# A Budget of Paradoxes

AUGUSTUS DE MORGAN
EDITED BY SOPHIA DE MORGAN

CAMBRIDGE
UNIVERSITY PRESS

# CAMBRIDGE
## UNIVERSITY PRESS

University Printing House, Cambridge, CB2 8BS, United Kingdom

Cambridge University Press is part of the University of Cambridge.

It furthers the University's mission by disseminating knowledge in the pursuit of
education, learning and research at the highest international levels of excellence.

www.cambridge.org
Information on this title: www.cambridge.org/9781108083201

This edition first published 1872
This digitally printed version 2015

ISBN 978-1-108-08320-1 Paperback

# CAMBRIDGE LIBRARY COLLECTION

*Books of enduring scholarly value*

## Mathematics

From its pre-historic roots in simple counting to the algorithms powering modern desktop computers, from the genius of Archimedes to the genius of Einstein, advances in mathematical understanding and numerical techniques have been directly responsible for creating the modern world as we know it. This series will provide a library of the most influential publications and writers on mathematics in its broadest sense. As such, it will show not only the deep roots from which modern science and technology have grown, but also the astonishing breadth of application of mathematical techniques in the humanities and social sciences, and in everyday life.

## A Budget of Paradoxes

An important figure in the development of modern mathematical logic and abstract algebra, Augustus De Morgan (1806–71) was also a witty writer who made a hobby of collecting evidence of paradoxical and illogical thinking from historical sources as well as contemporary pamphlets and periodicals. Based on articles that had appeared in *The Athenaeum* during his lifetime, this work was edited by his widow and published in book form in 1872. It parades all varieties of crackpot from circle-squarers to inventors of perpetual motion machines, all for the reader's entertainment and education. Filled with anecdotes, personal opinions and 'squibs' of every kind, the book remains enjoyable reading for those who are amused rather than appalled by the human condition. Also reissued in the Cambridge Library Collection are the *Memoir of Augustus De Morgan* (1882), prepared by his wife, and his ambitious *Formal Logic* (1847).

Cambridge University Press has long been a pioneer in the reissuing of out-of-print titles from its own backlist, producing digital reprints of books that are still sought after by scholars and students but could not be reprinted economically using traditional technology. The Cambridge Library Collection extends this activity to a wider range of books which are still of importance to researchers and professionals, either for the source material they contain, or as landmarks in the history of their academic discipline.

Drawing from the world-renowned collections in the Cambridge University Library and other partner libraries, and guided by the advice of experts in each subject area, Cambridge University Press is using state-of-the-art scanning machines in its own Printing House to capture the content of each book selected for inclusion. The files are processed to give a consistently clear, crisp image, and the books finished to the high quality standard for which the Press is recognised around the world. The latest print-on-demand technology ensures that the books will remain available indefinitely, and that orders for single or multiple copies can quickly be supplied.

The Cambridge Library Collection brings back to life books of enduring scholarly value (including out-of-copyright works originally issued by other publishers) across a wide range of disciplines in the humanities and social sciences and in science and technology.

A

# BUDGET OF PARADOXES.

LONDON : PRINTED BY
SPOTTISWOODE AND CO., NEW-STREET SQUARE
AND PARLIAMENT STREET

A

# BUDGET OF PARADOXES.

BY

## AUGUSTUS DE MORGAN,

F.R.A.S. & C.P.S.

OF TRINITY COLLEGE, CAMBRIDGE.

[*REPRINTED, WITH THE AUTHOR'S ADDITIONS, FROM THE 'ATHENÆUM.'*]

----

'Ut agendo surgamus arguendo gustamus.'

PTOCHODOKIARCHUS ANAGRAMMATISTES.

----

LONDON:

LONGMANS, GREEN, AND CO.

1872.

# EDITOR'S PREFACE.

It is not without hesitation that I have taken upon myself the editorship of a work left avowedly imperfect by the author, and, from its miscellaneous and discursive character, difficult of completion with due regard to editorial limitations by a less able hand.

Had the author lived to carry out his purpose he would have looked through his Budget again, amplifying and probably rearranging some of its contents. He had collected materials for further illustration of Paradox of the kind treated of in this book ; and he meant to write a second part, in which the contradictions and inconsistencies of orthodox learning would have been subjected to the same scrutiny and castigation as heterodox ignorance had already received.

It will be seen that the present volume contains more than the *Athenæum* Budget. Some of the additions formed a Supplement to the original articles. These supplementary paragraphs were, by the author, placed after those to which they respectively referred, being distinguished from the rest of the text by brackets. I have omitted these brackets as useless, except where they were needed to indicate subsequent writing.

Another and a larger portion of the work consists of discussion of matters of contemporary interest, for the Budget was in some degree a receptacle for the author's thoughts on any literary, scientific, or social question. Having grown thus gradually to its present size, the book as it was left was not quite in a fit condition for publication, but the alterations which have been made are slight and few, being in most cases verbal and such as the sense absolutely required, or transpositions of sentences to secure coherence with the rest, in places where the author, in his more recent insertion of them, had overlooked the connexion in which they stood. In no case has the meaning been in any degree modified or interfered with.

One rather large omission must be mentioned here. It is an account of the quarrel between Sir James South and Mr. Troughton on the mounting, &c. of the equatorial telescope at Campden Hill. At some future time when the affair has passed entirely out of the memory of living Astronomers, the appreciative sketch, which is omitted in this edition of the Budget, will be an interesting piece of history and study of character.

A very small portion of Mr. James Smith's circle-squaring has been left out, with a still smaller portion of Mr. De Morgan's answers to that Cyclometrical Paradoxer.

In more than one place repetitions, which would have disappeared under the author's revision, have been allowed to remain, because they could not have been taken away without leaving a hiatus, not easy to fill up without damage to the author's meaning.

I give these explanations in obedience to the rules laid down for the guidance of editors at page 11. If any apology for the fragmentary character of the book be thought necessary, it may be found in the author's own words at page 438.

The publication of the Budget could not have been delayed without lessening the interest attaching to the writer's thoughts upon questions of our own day. I trust that, incomplete as the work is compared with what it might have been, I shall not be held mistaken in giving it to the world. Rather let me hope that it will be welcomed as an old friend returning under great disadvantages, but bringing a pleasant remembrance of the amusement which its weekly appearance in the *Athenæum* gave to both writer and reader.

The Paradoxes are dealt with in chronological order. This will be a guide to the reader, and with the alphabetical Index of Names, &c., will, I trust, obviate all difficulty of reference.

SOPHIA DE MORGAN.

6 MERTON ROAD, PRIMROSE HILL.

*Erratum.*

Page 40, line 27, *for* Litchfield *read* Lichfield.

# A BUDGET

OF

# PARADOXES.

## INTRODUCTORY.

IF I had before me a fly and an elephant, having never seen more than one such magnitude of either kind; and if the fly were to endeavour to persuade me that he was larger than the elephant, I might by possibility be placed in a difficulty. The apparently little creature might use such arguments about the effect of distance, and might appeal to such laws of sight and hearing as I, if unlearned in those things, might be unable wholly to reject. But if there were a thousand flies, all buzzing, to appearance, about the great creature; and, to a fly, declaring, each one for himself, that he was bigger than the quadruped; and all giving different and frequently contradictory reasons; and each one despising and opposing the reasons of the others—I should feel quite at my ease. I should certainly say, My little friends, the case of each one of you is destroyed by the rest. I intend to show flies in the swarm, with a few larger animals, for reasons to be given.

In every age of the world there has been an established system, which has been opposed from time to time by isolated and dissentient reformers. The established system has sometimes fallen, slowly and gradually: it has either been upset by the rising influence of some one man, or it has been sapped by gradual change of opinion in the many.

I have insisted on the isolated character of the dissentients, as an element of the à priori probabilities of the case. Show me a schism, especially a growing schism, and it is another thing. The homœopathists, for instance, shall be, if any one so think, as

*B

wrong as St. John Long; but an organised opposition, supported by the efforts of many acting in concert, appealing to common arguments and experience, with perpetual succession and a common seal, as the Queen says in the charter, is, be the merit of the schism what it may, a thing wholly different from the case of the isolated opponent in the mode of opposition to it which reason points out.

During the last two centuries and a half, physical knowledge has been gradually made to rest upon a basis which it had not before. It has become *mathematical.* The question now is, not whether this or that hypothesis is better or worse to the pure thought, but whether it accords with observed phenomena in those consequences which can be shown necessarily to follow from it, if it be true. Even in those sciences which are not yet under the dominion of mathematics, and perhaps never will be, a working copy of the mathematical process has been made. This is not known to the followers of those sciences who are not themselves mathematicians, and who very often exalt their horns against the mathematics in consequence. They might as well be squaring the circle, for any sense they show in this particular.

A great many individuals, ever since the rise of the mathematical method, have, each for himself, attacked its direct and indirect consequences. I shall not here stop to point out how the very accuracy of exact science gives better aim than the preceding state of things could give. I shall call each of these persons a *paradoxer,* and his system a *paradox.* I use the word in the old sense: a paradox is something which is apart from general opinion, either in subject-matter, method, or conclusion.

Many of the things brought forward would now be called *crotchets,* which is the nearest word we have to old *paradox.* But there is this difference, that by calling a thing a *crotchet* we mean to speak lightly of it; which was not the necessary sense of *paradox.* Thus in the sixteenth century many spoke of the earth's motion as the *paradox of Copernicus,* who held the ingenuity of that theory in very high esteem, and some, I think, who even inclined towards it. In the seventeenth century, the depravation of meaning took place, in England at least. Phillips says paradox is ʻa thing which seemeth strange ʼ—here is the old meaning: after a colon, he proceeds—ʻand absurd, and is contrary to common opinion,ʼ which is an addition due to his own time.

Some of my readers are hardly inclined to think that the word *paradox* could once have had no disparagement in its meaning; still less that persons could have applied it to themselves. I chance to have met with a case in point against them. It is Spinoza's 'Philosophia Scripturæ Interpres, Exercitatio Paradoxa,' printed anonymously at Eleutheropolis, in 1666. This place was one of several cities in the clouds, to which the cuckoos resorted who were driven away by the other birds ; that is, a feigned place of printing, adopted by those who would have caught it if orthodoxy could have caught them. Thus, in 1656, the works of Socinus could only be printed at Irenopolis. The author deserves his self-imposed title, as in the following :—

Quanto sane satius fuisset illam [Trinitatem] pro mysterio non habuisse, et Philosophiæ ope, antequam quod esset statuerent, secundum veræ logices præcepta quid esset cum Cl. Keckermanno investigasse ; tanto fervore ac labore in profundissimas speluncas et obscurissimos metaphysicarum speculationum atque fictionum recessus se recipere ut ab adversariorum telis sententiam suam in tuto collocarent. Profecto magnus ille vir . . . dogma illud, quamvis apud theologos eo nomine non multum gratiæ iniverit, ita ex immotis Philosophiæ fundamentis explicat ac demonstrat, ut paucis tantum immutatis, atque additis, nihil amplius animus veritate sincere deditus desiderare possit.

This is properly paradox, though also heterodox. It supposes, contrary to all opinion, orthodox and heterodox, that philosophy can, with slight changes, explain the Athanasian doctrine so as to be at least compatible with orthodoxy. The author would stand almost alone, if not quite ; and this is what he meant. I have met with the counter-paradox. I have heard it maintained that the doctrine as it stands, in all its mystery, is à priori more likely than any other to have been Revelation, if such a thing were to be ; and that it might almost have been predicted.

After looking into books of paradoxes for more than thirty years, and holding conversation with many persons who have written them, and many who might have done so, there is one point on which my mind is fully made up. The manner in which a paradoxer will show himself, as to sense or nonsense, will not depend upon what he maintains, but upon whether he has or has not made a sufficient knowledge of what has been done by others, *especially as to the mode of doing it*, a preliminary to inventing knowledge for himself. That a little knowledge is a dangerous thing is one of the most fallacious of proverbs. A person of small knowledge is in danger of trying to make his *little* do the work of *more* ; but a person without any is in more danger of making his *no* knowledge do the work of *some*. Take the speculations on the tides as an instance. Persons with nothing but a little geometry have certainly exposed themselves in their modes of objecting to results which require the higher mathematics to be known before an independent opinion can be formed on sufficient grounds. But persons with no geometry at all have done the same thing much more completely.

There is a line to be drawn which is constantly put aside in the arguments held by parodoxers in favour of their right to instruct the world. Most persons must, or at least will, like the lady in Cadogan Place,[1] form and express an immense variety of opinions on an immense variety of subjects; and all persons must be their own guides in many things. So far all is well. But there are many who, in carrying the expression of their own opinions beyond the usual tone of private conversation, whether they go no further than attempts at oral proselytism, or whether they commit themselves to the press, do not reflect that they have ceased to stand upon the ground on which their process is defensible. Aspiring to lead *others*, they have never given themselves the fair chance of being first led by *other* others into something better than they can start for themselves; and that they should first do this is what both those classes of others have a fair right to expect. New knowledge, when to any purpose, must come by contemplation of old knowledge, in every matter which concerns thought; mechanical contrivance sometimes, not very often, escapes this rule. All the men who are now called discoverers, in every matter ruled by thought, have been men versed in the minds of their predecessors, and learned in what had been before them. There is not one exception. I do not say that every man has made direct acquaintance with the whole of his mental ancestry; many have, as I may say, only known their grandfathers by the report of their fathers. But even on this point it is remarkable how many of the greatest names in all departments of knowledge have been real antiquaries in their several subjects.

I may cite, among those who have wrought strongly upon opinion or practice in science, Aristotle, Plato, Ptolemy, Euclid, Archimedes, Roger Bacon, Copernicus, Francis Bacon, Ramus, Tycho Brahé, Galileo, Napier, Descartes, Leibnitz, Newton, Locke. I take none but names known out of their fields of work; and all were learned as well as sagacious. I have chosen my instances: if any one will undertake to show a person of little or no knowledge who has established himself in a great matter of pure thought, let him bring forward his man, and we shall see.

This is the true way of putting off those who plague others

---

[1] Mrs. Wititterly, in *Nicholas Nickleby*.

with their great discoveries. The first demand made should be —Mr. Moses, before I allow you to lead me over the Red Sea, I must have you show that you are learned in all the wisdom of the Egyptians upon your own subject. The plea that it is unlikely that this or that unknown person should succeed where Newton, &c. have failed, or should show Newton, &c. to be wrong, is utterly null and void. It was worthily versified by Sylvanus Morgan (the great herald who in his ' Sphere of Gentry' gave coat armour to ' *Gentleman Jesus*,' as he said), who sang of Copernicus as follows (1652):—

> If Tellus winged be,
> The earth a motion round ;
> Then much deceived are they
> Who nere before it found.
> Solomon was the wisest,
> His wit nere this attained ;
> Cease, then, Copernicus,
> Thy hypothesis vain.

Newton, &c. were once unknown ; but they made themselves known by what they knew, and then brought forward what they could do ; which I see is as good verse as that of Herald Sylvanus. The demand for previous knowledge disposes of twenty-nine cases out of thirty, and the thirtieth is worth listening to.

I have not set down Copernicus, Galileo, &c. among the para-doxers, merely because everybody knows them ; if my list were quite complete, they would have been in it. But the reader will find Gilbert, the great precursor of sound magnetical theory ; and several others on whom no censure can be cast, though some of their paradoxes are inadmissible, some unproved, and some capital jokes, true or false : the author of the 'Vestiges of Creation' is an instance. I expect that my old correspondent, General Perronet Thompson, will admit that his geometry is part and parcel of my plan ; and also that, if that plan embraced politics, he would claim a place for his ' Catechism on the Corn Laws,' a work at one time paradoxical, but which had more to do with the abolition of the bread-tax than Sir Robert Peel.

My intention in publishing this Budget in the *Athenœum* is *to enable those who have been puzzled by one or two discoverers to see how they look in the lump.* The only question is, has the selection been fairly made ? To this my answer is, that no selec-tion at all has been made. The books are, without exception, those which I have in my own library ; and I have taken *all*—I mean all of the kind : Heaven forbid that I should be supposed to have no other books! But I may have been a collector, influenced in choice by bias? I answer that I never have collected books of this sort—that is, I have never searched for them, never made up my mind to look out for this book or that. I have bought what happened to come in my way at shop or auction; I have retained what came in as part of the *undescribed* portion of miscellaneous auction lots; I have received a few from friends who found them among what they called their rubbish; and I have preserved books sent to me for review. In not a few instances the books have been bound up with others, unmentioned at the back; and for years I knew no more I had them than I knew I had Lord Macclesfield's speech on moving the change of Style, which, after I had searched shops, &c. for it in vain, I found had been reposing on my own shelves for many years, at the end of a summary of Leibnitz's philosophy. Consequently, I may positively affirm that the following list is formed by accident and circumstance alone, and that it truly represents the casualties of about a third of a century. For instance, the large proportion of works on the quadrature of the circle is not my doing : it is the natural share of this subject in the actual run of events.

[I keep to my plan of inserting only such books as I possessed in 1863, except by casual notice in aid of my remarks. I have found several books on my shelves which ought to have been inserted. These have their titles set out at the commencement of their articles, in leading paragraphs; the casuals are without this formality.[1]]

Before proceeding to open the Budget, I say something on my personal knowledge of the class of discoverers who square the circle, upset Newton, &c. I suspect I know more of the English class than any man in Britain. I never kept any reckoning; but I know that one year with another—and less of late years than in earlier time—I have talked to more than five in each year, giving more than a hundred and fifty specimens. Of this I am sure, that it is my own fault if they have not been a thousand. Nobody knows how they swarm, except those to whom they naturally resort. They are in all ranks and occupations, of all ages and characters. They are very earnest people, and their purpose is *bonâ fide* the dissemination of their paradoxes. A great many—the mass, indeed—are illiterate, and a great many waste their means, and are in or approaching penury. But I must say that never, in any one instance, has the quadrature of the circle, of

---

[1] The brackets mean that the paragraph is substantially from some one of the *Athenæum* Supplements.—(ED.)

the like, been made a pretext for begging; even to be asked to purchase a book is of the very rarest occurrence—it has happened, and that is all.

These discoverers despise one another: if there were the concert among them which there is among foreign mendicants, a man who admitted one to a conference would be plagued to death. I once gave something to a very genteel French applicant, who overtook me in the street, at my own door, saying he had picked up my handkerchief: whether he picked it up in my pocket for an introduction, I know not. But that day week came another Frenchman to my house, and that day fortnight a French lady; both failed, and I had no more trouble. The same thing happened with Poles. It is not so with circle-squarers, &c.: they know nothing of each other. Some will read this list, and will say I am right enough, generally speaking, but that there *is* an exception, if I could but see it.

I do not mean, by my confession of the manner in which I have sinned against the twenty-four hours, to hold myself out as accessible to personal explanation of new plans. Quite the contrary: I consider myself as having made my report, and being discharged from further attendance on the subject. I will not, from henceforward, talk to any squarer of the circle, trisector of the angle, duplicator of the cube, constructor of perpetual motion, subverter of gravitation, stagnator of the earth, builder of the universe, &c. I will receive any writings or books which require no answer, and read them when I please : I will certainly preserve them—this list may be enlarged at some future time.

There are three subjects which I have hardly anything upon; astrology, mechanism, and the infallible way of winning at play. I have never cared to preserve astrology. The mechanists make models, and not books. The infallible winners—though I have seen a few—think their secret too valuable, and prefer *mutare quadrata rotundis*—to turn dice into coin—at the gaming-house : verily they have their reward.

I shall now select, to the mystic number seven, instances of my personal knowledge of those who think they have discovered, in illustration of as many misconceptions.

1. *Attempt by help of the old philosophy, the discoverer not being in possession of modern knowledge.* A poor schoolmaster, in rags, introduced himself to a scientific friend with whom I was talking, and announced that he had found out the composition of the sun. ' How was that done? '—' By consideration of the four elements.'—' What are they ? '—' Of course, fire, air, earth, and

water.'—' Did you not know that air, earth, and water, have long
been known to be no elements at all, but compounds ? '—' What
do you mean, sir ?   Who ever heard of such a thing ? '

2. *The notion that difficulties are enigmas, to be overcome in
a moment by a lucky thought.*   A nobleman of very high rank,
now long dead, read an article by me on the quadrature, in an
early number of the *Penny Magazine.*   He had, I suppose, school
recollections of geometry.   He put pencil to paper, drew a circle,
and constructed what seemed likely to answer, and, indeed, was—
as he said— certain, if only this bit were equal to that; which of
course it was not.   He forwarded his diagram to the Secretary of
the Diffusion Society, to be handed to the author of the article, in
case the difficulty should happen to be therein overcome.

3. *Discovery at all hazards, to get on in the world.*   Thirty
years ago, an officer of rank, just come from foreign service, and
trying for a decoration from the Crown, found that his claims were
of doubtful amount, and was told by a friend that so and so, who
had got the order, had the additional claim of scientific distinction.   Now this officer, while abroad, had bethought himself one
day, that there really could be no difficulty in finding the circumference of a circle: if a circle were rolled upon a straight line
until the undermost point came undermost again, there would be
the straight line equal to the circle.   He came to me, saying that
he did not feel equal to the statement of his claim in this respect,
but that if some clever fellow would put the thing in a proper
light, he thought his affair might be managed.   I was clever
enough to put the thing in a proper light to himself, to this extent
at least, that, though perhaps they were wrong, the advisers of the
Crown would never put the letters K.C.B. to such a circle as his.

4. *The notion that mathematicians cannot find the circle for
common purposes.*   A working man measured the altitude of a
cylinder accurately, and—I think the process of Archimedes was
one of his proceedings—found its bulk.   He then calculated the
ratio of the circumference to the diameter, and found it answered
very well on other modes of trial.   His result was about 3·14.
He came to London, and somebody sent him to me.   Like many
others of his pursuit, he seemed to have turned the whole force of
his mind upon one of his points, on which alone he would be open
to refutation.   He had read some of Kater's experiments, and had
got the Act of 1825 on weights and measures.   Say what I would,
he had for a long time but one answer—' Sir ! I go upon Captain
Kater and the Act of Parliament.'   But I fixed him at last.   I
happened to have on the table a proof-sheet of the *Astronomical*

*Memoirs*, in which were a large number of observed places of the planets compared with prediction, and asked him whether it could be possible that persons who did not know the circle better than he had found it could make the calculations, of which I gave him a notion, so accurately? He was perfectly astonished, and took the titles of some books which he said he would read.

5. *Application for the reward from abroad.* Many years ago, about twenty-eight, I think, a Jesuit came from South America, with a quadrature, and a cutting from a newspaper, announcing that a reward was ready for the discovery in England. On this evidence he came over. After satisfying him that nothing had ever been offered here, I discussed his quadrature, which was of no use. I succeeded better when I told him of Richard White, also a Jesuit, and author of a quadrature published before 1648, under the name of *Chrysœspis*, of which I can give no account, having never seen it. This White (Albius) is the only quadrator who was ever convinced of his error. My Jesuit was struck by the instance, and promised to read more geometry—he was no Clavius—before he published his book. He relapsed, however, for I saw his book advertised in a few days. I may say, as sufficient proof of my being no collector, that I had not the curiosity to buy this book ; and my friend the Jesuit did not send me a copy, which he ought to have done, after the hour I had given him.

6. *Application for the reward at home.* An agricultural labourer squared the circle, and brought the proceeds to London. He left his papers with me, one of which was the copy of a letter to the Lord Chancellor, desiring his Lordship to hand over forthwith 100,000*l.*, the amount of the alleged offer of reward. He did not go quite so far as M. de Vausenville, who, I think in 1778, brought an action against the Academy of Sciences to recover a reward to which he held himself entitled. I returned the papers, with a note, stating that he had not the knowledge requisite to see in what the problem consisted. I got for answer a letter in which I was told that a person who could not see that he had done the thing should ' change his business, and appropriate his time and attention to a Sunday-school, to learn what he could, and keep the *litle* children from *durting* their *close.*' I also received a letter from a friend of the quadrator, informing me that I knew his friend had succeeded, and had been heard to say so. These letters were printed—without the names of the writers—for the amusement of the readers of *Notes and Queries*, First Series, xii. 57, and they will appear again in the sequel.

[There are many who have such a deep respect for any attempt at thought that they are shocked at ridicule even of those who have made themselves conspicuous by pretending to lead the world in matters which they have not studied. Among my anonymes is a gentleman who is angry at my treatment of the 'poor but thoughtful' man who is described in my introduction as recommending me to go to a Sunday-school because I informed him that he did not know in what the difficulty of quadrature consisted. My impugner quite forgets that this man's 'thoughtfulness' chiefly consisted in his demanding a hundred thousand pounds from the Lord Chancellor for his discovery; and I may add, that his greatest stretch of invention was finding out that 'the clergy' were the means of his modest request being unnoticed. I mention this letter because it affords occasion to note a very common error, namely, that men unread in their subjects have, by natural wisdom, been great benefactors of mankind. My critic says, 'Shakspeare, whom the Pro$^r$ (*sic*) may admit to be a wisish man, though an object of contempt as to learning. . . .' Shakspeare an object of contempt as to learning! Though not myself a thoroughgoing Shakspearean— and adopting the first half of the opinion given by George III., 'What! is there not sad stuff? only one must not say so'—I am strongly of opinion that he throws out the masonic signs of learning in almost every scene, to all who know what they are. And this over and above every kind of direct evidence. First, foremost, and enough, the evidence of Ben Jonson that he had 'little Latin and less Greek;' then Shakspeare had as much Greek as Jonson would call *some*, even when he was depreciating. To have any Greek at all was in those days exceptional. In Shakspeare's youth St. Paul's and Merchant Taylors' schools were to have masters learned in good and clean Latin literature, *and also in Greek if such may be gotten.* When Jonson spoke as above, he intended to put Shakspeare low among the learned, but not out of their pale; and he spoke as a rival dramatist, who was proud of his own learned sock; and it may be a subject of inquiry how much Latin *he* would call *little*. If Shakspeare's learning on certain points be very much less visible than Jonson's, it is partly because Shakspeare's writings hold it in chemical combination, Jonson's in mechanical aggregation.]

7. An elderly man came to me, to show me how the universe was created. There was one molecule, which by vibration became —Heaven knows how!—the Sun. Further vibration produced Mercury, and so on. I suspect the nebular hypothesis had got

into the poor man's head by reading, in some singular mixture with what it found there. Some modifications of vibration gave heat, electricity, &c. I listened until my informant ceased to vibrate—which is always the shortest way—and then said, ' Our knowledge of elastic fluids is imperfect.' 'Sir!' said he, ' I see you perceive the truth of what I have said, and I will reward your attention by telling you what I seldom disclose, never, except to those who can receive my theory—the little molecule whose vibrations have given rise to our solar system is the Logos of St. John's Gospel!' He went away to Dr. Lardner, who would not go into the solar system at all—the first molecule settled the question. So hard upon poor discoverers are men of science who are not antiquaries in their subject! On leaving, he said, ' Sir, Mr. De Morgan received me in a very different way ; he heard me attentively, and I left him perfectly satisfied of the truth of my system.' I have had much reason to think that many discoverers, of all classes, believe they have convinced every one who is not peremptory to the verge of incivility.

My list is given in chronological order. My readers will understand that my general expressions, where slighting or contemptuous, refer to the ignorant, who teach before they have learnt. In every instance, those of whom I am able to speak with respect, whether as right or wrong, have sought knowledge in the subject they were to handle before they completed their speculations. I shall further illustrate this at the conclusion of my list.

Before I begin the list, I give prominence to the following letter, addressed by me to the *Correspondent* of October 28, 1865. Some of my paradoxers attribute to me articles in this or that journal; and others may think—I know some do think—they know me as the writer of reviews of some of the very books noticed here. The following remarks will explain the way in which they may be right, and in which they may be wrong :—

THE EDITORIAL SYSTEM.

SIR,—I have reason to think that many persons have a very inaccurate notion of the *Editorial system*. What I call by this name has grown up in the last *centenary*—a word I may use to signify the hundred years now ending, and to avoid the ambiguity of *century*. It cannot conveniently be explained by editors themselves, and *edited* journals generally do not like to say much about it. In *your* paper perhaps, in which editorial duties differ somewhat from those of ordinary journals, the common system may be freely spoken of.

When a reviewed author, as very often happens, writes to the editor of the reviewing journal to complain of what has been said of him, he frequently—even more often than not—complains of 'your reviewer.' He sometimes presumes that 'you' have, 'through inadvertence' in this instance, 'allowed some incompetent person to lower the character of your usually accurate pages.' Sometimes he talks of 'your scribe,' and, in extreme cases, even of 'your hack.' All this shows perfect ignorance of the journal system, except where it is done under the notion of letting the editor down easy. But the editor never accepts the mercy.

All that is in a journal, except what is marked as from a correspondent, either by the editor himself or by the correspondent's real or fictitious signature, is published entirely on editorial responsibility, as much as if the editor had written it himself. The editor, therefore, may claim, and does claim and exercise, unlimited right of omission, addition, and alteration. This is so well understood that the editor performs his last function on the last revise without the 'contributor' knowing what is done. The word *contributor* is the proper one: it implies that he furnishes materials without stating what he furnishes or how much of it is accepted, or whether he be the only contributor. All this applies both to political and literary journals. No editor acknowledges the right of a contributor to withdraw an article, if he should find alterations in the proof sent to him for correction which would make him wish that the article should not appear. If the *demand* for suppression were made—I say nothing about what might be granted to *request*—the answer would be, 'It is not your article, but mine; I have all the responsibility; if it should contain a libel, I could not give you up, even at your own desire. You have furnished me with materials, on the known and common understanding that I was to use them at my discretion, and you have no right to impede my operations by making the appearance of the article depend on your approbation of my use of your materials.'

There is something to be said for this system, and something against it—I mean simply on its own merits. But the all-conquering argument in its favour is, that the only practicable alternative is the modern French plan of no articles without the signature of the writers. I need not discuss this plan; there is no collective party in favour of it. Some may think it is not the only alternative; they have not produced any intermediate proposal in which any dozen of persons have concurred. Many will say, Is not all this, though perfectly correct, well known to be matter of form? Is it not practically the course of events that an engaged contributor writes the article, and sends it to the editor, who admits it as written—substantially, at least? And is it not often very well known, by style and in other ways, who it was wrote the article? This system is matter of form just as much as loaded pistols are matter of form so long as the wearer is not assailed; but matter of form takes the form of matter in the pulling of a trigger, so soon as the need arises. Editors and contributors who can work together find each other out by elective affinity, so that the common run of events settles down into most articles appearing much as they are written. And there are two safety-valves; that is, when judicious persons come together. In the first place, the editor himself, when he has selected his contributor, feels that the contributor is likely to know his business better than an editor can teach him; in fact, it is on that principle that the selection is made. But he feels that he is more competent than the writer to judge questions of strength and of tone, especially when the general purpose of the journal is considered, of which the editor is the judge without appeal. An editor who meddles with substantive matter is likely to be wrong, even when he knows the subject; but one who prunes what he deems excess, is likely to be right, even when he does not know the subject. In the second place, a contributor knows that he is supplying an editor, and learns, without suppressing truth or suggesting falsehood, to make the tone of his communications suit the periodical in which they are to appear. Hence it very often arises that a reviewed author, who thinks he knows the name of his reviewer, and proclaims it with expressions of dissatisfaction, is only wrong in supposing that his critic has given all his mind. It has happened to myself, more than once, to be announced as the author of articles which I could not have signed, because they did not go far enough to warrant my affixing my name to them as to a sufficient expression of my own opinion.

There are two other ways in which a reviewed author may be wrong about his critic. At editor frequently makes slight insertions or omissions—I mean slight in quantity of type—as he goes over the last proof; this he does in a comparative hurry, and it may chance that he does not know the full sting of his little alteration. The very bit which the writer of the book most complains of may not have been seen by the person who is called the writer of the article until after the appearance of the journal; nay, if he be one of those—few, I daresay—who do not read their own articles, may never have been seen by him at all. Possibly, the insertion or omission would not have been made if the editor could have had one minute's conversation with his contributor. Sometimes it actually contradicts something which is allowed to remain in another part of the article; and sometimes, especially in the case of omission, it renders other parts of the article unintelligible. These are disadvantages of the system, and a judicious editor is not very free with his *unus et alter pannus*. Next, readers in general, when they see the pages of a journal with the articles so nicely fitting, and so many ending with the page or column, have very little notion of the cutting and carving which goes to the process. At the very last moment arises the necessity of some trimming of this kind; and the editor, who would gladly call the writer to counsel if he could, is obliged to strike out ten or twenty lines. He must do his best, but it may chance that the omission selected would take from the writer the power of owning the article. A few years ago, an able opponent of mine wrote to a journal some criticisms upon an article which he expressly attributed to me. I replied as if I were the writer, which, in a sense, I was. But if any one had required of me an unmodified 'Yes' or 'No' to the question whether I wrote the article, I must, of two falsehoods, have chosen 'No :' for certain omissions, dictated by the necessities of space and time, would have amounted, had my signature been affixed, to a silent surrender of points which, in my own character, I must have strongly insisted on, unless I had chosen to admit certain inferences against what I had previously published in my own name. I may here add that the forms of journalism obliged me in this case to remind my opponent that it could not be permitted to me, *in that journal*, either to acknowledge or deny the authorship of the articles. The cautions derived from the above remarks are particularly wanted with reference to the editorial comments upon letters of complaint. There is often no time to send these letters to the contributor, and even when this can be done, an editor is—and very properly— never of so editorial a mind as when he is revising the comments of a contributor upon an assailant of the article. He is then in a better position as to information, and a more critical position as to responsibility. Of course, an editor never meddles, except under notice, with the letter of a correspondent, whether of a complainant, of a casual informant, or of a contributor who sees reason to become a correspondent. Omissions must sometimes be made when a grievance is too highly spiced. It did once happen to me that a waggish editor made an insertion without notice in a letter signed by me with some fiction, which insertion contained the name of a friend of mine, with a satire which I did not believe, and should not have written if I had. To my strong rebuke, he replied—'I know it was very wrong; but human nature could not resist.' But this was the only occasion on which such a thing ever happened to me.

I daresay what I have written may give some of your readers to understand some of the *pericula et commoda* of modern journalism. I have known men of deep learning and science as ignorant of the prevailing system as any uneducated reader of a newspaper in a country town. I may, perhaps, induce some writers not to be too sure about this, that, or the other person. They may detect their reviewer, and they may be safe in attributing to him the general matter and tone of the article. But about one and another point, especially if it be a short and stinging point, they may very easily chance to be wrong. It has happened to myself, and within a few weeks to publication, to be wrong in two ways in reading a past article—to attribute to editorial insertion what was really my own, and to attribute to myself what was really editorial insertion.

What is a man to do who is asked whether he wrote an article. He may, of course, refuse to answer; which is regarded as an admission. He may say, as Swift did to Serjeant Bettesworth, 'Sir, when I was a young man, a friend of mine advised me, whenever I was asked whether I had written a certain paper, to deny it ; and I accordingly tell you that I did *not* write it.' He may say, as I often do, when charged with having invented a joke, story, or epigram, ' I want all the credit I can get, and therefore I always acknowledge all that is attributed to me, truly or not ; the story, &c. *is* mine. But for serious earnest, in the matter of imputed criticism, the answer may be, ' That article was of my material, but the editor has not let it stand as I gave it ; I cannot own it as a whole.' He may then refuse to be particular as to the amount of the editor's interference. Of this there are two extreme cases. The editor may have expunged nothing but a qualifying adverb. Or he may have done as follows. We all remember the account of Adam which satirizes woman, but eulogizes her if every second and third line be transposed. As in—

> Adam could find no solid peace
> When Eve was given him for a mate,
> Till he beheld a woman's face,
> Adam was in a happy state.

If this had been the article, and a gallant editor had made the transpositions, the author could not with truth acknowledge. If the alteration were only an omitted adverb, or a few things of the sort, the author could not with truth deny. In all that comes between, every man must be his own casuist. I stared, when I was a boy, to hear grave persons approve of Sir Walter Scott's downright denial that he was the author of Waverley, in answer to the Prince Regent's downright question. If I remember rightly, Samuel Johnson would have approved of the same course.

It is known that, whatever the law gives, it also gives all that is necessary to full possession ; thus a man whose land is environed by the land of others has a right of way over the land of these others. By analogy, it is argued that when a man has a right to his secret, he has a right to all that is necessary to keep it, and that is not unlawful. If, then, he can only keep his secret by denial, he has a right to denial. This I admit to be an answer as against all men except the denier himself ; if conscience and self respect will allow it, no one can impeach it. But the question cannot be solved on a case. That question is, A lie, is it *malum in se*, without reference to meaning and circumstances ? This is a question with two sides to it. Cases may be invented in which a lie is the only way of preventing a murder, or in which a lie may otherwise save a life. In these cases it is difficult to acquit, and almost impossible to blame; discretion introduced, the line becomes very hard to draw.

I know but one work which has precisely—as at first appears—the character and object of my Budget. It is the 'Review of the Works of the Royal Society of London,' by Sir John Hill, M.D. (1751 and 1780, 4to.) This man offended many: the Royal Society, by his work; the medical profession, by inventing and selling extra-pharmacopœian doses; Garrick, by resenting the rejection of a play. So Garrick wrote:

> For physic and farces his equal there scarce is;
> His farces are physic; his physic a farce is.

I have fired at the Royal Society and at the medical profession, but I have given a wide berth to the drama and its wits; so there is no epigram out against me, as yet. He was very able and very eccentric. Dr. Thomson (*Hist. Roy. Soc.*) says he has no humour, but Dr. Thomson was a man who never would have discovered humour.

Mr. Weld (*Hist. Roy. Soc.*) backs Dr. Thomson, but with a remarkable addition. Having followed his predecessor in observing that the *Transactions* in Martin Folkes's time have an unusual proportion of trifling and puerile papers, he says that Hill's book is a poor attempt at humour, and glaringly exhibits the feelings of a disappointed man. It is probable, he adds, that the points told with some effect on the Society; for shortly after its publication the *Transactions* possess a much higher scientific value.

I copy an account which I gave elsewhere.

When the Royal Society was founded, the Fellows set to work to prove all things, that they might hold fast that which was good. They bent themselves to the question whether sprats were young herrings. They made a circle of the powder of a unicorn's horn, and set a spider in the middle of it; 'but it immediately ran out.' They tried several times, and the spider 'once made some stay in the powder.' They enquired into Kenelm Digby's sympathetic powder. 'Magnetical cures being discoursed of, Sir Gilbert Talbot promised to communicate what he knew of sympathetical cures; and those members who had any of the powder of sympathy, were desired to bring some of it at the next meeting.' June 21, 1661, certain gentlemen were appointed 'curators of the proposal of tormenting a man with the sympathetic powder;' I cannot find any record of the result. And so they went on until the time of Sir John Hill's satire, in 1751. This once well-known work is, in my judgment, the greatest compliment the Royal Society ever received. It brought forward a number of what are now feeble and childish researches in the Philosophical Transactions. It showed that the inquirers had actually been inquiring; and that they did not pronounce decision about 'natural *knowledge*' by help of '*natural* knowledge.' But for this, Hill would neither have known what to assail, nor how. Matters are now entirely changed. The scientific bodies are far too well established to risk themselves. *Ibit qui zonam perdidit*—

Let him take castles who has ne'er a groat.

These great institutions are now without any collective purpose, except that of promoting individual energy ; they print for their contributors, and guard themselves by a general declaration that they will not be answerable for the things they print. Of course they will not put forward anything for everybody; but a writer of a certain reputation, or matter of a certain look of plausibility and safety, will find admission. This is as it should be ; the pasturer of flocks and herds and the hunters of wild beasts are two very different bodies, with very different policies. The scientific academies are what a spiritualist might call 'publishing mediums,' and *their* spirits fall occasionally into writing which looks as if minds in the higher state were not always impervious to nonsense.

The following joke is attributed to Sir John Hill. I cannot honestly say I believe it ; but it shows that his contemporaries did not believe he had no humour. Good stories are always in some sort of keeping with the characters on which they are fastened. Sir John Hill contrived a communication to the Royal Society from Portsmouth, to the effect that a sailor had broken his leg in a fall from the mast-head ; that bandages and a plentiful application of tarwater had made him, in three days, able to use his leg as well as ever. While this communication was under grave discussion—it must be remembered that many then thought tarwater had extraordinary remedial properties—the joker contrived that a second letter should be delivered, which stated that the writer had forgotten, in his previous communication, to mention that the leg was a wooden leg ! Horace Walpole told this story, I suppose for the first time ; he is good authority for the fact of circulation, but for nothing more.

Sir John Hill's book is droll and cutting satire. Dr. Maty, (Sec. Royal Society) wrote thus of it in the *Journal Britannique* (Feb. 1751), of which he was editor:

Il est fâcheux que cet ingénieux Naturaliste, qui nous a déjà donné et qui nous prépare encore des ouvrages plus utiles, emploie à cette odieuse tâche une plume qu'il trempe dans le fiel et dans l'absinthe. Il est vrai que plusieurs de ses remarques sont fondées, et qu'à l'erreur qu'il indique, il joint en même tems la correction. Mais il n'est pas toujours équitable, et ne manque jamais d'insulter. Que peut après tout prouver son livre, si ce n'est que la quarante-cinquième partie d'un très-ample et très-utile Recueil n'est pas exempte d'erreurs ? Devoit-il confondre avec des Écrivains superficiels, dont la Liberté du Corps ne permet pas de restreindre la fertilité, cette foule de savans du Premier ordre, dont les Écrits ont orné et ornent encore les Transactions ? A-t-il oublié qu'on y a vu fréquemment les noms des Boyle, des Newton, des Halley, des De Moivres, des Hans Sloane, etc.? Et qu'on y trouve encore ceux des Ward, des Bradley, des Graham, des Ellicot, des Watson, et d'un Auteur que Mr. Hill préfère à tous les autres, je veux dire de Mr. Hill lui-même ?

This was the only answer; but it was no answer at all. Hill's object was to expose the absurdities; he therefore collected the absurdities. I feel sure that Hill was a benefactor of the Royal Society; and much more than he would have been if he had softened their errors and enhanced their praises. No reviewer will object to me that I have omitted Young, Laplace, &c. But then my book has a true title. Hill should not have called his a review of the 'Works.'

It was charged against Sir John Hill that he had tried to become a Fellow of the Royal Society and had failed. This he denied, and challenged the production of the certificate which a candidate always sends in, and which is preserved. But perhaps he could not get so far as a certificate—that is, could not find any one to recommend him; he was a likely man to be in such a predicament. As I have myself run foul of the Society on some little points, I conceive it possible that I may fall under a like suspicion. Whether I could have been a Fellow, I cannot know; as the gentleman said who was asked if he could play the violin, I never tried. I have always had a high opinion of the Society upon its whole history. A person used to historical inquiry learns to look at wholes; the Universities of Oxford and Cambridge, the College of Physicians, &c. are taken in all their duration. But those who are not historians—I mean not possessed of the habit of history—hold a mass of opinions about current things which lead them into all kinds of confusion when they try to look back. Not to give an instance which will offend any set of existing men—this merely because I can do without it—let us take the country at large. Magna Charta for ever!

glorious safeguard of our liberties ! *Nullus liber homo capiatur aut imprisonetur. . . . aut aliquo modo destruatur, nisi per judicium parium. . . . Liber homo; frank home* ; a capital thing for him—but how about the *villeins?* Oh, there are none *now*! But there were. Who cares for villains, or barbarians, or helots ? And so England, and Athens, and Sparta, were free States : all the freemen in them were free. Long after Magna Charta, villains were sold with their ‘chattels and offspring,’ named in that order. Long after Magna Charta, it was law that ‘ Le Seigniour poit rob, naufrer, et chastiser son villein a son volunt, salve que il ne poit luy maim.’

The Royal Society was founded as a co-operative body, and co-operation was its purpose. The early charters, &c. do not contain a trace of the intention to create a *scientific distinction*, a kind of Legion of Honour. It is clear that the qualification was ability and willingness to do good work for the promotion of natural knowledge, no matter in how many persons, nor of what position in society. Charles II. gave a smart rebuke for exclusiveness, as elsewhere mentioned. In time arose, almost of course, the idea of distinction attaching to the title ; and when I first began to know the Society, it was in this state. Gentlemen of good social position were freely elected if they were really educated men ; but the moment a claimant was announced as resting on his science, there was a disposition to inquire whether he was scientific enough. The maxim of the poet was adopted ; and the Fellows were practically divided into *Drink-deeps* and *Taste-nots.*

I was, in early life, much repelled by the tone taken by the Fellows of the Society with respect to their very mixed body. A man high in science—some thirty-seven years ago (about 1830)— gave me some encouragement, as he thought. ‘ We shall have you a Fellow of the Royal Society in time,’ said he. Umph ! thought I : for I had that day heard of some recent elections, the united science of which would not have demonstrated I. 1, nor explained the action of a pump. Truly an elevation to look up at ! It came, further, to my knowledge that the Royal Society—if I might judge by the claims made by very influential Fellows—considered itself as entitled to the best of everything : second-best being left for the newer bodies. A secretary, in returning thanks for the Royal at an anniversary of the Astronomical, gave rather a lecture to the company on the positive duty of all present to send the very best to the old body, and the absolute right of the old body to expect it. An old friend of mine, on a similar occasion, stated as a fact that the thing was always done, as well as that it ought to be done.

Of late years this pretension has been made by a President of the Society.   In 1855, Lord Rosse presented a confidential memorandum to the Council on the expediency of enlarging their number.   He says, ' In a Council so small it is impossible to secure a satisfactory representation of the leading scientific Societies, and it is scarcely to be expected that, under such circumstances, they will continue to publish inferior papers while they send the best to our *Transactions.*'

And, again, with all the Societies represented on the Council, ' even if every Science had its Society, and if they published everything, withholding their best papers [i.e. from the Royal Society], which they would not be likely to do, still there would remain to the Royal Society . . .' Lord Rosse seems to imagine that the minor Societies themselves transfer their best papers to the Royal Society ; that if, for instance, the Astronomical Society were to receive from A. B. a paper of unusual merit, the Society would transfer it to the Royal Society.   This is quite wrong : any preference of the Royal to another Society is the work of the contributor himself.   But it shows how well hafted is the Royal Society's claim, that a President should acquire the notion that it is acknowledged and acted upon by the other Societies, in their joint and corporate capacities.   To the pretension thus made I never could give any sympathy.   When I first heard Mr. Christie, Sec. R. S., set it forth at the anniversary dinner of the Astronomical Society, I remembered the Baron in Walter Scott—

> Of Gilbert the Galliard a heriot he sought,
> Saying, Give thy best steed as a vassal ought.

And I remembered the answer—

> Lord and Earl though thou be, I trow
> I can rein Buck's-foot better than thou.

Fully conceding that the Royal Society is entitled to preeminent rank and all the respect due to age and services, I could not, nor can I now, see any more obligation in a contributor to send his best to that Society than he can make out to be due to himself.   This pretension, in my mind, was hooked on, by my historical mode of viewing things already mentioned, to my knowledge of the fact that the Royal Society—the chief fault, perhaps, lying with its President, Sir Joseph Banks—had sternly set itself against the formation of other societies ; the Geological and Astronomical, for instance, though it must be added that the chief rebels came out of the Society itself. And so a certain not very defined dislike was generated in my mind—an anti-aristocratic affair—to the body which seemed to me a little too uplifted. This would, I daresay, have worn off; but a more formidable objection arose. My views of physical science gradually arranged themselves into a form which would have rendered F.R.S., as attached to my name, a false representation symbol. The Royal Society is the great fortress of general physics: and in the philosophy of our day, as to general physics, there is something which makes the banner of the R.S. one under which I cannot march. Everybody who saw the three letters after my name would infer certain things as to my mode of thought which would not be true inference. It would take much space to explain this in full. I may hereafter, perhaps, write a budget of collected results of the *à priori philosophy*, the nibbling at the small end of omniscience, and the effect it has had on common life, from the family parlour to the jury-box, from the girls'-school to the vestry-meeting. There are in the Society those who would, were there no others, prevent my criticism, be its conclusions true or false, from having any basis; but they are in the minority.

There is no objection to be made to the principles of philosophy in vogue at the Society, when they are stated as principles; but there is an omniscience in daily practice which the principles repudiate. In like manner, the most retaliatory Christians have a perfect form of round words about behaviour to those who injure them: none of them are as candid as a little boy I knew, who, to his mother's admonition, You should love your enemies, answered—Catch me at it!

Years ago, a change took place which would alone have put a sufficient difficulty in the way. The co-operative body got tired of getting funds from and lending name to persons who had little or no science, and wanted F.R.S. to be in every case a Fellow Really Scientific. Accordingly, the number of yearly elections was limited to fifteen recommended by the Council, unless the general body should choose to elect more; which it does not do. The election is now a competitive examination: it is no longer—Are you able and willing to promote natural knowledge; it is—Are you one of the upper fifteen of those who make such claim. In the list of candidates—a list rapidly growing in number—each year shows from thirty to forty of those whom Newton and Boyle would have gladly welcomed as fellow-labourers. And though the rejected of one year may be the accepted of the next—or of the next but one, or but two, if self-respect will permit the candidate to hang on—yet the time is clearly coming when many of those who ought to be welcomed will be excluded for life, or else shelved at last, when past work, with a scientific peerage. Coupled with this attempt to create a kind of order of knighthood is an absurdity so glaring that it should always be kept before the general eye. This distinction, this mark set by science upon successful investigation, is of necessity a class-distinction. Rowan Hamilton, one of the greatest names of our day in mathematical science, never could attach F.R.S. to his name—*he could not afford it.* There is a condition precedent—Four Red Sovereigns. It is four pounds a year, or—to those who have contributed to the Transactions—forty pounds down. This is as it should be : the Society must be supported. But it is not as it should be that a kind of title of honour should be forged, that a body should take upon itself to confer distinctions *for* science, when it is in the background—and kept there when the distinction is trumpeted— that the wearer is a man who can spare four pounds a year. I am well aware that in England a person who is not gifted, either by nature or art, with this amount of money power, is, with the mass, a very second-rate sort of Newton, whatever he may be in the field of investigation. Even men of science, so called, have this feeling. I know that the *scientific advisers* of the Admiralty, who, years ago, received 100*l.* a year each for his trouble, were sneered at by a wealthy pretender as ' fellows to whom a hundred a year is an object.' Dr. Thomas Young was one of them. To a bookish man—I mean a man who can manage to collect books— there is no tax. To myself, for example, 40*l.* worth of books deducted from my shelves, and the life-use of the Society's splendid library instead, would have been a capital exchange. But there may be, and are, men who want books, and cannot pay the Society's price. The Council would be very liberal in allowing their books to be consulted. I have no doubt that if a known investigator were to call and ask to look at certain books, the Assistant-Secretary would forthwith seat him with the books before him, absence of F.R.S. not in any wise withstanding. But this is not like having the right to consult any book on any day, and to take it away, if farther wanted.

So much for the Royal Society as concerns myself. I must add, that there is not a spark of party feeling against those who wilfully remain outside. The better minds of course know better; and the smaller *savants* look complacently on the idea of an outer world which makes *élite* of them. I have done such a thing as serve on a committee of the Society, and report on a paper: they had the sense to ask, and I had the sense to see that none of my opinions were compromised by compliance. And I will be of any use which does not involve the status of *homo trium literarum*; as I have elsewhere explained, I would gladly be *Fautor Realis Scientiæ*, but I would not be taken for *Falsæ Rationis Sacerdos*.

Nothing worse will ever happen to me than the smile which individuals bestow on a man who does not *groove*. Wisdom, like religion, belongs to majorities; who can wonder that it should be so thought, when it is so clearly pictured in the New Testament from one end to the other?

The counterpart of *paradox*, the isolated opinion of one or of few, is the general opinion held by all the rest; and the counterpart of false and absurd paradox is what is called the 'vulgar error,' the *pseudodox*. There is one great work on this last subject, the *Pseudodoxia Epidemica* of Sir Thomas Browne, the famous author of the *Religio Medici*; it usually goes by the name of Browne 'On Vulgar Errors' (1st ed. 1646; 6th, 1672). A careful analysis of this work would show that vulgar errors are frequently opposed by scientific errors; but good sense is always good sense, and Browne's book has a vast quantity of it.

As an example of bad philosophy brought against bad observation. The Amphisbæna serpent was supposed to have two heads, one at each end; partly from its shape, partly because it runs backwards as well as forwards. On this Sir Thomas Browne makes the following remarks:—

And were there any such species or natural kind of animal, it would be hard to make good those six positions of body which, according to the three dimensions, are ascribed unto every Animal; that is, *infra, supra, ante, retro, dextrosum, sinistrosum*: for if (as it is determined) that be the anterior and upper part wherein the senses are placed, and that the posterior and lower part which is opposite thereunto, there is no inferior or former part in this Animal; for the senses, being placed at both extreams, doth make both ends anterior, which is impossible; the terms being Relative, which mutually subsist, and are not without each other. And therefore this duplicity was ill contrived to place one head at both extreams, and had been more tolerable to have settled three or four at one. And therefore also Poets have been more reasonable than Philosophers, and *Geryon* or *Cerberus* less monstrous than *Amphisbæna*.

There may be paradox upon paradox: and there is a good instance in the eighth century in the case of Virgil, an Irishman,

Bishop of Salzburg and afterwards Saint, and his quarrels with
Boniface, an Englishman, Archbishop of Mentz, also afterwards
Saint. All we know about the matter is, that there exists a
letter of 748 from Pope Zachary, citing Virgil—then, it seems, at
most a simple priest, though the Pope was not sure even of that
—to Rome to answer the charge of maintaining that there is
another world (*mundus*) under our earth (*terra*), with another
sun and another moon. Nothing more is known : the letter
contains threats in the event of the charge being true ; and there
history drops the matter. Since Virgil was afterwards a Bishop
and a Saint, we may fairly conclude that he died in the full
flower of orthodox reputation. It has been supposed — and it
seems probable—that Virgil maintained that the earth is peopled
all the way round, so that under some spots there are antipodes ;
that his contemporaries, with very dim ideas about the roundness
of the earth, and most of them with none at all, interpreted him
as putting another earth under ours—turned the other way,
probably, like the second piece of bread-and-butter in a sandwich,
with a sun and moon of its own. In the eighth century this
would infallibly have led to an underground Gospel, an underground Pope, and an underground Avignon for him to live in.
When, in later times, the idea of inhabitants for the planets
was started, it was immediately asked whether they had sinned,
whether Jesus Christ died for *them*, whether their wine and their
water could be lawfully used in the sacraments, &c.

On so small a basis as the above has been constructed a companion case to the persecution of Galileo. On one side the
positive assertion, with indignant comment, that Virgil was
deposed for antipodal heresy, on the other, serious attempts at
justification, palliation, or mystification. Some writers say that
Virgil was found guilty ; others that he gave satisfactory explanation, and became very good friends with Boniface : for all
which see Bayle. Some have maintained that the antipodist was
a different person from the canonised bishop : there is a second
Virgil, made to order. When your shoes pinch, and will not
stretch, always throw them away and get another pair : the same
with your facts. Baronius was not up to the plan of a substitute :
his commentator Pagi (probably writing about 1690) argues for
it in a manner which I think Baronius would not have approved.
This Virgil was perhaps a slippery fellow. The Pope says he
hears that Virgil pretended licence from him to claim one of
some new bishoprics : this he declares is totally false. It is part
of the argument that such a man as this could not have been

created a Bishop and a Saint: on this point there will be opinions and opinions.[1] Lactantius, four centuries before, had laughed at the antipodes in a manner which seems to be ridicule thrown on the idea of the earth's roundness. Ptolemy, without reference to the antipodes, describes the extent of the inhabited part of the globe in a way which shows that he could have had no objection to men turned opposite ways. Probably, in the eighth century, the roundness of the earth was matter of thought only to astronomers. It should always be remembered, especially by those who affirm persecution of a true opinion, that but for our knowing from Lactantius that the antipodal notion had been matter of assertion and denial among theologians, we could never have had any great confidence in Virgil really having maintained the simple theory of the existence of antipodes. And even now we are not entitled to affirm it as having historical proof: the evidence goes to Virgil having been charged with very absurd notions, which it seems more likely than not were the absurd constructions which ignorant contemporaries put upon sensible opinions of his.

One curious part of this discussion is, that neither side has allowed Pope Zachary to produce evidence to character. He shall have been an Urban, say the astronomers; an Urban he ought to have been, say the theologians. What sort of man was Zachary? He was eminently sensible and conciliatory; he contrived to make northern barbarians hear reason in a way which puts him high among that section of the early popes who had the knack of managing uneducated swordsmen. He kept the peace in Italy to an extent which historians mention with admiration. Even Bale, that Maharajah of pope-haters, allows himself to quote in favour of Zachary, that 'multa Papalem dignitatem decentia, eademque præclara (scilicet) opera confecit.' And this, though so willing to find fault that, speaking of Zachary putting a little geographical description of the earth on the portico of the Lateran Church, he insinuates that it was intended to affirm that the Pope was lord of the whole. Nor can he say how long Zachary held the see, except by announcing his death in 752, 'cum decem annis pestilentiæ sedi præfuisset.'

[1] An Irish antiquary informs me that Virgil is mentioned in annals, at A.D. 784, as Verghil, i.e. the geometer, Abbot of Achadhbo [and Bishop of Saltzburg], died in Germany in the thirteenth year of his bishoprick.' No allusion is made to his opinions; but it seems he was, by tradition, a mathematician. The Abbot of Aghabo (Queen's County) was canonised by Gregory IX., in 1233. The story of the second, or scapegoat, Virgil would be much damaged by the character given to the real bishop, if there were anything in it to dilapidate.

There was another quarrel between Virgil and Boniface which is an illustration. An ignorant priest had baptised ' in nomine Patria, et Filia, et Spiritua Sancta.' Boniface declared the rite null and void ; Virgil maintained the contrary ; and Zachary decided in favour of Virgil, on the ground that the absurd form was only ignorance of Latin, and not heresy. It is hard to believe that this man deposed a priest for asserting the whole globe to be inhabited. To me the little information that we have seems to indicate—but not with certainty—that Virgil maintained the antipodes : that his ignorant contemporaries travestied his theory into that of an underground cosmos ; that the Pope cited him to Rome to explain his system, which, as reported, looked like what all would then have affirmed to be heresy ; that he gave satisfactory explanations, and was dismissed with honour. It may be that the educated Greek monk, Zachary, knew his Ptolemy well enough to guess what the asserted heretic would say ; we have seen that he seems to have patronised geography. The *description* of the earth, according to historians, was a *map* ; this Pope may have been more ready than another to prick up his ears at any rumour of geographical heresy, from hope of information. And Virgil, who may have entered the sacred presence as frightened as Jacquard, when Napoleon I. sent for him and said, with a stern voice and threatening gesture, ' You are the man who can tie a knot in a stretched string,' may have departed as well pleased as Jacquard with the riband and pension which the interview was worth to him.

A word more about Baronius. If he had been pope, as he would have been but for the opposition of the Spaniards, and if he had lived ten years longer than he did, and if Clavius, who would have been his astronomical adviser, had lived five years longer than he did, it is probable, nay almost certain, that the great exhibition, the proceeding against Galileo, would not have furnished a joke against theology in all time to come. For Baronius was sensible and witty enough to say that in the Scriptures the Holy Spirit intended to teach how to go to Heaven, not how Heaven goes ; and Clavius, in his last years, confessed that the whole system of the heavens had broken down, and must be mended.

The manner in which the Galileo case, a reality, and the Virgil case, a fiction, have been hawked against the Roman see are enough to show that the Pope and his adherents have not cared much about physical philosophy. In truth, orthodoxy has always had other fish to fry. Physics, which in modern times has almost usurped the name *philosophy*, in England at least, has felt a little disposed to clothe herself with all the honours of persecution which belong to the real owner of the name. But the bishops, &c. of the middle ages knew that the contest between nominalism and realism, for instance, had a hundred times more bearing upon orthodoxy than anything in astronomy, &c. A wrong notion about *substance* might play the mischief with *transubstantiation*.

The question of the earth's motion was the single point in which orthodoxy came into real contact with science. Many students of physics were suspected of magic, many of atheism : but, stupid as the mistake may have been, it was *bonâ fide* the magic or the atheism, not the physics, which was assailed. In the astronomical case it was the very doctrine, as a doctrine, independently of consequences, which was the *corpus delicti* : and this because it contradicted the Bible. And so it did ; for the stability of the earth is as clearly assumed from one end of the Old Testament to the other as the solidity of iron. Those who take the Bible to be *totidem verbis* dictated by the God of Truth can refuse to believe it ; and they make strange reasons. They undertake, *à priori*, to settle Divine intentions. The Holy Spirit did not *mean* to teach natural philosophy : this they know beforehand ; or else they infer it from finding that the earth does move, and the Bible says it does not. Of course, ignorance apart, every word is truth, or the writer did not mean truth. But this puts the whole book on its trial : for we never can find out what the writer meant, until we otherwise find out what is true. Those who like may, of course, declare for an inspiration over which they are to be viceroys ; but common sense will either accept verbal meaning or deny verbal inspiration.

Questiones Morales, folio, 1489 [Paris]. By T. Buridan.

This is the title from the Hartwell Catalogue of Law Books. I suppose it is what is elsewhere called the 'Commentary on the Ethics of Aristotle,' printed in 1489. Buridan (died about 1358) is the creator of the famous ass which, as *Burdin's* ass, was current in Burgundy, perhaps is, as a vulgar proverb. Spinoza says it was a jenny ass, and that a man would not have been so foolish; but whether the compliment is paid to human or to masculine character does not appear—perhaps to both in one. The story *told* about the famous paradox is very curious. The Queen of France, Joanna or Jeanne, was in the habit of sewing her lovers up in sacks, and throwing them into the Seine; not for blabbing, but that they might not blab—certainly the safer plan. Buridan was exempted, and, in gratitude, invented the sophism. What it has to do with the matter has never been explained. Assuredly *qui facit per alium facit per se* will convict Buridan of prating. The argument is as follows, and is seldom told in full. Buridan was for free-will—that is, will which determines conduct, let motives be ever so evenly balanced. An ass is *equally* pressed by hunger and by thirst; a bundle of hay is on one side, a pail of water on the other. Surely, you will say, he will not be ass enough to die for want of food or drink; he will then make a choice—that is, will choose between alternatives of equal force. The problem became famous in the schools; some allowed the poor donkey to die of indecision; some denied the possibility of the balance, which was no answer at all.

The following question is more difficult, and involves free-will to all who answer—'Which you please.' If the northern hemisphere were land, and all the southern hemisphere water, ought we to call the northern hemisphere an island, or the southern hemisphere a lake? Both the questions would be good exercises for paradoxers who must be kept employed, like Michael Scott's devils. The wizard knew nothing about squaring the circle, &c., so he set them to make ropes out of sea sand, which puzzled them. Stupid devils! much of our glass is sea sand, and it makes beautiful thread. Had Michael set them to square the circle or to find a perpetual motion, he would have done his work much better. But all this is conjecture: who knows that I have not hit on the very plan he adopted? Perhaps the whole race of paradoxers on hopeless subjects are Michael's subordinates, condemned to transmigration after transmigration, until their task is done.

The above was not a bad guess. A little after the time when the famous Pascal papers were produced, I came into possession of a correspondence which, but for these papers, I should have held too incredible to be put before the world. But when one sheep leaps the ditch, another will follow : so I gave the following account in the *Athenæum* of October 5, 1867 :—

The recorded story is that Michael Scott, being bound by contract to procure perpetual employment for a number of young demons, was worried out of his life in inventing jobs for them, until at last he set them to make ropes out of sea sand, which they never could do. We have obtained a very curious correspondence between the wizard Michael and his demon-slaves ; but we do not feel at liberty to say how it came into our hands. We much regret that we did not receive it in time for the British Association. It appears that the story, true as far as it goes, was never finished. The demons easily conquered the rope difficulty, by the simple process of making the sand into glass, and spinning the glass into thread, which they twisted. Michael, thoroughly disconcerted, hit upon the plan of setting some to square the circle, others to find the perpetual motion, &c. He commanded each of them to transmigrate from one human body into another, until their tasks were done. This explains the whole succession of cyclometers, and all the heroes of the Budget. Some of this correspondence is very recent; it is much blotted, and we are not quite sure of its meaning : it is full of figurative allusions to driving something illegible down a steep into the sea. It looks like a humble petition to be allowed some diversion in the intervals of transmigration ; and the answer is—

Rumpat et serpens iter institutum,

—a line of Horace, which the demons interpret as a direction to come athwart the proceedings of the Institute by a sly trick. Until we saw this, we were suspicious of M. Libri : the unvarying blunders of the correspondence look like knowledge. To be always out of the road requires a map : genuine ignorance occasionally lapses into truth. We thought it possible M. Libri might have played the trick to show how easily the French are deceived ; but with our present information, our minds are at rest on the subject. We see M. Chasles does not like to avow the real source of information : he will not confess himself a spiritualist.

Philo of Gadara is asserted by Montucla, on the authority of Eutocius, the commentator on Archimedes, to have squared the circle within the *ten-thousandth* part of a unit, that is, to *four* places of decimals. A modern classical dictionary represents it as done by Philo to *ten thousand* places of decimals. Lacroix comments on Montucla to the effect that *myriad* (in Greek *ten thousand*) is here used as we use it, vaguely, for an immense number. On looking into Eutocius, I find that not one definite word is said about the extent to which Philo carried the matter. I give a translation of the passage :—

We ought to know that Apollonius Pergæus, in his Ocytocium [this work is lost], demonstrated the same by other numbers, and came nearer, which seems more accurate, but has nothing to do with Archimedes; for, as before said, he aimed only at going near enough for the wants of life. Neither is Porus of Nicæa fair when he takes Archimedes to task for not giving a line accurately equal to the circumference. He says in his Cerii that his teacher, Philo of Gadara, had given a more accurate approximation (εἰς ἀκριβεστέρους ἀριθμοὺς ἀγαγειν) than that of Archimedes, or than 7 to 22. But all these [the rest as well as Philo] miss the intention. They multiply and divide by *tens of thousands*, which no one can easily do, unless he be versed in the logistics [fractional computation] of Magnus [now unknown].

Montucla, or his source, ought not to have made this mistake. He had been at the Greek to correct Philo *Gadetanus*, as he had often been called, and he had brought away and quoted ἀπὸ Γαδαρων. Had he read two sentences further, he would have found the mistake.

We here detect a person quite unnoticed hitherto by the moderns, Magnus the arithmetician. The phrase is ironical; it is as if we should say, ' To do this a man must be deep in Cocker.' Accordingly, Magnus, Baveme, and Cocker, are three personifica-. tions of arithmetic; and there may be more.

Aristotle, treating of the category of relation, denies that the quadrature has been found, but appears to assume that it can be done. Boethius, in his comment on the passage, says that it has been done since Aristotle, but that the demonstration is too long for him to give. Those who have no notion of the quadrature question may look at the *English Cyclopædia*, art. 'Quadrature of the Circle.'

Tetragonismus. Id est circuli quadratura per Campanum, Archimedem Syracusanum, atque Boetium mathematicæ perspicacissimos adinventa.—At the end, Impressum Venetiis per Ioan. Bapti. Sessa. Anno ab incarnatione Domini, 1503. Die 28 Augusti.

This book has never been noticed in the history of the subject, and I cannot find any mention of it. The quadrature of Campanus takes the ratio of Archimedes, 7 to 22, to be absolutely correct; the account given of Archimedes is not a translation of his book; and that of Boetius has more than is in Boet*h*ius. This book must stand, with the next, as the earliest in print on the subject, until further showing: Murhard and Kastner have nothing so early. It is edited by Lucas Gauricus, who has given a short preface. Luca Gaurico, Bishop of Civita Ducale, an astrologer of astrologers, published this work at about thirty years of age, and lived to eighty-two. His works are collected in folios, but I do not know whether they contain this production. The poor fellow could never tell his own fortune, because his father neglected to note the hour and minute of his birth. But if there had been anything in astrology, he could have worked back, as Adams and Leverrier did when they caught Neptune: at sixty he could have examined every minute of his day of birth, by the events of his life, and so would have found the right minute. He could then have gone on, by rules of prophecy. Gauricus was the mathematical teacher of Joseph Scaliger, who did him no credit, as we shall see.

In hoc opere contenta Epitome. . . . . Liber de quadratura Circuli. . . . . Paris, 1503, folio.

The quadrator is Charles Bovillus, who adopted the views of Cardinal Cusa, presently mentioned. Montucla is hard on his compatriot, who, he says, was only saved from the laughter of geometers by his obscurity. Persons must guard against most historians of mathematics in one point: they frequently attribute to *his own* age the obscurity which a writer has in *their own* time. This tract was printed by Henry Stephens, at the instigation of Faber Stapulensis, and is recorded by Dechales, &c. It was also introduced into the 'Margarita Philosophica' of 1815, in the same appendix with the new perspective from Viator. This is not extreme obscurity, by any means. The quadrature deserved it; but that is another point.

It is stated by Montucla that Bovillus makes $\pi = \sqrt{10}$. But-Montucla cites a work of 1507, *Introductorium Geometricum*, which I have never seen. He finds in it an account which Bovillus gives of the quadrature of the peasant labourer, and describes it as agreeing with his own. But the description makes $\pi = 3\frac{1}{8}$, which it thus appears Bovillus could not distinguish from $\sqrt{10}$. It seems also that this $3\frac{1}{8}$, about which we shall see so much in the sequel, takes its rise in the thoughtful head of a poor labourer. It does him great honour, being so near the truth, and he having no means of instruction. In our day, when an ignorant person chooses to bring his fancy forward in opposition to demonstration which he will not study, he is deservedly laughed at.

Mr. James Smith, of Liverpool—hereinafter notorified—attributes the first announcement of $3\frac{1}{8}$ to M. Joseph Lacomme, a French well-sinker, of whom he gives the following account:—

In the year 1836, at which time Lacomme could neither read nor write, he had constructed a circular reservoir and wished to know the quantity of stone that would be required to pave the bottom, and for this purpose called on a professor of mathematics. On putting his question and giving the diameter, he was surprised at getting the following answer from the Professor—' *Qu'il lui était impossible de le lui dire au juste, attendu que personne n'avait encore pu trouver d'une manière exacte le rapport de la circonférence au diamètre.*' From this he was led to attempt the solution of the problem. His first process was purely mechanical, and he was so far convinced he had made the discovery that he took to educating himself, and became an expert arithmetician, and then found that arithmetical results agreed with his mechanical experiments. He appears to have eked out a bare existence for many years by teaching arithmetic, all the time struggling to get a hearing from some of the learned societies, but without success. In the year 1855 he found his way to Paris, where, as if by accident, he made the acquaintance of a young gentleman, son of M. Winter, a commissioner of police, and taught him his peculiar methods of calculation. The young man was so enchanted that he strongly recommended Lacomme to his father, and subsequently through M. Winter he obtained an introduction to the President of the Society of Arts and Sciences of Paris. A committee of the society was appointed to examine and report upon his discovery, and the society at its *séance* of March 17, 1856, awarded a silver medal of the first class to M. Joseph Lacomme for his discovery of the true ratio of diameter to circumference in a circle. He subsequently received three other medals from other societies. While writing this I have his likeness before me, with his-medals on his breast, which stands as a frontispiece to a short biography of this extraordinary man, for which I am indebted to the gentleman who did me the honour to publish a French translation of the pamphlet I distributed at the meeting of the British Association for the Advancement of Science, at Oxford, in 1860.—*Correspondent*, May 3, 1866.

My inquiries show that the story of the medals is not incredible. There are at Paris little private societies which have not so much claim to be exponents of scientific opinion as our own Mechanics' Institutes. Some of them were intended to give a false lustre: as the 'Institut Historique,' the members of which are 'Membre de l'Institut Historique.' That M. Lacomme should have got four medals from societies of this class is very possible: that he should have received one from any society at Paris which has the least claim to give one is as yet simply incredible.

Nicolai de Cusa Opera Omnia. Venice, 1514. 3 vols. folio.

The real title is 'Hæc accurata recognitio trium voluminum operum clariss. P. Nicolai Cusæ . . . proxime sequens pagina monstrat.' Cardinal Cusa, who died in 1464, is one of the earliest modern attempters. His quadrature is found in the second volume, and is now quite unreadable. In these early days every quadrator found a geometrical opponent, who finished him. Regiomontanus did this office for the Cardinal.

De Occulta Philosophia libri III. By Henry Cornelius Agrippa. Lyons, 1550, 8vo.
De incertitudine et vanitate scientiarum. By the same. Cologne, 1531, 8vo.

The first editions of these works were of 1530, as well as I can make out; but the first was in progress in 1510. In the second work Agrippa repents of having wasted time on the magic of the first; but all those who actually deal with demons are destined to eternal fire with Jamnes and Mambres and Simon Magus. This means, as is the fact, that his occult philosophy did not actually enter upon *black* magic, but confined itself to the power of the stars, of numbers, &c. The fourth book, which appeared after the death of Agrippa, and really concerns dealing with evil spirits, is undoubtedly spurious. It is very difficult to make out what Agrippa really believed on the subject. I have introduced his books as the most marked specimens of treatises on magic, a paradox of our day, though not far from orthodoxy in his; and here I should have ended my notice, if I had not casually found something more interesting to the reader of our day.

D

Walter Scott, it is well known, was curious on all matters connected with magic, and has used them very widely. But it is hardly known how much pains he has taken to be correct, and to give the real thing. The most decided detail of a magical process which is found in his writings is that of Dousterswivel in 'The Antiquary'; and it is obvious, by his accuracy of process, that he does not intend the adept for a mere impostor, but for one who had a lurking belief in the efficacy of his own processes, coupled with intent to make a fradulent use of them. The materials for the process are taken from Agrippa. I first quote Mr. Dousterswivel:

> . . . I take a silver plate when she [the moon] is in her fifteenth mansion, which mansion is in de head of *Libra,* and I engrave upon one side de worts *Schedbarschemoth Schartachan* [*ch* should be *t*]—dat is, de Intelligence of de Intelligence of de moon—and I make his picture like a flying serpent with a turkey-cock's head—vary well—Then upon this side I make de table of de moon, which is a square of nine, multiplied into itself, with eighty-one numbers [nine] on every side, and diameter nine. . . .

In the 'De Occulta Philosophia,' p. 290, we find that the fifteenth mansion of the moon *incipit capite Libræ,* and is good *pro extrahendis thesauris,* the object being to discover hidden treasure. In p. 246, we learn that a *silver* plate must be used with the moon. In p. 248, we have the words which denote the Intelligence, &c. But, owing to the falling of a number into a wrong line, or the misplacement of a line, one or other—which takes place in all the editions I have examined—Scott has, sad to say, got hold of the wrong words; he has written down the *demon of the demons* of the moon. Instead of the gibberish above, it should have been *Malcha betharsisim hed beruah schehakim.* In p. 253, we have the magic square of the moon, with eighty-one numbers, and the symbol for the Intelligence, which Scott likens to a flying serpent with a turkey-cock's head. He was obliged to say something; but I will stake my character—and so save a woodcut—on the scratches being more like a pair of legs, one shorter than the other, without a body, jumping over a six-barred gate placed side uppermost. Those who thought that Scott forged his own nonsense, will henceforth stand corrected. As to the spirit Peolphan, &c., no doubt Scott got it from the authors he elsewhere mentions, Nicolaus Remigius and Petrus Thyracus; but this last word should be Thyræus.

The tendency of Scott's mind towards prophecy is very marked, and it is always fulfilled. Hyder, in his disguise, calls out to Tippoo—' Cursed is the prince who barters justice for lust; he shall die in the gate by the sword of the stranger.' Tippoo was killed in a gateway at Seringapatam.

Orontii Finaei. . . Quadratura Circuli. Paris, 1544, 4to.

Orontius squared the circle out of all comprehension; but he was killed by a feather from his own wing. His former pupil, John Buteo, the same who—I believe for the first time—calculated the question of Noah's ark, as to its power to hold all the animals and stores, unsquared him completely. Orontius was the author of very many works, and died in 1555. Among the laudatory verses which, as was usual, precede this work, there is one of a rare character : a congratulatory ode to the wife of the author. The French now call this writer Oronce Finée ; but there is much difficulty about delatinisation. Is this more correct than Oronce Fine, which the translator of De Thou uses? Or than Horonce Phine, which older writers give? I cannot understand why M. de Viette should be called Viète, because his Latin name is Vieta. It is difficult to restore Buteo ; for not only now is *butor* a block-head as well as a bird, but we really cannot know what kind of bird Buteo stood for. We may be sure that Madame Fine was Denise Blanche ; for Dionysia Candida can mean nothing else. Let her shade rejoice in the fame which Hubertus Sussannæus has given her.

I ought to add that the quadrature of Orontius, and solutions of all the other difficulties, were first published in ' De Rebus Mathematicis Hactenus Desideratis,' of which I have not the date.

Nicolai Raymari Ursi Dithmarsi Fundamentum Astronomicum, id est, nova doctrina sinuum et triangulorum. . . . Strasburg, 1588, 4to.

People choose the name of this astronomer for themselves: I take *Ursus*, because he *was* a bear. This book gave the quadra-ture of Simon Duchesne, or à Quercu, which excited Peter Metius, as presently noticed. It also gave that unintelligible reference to Justus Byrgius which has been used in the discussion about the invention of logarithms.

The real name of Duchesne is Van der Eycke. I have met with a tract in Dutch, *Letterkundige Aanteekeningen,* upon Van Eycke, Van Ceulen, &c., by J. J. Dodt van Flensburg, which I make out to be since 1841 in date. I should much like a trans-

lation of this tract to be printed, say in the *Phil. Mag.*   Dutch
would be clear English if it were properly spelt.   For example,
*learn-master* would be seen at once to be *teacher*; but they will
spell it *leermeester*.   *Of these* they write as *van deze; widow*
they make *weduwe*.   All this is plain to me, who never saw a
Dutch dictionary in my life; but many of their mispellings are
quite unconquerable.

> Jacobus Falco Valentinus, miles Ordinis Montesiani, hanc circuli
> quadraturam invenit.   Antwerp, 1589, 4to.

The attempt is more than commonly worthless; but as Montucla and others have referred to the verses at the end, and as the
tract is of the rarest, I will quote them:—

*Circulus loquitur.*

Vocabar ante circulus
Eramque curvus undique
Ut alta solis orbita
Et arcus ille nubium.
Eram figura nobilis
Carensque sola origine
Carensque sola termino.
Modo indecora prodeo
Novisque fœdor angulis.
Nec hoc peregit Archytas
Neque Icari pater neque
Tuus Iapete filius.
Quis ergo casus aut Deus
Meam quadravit aream?

*Respondet auctor.*

Ad alta Turiæ ostia
Lacumque limpidissimum
Sita est beata civitas
Parum Saguntus abfuit
Abestque Sucro plusculum.
Hic est poeta quispiam
Libenter astra consulens
Sibique semper arrogans
Negata doctioribus.
Senex ubique cogitans
Sui frequenter immemor
Nec explicare circinum

Nec exarare lineas
Sciens ut ipse præedicat.
Hic ergo bellus artifex
Tuam quadravit aream.

Falco's verses are pretty, if the ˘ ¯ mysteries be correct; but of these things I have forgotten—what I knew. [One mistake has been pointed out to me: it is Archȳtas].

As a specimen of the way in which history is written, I copy the account which Montucla—who is accurate when he writes about what he has seen—gives of these verses. He gives the date 1587; he places the verses at the beginning instead of the end; he says the circle thanks its quadrator affectionately; and he says the good and modest chevalier gives all the glory to the patron saint of his order. All of little consequence, as it happens; but writing at second-hand makes as complete mistakes about more important matters.

> Petri Bungi Bergomatis Numerorum mysteria. Bergomi [Bergamo], 1591, 4to. Second Edition.

The first edition is said to be of 1585; the third, Paris, 1618. Bungus is not for my purpose on his own score, but those who gave the numbers their mysterious characters: he is but a collector. He quotes or uses 402 authors, as we are informed by his list: this just beats Warburton, whom some eulogist or satirist, I forget which, holds up as having used 400 authors in some one work. Bungus goes through 1, 2, 3, &c., and gives the account of everything remarkable in which each number occurs; his accounts not being always mysterious. The numbers which have nothing to say for themselves are omitted: thus there is a gap between 50 and 60. In treating 666, Bungus, a good Catholic, could not compliment the Pope with it, but he fixes it on Martin Luther with a little forcing. If from A to I represent 1–10, from K to S 10–90, and from T to Z 100–500, we see—

$$\begin{array}{cccccc} \text{M} & \text{A} & \text{R} & \text{T} & \text{I} & \text{N} \\ 30 & 1 & 80 & 100 & 9 & 40 \end{array} \qquad \begin{array}{cccccc} \text{L} & \text{U} & \text{T} & \text{E} & \text{R} & \text{A} \\ 20 & 200 & 100 & 5 & 80 & 1 \end{array}$$

which gives 666. Again, in Hebrew, *Lulter* does the same :—

$$\begin{array}{ccccc} \text{ר} & \text{ת} & \text{ל} & \text{ו} & \text{ל} \\ 200 & 400 & 30 & 6 & 30 \end{array}$$

And thus two can play at any game. The second is better than the first: to Latinise the surname and not the Christian name is very unscholarlike. The last number mentioned is a thousand millions; all greater numbers are dismissed in half a page. Then follows an accurate distinction between *number* and *multitude*—a thing much wanted both in arithmetic and logic.

What may be the use of such a book as this? The last occasion on which it was used was the following. Fifteen or sixteen years ago the Royal Society determined to restrict the number of yearly admissions to fifteen men of science, and noblemen *ad libitum*; the men of science being selected and recommended by the Council, with a power, since practically surrendered, to the Society to elect more. This plan appears to me to be directly against the spirit of their charter, the true intent of which is, that all who are fit should be allowed to promote natural knowledge in association, from and after the time at which they are both fit and willing. It is also working more absurdly from year to year; the tariff of fifteen per annum will soon amount to the practical exclusion of many who would be very useful. This begins to be felt already, I suspect. But, as appears above, the body of the Society has the remedy in its own hands. When the alteration was discussed by the Council, my friend the late Mr. Galloway, then one of the body, opposed it strongly, and inquired particularly into the reason why *fifteen*, of all numbers, was the one to be selected. Was it because fifteen is seven and eight, typifying the Old Testament Sabbath, and the New Testament day of the resurrection following? Was it because Paul strove fifteen days against Peter, proving that he was a doctor both of the Old and New Testament? Was it because the prophet Hosea bought a lady for fifteen pieces of silver? Was it because, according to Micah, seven shepherds and eight chiefs should waste the Assyrians? Was it because Ecclesiastes commands equal reverence to be given to both Testaments—such was the interpretation—in the words ' Give a portion to seven, and also to eight'? Was it because the waters of the Deluge rose fifteen cubits above the mountains?—or because they lasted fifteen decades of days? Was it because Ezekiel's temple had fifteen steps? Was it because Jacob's ladder has been supposed to have had fifteen steps? Was it because fifteen years were added to the life of Hezekiah? Was it because the feast of unleavened bread was on the fifteenth day of the month? Was it because the scene of the Ascension was fifteen stadia from Jerusalem? Was it because the stone-masons and porters employed in Solomon's temple amounted to fifteen myriads? &c. The Council were amused and astounded by the volley of fifteens which was fired at them; they knowing nothing about Bungus, of which Mr. Galloway—who did not, as the French say, indicate his sources—possessed the copy now before me. In giving this anecdote I give a specimen of the book, which is exceedingly rare. Should another edition ever appear, which is not very probable, he would be but a bungling Bungus who should forget the *fifteen* of the Royal Society.

[I make a remark on the different colours which the same person gives to one story, according to the bias under which he tells it. My friend Galloway told me how he had quizzed the Council of the Royal Society, to my great amusement. Whenever I am struck by the words of any one, I carry away a vivid recollection of position, gestures, tones, &c. I do not know whether this be common or uncommon. I never recall this joke without seeing before me my friend, leaning against his bookcase, with Bungus open in his hand, and a certain half-depreciatory tone which he often used when speaking of himself. Long after his death, an F.R.S. who was present at the discussion, told me the story. I did not say I had heard it, but I watched him, with Galloway at the bookcase before me. I wanted to see whether the two would agree as to the fact of an enormous budget of fifteens having been fired at the Council, and they did agree perfectly. But when the paragraph of the Budget appeared in the *Athenæum,* my friend, who seemed rather to object to the *shewing-up,* assured me that the thing was grossly exaggerated; there was indeed a fifteen or two, but nothing like the number I had given. I had, however, taken sharp note of the previous narration.

I will give another instance. An Indian officer gave me an account of an elephant, as follows. A detachment was on the march, and one of the gun-carriages got a wheel off the track, so that it was also off the ground, and hanging over a precipice. If the bullocks had moved a step, carriages, bullocks, and all must have been precipitated. No one knew what could be done until some one proposed to bring up an elephant, and let him manage it his own way. The elephant took a moment's survey of the fix, put his trunk under the axle of the free wheel, and waited. The surrounders, who saw what he meant, moved the bullocks gently forward, the elephant followed, supporting the axle, until there was ground under the wheel, when he let it quietly down. From all I had heard of the elephant, this was not too much to believe. But when, years afterwards, I reminded my friend

of his story, he assured me that I had misunderstood him, that
the elephant was *directed* to put his trunk under the wheel, and
saw in a moment why. This is reasonable sagacity, and very
likely the correct account; but I am quite sure that, in the fit
of elephant-worship under which the story was first told, it was
told as I have first stated it.]

[Jordani Bruni Nolani de Monade, Numero et Figura . . . item de
Innumerabilibus, Immenso, et Infigurabili. . . Frankfort, 1591,
8vo.

I cannot imagine how I came to omit a writer whom I have
known so many years, unless the following story will explain it.
The officer reproved the boatswain for perpetual swearing; the
boatswain answered that he heard the officers swear. 'Only in
an emergency,' said the officer. 'That's just it,' replied the
other; 'a boatswain's life is a life of 'mergency.' Giordano
Bruno was all paradox; and my mind was not alive to his
paradoxes, just as my ears might have become dead to the boatswain's oaths. He was, as has been said, a vorticist before
Descartes, an optimist before Leibnitz, a Copernican before
Galileo. It would be easy to collect a hundred strange opinions
of his. He was born about 1550, and was roasted alive at Rome,
February 17, 1600, for the maintenance and defence of the holy
Church, and the rights and liberties of the same. These last
words are from the writ of our own good James I., under which
Leggatt was roasted at Smithfield, in March 1612; and if I had a
copy of the instrument under which Wightman was roasted at
Litchfield, a month afterwards, I daresay I should find something
quite as edifying. I extract an account which I gave of Bruno
in the *Comp. Alm.* for 1855 :—

He was first a Dominican priest, then a Calvinist; and was roasted
alive at Rome, in 1600, for as many heresies of opinion, religious and
philosophical, as ever lit one fire. Some defenders of the papal cause
have at least worded their accusations so to be understood as imputing
to him villainous actions. But it is positively certain that his death
was due to opinions alone, and that retractation, even after sentence,
would have saved him. There exists a remarkable letter, written from
Rome on the very day of the murder, by Scioppius (the celebrated
scholar, a waspish convert from Lutheranism, known by his hatred to
Protestants and Jesuits) to Rittershusius, a well-known Lutheran
writer on civil and canon law, whose works are in the index of prohibited books. This letter has been reprinted by Libri (vol. iv. p. 407).
The writer informs his friend (whom he wished to convince that even a
Lutheran would have burnt Bruno) that all Rome would tell him that

Bruno died for Lutheranism; but this is because the Italians do not know the difference between one heresy and another, in which simplicity (says the writer) may God preserve them. That is to say, they knew the difference between a live heretic and a roasted one by actual inspection, but had no idea of the difference between a Lutheran and a Calvinist. The countrymen of Boccaccio would have smiled at the idea which the German scholar entertained of them. They said Bruno was burnt for Lutheranism, a name under which they classed all Protestants: and they are better witnesses than Schopp, or Scioppius. He then proceeds to describe to his Protestant friend (to whom he would certainly not have omitted any act which both their Churches would have condemned) the mass of opinions with which Bruno was charged; as that there are innumerable worlds, that souls migrate, that Moses was a magician, that the Scriptures are a dream, that only the Hebrews descended from Adam and Eve, that the devils would be saved, that Christ was a magician and deservedly put to death, &c. In fact, says he, Bruno has advanced all that was ever brought forward by all heathen philosophers, and by all heretics, ancient and modern. A time for retractation was given, both before sentence and after, which should be noted, as well for the wretched palliation which it may afford, as for the additional proof it gives that opinions, and opinions only, brought him to the stake. In this medley of charges the Scriptures are a dream, while Adam, Eve, devils, and salvation are truths, and the Saviour a deceiver. We have examined no work of Bruno except the *De Monade*, &c., mentioned in the text. A strong though strange *theism* runs through the whole, and Moses, Christ, the Fathers, &c., are cited in a manner which excites no remark either way. Among the versions of the cause of Bruno's death is *atheism* : but this word was very often used to denote rejection of revelation, not merely in the common course of dispute, but by such writers, for instance, as Brucker and Morhof. Thus Morhof says of the *De Monade*, &c., that it exhibits no manifest signs of atheism. What he means by the word is clear enough, when he thus speaks of a work which acknowledges God in hundreds of places, and rejects opinions as blasphemous in several. The work of Bruno in which his astronomical opinions are contained is *De Monade*, &c. (Frankfort, 1591, 8vo). He is the most thorough-going Copernican possible, and throws out almost every opinion, true or false, which has ever been discussed by astronomers, from the theory of innumerable inhabited worlds and systems to that of the planetary nature of comets. Libri (vol. iv.) has reprinted the most striking part of his expressions of Copernican opinion.

The Satanic doctrine that a Church may employ force in aid of its dogma is supposed to be obsolete in England, except as an individual paradox ; but this is difficult to settle. Opinions are much divided as to what the Roman Church would do in

England, if she could : any one who doubts that she claims the right does not deserve an answer. When the hopes of the Tractarian section of the High Church were in bloom, before the most conspicuous intellects among them had *transgressed* their ministry, that they might go to their own place, I had the curiosity to see how far it could be ascertained whether they held the only doctrine which makes me the personal enemy of a sect. I found in one of their tracts the assumption of a right to persecute, modified by an asserted conviction that force was not efficient. I cannot now say that this tract was one of the celebrated ninety ; and on looking at the collection I find it so poorly furnished with contents, &c., that nothing but searching through three thick volumes would decide. In these volumes I find, augmenting as we go on, declarations about the character and power of 'the Church' which have a suspicious appearance. The suspicion is increased by that curious piece of sophistry, No. 87, on religious reserve. The queer paradoxes of that tract leave us in doubt as to everything but this, that the church(man) is not bound to give his whole counsel in all things, and not bound to say what the things are in which he does not give it. It is likely enough that some of the 'rights and liberties' are but scantily described. There is now no fear ; but the time was when, if not fear, there might be a looking for of fear to come ; nobody could then be so sure as we now are that the lion was only asleep. There was every appearance of a harder fight at hand than was really found needful.

Among other exquisite quirks of interpretation in the No. 87 above mentioned is the following. God himself employs reserve ; he is said to be decked with light as with a garment (the old or prayer-book version of Psalm civ. 2). To an ordinary apprehension this would be a strong image of display, manifestation, revelation ; but there is something more. 'Does not a garment veil in some measure that which it clothes? Is not that very light concealment ?'

This No. 87, admitted into a series, fixes upon the managers of the series, who permitted its introduction, a strong presumption of that underhand intent with which they were charged. At the same time it is honourable to our liberty that this series could be published : though its promoters were greatly shocked when the Essayists and Bishop Colenso took a swing on the other side. When No. 90 was under discussion, Dr. Maitland, the librarian at Lambeth, asked Archbishop Howley a question about No. 89. ' I did not so much as know there *was* a No. 89,' was the answer. I am almost sure I have seen this in print, and quite sure that Dr. Maitland told it to me. It is creditable that there was so much freedom; but No. 90 was *too bad*, and was stopped.

The Tractarian mania has now (October 1866) settled down into a chronic vestment disease, complicated with fits of transubstantiation, which has taken the name of *Ritualism*. The common sense of our national character will not put up with a continuance of this grotesque folly; millinery in all its branches will at last be advertised only over the proper shops. I am told that the Ritualists give short and practical sermons; if so, they may do good in the end. The English Establishment has always contained those who want an excitement; the New Testament, in its plain meaning, can do little for them. Since the Revolution, Jacobitism, Wesleyanism, Evangelicism, Puseyism, and Ritualism, have come on in turn, and have furnished hot water for those who could not wash without it. If the Ritualists should succeed in substituting short and practical teaching for the high-spiced lectures of the doctrinalists, they will be remembered with praise. John the Baptist would perhaps not have brought all Jerusalem out into the wilderness by his plain and good sermons: it was the camel's hair and the locusts which got him a congregation, and which, perhaps, added force to his precepts. When at school I heard a dialogue, between an usher and the man who cleaned the shoes, about Mr. ——, a minister, a very corporate body with due area of waistcoat. 'He is a man of great erudition,' said the first. 'Ah, yes, sir,' said Joe; 'anyone can see that who looks at that silk waistcoat.']

[When I said at the outset that I had only taken books from my own store, I should have added that I did not make any search for information given as *part* of a work. Had I looked *through* all my books, I might have made some curious additions. For instance, in Schott's *Magia Naturalis* (vol. iii. pp. 756–778) is an account of the quadrature of Gephyrauder, as he is misprinted in Montucla. He was Thomas Gephyrander Salicetus; and he published two editions, in 1608 and 1609: I never even heard of a copy of either. His work is of the extreme of absurdity: he makes a distinction between geometrical and arithmetical fractions, and evolves theorems from it. More curious than his quadrature is his name; what are we to make of it? If a German, he is probably a German form of *Bridgeman*, and Salicetus refers him to *Weiden*. But *Thomas* was hardly a German Christian name of his time ; of 526 German philosophers, physicians, lawyers, and theologians who were biographed by Melchior Adam, only two are of this name. Of these one is Thomas Erastus, the physician whose theological writings against the Church as a separate power have given the name of Erastians to those who follow his doctrine, whether they have heard of him or not. Erastus is little known ; accordingly, some have supposed that he must be Erastus, the friend of St. Paul and Timothy (Acts xix. 22 ; 2 Tim. iv. 20 ; Rom. xvi. 23), but what this gentleman did to earn the character is not hinted at. Few words would have done : Gaius (Rom. xvi. 23) has an immortality which many more noted men have missed, given by John Bunyan, out of seven words of St. Paul. I was once told that the Erastians got their name from *Blastus*, and I could not solve *bl* = *er* : at last I remembered that Blastus was a *chamberlain* as well as Erastus ; hence the association which caused the mistake. The real heresiarch was a physician who died in 1583 ; his heresy was promulgated in a work, published immediately after his death by his widow, *De Excommunicatione Ecclesiastica.* He denied the power of excommunication on the principle above stated ; and was answered by Beza. The work was translated by Dr. R. Lee (Edinb. 1844, 8vo). The other is Thomas Grynæus, a theologian, nephew of Simon, who first printed Euclid in Greek ; of him Adam says that of works he published none, of learned sons four. If Gephyrander were a Frenchman, his name is not so easily guessed at ; but he must have been of La Saussaye. The account given by Schott is taken from a certain Father Philip Colbinus, who wrote against him.

In some manuscripts lately given to the Royal Society, David Gregory, who seems to have seen Gephyrander's work, calls him Salicetus *Westphalus*, which is probably on the title-page. But the only Weiden I can find is in Bavaria. Murhard has both editions in his Catalogue, but had plainly never seen the books : he gives the author as Thomas Gep. Hyandrus, Salicettus Westphalus. Murhard is a very old referee of mine ; but who the *non nominandus* was to see Montucla's *Gephyrauder* in Murhard's *Gep. Hyandrus*, both writers being usually accurate ?]

> A. plain discoverie of the whole Revelation of St. John . . .
> whereunto are annexed certain oracles of Sibylla . . . Set Foorth
> by John Napeir L. of Marchiston. London, 1611, 4to.

The first edition was Edinburgh, 1593, 4to. Napier always
believed that his great mission was to upset the Pope, and that
logarithms, and such things, were merely episodes and relaxations.
It is a pity that so many books have been written about this
matter, while Napier, as good as any, is forgotten and unread.
He is one of the first who gave us the six thousand years. 'There
is a sentence of the house of Elias reserved in all ages, bearing
these words : The world shall stand six thousand years, and then
it shall be consumed by fire : two thousand yeares voide or without
lawe, two thousand yeares under the law, and two thousand
yeares shall be the daies of the Messias. . . .'

I give Napier's parting salute : it is a killing dilemma :—

> In summar conclusion, if thou o *Rome* aledges thyselfe reformed, and
> to beleeue true Christianisme, then beleeue Saint *John* the Disciple,
> whome Christ loued, publikely here in this Reuelation proclaiming thy
> wracke, but if thou remain Ethnick in thy priuate thoghts, beleeuing
> the old Oracles of the *Sibyls* reuerently keeped somtime in thy *Capitol* :
> then doth here this *Sibyll* proclame also thy wracke. Repent therefore
> alwayes, in this thy latter breath, as thou louest thine Eternall salvation.
> *Amen.*

—Strange that Napier should not have seen that this appeal could
not succeed, unless the prophecies of the Apocalypse were no true
prophecies at all.

> De Magnete magneticisque corporibus, et de magno magnete
> tellure. By William Gilbert. London, 1600, folio.—There is
> a second edition ; and a third, according to Watt.

Of the great work on the magnet there is no need to speak,
though it was a paradox in its day. The posthumous work of
Gilbert, 'De Mundo nostro sublunari philosophia nova' (Amsterdam, 1651, 4to) is, as the title indicates, confined to the
physics of the globe and its atmosphere. It has never excited
attention : I should hope it would be examined with our present
lights.

> Elementorum Curvilineorum Libri tres. By John Baptista Porta.
> Rome, 1610, 4to.

This is a ridiculous attempt, which defies description, except
that it is all about lunules. Porta was a voluminous writer.

His printer announces fourteen works printed, and four to come, besides thirteen plays printed, and eleven waiting. His name is, and will be, current in treatises on physics for more reasons than one.

> Trattato della quadratura del cerchio. Di Pietro Antonio Cataldi. Bologna, 1612, folio.

Rheticus, Vieta, and Cataldi are the three untiring computers of Germany, France, and Italy; Napier in Scotland, and Briggs in England, come just after them. This work claims a place as beginning with the quadrature of Pellegrino Borello of Reggio, who will have the circle to be exactly 3 diameters and $\frac{69}{484}$ of a diameter. Cataldi, taking Van Ceulen's approximation, works hard at the finding of integers which nearly represent the ratio. He had not then the *continued fraction*, a mode of representation which he gave the next year in his work on the square root. He has but twenty of Van Ceulen's thirty places, which he takes from Clavius: and anyone might be puzzled to know whence the Italians got the result; Van Ceulen, in 1612, not having been translated from Dutch. But Clavius names his comrade Gruenberger, and attributes the approximation to them jointly; 'Lud. a Collen et Chr. Gruenbergerus invenerunt,' which he had no right to do, unless, to his private knowledge, Gruenberger had verified Van Ceulen. And Gruenberger only handed over twenty of the places. But here is one instance, out of many, of the polyglot character of the Jesuit body, and its advantages in literature.

> Philippi Lausbergii Cyclometriæ Novæ Libri Duo. Middleburg, 1616, 4to.

This is one of the legitimate quadratures, on which I shall here only remark that by candlelight it is quadrature under difficulties, for all the diagrams are in red ink.

> Recherches Curieuses des Mesures du Monde. By S. C. de V. Paris, 1626, 8vo. (pp. 48).

It is written by some Count for his son; and if all the French nobility would have given their sons the same kind of instruction about rank, the old French aristocracy would have been as prosperous at this moment as the English peerage and squireage. I sent the tract to Capt. Speke, shortly after his arrival in England, thinking he might like to see the old names of the Ethiopian provinces. But I first made a copy of all that relates to Prester John, himself a paradox. The tract contains, *inter alia*, an account of

the four empires; of the great Turk, the great Tartar, the great
Sophy, and the great Prester John. This word *great* (*grand*),
which was long used in the phrase 'the great Turk,' is a generic
adjunct to an emperor. Of the Tartars it is said that 'c'est vne
nation prophane et barbaresque, sale et vilaine, qui mangent la
chair demie cruë, qui boiuent du laict de jument, et qui n'vsent
de nappes et seruiettes que pour essuyer leurs bouches et leurs
mains.' Many persons have heard of Prester John, and have
a very indistinct idea of him. I give all that is said about him,
since the recent discussions about the Nile may give an interest
to the old notions of geography.—

Le grand Prestre Jean qui est le quatriesme en rang, est Empereur
d'Ethiopie, et des Abyssins, et se vante d'estre issu de la race de Dauid,
comme estant descendu de la Royne de Saba, Royne d'Ethiopie, laquelle
estant venuë en Hierusalem pour voir la sagesse de Salomon, enuiron
l'an du monde 2952, s'en retourna grosse d'vn fils qu'ils nomment
Moylech, duquel ils disent estre descendus en ligne directe. Et ainsi
il se glorifie d'estre le plus ancien Monarque de la terre, disant que son
Empire a duré plus de trois mil ans, ce que nul autre Empire ne peut
dire. Aussi met-il en ses tiltres ce qui s'ensuit : Nous, N. Souuerain
en mes Royaumes, vniquement aymé de Dieu, colomne de la foy, sorty
de la race de Iuda, &c. Les limites de cet Empire touchent à la mer
Rouge, et aux montagnes d'Azuma vers l'Orient, et du costé de
l'Occident, il est borné du fleuue du Nil, qui le separe de la Nubie, vers
le Septentrion il a l'Ægypte, et au Midy les Royaumes de Congo, et de
Mozambique, sa longueur contenant quarante degré, qui font mille
vingt cinq lieuës, et ce depuis Congo ou Mozambique qui sont au Midy,
iusqu'en Ægypte qui est au Septentrion, et sa largeur contenant depuis
le Nil qui est à l'Occident, iusqu'aux montagnes d'Azuma, qui sont à
l'Orient, sept cens vingt cinq lieues, qui font vingt neuf degrez. Cét
empire a sous soy trente grandes Prouinces, sçauoir, Medra, Gaga,
Alchy, Cedalon, Mantro, Finazam, Barnaquez, Ambiam, Fungy,
Angoté, Cigremaon, Gorga Cafatez, Zastanla, Zeth, Barly, Belangana,
Tygra, Gorgany, Barganaza, d'Ancut, Dargaly Ambiacatina, Cara-
cogly, Amara . Maon (*sic*), Guegiera, Bally, Dobora et Macheda.
Toutes ces Prouinces cy dessus sont situées iustement sous la ligne
equinoxiale, entres les Tropiques de Capricorne, et de Cancer. Mais
elles s'approchent de nostre Tropique, de deux cens cinquante lieuës
plus qu'elles ne font de l'autre Tropique. Ce mot de Prestre Jean
signifie grand Seigneur, et n'est pas Prestre comme plusieurs pense, il a
esté tousiours Chrestien, mais souuent Schismatique : maintenant il est
Catholique, et reconnaist le Pape pour Souuerain Pontife. I'ay veu
quelqu'vn des ses Euesques, estant en Hierusalem, auec lequel i'ay
conferé souuent par le moyen de nostre trucheman : il estoit d'vn port
graue et serieux, succiur (*sic*) en son parler, mais subtil à merueilles
en tout ce qu'il disoit. Il prenoit grand plaisir au recit que ie luy

faisais de nos belles ceremonies, et de la grauité de nos Prelats en leurs habits Pontificaux, et autres choses que je laisse pour dire, que l'Ethiopien est ioyoux et gaillard, ne ressemblant en rien à la saleté du Tartare, ny à l'affreux regard du miserable Arabe, mais ils sont fins et cauteleux, et ne se fient en personne, soupçonneux à merueilles, et fort devotieux, ils ne sont du tout noirs comme l'on croit, i'entens parler de ceux qui ne sont pas sous la ligne Equinoxiale, ny trop proches d'icelle, car ceux qui sont dessous sont les Mores que nous voyons.

It will be observed that the author speaks of his conversation with an Ethiopian bishop, about that bishop's sovereign. Something must have passed between the two which satisfied the writer that the bishop acknowledged his own sovereign under some title answering to Prester John.

> De Cometa anni 1618 dissertationes Thomæ Fieni et Liberti Fromondi. . . Equidem Thomæ Fieni epistolica quæstio, An verum sit Cœlum moveri et Terram quiescere ? London, 1670, 8vo.

This tract of Fienus against the motion of the earth is a reprint of one published in 1619. I have given an account of it as a good summary of arguments of the time, in the *Companion to the Almanac* for 1836.

> Willebrordi Snellii. R. F. Cyclometricus. Leyden, 1621, 4to.

This is a celebrated work on the approximative quadrature, which, having the suspicious word *cyclometricus*, must be noticed here for distinction.

1620. In this year, Francis Bacon published his 'Novum Organum,' which was long held in England—but not until the last century—to be the work which taught Newton and all his successors how to philosophise. That Newton never mentions Bacon, nor alludes in any way to his works, passed for nothing. Here and there a parodoxer ventured not to find all this teaching in Bacon, but he was pronounced blind. In our day it begins to be seen that, great as Bacon was, and great as his book really is, he is not the philosophical father of modern discovery.

But old prepossession will find reason for anything. A learned friend of mine wrote to me that he had discovered proof that Newton owned Bacon for his master : the proof was that Newton, in some of his earlier writings, used the phrase *experimentum crucis*, which is Bacon's. Newton may have read some of Bacon, though no proof of it appears. I have a dim idea that I once saw the two words attributed to the alchemists : if so, there is another explanation ; for Newton was deeply read in the alchemists.

I subjoin a review which I wrote of the splendid edition of Bacon by Spedding, Ellis, and Heath. All the opinions therein expressed had been formed by me long before : most of the materials were collected for another purpose.

> The Works of Francis Bacon. Edited by James Spedding, R. Leslie Ellis, and Douglas D. Heath. 5 vols.

No knowledge of nature without experiment and observation : so said Aristotle, so said Bacon, so acted Copernicus, Tycho Brahé, Gilbert, Kepler, Galileo, Harvey, &c., before Bacon wrote. No derived knowledge *until* experiment and observation are concluded : so said Bacon, and no one else. We do not mean to say that he laid down his principle in these words, or that he carried it to the utmost extreme : we mean that Bacon's ruling idea was the collection of enormous masses of facts, and then digested processes of arrangement and elimination, so artistically contrived, that a man of common intelligence, without any unusual sagacity, should be able to announce the truth sought for. Let Bacon speak for himself, in his editor's English :—

But the course I propose for the discovery of sciences is such as leaves but little to the acuteness and strength of wits, but places all wits and understandings nearly on a level. For, as in the drawing of a straight line or a perfect circle, much depends on the steadiness and practice of the hand, if it be done by aim of hand only, but if with the aid of rule or compass little or nothing, so it is exactly with my plan. . . For my way of discovering sciences goes far to level men's wits, and leaves but little to individual excellence ; because it performs everything by the surest rules and demonstrations.

To show that we do not strain Bacon's meaning, we add what is said by Hooke, whom we have already mentioned as his professed disciple, and, we believe, his only disciple of the day of Newton. We must, however, remind the reader that Hooke was very little of a mathematician, and spoke of algebra from his own idea of what others had told him :—

The intellect is not to be suffered to act without its helps, but is continually to be assisted by some method or engine, which shall be as a guide to regulate its actions, so as that it shall not be able to act amiss. Of this engine, no man except the incomparable Verulam hath had any thoughts, and he indeed hath promoted it to a very good pitch ; but there is yet somewhat more to be added, which he seemed to want time to complete. By this, as by that art of algebra in geometry, 'twill be very easy to proceed in any natural inquiry, regularly and certainly. . . For as 'tis very hard for the most acute wit to find out any difficult problem in geometry without the help of algebra . . . and altogether as easy for the meanest capacity acting by that method to complete and perfect it, so will it be in the inquiry after natural knowledge.

Bacon did not live to mature the whole of this plan. Are we really to believe that if he had completed the 'Instauratio' we who write this—and who feel ourselves growing bigger as we write it—should have been on a level with Newton in physical discovery? Bacon asks this belief of us, and does not get it. But it may be said, Your business is with what he *did* leave, and with its consequences. Be it so. Mr. Ellis says: 'That his method is impracticable cannot, I think, be denied, if we reflect not only that it never has produced any result, but also that the process by which scientific truths have been established cannot be so presented as even to appear to be in accordance with it.' That this is very true is well known to all who have studied the history of discovery: those who deny it are bound to establish either that some great discovery has been made by Bacon's method—we mean by the part peculiar to Bacon—or, better still, to show that some new discovery can be made, by actually making it. No general talk about *induction*: no reliance upon the mere fact that certain experiments or observations have been made; let us see where *Bacon's induction* has been actually used or can be used. Mere induction, *enumeratio simplex*, is spoken of by himself with contempt, as utterly incompetent. For Bacon knew well that a thousand instances may be contradicted by the thousand and first: so that no enumeration of instances, however large, is 'sure demonstration,' so long as any are left.

The immortal Harvey, who was *inventing*—we use the word in its old sense—the circulation of the blood, while Bacon was in the full flow of thought upon his system, may be trusted to say whether, when the system appeared, he found any likeness in it to his own processes, or what would have been any help to him, if he had waited for the 'Novum Organum.' He said of Bacon, 'He writes philosophy like a Lord Chancellor.' This has been generally supposed to be only a sneer at the *sutor ultra crepidam*; but we cannot help suspecting that there was more intended by it. To us, Bacon is eminently the philosopher of *error prevented*, not of *progress facilitated*. When we throw off the idea of being *led right*, and betake ourselves to that of being *kept from going wrong*, we read his writings with a sense of their usefulness, his genius, and their probable effect upon purely experimental science, which we can be conscious of upon no other supposition. It amuses us to have to add that the part of Aristotle's logic of which he saw the value was the book on *refutation of fallacies*. Now is this not the notion of things to which the bias of a practised lawyer might lead him? In the case which is before the Court, generally speaking, truth lurks somewhere about the facts, and the elimination of all error will show it in the residuum. The two senses of the word *law* come in so as to look almost like a play upon words. The judge can apply the law so soon as the facts are settled : the physical philosopher has to deduce the law from the facts. Wait, says the judge, until the facts are determined : did the prisoner take the goods with felonious intent? did the defendant give what amounts to a warranty? or the like. Wait, says Bacon, until all the facts, or all the obtainable facts, are brought in : apply my rules of separation to the facts, and the result shall come out as easily as by ruler and compasses. We think it possible that Harvey might allude to the legal character of Bacon's notions: we can hardly conceive so acute a man, after seeing what manner of writer Bacon was, meaning only that he was a lawyer and had better stick to his business. We do ourselves believe that Bacon's philosophy more resembles the action of mind of a common-law judge—not a Chancellor—than that of the physical inquirers who have been supposed to follow in his steps. It seems to us that Bacon's argument is, there can be nothing of law but what must be either perceptible, or mechanically deducible, when all the results of law, as exhibited in phenomena, are before us. Now the truth is, that the physical philosopher has frequently to conceive law which never was in his previous thought—to educe the unknown, not to choose among the known. Physical discovery would be very easy work if the inquirer could lay down his this, his that, and his t'other, and say, 'Now, one of these it must be; let us proceed to try which.' Often has he done this, and failed; often has the truth turned out to be neither this, that, nor t'other. Bacon seems to us to think that the philosopher is a judge who has to choose, upon ascertained facts, which of known statutes is to rule the decision : he appears to us more like a person who is to write the statute-book, with no guide except cases and decisions presented in all their confusion and all their conflict.

Let us take the well-known first aphorism of the 'Novum Organum :'

Man being the servant and interpreter of nature, can do and understand so much, and so much only, as he has observed in fact or in thought of the course of nature : beyond this he neither knows anything nor can do anything.

This aphorism is placed by Sir John Herschel at the head of his ' Discourse on the Study of Natural Philosophy:' a book containing notions of discovery far beyond any of which Bacon `ever dreamed; and this because it was written after discovery, instead of before. Sir John Herschel, in his version, has avoided the translation of *re vel mente observaverit*, and gives us only ' by his observation of the order of nature.' In making this the opening of an excellent sermon, he has imitated the theologians, who often employ the whole time of the discourse in stuffing matter into the text, instead of drawing matter out of it. By *observation* he (Herschel) means the whole course of discovery, observation, hypothesis, deduction, comparison, &c. The type of the Baconian philosopher, as it stood in his mind, had been derived from a noble example, his own father, William Herschel, an inquirer whose processes would have been held by Bacon to have been vague, insufficient, compounded of chance work and sagacity, and too meagre of facts to deserve the name of induction. In another work, his treatise on Astronomy, Sir John Herschel, after noting that a popular account can only place the reader on the threshold, proceeds to speak as follows of all the higher departments of science. The italics are his own :—

Admission to its sanctuary, and to the privileges and feelings of a votary, is only to be gained by one means—*sound and sufficient knowledge of mathematics, the great instrument of all exact inquiry, without which no man can ever make such advances in this or any other of the higher departments of science as can entitle him to form an independent opinion on any subject of discussion within their range.*

How is this? Man can know no more than he gets from observation, and yet mathematics is the great instrument of all exact inquiry. Are the results of mathematical deduction results of observation? We think it likely that Sir John Herschel would reply that Bacon, in coupling together *observare re* and *observare mente*, has done what some wags said Newton afterwards did in his study-door—cut a large hole of exit for the large cat, and a little hole for the little cat. But Bacon did no such thing: he never included any deduction under observation. To mathematics he had a dislike. He averred that logic and mathematics should be the handmaids, not the mistresses, of philosophy. He meant that they should play a subordinate and subsequent part in the dressing of the vast mass of facts by which discovery was to be rendered equally accessible to Newton and to us. Bacon himself was very ignorant of all that had been done by mathematics ; and, strange to say, he especially objected to astronomy being handed over to the mathematicians. Leverrier and Adams, calculating an unknown planet into visible existence by enormous heaps of algebra, furnish the last comment of note on this specimen of the goodness of Bacon's views. The following account of his knowledge of what had been done in his own day or before it, is Mr. Spedding's collection of casual remarks in Mr. Ellis's several prefaces :—

Though he paid great attention to astronomy, discussed carefully the methods in which it ought to be studied, constructed for the satisfaction of his own mind an elaborate theory of the heavens, and listened eagerly for the news from the stars brought by Galileo's telescope, he appears to have been utterly ignorant of the discoveries which had just been made by Kepler's calculations. Though he complained in 1623 of the want of compendious methods for facilitating arithmetical computations, especially with regard to the doctrine of Series, and fully recognized the importance of them as an aid to physical inquiries— he does not say a word about Napier's Logarithms, which had been published only nine years before and reprinted more than once in the interval. He complained that no considerable advance had been made in geometry beyond Euclid, without taking any notice of what had been done by Archimedes and Apollonius. He saw the importance of determining accurately the specific gravities of different substances, and himself attempted to form a table of them by a rude process of his own, without knowing of the more scientific though still imperfect methods previously employed by Archimedes, Ghetaldus, and Porta. He speaks of the εὕρηκα of Archimedes in a manner which implies that he did not clearly apprehend either the nature of the problem to be solved or the principles upon which the solution depended. In reviewing the progress of mechanics, he makes no mention of Archimedes himself, or of Stevinus, Galileo, Guldinus, or Ghetaldus. He makes no allusion to the theory of equilibrium. He observes that a ball of one pound weight will fall nearly as fast through the air as a ball of two, without alluding to the theory of the acceleration of falling bodies,. which had been made known by Galileo more than thirty years before. He proposes an inquiry with regard to the lever—namely, whether in a balance with arms of different length but equal weight the distance from the fulcrum has any effect upon the inclination,—though the theory of the lever was as well understood in his own time as it is now. In making an experiment of his own to ascertain the cause of the motion of a windmill, he overlooks an obvious circumstance which makes the experiment inconclusive, and an equally obvious variation of the same experiment which would have shown him that his theory was false. He speaks of the poles of the earth as fixed, in a manner which seems to imply that he was not acquainted with the precession of the equinoxes; and in another place, of the north pole being above and the south pole below, as a reason why in our hemisphere the north winds predominate over the south.

Much of this was known before, but such a summary of Bacon's want of knowledge of the science of his own time was never yet collected in one place. We may add, that Bacon seems to have been as ignorant of Wright's memorable addition to the resources of navigation as of Napier's addition to the means of calculation. Mathematics was beginning to be the great instrument of exact inquiry: Bacon threw the science aside, from ignorance, just at the time when his enormous sagacity, applied to knowledge, would have made him see the part it was to play. If Newton had taken Bacon for his master, not he, but somebody else, would have been Newton.

There is an attempt at induction going on, which has yielded little or no fruit, the observations made in the meteorological observatories. This attempt is carried on in a manner which would have caused Bacon to dance for joy; for he lived in times when Chancellors did dance. Russia, says M. Biot, is covered by an army of meteorographs, with generals, high officers, subalterns, and privates with fixed and defined duties of observation. Other countries have also their systematic observations. And what has come of it? Nothing, says M. Biot, and nothing will ever come of it: the veteran mathematician and experimental philosopher declares, as does Mr. Ellis, that no single branch of science has ever been fruitfully explored in this way. There is no *special object*, he says. Any one would suppose that M. Biot's opinion, given to the French Government upon the proposal to construct meteorological observatories in Algeria (*Comptes Rendus*, vol. xli, Dec. 31, 1855), was written to support the mythical Bacon, modern physics, against the real Bacon of the 'Novum Organum.' There is no *special object*. In these words lies the difference between the two methods.

[In the report to the Greenwich Board of Visitors for 1867, Mr. Airy, speaking of the increase of meteorological observatories, remarks 'Whether the effect of this movement will be that millions of useless observations will be added to the millions that already exist, or whether something may be expected to result which will lead to a meteorological theory, I cannot hazard a conjecture.' This *is* a conjecture, and a very obvious one: if

Mr. Airy would have given $2\frac{3}{4}d$. for the chance of a meteorological theory formed by masses of observations, he would never have said what I have quoted.]

Modern discoveries have not been made by large collections of facts, with subsequent discussion, separation, and resulting deduction of a truth thus rendered perceptible. A few facts have suggested an *hypothesis*, which means a *supposition*, proper to explain them. The necessary results of this supposition are worked out, and then, and not till then, other facts are examined to see if these ulterior results are found in nature. The trial of the hypothesis is the *special object*: prior to which, hypothesis must have been started, not by rule, but by that sagacity of which no description can be given, precisely because the very owners of it do not act under laws perceptible to themselves. The inventor of hypothesis, if pressed to explain his method, must answer as did Zerah Colburn, when asked for his mode of instantaneous calculation. When the poor boy had been bothered for some time in this manner, he cried out in a huff, 'God put it into my head, and I can't put it into yours.' Wrong hypotheses, rightly worked from, have produced more useful results than unguided observation. But this is not the Baconian plan. Charles the Second, when informed of the state of navigation, founded a Baconian observatory at Greenwich, to observe, observe, observe away at the moon, until her motions were known sufficiently well to render her useful in guiding the seaman. And no doubt Flamsteed's observations, twenty or thirty of them at least, were of signal use. But how? A somewhat fanciful thinker, one Kepler, had hit upon the approximate orbits of the planets by trying one hypothesis after another: he found the *ellipse*, which the Platonists, well despised of Bacon, and who would have despised him as heartily if they had known him, had investigated and put ready to hand nearly 2,000 years before. The sun in the focus, the motions of the planet more and more rapid as they approach the sun, led Kepler—and Bacon would have reproved him for his rashness—to imagine that a force residing in the sun might move the planets, a force inversely as the distance. Bouillaud, upon a fanciful analogy, rejected the inverse distance, and, rejecting the force altogether, declared that if such a thing there were, it would be as the inverse *square* of the distance. Newton, ready prepared with the mathematics of the subject, tried the fall of the moon towards the earth, away from her tangent, and found that, as compared with the fall of a stone, the law of the inverse square did hold for the moon. He deduced the ellipse, he proceeded to deduce the effect of the disturbance of the sun upon the moon, upon the assumed theory of *universal* gravitation. He found result after result of his theory in conformity with observed fact: and, by aid of Flamsteed's observations, which amended what mathematicians call his *constants*, he constructed his lunar theory. Had it not been for Newton, the whole dynasty of Greenwich astronomers, from Flamsteed of happy memory, to Airy whom Heaven preserve, might have worked away at nightly observation and daily reduction, without any remarkable result: looking forward, as to a millennium, to the time when any man of moderate intelligence was to see the whole explanation. What are large collections of facts for? To make theories *from*, says Bacon: to try ready-made theories *by*, says the history of discovery: it's all the same, says the idolater: nonsense, say we!

Time and space run short: how odd it is that of the three leading ideas of mechanics, time, space, and matter, the first two should always fail a reviewer before the third. We might dwell upon many points, especially if we attempted a more descriptive account of the valuable edition before us. No one need imagine that the editors, by their uncompromising attack upon the notion of Bacon's influence common even among mathematicians and experimental philosophers, have lowered the glory of the great man whom it was, many will think, their business to defend through thick and thin. They have given a clearer notion of his excellencies, and a better idea of the power of his mind, than ever we saw given before. Such a correction as theirs must have come, and soon, for as Hallam says—after noting that the 'Novum Organum' was *never published separately in England*, Bacon has probably been more read in the last thirty years—now forty—than in the two hundred years which preceded. He will now be more read than ever he was. The history of the intellectual world is the history of the worship of one idol after another. No sooner is it clear that a Hercules has appeared among men, than all that imagination can conceive of strength is attributed to him, and his labours are recorded in the heavens. The time arrives when, as in the case of Aristotle, a new deity is found, and the old one is consigned to shame and reproach. A reaction may afterwards take place, and this is now happening in the case of the Greek philosopher. The end of the process is, that the opposing deities take their places, side by side, in a Pantheon dedicated not to gods, but to heroes.

Passing over the success of Bacon's own endeavours to improve the details of physical science, which was next to nothing, and of his method as a whole, which has never been practised, we might say much of the good influence of his writings. Sound wisdom, set in sparkling wit, must instruct and amuse to the end of time : and, as against error, we repeat that Bacon is soundly wise, so far as he goes. There is hardly a form of human error within his scope which he did not detect, expose, and attach to a satirical metaphor which never ceases to sting. He is largely indebted to a very extensive reading ; but the thoughts of others fall into his text with such a close-fitting compactness that he can make even the words of the Sacred Writers pass for his own. A saying of the prophet Daniel, rather a hackneyed quotation in our day, *Multi pertransibunt, et augebitur scientia,* stands in the title-page of the first edition of Montucla's 'History of Mathematics' as a quotation from Bacon—and it is not the only place in which this mistake occurs. When the truth of the matter, as to Bacon's system, is fully recognized, we have little fear that there will be a reaction against the man. First, because Bacon will always live to speak for himself, for he will not cease to be read : secondly, because those who seek the truth will find it in the best edition of his works, and will be most ably led to know what Bacon *was,* in the very books which first showed at large what he *was not.*

In this year (1620) appeared the corrections under which the Congregation of the Index — i.e. the Committee of Cardinals which superintended the *Index* of forbidden books—proposed to allow the work of Copernicus to be read. I insert these conditions in full, because they are often alluded to, and I know of no source of reference accessible to a twentieth part of those who take interest in the question.

By a decree of the Congregation of the Index, dated March 5, 1616, the work of Copernicus, and another of Didacus Astunica, are suspended *donec corrigantur,* as teaching :

' Falsam illam doctrinam Pythagoricam,divinæ que Scripturæ omnino adversantem, de mobilitate Terræ et immobilitate Solis.'

But a work of the Carmelite Foscarini is:

' Omnino prohibendum atque damnandum,'because 'ostendere conatur præfatam doctrinam . . . . consonam esse veritati et non adversari Sacræ Scripturæ.'

Works which teach the false doctrine of the earth's motion

are to be corrected ; those which declare the doctrine conformable to Scripture are to be utterly prohibited.

In a 'Monitum ad Nicolai Copernici lectorem, ejusque emendatio, permissio, et correctio,' dated 1620 without the month or day, permission is given to reprint the work of Copernicus with certain alterations ; and, by implication, to read existing copies after correction in writing. In the preamble the author is called *nobilis astrologus*; not a compliment to his birth, which was humble, but to his fame. The suspension was because :

'Sacræ Scripturæ, ejusque veræ et Catholicæ interpretationi repugnantia (quod in homine Christiano minime tolerandum) non *per hypothesin* tractare, sed *ut verissima* adstruere non dubitat !

And the corrections relate :

'Locis in quibus non *ex hypothesi*, sed *asserendo* de situ et motu Terræ disputat.'

That is, the earth's motion may be an hypothesis for elucidation of the heavenly motions, but must not be asserted as a fact.

(In Pref. circa finem.) '*Copernicus.* Si fortasse erunt ματαιόλογοι, qui cum omnium Mathematum ignari sint, tamen de illis judicium sibi summunt, propter aliquem locum scripturæ, male ad suum propositum detortum, ausi fuerint meum hoc institutum reprehendere ac insectari: illos nihil moror adeo ut etiam illorum judicium tanquam temerarium contemnam. Non enim obscurum est Lactantium, celebrem alioqui scriptorem, sed Mathematicum parum, admodum pueriliter de forma terræ loqui, cum deridet eos, qui terram globi formam habere prodiderunt. Itaque non debet mirum videri studiosis, si qui tales nos etiam videbunt. Mathemata Mathematicis scribuntur, quibus et hi nostri labores, si me non fallit opinio, videbuntur etiam Reipub. ecclesiasticæ conducere aliquid . . . *Emend.* Ibi *si fortasse* dele omnia, usque ad verbum *hi nostri labores* et sic accommoda—*Cœterum hi nostri labores.*'

All the allusion to Lactantius, who laughed at the notion of the earth being round, which was afterwards found true, is to be struck out.

(Cap. 5. lib. i. p. 3.) '*Copernicus.* Si tamen attentius rem consideremus, videbitur hæc quæstio nondum absoluta, et idcirco minime contemnenda. *Emend.* Si tamen attentius rem consideremus, nihil refert an Terram in medio Mundi, an extra Medium existere, quoad solvendas cœlestium motuum apparentias existimemus.'

We must not say the question is not yet settled, but only that it may be settled either way, so far as mere explanation of the celestial motions is concerned.

(Cap. 8. lib. i.) ' Totum hoc caput potest expungi, quia ex professo tractat de veritate motus Terræ, dum solvit veterum rationes probantes ejus quietem. Cum tamen problematice videatur loqui ; ut studiosis satisfiat, seriesque et ordo libri integer maneat ; emendetur ut infra.'

A chapter which seems to assert the motion should perhaps be expunged ; but it may perhaps be problematical; and, not to break up the book, must be amended as below.

(p. 6.) ' *Copernicus*. Cui ergo hesitamus adhuc, nobilitatem illi formæ suæ a natura congruentem concedere, magisquam quod totus labatur mundus, cujus finis ignoratur, scirique nequit, neque fateamur ipsius cotidianæ revolutionis in cœlo apparentiam esse, et in terra veritatem ? Et hæc perinde se habere, ac si diceret Virgilianus Æneas : Provehimur portu . . . . *Emend*. Cur ergo non possum mobilitatem illi formæ suæ concedere, magisque quod totus labatur mundus, cujus finis ignoratur scirique nequit, et quæ apparent in cœlo, perinde se habere, ac si . . . .'

' Why should we hesitate to allow the earth's motion,' must be altered into ' I cannot concede the earth's motion.'

(p. 7.) ' *Copernicus*. Addo etiam, quod satis absurdum videretur, continenti sive locanti motum adscribi, et non potius contento et locato, quod est terra. *Emend*. Addo etiam difficilius non esse contento et locato, quod est Terra, motum adscribere, quam continenti.'

We must not say it is absurd to refuse motion to the *contained* and *located*, and to give it to the containing and locating; say that neither is more difficult than the other.

(p. 7.) ' *Copernicus*. Vides ergo quod ex his omnibus probabilior sit mobilitas Terræ, quam ejus quies, præsertim in cotidiana revolutione, tanquam terræ maxime propria. *Emend*. *Vides* . . . delendus est usque ad finem capitis.'

Strike out the whole of the chapter from this to the end; it says that the motion of the earth is the most probable hypothesis.

(Cap. 9. lib. i. p. 7.) ' *Copernicus*. Cum igitur nihil prohibeat mobilitatem Terræ, videndum nunc arbitror, an etiam plures illi motus conveniant, ut possit una errantium syderum existimari. *Emend*. Cum igitur Terram moveri assumpserim, videndum nunc arbitror, an etiam illi plures possint convenire motus.'

We must not say that nothing prohibits the motion of the earth, only that having *assumed* it, we may inquire whether our explanations require several motions.

(Cap. 10. lib. 1. p. 9.) '*Copernicus*. Non pudet nos fateri . . . . hoc potius in mobilitate terræ verificari. *Emend.* Non pudet nos assumere . . . . hoc consequenter in mobilitate verificari.'

(Cap. 10. lib. i. p. 10.) '*Copernicus*. Tanta nimirum est divina hæc Opt. Max. fabrica. *Emend.* Dele illa verba postrema.'

(Cap. ii. lib. i.) '*Copernicus*. De triplici motu telluris demonstratio. *Emend.* De hypothesi triplicis motus Terræ, ejusque demonstratione.'

(Cap. 10. lib. iv. p. 122.) '*Copernicus*. De magnitudine horum trium siderum, Solis, Lunæ, et Terræ. *Emend.* Dele verba *horum trium siderum*, quia terra non est sidus, ut facit eam Copernicus.'

We must not say we are not ashamed to *acknowledge; assume* is the word. We must not call this assumption a *Divine work.* A chapter must not be headed *demonstration*, but *hypothesis.* The earth must not be called a *star*; the word implies motion.

It will be seen that it does not take much to reduce Copernicus to pure hypothesis. No personal injury being done to the author — who indeed had been 17 years out of reach—the treatment of his book is now an excellent joke. It is obvious that the Cardinals of the Index were a little ashamed of their position, and made a mere excuse of a few corrections. Their mode of dealing with chap. 8, this *problematice videtur loqui, ut studiosis satisfiat*, is an excuse to avoid corrections. But they struck out the stinging allusion to Lactantius in the preface, little thinking, honest men, for they really believed what they said—that the light of Lactantius would grow dark before the brightness of their own.

1622. I make no reference to the case of Galileo, except this. I have pointed out (*Penny Cycl. Suppl.* 'Galileo;' *Engl. Cycl.* 'Motion of the Earth') that it is clear the absurdity was the act of the *Italian* Inquisition—for the private and personal pleasure of the Pope, who *knew* that the course he took would not commit him as *Pope*—and not of the body which calls itself the *Church.* Let the dirty proceeding have its right name. The Jesuit Riccioli, the stoutest and most learned Anti-Copernican in Europe, and the Puritan Wilkins, a strong Copernican and Pope-hater, are equally positive that the Roman *Church* never pronounced any decision : and this in the time immediately following the ridiculous proceeding of the Inquisition. In like manner a decision of the Convocation of Oxford is not a law of the *English* Church; which is fortunate, for that Convocation, in 1622, came to a decision quite as absurd, and a great deal more wicked than the declaration against the motion of the earth. The second was a foolish mistake : the first was a disgusting surrender of right feeling. The story is told without disapprobation by Anthony Wood, who never exaggerated anything against the university of which he is writing eulogistic history.

In 1622, one William Knight put forward in a sermon preached before the University certain theses which, looking at the state of the times, may have been improper and possibly of seditious intent. One of them was that the bishop might excommunicate the civil magistrate : this proposition the clerical body could not approve, and designated it by the term *erronea*, the mildest going. But Knight also declared as follows—

Subditis mere privatis, si Tyrannus tanquam latro aut stuprator in ipsos faciat impetum, et ipsi nec potestatem ordinariam implorare, nec alia ratione effugere periculum possint, in presenti periculo se et suos contra tyrannum, sicut contra privatum grassatorem, defendere licet.

That is, a man may defend his purse or a woman her honour, against the personal attack of a king, as against that of a private person, if no other means of safety can be found. The Convocation sent Knight to prison, declared the proposition '*falsa, periculosa, et impia*,' and enacted that all applicants for degrees should subscribe this censure, and make oath that they would neither hold, teach, nor defend Knight's opinions.

The thesis, in the form given, was unnecessary and improper. Though strong opinions of the king's rights were advanced at the time, yet no one ventured to say that, ministers and advisers apart, the king might *personally* break the law ; and we know that the first and only attempt which his successor made brought on the crisis which cost him his throne and his head. But the declaration that the proposition was *false* far exceeds in all that is disreputable the decision of the Inquisition against the earth's motion. We do not mention this little matter in England. Knight was a Puritan, and Neal gives a short account of his sermon. From comparison with Wood, I judge that the theses, as given, were not Knight's words, but the digest which it was customary to make in criminal proceedings against opinion. This heightens the joke, for it appears that the qualifiers of the Convocation took pains to present their condemnation of Knight in the terms which would most unequivocally make their censure condemn themselves. This proceeding took place in the interval between the two proceedings against Galileo : it is left undetermined whether we must say pot-kettle-pot or kettle-pot-kettle.

> Liberti Fromondi . . '. Ant-Aristarchus, sive orbis terræ immo-
> bilis. Antwerp, 1631, 8vo.

This book contains the evidence of an ardent opponent of Galileo to the fact, that Roman Catholics of the day did not consider the decree of the *Index* or of the *Inquisition* as a declaration of their *Church*. Fromond would have been glad to say as much, and tries to come near it, but confesses he must abstain. See *Penny Cyclop. Suppl.* 'Galileo,' and *Eng. Cycl.* 'Motion of the Earth.' The author of a celebrated article in the *Dublin Review*, in defence of the Church of Rome, seeing that Drinkwater Bethune makes use of the authority of Fromondus, but for another purpose, sneers at him for bringing up a ' musty old Professor.' If he had known Fromondus, and used him he would have helped his own case, which is very meagre for want of knowledge.[1]

> Advis à Monseigneur l'eminentissime Cardinal Duc de Richelieu,
> sur la Proposition faicte par le Sieur Morin pour l'invention des
> longitudes. Paris, 1634, 8vo.

This is the Official Report of the Commissioners appointed by the Cardinal, of whom Pascal is the one now best known, to consider Morin's plan. See the full account in Delambre, *Hist. Astr. Mod.* ii. 236, &c.

> Arithmetica et Geometria practica. By Adrian Metius. Ley-
> den, 1640, 4to.

This book contains the celebrated approximation *guessed at* by his father, Peter Metius, namely, that the diameter is to the circumference as 113 to 355. The error is at the rate of about a foot in 2,000 miles. Peter Metius, having his attention called to the subject by the false quadrature of Duchesne, found that the ratio lay between $\frac{333}{106}$ and $\frac{377}{120}$. He then took the liberty of taking the mean of both numerators and denominators, giving $\frac{355}{113}$. He had no right to presume that this mean was better than either of the extremes; nor does it appear positively that he did so. He published nothing: but his son Adrian, when Van Ceulen's work showed how near his father's result came to the truth, first made it known in the work above. (See *Eng. Cyclop.* art. ' Quadrature.')

> A discourse concerning a new world and another planet, in two
> books. London, 1640, 8vo.
> Cosmotheoros: or conjectures concerning the planetary worlds

---

[1] The article referred to is about thirty years old : since it appeared another has been given (*Dubl. Rev.* Sept. 1865) which is of much greater depth. In it will also be found the Roman view of Bishop Virgil (*ante*, p. 24).

and their inhabitants. Written in Latin, by Christianus Huy-ghens. This translation was first published in 1698. Glasgow 1757, 8vo. [The original is also of 1698.]

The first work is by Bishop Wilkins, being the third edition, [first in 1638] of the first book, 'That the Moon may be a Planet;' and the first edition of the second work, 'That the Earth may be a Planet.' [See more under the reprint of 1802.] Whether other planets be inhabited or not, that is, crowded with organisations, some of them having consciousness, is not for me to decide ; but I should be much surprised if, on going to one of them, I should find it otherwise. The whole dispute tacitly assumes that, if the stars and planets be inhabited, it must be by things of which we can form some idea. But for aught we know, what number of such bodies there are, so many organisms may there be, of which we have no way of thinking nor of speaking. This is seldom re-membered. In like manner it is usually forgotten that the *matter* of other planets may be of different chemistry from ours. There may be no oxygen and hydrogen in Jupiter, which may have *gens* of its own. But this must not be said : it would limit the omni-science of the *à priori* school of physical inquirers, the larger half of the whole, and would be very *unphilosophical*. Nine-tenths of my best paradoxers come out from among this larger half, because they are just a little more than of it at their entrance.

There was a discussion on the subject some years ago, which began with—

> The plurality of worlds: an Essay. London, 1853, 8vo. [By Dr. Wm. Whewell, Master of Trinity College, Cambridge]. A dialogue on the plurality of worlds, being a supplement to the Essay on that subject. [First found in the second edition, 1854; removed to the end in subsequent editions, and separate copies issued.]

A work of sceptical character, insisting on analogies which pro-hibit the positive conclusion that the planets, stars, &c., are what we should call *inhabited* worlds. It produced several works and a large amount of controversy in reviews. The last predecessor of whom I know was—

> Plurality of Worlds. . . . By Alexander Maxwell. Second Edition. London, 1820, 8vo.

This work is directed against the plurality by an author who does not admit modern astronomy. It was occasioned by Dr. Chal-mers's celebrated discourses on religion in connexion with astro-nomy. The notes contain many citations on the gravity controversy, from authors now very little read : and this is its present value. I find no mention of Maxwell, not even in Watt. He communicated with mankind without the medium of a publisher ; and, from Vieta till now, this method has always been favourable to loss of books.

A correspondent informs me that Alex. Maxwell, who wrote on the plurality of worlds, in 1820, was a law-bookseller and publisher (probably his own publisher) in Bell Yard. He had peculiar notions, which he was fond of discussing with his customers. He was a bit of a Swedenborgian.

There is a class of hypothetical creations which do not belong to my subject, because they are *acknowledged* to be fictions, as those of Lucian, Rabelais, Swift, Francis Godwin, Voltaire, &c. All who have more positive notions as to either the composition or organisation of other worlds, than the reasonable conclusion that our Architect must be quite able to construct millions of other buildings on millions of other plans, ought to rank with the writers just mentioned, in all but self-knowledge. Of every one of their systems I say, as the Irish Bishop said of Gulliver's book,—I don't believe half of it. Huyghens had been preceded by Fontenelle, who attracted more attention. Huyghens is very fanciful and very positive ; but he gives a true account of his method. ' But since there's no hopes of a Mercury to carry us such a journey, we shall e'en be contented with what's in our power : we shall suppose ourselves there. . . .' And yet he says, —'We have proved that they live in societies, have hands and feet. . . .' Kircher had gone to the stars before him, but would not find any life in them, either animal or vegetable.

The question of the inhabitants of a particular planet is one which has truth on one side or the other : either there are some inhabitants, or there are none. Fortunately, it is of no consequence which is true. But there are many cases, where the balance is equally one of truth and falsehood, in which the choice is a matter of importance. My work selects, for the most part, sins against demonstration : but the world is full of questions of fact or opinion, in which a struggling minority will become a majority, or else will be gradually annihilated : and each of the cases subdivides into results of good, and results of evil. What is to be done ?

> Periculosum est credere et non credere ;
> Hippolitus obiit quia novercæ creditum est ;
> Cassandræ quia non creditum ruit Ilium :
> Ergo exploranda est veritas multum prius
> Quam stulta prove judicet sententia.

Nova Demonstratio immobilitatis terræ petita ex virtute mag-
netica. By Jacobus Grandamicus. Flexiæ (La Flèche), 1645,
4to.

No magnetic body can move about its poles : the earth is a
magnetic body, therefore, &c.   The iron and its magnetism are
typical of two natures in one person ; so it is said, ' Si exaltatus
fuero à terra, omnia traham ad me ipsum.'

Le glorie degli incogniti, o vero gli huomini illustri dell' ac-
cademia de' signori incogniti di Venetia.   Venice, 1647, 4to.

This work is somewhat like a part of my own : it is a budget
of Venetian nobodies who wished to be somebodies ; but paradox
is not the only means employed.   It is of a serio-comic character,
gives genuine portraits in copper-plate, and grave lists of works ;
but satirical accounts.   The astrologer Andrew Argoli is there,
and his son ; both of whom, with some of the others, have place in
modern works on biography.   Argoli's discovery that logarithms
facilitate easy processes, but increase the labour of difficult ones,
is worth recording.

Controversiæ de vera circuli mensura . . . inter . . . . . . . .
C. S. Longomontanum et Jo. Pellium. Amsterdam, 1647, 4to.

Longomontanus, a Danish astronomer of merit, squared the
circle in 1644 : he found out that the diameter 43 gives the square
root of 18252 for the circumference ; which gives 3·14185 . . .
for the ratio.   Pell answered him, and being a kind of circulating
medium, managed to engage in the controversy names known and
unknown, as Roberval, Hobbes, Carcavi, Lord Charles Cavendish,
Pallieur, Mersenne, Tassius, Baron Wolzogen, Descartes, Cavalieri
and Golius.   Among them, of course, Longomontanus was made
mincemeat : but he is said to have insisted on the discovery in his
epitaph.

The great circulating mediums, who wrote to everybody, heard
from everybody, and sent extracts to everybody else, have been
Father Mersenne, John Collins, and the late Prof. Schumacher :
all 'late' no doubt, but only the last recent enough to be so
styled.   If M.C.S. should ever again stand for ' Member of the
Corresponding Society,' it should raise an acrostic thought of the
three.   There is an allusion to Mersenne's occupation in Hobbes's
reply to him.   He wanted to give Hobbes, who was very ill at
Paris, the Roman Eucharist : but Hobbes said, 'I have settled all

F

that long ago; when did you hear from Gassendi?' We are reminded of William's answer to Burnet. John Collins disseminated
Newton, among others. Schumacher ought to have been called
the postmaster-general of astromony, as Collins was called the
attorney-general of mathematics.

> A late discourse. . . . by Sir Kenelme Digby . . . Rendered into
> English by R. White. London, 1658, 12mo.

On this work see *Notes and Queries*, 2nd series, vii. 231, 299,
445, viii. 190. It contains the celebrated sympathetic powder. I
am still in much doubt as to the connexion of Digby with this
tract. Without entering on the subject here, I observe that in
Birch's 'History of the Royal Society,' to which both Digby and
White belonged, Digby, though he brought many things before
the Society, never mentioned the powder, which is connected only
with the names of Evelyn and Sir Gilbert Talbot. The sympathetic powder was that which cured by anointing the weapon
with its salve instead of the wound. I have long been convinced
that it was efficacious. The directions were to keep the wound
clean and cool, and to take care of diet, rubbing the salve on the
knife or sword. If we remember the dreadful notions upon drugs
which prevailed, both as to quantity and quality, we shall readily
see that any way of *not* dressing the wound would have been useful. If the physicians had taken the hint, had been careful of
diet, &c., and had poured the little barrels of medicine down the
throat of a practicable doll, *they* would have had their magical
cures as well as the surgeons. Matters are much improved now;
the quantity of medicine given, even by orthodox physicians,
would have been called infinitesimal by their professional ancestors. Accordingly, the College of Physicians has a right to
abandon its motto, which is *Ars longa, vita brevis*, meaning
*Practice is long, so life is short.*

> Examinatio et emendatio Mathematicæ Hodiernæ. By Thomas
> Hobbes. London, 1666, 4to.

In six dialogues: the sixth contains a quadrature of the circle.
But there is another edition of this work, without place or date
on the title-page, in which the quadrature is omitted. This
seems to be connected with the publication of another quadrature, without date, but about 1670, as may be judged from its
professing to answer a tract of Wallis, printed in 1669. The
title is ' Quadratura circuli, cubatio sphæræ, duplicatio cubi,' 4to.

Hobbes, who began in 1655, was very wrong in his quadrature; but, though not a Gregory St. Vincent, he was not the ignoramus in geometry that he is sometimes supposed. His writings, erroneous as they are in many things, contain acute remarks on points of principle. He is wronged by being coupled with Joseph Scaliger, as the two great instances of men of letters who have come into geometry to help the mathematicians out of their difficulty. I have never seen Scaliger's quadrature, except in the answers of Adrianus Romanus, Vieta and Clavius, and in the extracts of Kastner. Scaliger had no right to such strong opponents: Erasmus or Bentley might just as well have tried the problem, and either would have done much better in any twenty minutes of his life.

Scaliger inspired some mathematicians with great respect for his geometrical knowledge. Vieta, the first man of his time, who answered him, had such regard for his opponent as made him conceal Scaliger's name. Not that he is very respectful in his manner of proceeding: the following dry quiz on his opponent's logic must have been very cutting, being true. ' In grammaticis, dare navibus Austros, et dare naves Austris, sunt æque significantia. Sed in Geometricis, aliud est adsumpsisse circulum BCD non esse majorem triginta sex segmentis BCDF, aliud circulo BCD non esse majora triginta sex segmenta BCDF. Illa adsumptiuncula vera est, hæc falsa.' Isaac Casaubon, in one of his letters to De Thou, relates that, he and another paying a visit to Vieta, the conversation fell upon Scaliger, of whom the host said that he believed Scaliger was the only man who perfectly understood mathematical writers, especially the Greek ones: and that he thought more of Scaliger when wrong than of many others when right; pluris se Scaligerum vel errantem facere quam multos κατορθοῦντας. This must have been before Scaliger's quadrature (1594). There is an old story of some one saying, ' Mallem cum Scaligero errare, quam cum Clavio recte sapere.' This I cannot help suspecting to have been a version of Vieta's speech with Clavius satirically inserted, on account of the great hostility which Vieta showed towards Clavius in the latter years of his life.

Montucla could not have read with care either Scaliger's quadrature or Clavius's refutation. He gives the first a wrong date: he assures the world that there is no question about Scaliger's quadrature being wrong, in the eyes of geometers at least: and he states that Clavius mortified him extremely by showing that it made the circle less than its inscribed dodecagon, which is, of course, equivalent to asserting that a straight line is not always the shortest distance between two points. Did *Clavius* show this? No, it was Scaliger himself who showed it, boasted of it, and declared it to be a 'noble paradox' that a theorem false in geometry is true in arithmetic; a thing, he says with great triumph, not noticed by Archimedes himself! He says in so many words that the periphery of the dodecagon is greater than that of the circle; and that the more sides there are to the inscribed figure, the more does it exceed the circle in which it is. And here *are* the words, on the independent testimonies of Clavius and Kastner:—

Ambitus dodecagoni circulo inscribendi plus potest quam circuli ambitus. Et quanto deinceps plurium laterum fuerit polygonum circulo inscribendum, tanto plus poterit ambitus polygoni quam ambitus circuli.

There is much resemblance between Joseph Scaliger and William Hamilton, in a certain impetuosity of character, and inaptitude to think of quantity. Scaliger maintained that the arc of a circle is less than its chord in arithmetic, though greater in geometry; Hamilton arrived at two quantities which are identical, but the greater the one the less the other. But, on the whole, I liken Hamilton rather to Julius than to Joseph. On this last hero of literature I repeat Thomas Edwards, who says that a man is unlearned who, be his other knowledge what it may, does not understand the subject he writes about. And now one of many instances in which literature gives to literature character in science. Anthony Teissier, the learned annotator of De Thou's biographies, says of Finæus, 'Il se vanta sans raison avoir trouvé la quadrature du cercle; la gloire de cette admirable découverte était réservée à Joseph Scaliger, comme l'a écrit Scévole de St. Marthe.'

> Natural and Political Observations . . . upon the Bills of Mortality. By John Graunt, citizen of London. London, 1662, 4to.

This is a celebrated book, the first great work upon mortality. But the author, going *ultra crepidam*, has attributed to the motion of the moon in her orbit all the tremors which she gets from a shaky telescope. But there is another paradox about this book: the above absurd opinion is attributed to that excellent mechanist, Sir William Petty, who passed his days among the astronomers. Graunt did not write his own book! Anthony Wood hints that Petty 'assisted, or put into a way' his old benefactor: no doubt the two friends talked the matter over many a time.

Burnet and Pepys state that Petty wrote the book. It is enough for me that Graunt, whose honesty was never impeached, uses the plainest incidental professions of authorship throughout; that he was elected into the Royal Society because he was the author; that Petty refers to him as author in scores of places, and published an edition, as editor, after Graunt's death, with Graunt's name of course. The note on Graunt in the ' Biographia Britannica ' may be consulted; it seems to me decisive. Mr. C. B. Hodge, an able actuary, has done the best that can be done on the other side in the *Assurance Magazine*, viii. 234. If I may say what is in my mind, without imputation of disrespect, I suspect some actuaries have a bias: they would rather have Petty the greater for their Coryphæus than Graunt the less.

Pepys is an ordinary gossip: but Burnet's account has an animus which is of a worse kind. He talks of ' one Graunt, a Papist, under whose name Sir William Petty published his observations on the bills of mortality.' He then gives the cock without a bull story of Graunt being a trustee of the New River Company, and shutting up the cocks and carrying off their keys, just before the fire of London, by which a supply of water was delayed. It was one of the first objections made to Burnet's work, that Graunt was *not* a trustee at the time; and Maitland, the historian of London, ascertained from the books of the Company that he was not admitted until twenty-three days after the breaking out of the fire. Graunt's first admission to the Company took place on the very day on which a committee was appointed to inquire into the cause of the fire. So much for Burnet. I incline to the view that Graunt's setting London on fire strongly corroborates his having written on the bills of mortality : every practical man takes stock before he commences a grand operation in business.

> De Cometis: or a discourse of the natures and effects of Comets, as they are philosophically, historically, and astrologically considered. With a brief (yet full) account of the III late Comets, or blazing stars, visible to all Europe. And what (in a natural way of judicature) they portend. Together with some observations on the nativity of the Grand Seignior. By John Gadbury, Φιλομαϑηματικός. London, 1665, 4to.

Gadbury, though his name descends only in astrology, was a well-informed astronomer. D'Israeli sets down Gadbury, Lilly, Wharton, Booker, &c., as rank rogues : I think him quite wrong. The easy belief in roguery and intentional imposture which prevails in educated society is, to my mind, a greater presumption against the

honesty of mankind than all the roguery and imposture itself.
Putting aside mere swindling for the sake of gain, and looking at
speculation and paradox, I find very little reason to suspect wilful
deceit. My opinion of mankind is founded upon the mournful fact
that, so far as I can see, they find within themselves the means of
believing in a thousand times as much as there is to believe in,
judging by experience. I do not say anything against Isaac
D'Israeli for talking his time. We are all in the team, and we all
go the road, but we do not all draw.

> An essay towards a real character and a philosophical language.
> By John Wilkins [Dean of Ripon, afterwards Bishop of
> Chester]. London, 1668, folio.

This work is celebrated, but little known. Its object gives it
a right to a place among paradoxes. It proposes a language—if
that be the proper name—in which *things* and their relations
shall be denoted by signs, not *words* : so that any person, whatever may be his mother tongue, may read it in his own words.
This is an obvious possibility, and, I am afraid, an obvious impracticability. One man may construct such a system—Bishop
Wilkins has done it—but where is the man who will learn it?
The second tongue makes a language, as the second blow makes a
fray. There has been very little curiosity about his performance,
the work is scarce ; and I do not know where to refer the reader for
any account of its details, except to the partial reprint of Wilkins
presently mentioned under 1802, in which there is an unsatisfactory abstract. There is nothing in the ' Biographia Britannica,'
except discussion of Anthony Wood's statement that the hint was
derived from Dalgarno's book, ' De Signis,' 1661. Hamilton
(' Discussions,' Art. 5,' Dalgarno ') does not say a word on this point,
beyond quoting Wood ; and Hamilton, though he did now and
then write about his countrymen with a rough-nibbed pen, knew
perfectly well how to protect their priorities.

> Problema Austriacum. Plus ultra Quadratura Circuli. Auctore
> P. Gregorio a Sancto Vincentio Soc. Jesu., Antwerp, 1647,
> folio.—Opus Geometricum posthumum ad Mesolabium. By the
> same. Gandavi [Ghent], 1668, folio.

The first book has more than 1200 pages, on all kinds of
geometry. Gregory St. Vincent is the greatest of circle-squarers,
and his investigations led him into many truths : he found the
property of the area of the hyperbola which led to Napier's logarithms being called *hyperbolic*. Montucla says of him, with sly

truth, that no one has ever squared the circle with so much genius, or, excepting his principal object, with so much success. His reputation, and the many merits of his work, led to a sharp controversy on his quadrature, which ended in its complete exposure by Huyghens and others. He had a small school of followers, who defended him in print.

Renati Francisci Slusii Mesolabum. Leodii Eburonum [Liége], 1668, 4to.

The Mesolabum is the solution of the problem of finding two mean proportionals, which Euclid's geometry does not attain. Slusius is a true geometer, and uses the ellipse, &c.: but he is sometimes ranked with the trisectors, for which reason I place him here, with this explanation.

The finding of two mean proportionals is the preliminary to the famous old problem of the duplication of the cube, proposed by Apollo (not Apollonius) himself. D'Israeli speaks of the 'six follies of science,'—the quadrature, the duplication, the perpetual motion, the philosopher's stone, magic, and astrology. He might as well have added the trisection, to make the mystic number seven : but had he done so, he would still have been very lenient ; only seven follies in all science, from mathematics to chemistry ! Science might have said to such a judge—as convicts used to say who got seven years, expecting it for life, 'Thank you, my Lord, and may you sit there till they are over,'—may the Curiosities of Literature outlive the Follies of Science !

1668. In this year James Gregory, in his *Vera Circuli et Hyperbolæ Quadratura*, held himself to have proved that the *geometrical* quadrature of the circle is impossible. Few mathematicians read this very abstruse speculation, and opinion is somewhat divided. The regular circle-squarers attempt the *arithmetical* quadrature, which has long been proved to be impossible. Very few attempt the geometrical quadrature. One of the last is Malacarne, an Italian, who published his *Solution Géométrique*, at Paris, in 1825. His method would make the circumference less than three times the diameter.

La Géométrie Françoise, ou la Pratique aisée . . . La quadracture du cercle. Par le Sieur de Beaulieu, Ingénieur, Géographe du Roi . . . Paris, 1676, 8vo. [not Pontault de Beaulieu, the celebrated topographer ; he died in 1674].

If this book had been a fair specimen, I might have pointed to it in connection with contemporary English works, and made a scornful comparison. But it is not a fair specimen. Beaulieu was attached to the Royal Household, and throughout the century it may be suspected that the household forced a royal road to geometry. Fifty years before, Beaugrand, the king's secretary, made a fool of himself, and [so?] contrived to pass for a geometer. He had interest enough to get Desargues, the most powerful geometer of his time, the teacher and friend of Pascal, prohibited from lecturing. See some letters on the History of Perspective, which I wrote in the *Athenœum*, in October and November, 1861. Montucla, who does not seem to know the true secret of Beaugrand's greatness, describes him as 'un certain M. de Beaugrand, mathématicien, fort mal traité par Descartes, et à ce qu'il paroît avec justice.'

Beaulieu's quadrature amounts to a geometrical construction which gives $\pi = \sqrt{10}$. His depth may be ascertained from the following extracts. First, on Copernicus :—

Copernic, Allemand, ne s'est pas moins rendu illustre par ses doctes écrits ; et nous pourrions dire de luy, qu'il seroit le seul et unique en la force de ses Problèmes, si sa trop grande présomption ne l'avoit porté à avancer en cette Science une proposition aussi absurde, qu'elle est contre la Foy et raison, en faisant la circonférence d'un Cercle fixe, immobile, et le centre mobile, sur lequel principe Géométrique, il a avancé en son Traitté Astrologique le Soleil fixe, et la Terre mobile.

I digress here to point out that though our quadrators, &c., very often, and our historians sometimes, assert that men of the character of Copernicus, &c. were treated with contempt and abuse until their day of ascendancy came, nothing can be more incorrect. From Tycho Brahé to Beaulieu, there is but one expression of admiration for the genius of Copernicus. There is an exception, which, I believe, has been quite misunderstood. Maurolycus, in his 'De Sphæra,' written many years before its posthumous publication in 1575, and which it is not certain he would have published, speaking of the safety with which various authors may be read after his cautions, says, 'Toleratur et Nicolaus Copernicus qui Solem fixum et Terram *in girum circumverti* posuit: et scutica potius, aut flagello, quam reprehensione dignus est.' Maurolycus was a mild and somewhat contemptuous satirist, when expressing disapproval : as we should now say, he pooh-poohed his opponents ; but, unless the above be an instance, he was never savage nor impetuous. I am fully satisfied that the meaning of the sentence is, that Copernicus, who turned the earth like a boy's top, ought rather to have a whip given him wherewith to keep up his plaything than a serious refutation.   To speak of *tolerating* a person *as being* more worthy of a flogging than an argument, is almost a contradiction.

I will now extract Beaulieu's treatise on algebra, entire.

L'Algebre est la science curieuse des Sçavans et specialement d'un General d'Armée ou Capitaine, pour promptement ranger une Armée en bataille, et nombre de Mousquetaires et Piquiers qui composent les bataillons d'icelle, outre les figures de l'Arithmetique.   Cette science a 5 figures particulieres en cette sorte.   P signifie *plus* au commerce, et à l'Armée *Piquiers*.   M signifie *moins*, et *Mousquetaire* en l'Art des bataillons.   [It is quite true that P and M were used for *plus* and *minus* in a great many old works.]   R signifie *racine* en la mesure du Cube, et en l'Armée *rang*.   Q signifie *quaré* en l'un et l'autre usage.   C signifie *cube* en la mesure, et *Cavallerie* en la composition des bataillons et escadrons.   Quant à l'operation de cette science, c'est d'additionner un *plus* d'avec *plus*, la somme sera *plus*, et *moins* d'avec *plus*, on soustrait le moindre du *plus*, et la reste est la somme requise ou nombre trouvé.   Je dis seulement cecy en passant pour ceux qui n'en sçavent rien du tout.

This is the algebra of the Royal Household, seventy-three years after the death of Vieta.   Quære, is it possible that the fame of Vieta, who himself held very high stations in the household all his life, could have given people the notion that when such an officer chose to declare himself an algebraist, he must be one indeed ?   This would explain Beaugrand, Beaulieu, and all the *beaux*.   Beaugrand—not only secretary to the king, but ' mathematician ' to the Duke of Orleans—I wonder what his 'fool' could have been like, if indeed he kept the offices separate,—would have been in my list if I had possessed his *Geostatique*, published about 1638.   He makes bodies diminish in weight as they approach the earth, because the effect of a weight on a lever is less as it approaches the fulcrum.

> Remarks upon two late ingenious discourses . . . By Dr. Henry More.   London, 1676, 8vo.

In 1673 and 1675, Matthew Hale, then Chief Justice, published two tracts, an ' Essay touching Gravitation,' and ' Difficiles Nugæ' on the Torricellian experiment.   Here are the answers by the learned and voluminous Henry More.   The whole would be useful to any one engaged in research about ante-Newtonian notions of gravitation.

> Observations touching the principles of natural motions; and especially touching rarefaction and condensation . . . By the author of *Difficiles Nugæ*. London, 1677, 8vo.

This is another tract of Chief Justice Hale, published the year after his death. The reader will remember that *motion*, in old philosophy, meant any change from state to state : what we now describe as *motion* was *local motion*. This is a very philosophical book, about *flux* and *materia prima*, *virtus activa* and *essentialis*, and other fundamentals. I think Stephen Hales, the author of the 'Vegetable Statics,' has the writings of the Chief Justice sometimes attributed to him, which is very puny justice indeed. Matthew Hale died in 1676, and from his devotion to science it probably arose that his famous *Pleas of the Crown* and other law works did not appear until after his death. One of his contemporaries was the astronomer Thomas Street, whose *Caroline Tables* were several times printed : another contemporary was his brother judge, Sir Thomas Street. But of the astronomer absolutely nothing is known : it is very unlikely that he and the judge were the same person, but there is not a bit of positive evidence either for or against, so far as can be ascertained. Halley— no less a person—published two editions of the 'Caroline Tables,' no doubt after the death of the author : strange indeed that neither Halley nor any one else should leave evidence that Street was born or died.

Matthew Hale gave rise to an instance of the lengths a lawyer will go when before a jury who cannot detect him. Sir Samuel Shepherd, the Attorney General, in opening Hone's first trial, calls him ' one who was the most learned man that ever adorned the Bench, the most even man that ever blessed domestic life, *the most eminent man that ever advanced the progress of science*, and one of the [very moderate] best and most purely religious men that ever lived.

> Basil Valentine his triumphant Chariot of Antimony, with annotations of Theodore Kirkringius, M.D. With the true book of the learned Syncsius, a Greek abbot, taken out of the Emperour's library, concerning the Philosopher's Stone. London, 1678, 8vo.

There are said to be three Hamburg editions of the collected works of Valentine, who discovered the common antimony, and is said to have given the name *antimoine*, in a curious way. Finding that the pigs of his convent throve upon it, he gave it to his brethren, who died of it. The impulse given to chemistry by R. Boyle seems to have brought out a vast number of translations, as in the following tract :—

> *Collectanea Chymica* : A collection of ten several treatises in chymistry, concerning the liquor Alkehest, the Mercury of Philosophers, and other curiosities worthy the perusal. Written by Eir. Philaletha, Anonymus, J. B. Van-Helmont, Dr. Fr. Antonie, Bernhard Earl of Trevisan, Sir Geo. Ripley, Rog. Bacon, Geo. Starkie, Sir Hugh Platt, and the Tomb of Semiramis. See more in the contents. London, 1684, 8vo.

In the advertisements at the ends of these tracts there are upwards of a hundred English tracts, nearly all of the period, and most of them translations. Alchemy looks up since the chemists have found perfectly different substances composed of the same elements and proportions. It is true the chemists cannot yet *transmute* ; but they may in time : they poke about most assiduously. It seems, then, that the conviction that alchemy *must* be impossible was a delusion : but we do not mention it.

The astrologers and the alchemists caught it in company in the following, of which I have an unreferenced note.

> Mendacem et futilem hominem nominare qui volunt, calendariographum dicunt; at qui sceleratum simul ac impostorem, chimicum.
>
> Crede ratem ventis, corpus ne crede chimistis ;
> Est quævis chimica tutior aura fide.

Among the smaller paradoxes of the day is that of the *Times* newspaper, which always spells it *chymistry* : but so, I believe, do Johnson, Walker, and others. The Arabic word is very likely formed from the Greek : but it may be connected either with χημεια or with χυμεια.

> Lettre d'un gentil-homme de province à une dame de qualité, sur le sujet de la Comète. Paris, 1681, 4to.

An opponent of astrology, whom I strongly suspect to have been one of the members of the Academy of Sciences under the name of a country gentleman, writes very good sense on the tremors excited by comets.

> The Petitioning-Comet: or, a brief Chronology of all the famous
> Comets and their events, that have happened from the birth of
> Christ to this very day. Together with a modest enquiry into
> this present comet, London, 1681, 4to.

A satirical tract against cometic prophecy :—

'This present comet (it's true) is of a menacing aspect, but if the
*new parliament* (for whose convention so many good men pray) continue
long to sit, I fear not but the star will lose its virulence and malignancy,
or at least its portent be averted from this our nation; which being
the humble request to God of all good men, makes me thus entitle it,
a Petitioning-Comet.

The following anecdote is new to me :—

Queen Elizabeth (1558) being then at Richmond, and being
disswaded from looking on a comet which did then appear, made
answer, *jacta est alea*, the dice are thrown; thereby intimating that
the pre-order'd providence of God was above the influence of any star
or comet.

The argument was worth nothing : for the comet might have
been *on the dice* with the event; the astrologers said no more,
at least the more rational ones, who were about half of the
whole.

> An astrological and theological discourse upon this present
> great conjunction (the like whereof hath not (likely) been in
> some ages) ushered in by a great comet. London, 1682, 4to.
> By C. N.

The author foretells the approaching 'sabbatical jubilee,' but
will not fix the date: he recounts the failures of his predecessors.

> A judgment of the comet which became first generally visible
> to us in Dublin, December 13, about 15 minutes before 5 in the
> evening, A.D. 1680. By a person of quality. Dublin, 1682,
> 4to.

The author argues against cometic astrology with great ability.

> A prophecy on the conjunction of Saturn and Jupiter in this
> present year 1682. With some prophetical predictions of what
> is likely to ensue therefrom in the year 1684. By John Case,
> Student in physic and astrology. London, 1682, 4to.

According to this writer, great conjunctions of Jupiter and
Saturn occur 'in the fiery trigon,' about once in 800 years. Of

these there are to be seven: six happened in the several times of Enoch, Noah, Moses, Solomon, Christ, Charlemagne. The seventh, which is to happen at 'the lamb's marriage with the bride,' seems to be that of 1682 ; but this is only vaguely hinted.

> De Quadrature van de Circkel. By Jacob Marcelis. Amsterdam, 1698, 4to.
>
> Ampliatie en demonstratie wegens de Quadrature . . . By Jacob Marcelis. Amsterdam, 1699, 4to.
>
> Eenvoudig vertoog briev-wys geschrevem am J. Marcelis . . Amsterdam, 1702, 4to.
>
> De sleutel en openinge van de quadrature. . . . Amsterdam, 1704, 4to.

Who shall contradict Jacob Marcelis? He says the circumference contains the diameter exactly times

$$3\frac{100844908737754167989428218489 4}{699718363754081944003523927170 2}$$

But he does not come very near, as the young arithmetician will find.

> Theologiæ Christianæ Principia Mathematica. Auctore Johanne Craig. London, 1699, 4to.

This is a celebrated speculation, and has been reprinted abroad, and seriously answered. Craig is known in the early history of fluxions, and was a good mathematician. He professed to calculate, on the hypothesis that the suspicions against historical evidence increase with the square of the time, how long it will take the evidence of Christianity to die out. He finds, by formulæ, that had it been oral only, it would have gone out A.D. 800; but, by aid of the written evidence, it will last till A.D. 3150. At this period he places the second coming, which is deferred until the extinction of evidence, on the authority of the question ' When the Son of Man cometh, shall he find faith on the earth ? ' It is a pity that Craig's theory was not adopted : it would have spared a hundred treatises on the end of the world, founded on no better knowledge than his, and many of them falsified by the event. The most recent (October, 1863) is a tract in proof of Louis Napoleon being Antichrist, the Beast, the eighth Head, &c.; and the present dispensation is to close soon after 1864.

In order rightly to judge Craig, who added speculations on the variations of pleasure and pain treated as functions of time, it is necessary to remember that in Newton's day the idea of force, as a quantity to be measured, and as following a law of variation, was very new: so likewise was that of probability, or belief, as an object of measurement. The success of the 'Principia' of Newton put it into many heads to speculate about applying notions of quantity to other things not then brought under measurement. Craig imitated Newton's title, and evidently thought he was making a step in advance: but it is not every one who can plough with Samson's heifer.

It is likely enough that Craig took a hint, directly or indirectly, from Mahometan writers, who make a reply to the argument that the Koran has not the evidence derived from miracles. They say that, as evidence of Christian miracles is daily becoming weaker, a time must at last arrive when it will fail of affording assurance that they were miracles at all: whence would arise the necessity of another prophet and other miracles. Lee, the Cambridge orientalist, from whom the above words are taken, almost certainly never heard of Craig or his theory.

> Copernicans of all sorts convicted . . . to which is added a Treatise of the Magnet. By the Hon. Edw. Howard, of Berks. London, 1705, 8vo.

Not all the blood of all the Howards will gain respect for a writer who maintains that eclipses admit no possible explanation under the Copernican hypothesis, and who asks how a man can 'go 200 yards to any place if the moving superficies of the earth does carry it from him?' Horace Walpole, at the beginning of his 'Royal and Noble Authors,' has mottoed his book with the Cardinal's address to Ariosto, 'Dove diavolo, Messer Ludovico, avete pigliato tante coglionerie?' Walter Scott says you could hardly pick out, on any principle of selection—except badness itself, he means of course—the same number of plebeian authors whose works are so bad. But his implied satire on aristocratic writing forgets two points. First, during a large period of our history, when persons of rank condescended to write, they veiled themselves under 'a person of honour,' 'a person of quality,' and the like, when not wholly undescribed. Not one of these has Walpole got; he omits, for instance, Lord Brounker's translation of Descartes on Music. Secondly, Walpole only takes the heads of houses: this cuts both ways; he equally eliminates the Hon. Robert Boyle and the precious Edward Howard. This last writer is hardly out of the time in which aristocracy suppressed its names; the avowal was then usually meant to make the author's greatness useful to the book.   In our day, literary peers and honourables are very favourably known, and contain an eminent class.   They rough it like others, and if such a specimen as Edw. Howard were now to appear, he would be greeted with

> Hereditary noodle ! knowest thou not,
> Who would be wise, himself must make him so ?

A new and easy method to find the longitude at land or sea. London, 1710, 4to.

This tract is a little earlier than the great epoch of such publications (1714), and professes to find the longitude by the observed altitudes of the moon and two stars.

A new method for discovering the longitude both at sea and land, humbly proposed to the consideration of the public.   By Wm. Whiston and Humphry Ditton.   London, 1714, 8vo.

This is the celebrated tract, written by the two Arian heretics. Swift, whose orthodoxy was as undoubted as his meekness, wrote upon it the epigram  if, indeed, that be epigram of which the point is pious wish—which has been so often recited for the purity of its style, a purity which transcends modern printing. Perhaps some readers may think that Swift cared little for Whiston and Ditton, except as a chance hearing of their plan pointed them out as good marks.   But it was not so : the clique had their eye on the guilty pair before the publication of the tract.   The preface is dated July 7 ; and ten days afterwards Arbuthnot writes as follows to Swift :—

Whiston has at last published his project of the longitude ; the most ridiculous thing that ever was thought on.   But a pox on him ! he has spoiled one of my papers of Scriblerus, which was a proposition for the longitude not very unlike his, to this purpose ; that since there was no pole for east and west, that all the princes of Europe should join and build two prodigious poles, upon high mountains, with a vast lighthouse to serve for a polestar.   I was thinking of a calculation of the time, charges, and dimensions.   Now you must understand his project is by lighthouses, and explosion of bombs at a certain hour.

The plan was certainly impracticable ; but Whiston and Ditton might have retorted that they were nearer to the longitude than their satirist to the kingdom of heaven, or even to a bishopric. Arbuthnot, I think, here and elsewhere, reveals himself as the calculator who kept Swift right in his proportions in the matter of the Lilliputians, Brobdingnagians, &c. Swift was very ignorant about things connected with number.   He writes to Stella that he has discovered that leap-year comes every four years, and that all his life he had thought it came every three years.   Did he begin with the mistake of Cæsar's priests?   Whether or no, when I find the person who did not understand leap-year inventing satellites of Mars in correct accordance with Kepler's third law, I feel sure he must have had help.

> An essay concerning the late apparition in the heavens on the 6th of March.  Proving by mathematical, logical, and moral arguments, that it cou'd not have been produced meerly by the ordinary course of nature, but must of necessity be a prodigy. Humbly offered to the consideration of the Royal Society. London, 1716, 8vo.

The prodigy, as described, was what we should call a very decided and unusual aurora borealis.   The inference was, that men's sins were bringing on the end of the world.   The author thinks that if one of the old ' threatening prophets ' were then alive, he would give ' something like the following.'   I quote a few sentences of the notion which the author had of the way in which Ezekiel, for instance, would have addressed his Maker in the reign of George the First :—

> Begin !  Begin !  O Sovereign, for once, with an effectual clap of thunder. . . . O Deity ! either thunder to us no more, or when you thunder, do it home, and strike with vengeance to the mark. . . . 'Tis not enough to raise a storm, unless you follow it with a blow, and the thunder without the bolt, signifies just nothing at all. . . . Are then your lightnings of so short a sight, that they don't know how to hit, unless a mountain stands like a barrier in their way ?  Or perhaps so many eyes open in the firmament make you lose your aim when you shoot the arrow ?  Is it this ?  No ! but, my dear Lord, it is your custom never to take hold of your arms till you have first bound round your majestic countenance with gathered mists and clouds.

> The principles of the Philosophy of the Expansive and Contractive Forces. . . . By Robert Greene, M.A., Fellow of Clare Hall.  Cambridge, 1727, folio.

Sanderson writes to Jones : ' The gentleman has been reputed mad for these two years last past, but never gave the world such ample testimony of it before.'   This was said of a former work of Greene's, on solid geometry, published in 1712, in which he gives a quadrature. He gives the same or another, I do not know which, in the present work, in which the circle is 3⅕ diameters. This volume is of 981 good folio pages, and treats of all things, mental and material. The author is not at all mad, only wrong on many points. It is the weakness of the orthodox follower of any received system to impute insanity to the solitary dissentient : which is voted (in due time) a very wrong opinion about Copernicus, Columbus, or Galileo, but quite right about Robert Greene. If misconceptions, acted on by too much self-opinion, be sufficient evidence of madness, it would be a curious inquiry what is the least per-centage of the reigning school which has been insane at any one time. Greene is one of the sources for Newton being led to think of gravitation by the fall of an apple : his authority is the gossip of Martin Folkes. Probably Folkes had it from Newton's niece, Mrs. Conduitt, whom Voltaire acknowledges as *his* authority. It is in the draft found among Conduitt's papers of memoranda to be sent to Fontenelle. But Fontenelle, though a great retailer of anecdote, does not mention it in his *éloge* of Newton ; whence it may be suspected that it was left out in the copy forwarded to France. D'Israeli has got an improvement on the story : the apple ' struck him a smart blow on the head : ' no doubt taking him just on the organ of causality. He was ' surprised at the force of the stroke ' from so small an apple : but then the apple had a mission ; Homer would have said it was Minerva in the form of an apple. ' This led him to consider the accelerating motion of falling bodies,' which Galileo had settled long before : ' from whence he deduced the principle of gravity,' which many had considered before him, but no one had *deduced anything from it.* I cannot imagine whence D'Israeli got the rap on the head, I mean got it for Newton : this is very unlike his usual accounts of things. The story is pleasant and possible : its only defect is that various writings, well known to Newton, a very *learned* mathematician, had given more suggestion than a whole sack of apples could have done, if they had tumbled on that mighty head all at once. And Pemberton, speaking from Newton himself, says nothing more than that the idea of the moon being retained by the same force which causes the fall of bodies struck him for the first time while meditating in a garden. One particular tree at Woolsthorpe has been selected as the gallows of the apple-shaped goddess : it died in 1820, and Mr. Turnor kept the wood ; but Sir D. Brewster brought away a bit of root in 1814, and must have had it on his conscience for 43 years that he may have killed the tree. Kepler's suggestion of gravitation with the inverse distance, and Bouillaud's proposed substitution of the inverse square of the distance, are things which Newton knew better than his modern readers. I discovered two anagrams on his name, which are quite conclusive: the notion of gravitation was *not new*; but Newton *went on*. Some wandering spirit, probably, whose business it was to resent any liberty taken with Newton's name, put into the head of a friend of mine *eighty-one* anagrams on my own pair, some of which hit harder than any apple.

This friend, whom I must not name, has since made it up to about 800 anagrams on my name, of which I have seen about 650. Two of them I have joined in the title-page: the reader may find the sense. A few of the others are personal remarks.

Great gun! do us a sum!

is a sneer at my pursuits: but,

Go! great sum! $\int a^{u^n} du$

is more dignified.

Sunt agro! gaudemus,

is happy as applied to one of whom it may be said:

Ne'er out of town; 'tis such a horrid life:
But duly sends his family and wife.

Adsum, nugator, suge!

is addressed to a student who continues talking after the lecture has commenced: oh! the rascal!

Graduatus sum! nego

applies to one who declined to subscribe for an M.A. degree.

Usage mounts guard

symbolises a person of very fixed habits.

Gus! Gus! a mature don!
August man! sure, god!
And Gus must argue, O!
Snug as mud to argue,
Must argue on gauds.
A mad rogue stung us.
Gag a numerous stud.
Go! turn us! damage us!
Tug us! O drag us! Amen.
Grudge us! moan at us!
Daunt us! gag us more!
Dog-ear us, man! gut us!
D — us! a rogue tugs!

are addressed to me by the circle-squarers ; and,

O ! Gus ! tug a mean surd !

is smart upon my preference of an incommensurable value of $\pi$ to $3\frac{1}{5}$, or some such simple substitute. While,

Gus! Gus! at 'em a' round!

ought to be the backing of the scientific world to the author of the 'Budget of Paradoxes.'

The whole collection commenced existence in the head of a powerful mathematician during some sleepless nights. Seeing how large a number was practicable, he amused himself by inventing a digested plan of finding more.

Is there any one whose name cannot be twisted into either praise or satire ? I have had given to me,

Thomas Babington Macaulay
Mouths big : a Cantab anomaly.

A treatise of the system of the world. By Sir Isaac Newton. Translated into English. London, 1728, 8vo.

I think I have a right to one little paradox of my own : I greatly doubt that Newton wrote this book. Castiglione, in his 'Newtoni Opuscula,' gives it in the Latin which appeared in 1731, not for the first time ; he says *Angli omnes Newtono tribuunt.* It appeared just after Newton's death, without the name of any editor, or any allusion to Newton's recent departure, purporting to be that popular treatise which Newton, at the beginning of the third book of the 'Principia,' says he wrote, intending it to be the third book. It is very possible that some observant turn-penny might construct such a treatise as this from the third book, that it might be ready for publication the moment Newton could not disown it. It has been treated with singular silence : the name of the editor has never been given. Rigaud mentions it without a word : I cannot find it in Brewster's *Newton*, nor in the 'Biographia Britannica.' There is no copy in the Catalogue of the Royal Society's Library, either in English or Latin, except in Castiglione. I am open to correction ; but I think nothing from Newton's acknowledged works will prove—as laid down in the suspected work—that he took Numa's temple of Vesta, with a central fire, to be intended to symbolise the sun as the centre of our system, in the Copernican sense.

Mr. Edleston gives an account of the *lectures* ' de motu corporum,' and gives the corresponding pages of the *Latin* ' De Systemate

Mundi' of 1731. But no one mentions the *English* of 1728. This English seems to agree with the Latin ; but there is a mystery about it. The preface says, 'That this work as here published is genuine will so clearly appear by the intrinsic marks it bears, that it will be but losing words and the reader's time to take pains in giving him any other satisfaction.' Surely fewer words would have been lost if the prefator had said at once that the work was from the manuscript preserved at Cambridge. Perhaps it was a mangled copy clandestinely taken and interpolated.

> Lord Bacon not the author of 'The Christian Paradoxes,' being a reprint of 'Memorials of Godliness and Christianity,' by Herbert Palmer, B.D. With Introduction, Memoir, and Notes, by the Rev. Alexander B. Grosart, Kenross. (Private circulation, 1864).

I insert the above in this place on account of a slight connexion with the last. Bacon's Paradoxes,—so attributed—were first published as his in some asserted 'Remains,' 1648. They were admitted into his works in 1730, and remain there to this day. The title is 'The Character of a believing Christian, set forth in paradoxes and seeming contradictions.' The following is a specimen :—

> He believes three to be one and one to be three ; a father not to be older than his son ; a son to be equal with his father ; and one proceeding from both to be equal with both : he believes three persons in one nature, and two natures in one person. . . . . He believes the God of all grace to have been angry with one that never offended Him ; and that God that hates sin to be reconciled to himself though sinning continually, and never making or being able to make Him any satisfaction. He believes a most just God to have punished a most just person, and to have justified himself, though a most ungodly sinner. He believes himself freely pardoned, and yet a sufficient satisfaction was made for him.

Who can doubt that if Bacon had written this, it must have been wrong? Many writers, especially on the Continent, have taken him as sneering at (Athanasian) Christianity right and left. Many Englishmen have taken him to be quite in earnest, and to have produced a body of edifying doctrine. More than a century ago the Paradoxes were published as a penny tract; and, again, at the same price, in the 'Penny Sunday Reader,' vol. vi. No. 148, a few passages were omitted, as *too strong*. But all did not agree : in my copy of Peter Shaw's edition (vol. ii. p. 283) the Paradoxes have been cut out by the binder, who has left the backs of the leaves. I never had the curiosity to see whether other copies of the edition have been served in the same way. The Religious Tract Society republished them recently in 'Selections from the Writings of Lord Bacon,' (no date; bad plan; about 1863, I suppose). No omissions were made, so far as I find.

I never believed that Bacon wrote this paper; it has neither his *sparkle* nor his idiom. I stated my doubts even before I heard that Mr. Spedding, one of Bacon's editors, was of the same mind. (*Athenæum*, July 16, 1864). I was little moved by the wide consent of orthodox men: for I knew how Bacon, Milton, Newton, Locke, &c., were always claimed as orthodox until almost the present day. Of this there is a remarkable instance.

Among the books which in my younger day were in some orthodox publication lists—I think in the list of the Christian Knowledge Society, but I am not sure—was Locke's 'Reasonableness of Christianity.' It seems to have come down from the eighteenth century, when the battle was belief in Christ against unbelief, *simpliciter*, as the logicians say. Now, if ever there were a Socinian[1] book in the world, it is this work of Locke. 'These two,' says Locke, 'faith and repentance, i.e. believing Jesus to be the Messiah, and a good life, are the indispensable conditions of the new covenant, to be performed by all those who would obtain eternal life.' All the book is amplification of this doctrine. Locke, in this and many other things, followed Hobbes, whose doctrine, in the Leviathan, is *fidem, quanta ad salutem necessaria est, contineri in hoc articulo, Jesus est Christus.* For this Hobbes was called an atheist, which many still believe him to have been: some of his contemporaries called him, rightly,

---

[1] I use the word *Socinian* because it was so much used in Locke's time; it is used in our own day by the small fry, the unlearned clergy and their immediate followers, as a term of reproach for *all* Unitarians. I suspect they have a kind of liking for the *word*; it sounds like *so sinful*. The learned clergy and the higher laity know better: they know that the bulk of the modern Unitarians go farther than Socinus, and are not correctly named as his followers. The Unitarians themselves neither desire nor deserve a name which puts them one point nearer to orthodoxy than they put themselves. That point is the doctrine that direct prayer to Jesus Christ is lawful and desirable: this Socinus held, and the modern Unitarians do not hold. Socinus, in treating the subject in his own *Institutio*, an imperfect catechism which he left, lays much more stress on John xiv. 13 than on xv. 16 and xvi. 23. He is not disinclined to think that *Patrem* should be in the first citation, where some put it; but he says that to ask the Father in the name of the Son is nothing but praying to the Son in prayer to the Father. He labours the point with obvious wish to secure a conclusive sanction. In the Racovian Catechism, of which Faustus Socinus probably drew the first sketch, a clearer light is arrived at. The translation says: 'But wherein consists the divine honour due to Christ? In adoration likewise and invocation. For we ought at all times to adore Christ, and may in our necessities address our prayers to him as often as we please; and there are many reasons to induce us to do this freely.' There are some who like accuracy, even in aspersion.

a Socinian.    Locke was known for a Socinian as soon as his work
appeared :    Dr. John Edwards, his assailant, says he is ' Socin-
ianized all over.'    Locke, in his reply, says ' there is not one
word of Socinianism in it : ' and he was right : the positive
Socinian doctrine has *not one word of Socinianism in it* ; Socinianism consists in omissions.    Locke and Hobbes did not dare
*deny* the Trinity : for such a thing Hobbes might have been
roasted, and Locke might have been strangled.    Accordingly, the
well known way of teaching Unitarian doctrine was the collection
of the asserted essentials of Christianity, without naming the
Trinity, &c. This is the plan Newton followed, in the papers which
have at last been published.

So I, for one, thought little about the general tendency of
orthodox writers to claim Bacon by means of the Paradoxes.    I
knew that, in his ' Confession of Faith ' he is a Trinitarian of a
heterodox stamp.    His second Person takes human nature before
he took flesh, not for redemption, but as a condition precedent of
creation.    ' God is so holy, pure, and jealous, that it is impossible
for him to be pleased in any creature, though the work of his
own hands. . . . . [Genesis i. 10, 12, 18, 21, 25, 31, freely
rendered].    But—purposing to become a Creator, and to communicate to his creatures, he ordained in his eternal counsel that
one person of the Godhead should be united to one nature, and
to one particular of his creatures ; that so, in the person of
the Mediator, the true ladder might be fixed, whereby God
might descend to his creatures and his creatures might ascend
to God. · . . . . '

This is republished by the Religious Tract Society, and seems
to suit their theology, for they confess to having omitted some
things of which they disapprove.

In 1864, Mr. Grosart published his discovery that the Paradoxes
are by Herbert Palmer ; that they were first published surreptitiously, and immediately afterwards by himself, both in 1645 ;
that the ' Remains ' of Bacon did not appear until 1648 ; that
from 1645 to 1708, thirteen editions of the ' Memorials ' were
published, all containing the Paradoxes.    In spite of this, the
Paradoxes were introduced into Bacon's works in 1730, where they
have remained.

Herbert Palmer was of good descent, and educated as a Puritan.    He was an accomplished man, one of the few of his day who
could speak French as well as English.    He went into the Church,
and was beneficed by Laud, in spite of his puritanism ; he sat in
the Assembly of Divines, and was finally President of Queens'

College, Cambridge, in which post he died, August 13, 1647, in the 46th year of his age.

Mr. Grosart says, speaking of Bacon's ' Remains,' ' All who have had occasion to examine our early literature are aware that it was a common trick to issue imperfect, false, and unauthorised writings under any recently deceased name that might be expected to take. The Puritans, down to John Bunyan, were perpetually expostulating and protesting against such procedure.' I have met with instances of all this; but I did not know that there was so much of it: a good collection would be very useful. The work of 1728, attributed to Newton, is likely enough to be one of the class.

> Demonstration de l'immobilitez de la Terre. . . . Par M. de la Jonchere, Ingénieur Français. Londres, 1728, 8vo.

A synopsis which is of a line of argument belonging to the beginning of the preceding century.

> The Circle squared; together with the Ellipsis and several reflections on it. The finding two geometrical mean proportionals, or doubling the cube geometrically. By Richard Locke. . . . . London, no date, probably about 1730, 8vo.

According to Mr. Locke, the circumference is three diameters, three-fourths the difference of the diameter and the side of the inscribed equilateral triangle, and three-fourths the difference between seven-eighths of the diameter and the side of the same triangle. This gives, he says, 3·18897. There is an addition to this tract, being an appendix to a book on the longitude.

> The Circle squar'd. By Thos. Baxter, Crathorn, Cleaveland, Yorkshire. London, 1732, 8vo.

Here $\pi = 3\cdot0625$. No proof is offered.

> The longitude discovered by the Eclipses, Occultations, and Conjunctions of Jupiter's planets. By William Whiston. London, 1738.

This tract has, in some copies, the celebrated preface containing the account of Newton's appearance before the Parliamentary Committee on the longitude question, in 1714 (Brewster, ii. 257–266). This ' historical preface,' is an insertion, and is dated April 28, 1741, with four additional pages dated August 10, 1741. The short ' preface ' is by the publisher, John Whiston, the author's son.

> A description and draught of a new-invented machine for carrying
> vessels or ships out of, or into any harbour, port, or river,
> against wind and ti.'e, or in a calm.  For which, His Majesty
> has granted letters patent, for the sole benefit of the author, for
> the space of fourteen years.  By Jonathan Hulls.   London :
> printed for the author, 1737.   Price sixpence (folding plate and
> pp. 48, beginning from title).

(I ought to have entered this tract in its place.   It is so rare
that its existence was once doubted.   It is the earliest description
of steam-power applied to navigation.   The plate shows a barge,
with smoking funnel, and paddles at the stem, towing a ship of
war.   The engine, as described, is Newcomen's.

In 1855, John Sheepshanks, so well known as a friend of Art
and a public donor, reprinted this tract, in fac-simile, from his own
copy ; twenty-seven copies of the original 12mo. size, and twelve
on old paper, small 4to.   I have an original copy, wanting the
plate, and with ' Price sixpence' carefully erased, to the honour of
the book.

It is not known whether Hulls actually constructed a boat.   In
all probability his tract suggested to Symington, as Symington did
to Fulton.)

> Le vrai système de physique générale de M. Isaac Newton ex-
> posé et analysé en parallèle avec celui de Descartes.   By
> Louis Castel [Jesuit and F.R.S.].   Paris, 1743, 4to.

This is an elaborate correction of Newton's followers, and of
Newton himself, who it seems did not give his own views with
perfect fidelity.   Father Castel, for instance, assures us that Newton placed the sun *at rest* in the centre of the system.   Newton left
the sun to arrange that matter with the planets and the rest of the
universe.   In this volume of 500 pages there is right and wrong,
both clever.

> A dissertation on the Æther of Sir Isaac Newton.   By Bryan
> Robinson, M.D.   Dublin, 1743, 8vo.

A mathematical work, professing to prove that the assumed
ether causes gravitation.

> Mathematical principles of theology, or the existence of God
> geometrically demonstrated.   By Richard Jack, teacher of
> Mathematics.   London, 1747, 8vo.

Propositions arranged after the manner of Euclid, with beings
represented by circles and squares.   But these circles and squares

are logical symbols, not geometrical ones. I brought this book forward to the Royal Commission on the British Museum as an instance of the absurdity of attempting a *classed* catalogue from the *titles* of books. The title of this book sends it either to theology or geometry: when, in fact, it is a logical vagary. Some of the houses which Jack built were destroyed by the fortune of war in 1745, at Edinburgh : who will say the rebels did no good whatever? I suspect that Jack copied the ideas of J. B. Morinus, ' Quod Deus sit,' Paris, 1636, 4to., containing an attempt of the same kind, but not stultified with diagrams.

> Dissertation, découverte, et démonstrations de la quadrature mathématique du cercle. Par M. de Fauré, géomètre. [*s. l.*, probably Geneva] 1747, 8vo.
>
> Analyse de la Quadrature du Cercle. Par M. de Fauré, Gentilhomme Suisse. Hague, 1749, 4to.

According to this octavo geometer and quarto gentleman, a diameter of 81 gives a circumference of 256. There is an amusing circumstance about the quarto which has been overlooked, if indeed the book has ever been examined. John Bernoulli (the one of the day) and Koenig have both given an attestation : my mathematical readers may stare as they please, such is the fact. But, on examination, there will be reason to think the two sly Swiss played their countryman the same trick as the medical man played Miss Pickle, in the novel of that name. The lady only wanted to get his authority against sousing her little nephew, and said, ' Pray, doctor, is it not both dangerous and cruel to be the means of letting a poor tender infant perish by sousing it in water as cold as ice ?'—' Downright murder, I affirm,' said the doctor ; and certified accordingly. De Fauré had built a tremendous scaffolding of equations, quite out of place, and feeling cock-sure that his solutions, if correct, would square the circle, applied to Bernoulli and Koenig—who after his tract of two years before, must have known what he was at—for their approbation of the solutions. And he got it, as follows, well guarded :—

Suivant les suppositions posées dans ce Mémoire, il est si évident que $t$ doit être $= 34$, $y = 1$, et $z = 1$, que cela n'a besoin ni de preuve ni d'autorité pour être reconnu par tout le monde.

à Basle le 7e Mai 1749.                                    JEAN BERNOULLI.

Je souscris au jugement de Mr. Bernoulli, en conséquence de ces suppositions.

à la Haye le 21 Juin 1749.                                    S. KOENIG.

On which de Fauré remarks with triumph—as I have no doubt

it was intended he should do—' il conste clairement par ma présente Analyse et Démonstration, qu'ils y ont déja reconnu et approuvé parfaitement que la quadrature du cercle est mathématiquement démontrée.' It should seem that it is easier to square the circle than to get round a mathematician.

> An attempt to demonstrate that all the Phenomena in Nature may be explained by two simple active principles, Attraction and Repulsion, wherein the attractions of Cohesion, Gravity and Magnetism are shown to be one and the same. By Gowin Knight. London, 1748, 4to.

Dr. Knight was Mr. Panizzi's archetype, the first Principal Librarian of the British Museum. He was celebrated for his magnetical experiments. This work was long neglected; but is now recognised as of remarkable resemblance to modern speculations.

> An original theory or Hypothesis of the Universe. By Thomas Wright of Durham. London, 4to. 1750.

Wright is a speculator whose thoughts are now part of our current astronomy. He took that view—or most of it—of the milky way which afterwards suggested itself to William Herschel. I have given an account of him and his work in the *Philosophical Magazine* for April, 1848.

Wright was mathematical instrument maker to the King; and kept a shop in Fleet Street. Is the celebrated business of Troughton & Simms, also in Fleet Street, a lineal descendant of that of Wright? It is likely enough, more likely than that—as I find ʰim reported to have affirmed—Prester John was the descendant f Solomon and the Queen of Sheba. Having settled it thus, it struck me that I might apply to Mr. Simms, and he informs me that it is as I thought, the line of descent being Wright, Cole, John Troughton, Edward Troughton, Troughton & Simms.

> The theology and philosophy in Cicero's *Somnium Scipionis* explained. Or, a brief attempt to demonstrate, that the Newtonian system is perfectly agreeable to the notions of the wisest ancients: and that mathematical principles are the only sure ones. [By Bishop Horne, at the age of nineteen.] London, 1751, 8vo.

This tract, which was not printed in the collected works, and is now excessively rare, is mentioned in *Notes and Queries*, 1st S.,

v. 490, 573 ; 2nd S., ix. 15. The boyish satire on Newton is amusing. Speaking of old Benjamin Martin, he goes on as follows :—

But the most elegant account of the matter [attraction] is by that hominiform animal, Mr. Benjamin Martin, who having attended Dr. Desaguliers' fine, raree, gallanty shew for some years [Desaguliers was one of the first who gave public experimental lectures, before the saucy boy was born] in the capacity of a turnspit, has, it seems, taken it into his head to set up for a philosopher.

Thus is preserved the fact, unknown to his biographers, that Benj. Martin was an assistant to Desaguliers in his lectures. Hutton says of him, that ' he was well skilled in the whole circle of the mathematical and philosophical sciences, and wrote useful books on every one of them' : this is quite true ; and even at this day he is read by twenty where Horne is read by one ; see the stalls, *passim*. All that I say of him, indeed my knowledge of the tract, is due to this contemptuous mention of a more durable man than himself. My assistant secretary at the Astronomical Society, the late Mr. Epps, bought the copy at a stall because his eye was caught by the notice of 'Old Ben Martin,' of whom he was a great reader. Old Ben could not be a Fellow of the Royal Society, because he kept a shop : even though the shop sold nothing but philosophical instruments. Thomas Wright, similarly situated as to shop and goods, never was a Fellow. The Society of our day has greatly degenerated : those of the old time would be pleased, no doubt, that the glories of their day should be commemorated. In the early days of the Society, there was a similar difficulty about Graunt, the author of the celebrated work on mortality. But their royal patron, 'who never said a foolish thing,' sent them a sharp message, and charged them if they found any more such tradesmen, they should 'elect them without more ado.'

Horne's first pamphlet was published when he was but twenty-one years old. Two years afterwards, being then a Fellow of his college, and having seen more of the world, he seems to have felt that his manner was a little too pert. He endeavoured, it is said, to suppress his first tract : and copies are certainly of extreme rarity. He published the following as his maturer view :

> A fair, candid, and impartial state of the case between Sir
> Isaac Newton and Mr. Hutchinson. In which is shown how
> far a system of physics is capable of mathematical demonstra-
> tion; how far Sir Isaac's, as such a system, has that demon-
> stration ; and consequently, what regard Mr. Hutchinson's claim
> may deserve to have paid to it. By George Horne, M.A.
> Oxford, 1753, 8vo.

It must be remembered that the successors of Newton were
very apt to declare that Newton had demonstrated attraction as a
*physical* cause : he had taken reasonable pains to show that he
did not pretend to this. If any one had said to Newton, I hold
that every particle of matter is a responsible being of vast intellect, ordered by the Creator to move as it would do if every other
particle attracted it, and gifted with power to make its way in
true accordance with that law, as easily as a lady picks her way
across the street; what have you to say against it?—Newton
must have replied, Sir! if you really undertake to maintain this
as *demonstrable,* your soul had better borrow a little power from
the particles of which your body is made : if you merely ask me
to refute it, I tell you that I neither can nor need do it; for
whether attraction comes in this way or in any other, *it comes,*
and that is all I have to do with it.

The reader should remember that the word *attraction,* as used
by Newton and the best of his followers, only meant a *drawing
towards,* without any implication as to the cause. Thus whether
they said that matter attracts matter, or that young lady attracts
young gentleman, they were using one word in one sense. Newton
found that the law of the first is the inverse square of the distance : I am not aware that the law of the second has been
discovered ; if there be any chance, we shall see it at the year
1856 in this list.

In this point young Horne made a hit. He justly censures
those who fixed upon Newton a more positive knowledge of what
attraction is than he pretended to have. 'He has owned over
and over he did not know what he meant by it—it might be this,
or it might be that, or it might be anything, or it might be
nothing.' With the exception of the *nothing* clause, this is true,
though Newton might have answered Horne by 'Thou hast said
it,'

(I thought everybody knew the meaning of 'Thou hast said it :'
but I was mistaken. In three of the evangelists Συ λεγεις is the

answer to ' Art thou a king ? '  The force of this answer, as always understood, is ' That is your way of putting it.'  The Puritans, who lived in Bible phrases, so understood it: and Walter Scott, who caught all peculiarities of language with great effect, makes a marked instance, ' Were you armed ?—I was not—I went in my calling, as a preacher of God's word, to encourage them that drew the sword in His cause.  In other words, to aid and abet the rebels, said the Duke.  *Thou hast spoken it,* replied the prisoner.')

Again, Horne quotes Rowning as follows : —

Mr. Rowning, pt. 2 p. 5 in a note, has a very pretty conceit upon this same subject of attraction, about every particle of a fluid being intrenched in three spheres of attraction and repulsion, one within another, ' the innermost of which (he says) is a sphere of repulsion, which keeps them from approaching into contact; the next, a sphere of attraction, diffused around this of repulsion, by which the particles are disposed to run together into drops ; and the outermost of all, a sphere of repulsion, whereby they repel each other, when removed out of the attraction.'  So that between the *urgings*, and *sollicitations*, of one and t'other, a poor unhappy particle must ever be at his wit's end, not knowing which way to turn, or whom to obey first.

Rowning has here started the notion which Boscovich afterwards developed.

I may add to what precedes that it cannot be settled that, as Granger says, Desaguliers was the first who gave experimental lectures in London.  William Whiston gave some, and Francis Hauksbee made the experiments.  The prospectus, as we should now call it, is extant, a quarto tract of plates and descriptions, without date.  Whiston, in his life, gives 1714 as the first date of publication, and therefore, no doubt, of the lectures.  Desaguliers removed to London soon after 1712, and commenced his lectures soon after that.  It will be rather a nice point to settle which lectured first; probabilities seem to go in favour of Whiston.

An Essay to ascertain the value of leases, and annuities for years and lives.  By W[eyman] L[ee].  London, 1737, 8vo.

A valuation of Annuities and Leases certain, for a single life.  By Weyman Lee, Esq. of the Inner Temple.  London, 1751, 8vo.  Third edition, 1773.

Every branch of exact science has its paradoxer.  The world at large cannot tell with certainty who is right in such questions as squaring the circle, &c.  Mr. Weyman Lee was the assailant of what all who had studied called demonstration in the question of annuities.  He can be exposed to the world : for his error arose out of his not being able to see that the whole is the sum of all its parts.

By an annuity, say of 100*l.*, now bought, is meant that the buyer is to have for his money 100*l.* in a year, if he be then alive, 100*l.* at the end of two years, if then alive, and so on.  It is clear that he would buy a life annuity if he should buy the first 100*l.* in one office, the second in another, and so on.  All the difference between buying the whole from one office, and buying all the separate contingent payments at different offices, is immaterial to calculation.  Mr. Lee would have agreed with the rest of the world about the payments to be made to the several different offices, in consideration of their several contracts : but he differed from every one else about the sum to be paid to *one* office.  He contended that the way to value an annuity is to find out the term of years which the individual has an even chance of surviving, and to charge for the life annuity the value of an annuity certain for that term.

It is very common to say that Lee took the average life, or expectation, as it is wrongly called, for his term : and this I have done myself, taking the common story.  Having exposed the absurdity of this second supposition, taking it for Lee's, in my 'Formal Logic,' I will now do the same with the first.

A mathematical truth is true in its extreme cases.  Lee's principle is that an annuity on a life is the annuity made certain for the term within which it is an even chance the life drops.  If, then, of a thousand persons, 500 be sure to die within a year, and the other 500 be immortal, Lee's price of an annuity to any one of these persons is the present value of one payment : for one year is the term which each one has an even chance of surviving and not surviving.  But the true value is obviously half that of a perpetual annuity : so that at 5 per cent. Lee's rule would give less than the tenth of the true value.  It must be said for the poor circle-squarers, that they never err so much as this.

Lee would have said, if alive, that I have put an *extreme case* : but any *universal* truth is true in its extreme cases.  It is not fair to bring forward an extreme case against a person who is speaking as of usual occurrences : but it is quite fair when, as frequently happens, the proposer insists upon a perfectly general acceptance of his assertion.  And yet many who go the whole hog protest against being tickled with the tail.  Counsel in court are good instances : they are paradoxers by trade.  June 13, 1849, at

Hertford, there was an action about a ship, insured against a *total* loss : some planks were saved, and the underwriters refused to pay. Mr. Z. (for deft.) 'There can be no degrees of totality; and some timbers were saved.'—L. C. B. 'Then if the vessel were burned to the water's edge, and some rope saved in the boat, there would be no total loss.'—Mr. Z. 'This is putting a very extreme case.'— L. C. B. 'The argument would go that length.' What would *Judge Z.*—as he now is—say to the extreme case beginning some-where between six planks and a bit of rope ?

> Histoire des recherches sur la quadrature du cercle. . . . avec une addition concernant les problèmes de la duplication du cube et de la trisection de l'angle. Paris, 1754, 12mo. [By Montucla.]

This is *the* history of the subject. It was a little episode to the great history of mathematics by Montucla, of which the first edition appeared in 1758. There was much addition at the end of the fourth volume of the second edition ; this is clearly by Montucla, though the bulk of the volume is put together, with help from Montucla's papers, by Lalande. There is also a second edition of the history of the quadrature, Paris, 1831, 8vo, edited, I think, by Lacroix ; of which it is the great fault that it makes hardly any use of the additional matter just mentioned.

Montucla is an admirable historian when he is writing from his own direct knowledge : it is a sad pity that he did not tell us when he was depending on others. We are not to trust a quarter of his book, and we must read many other books to know which quarter. The fault is common enough, but Montucla's good three-quarters is so good that the fault is greater in him than in most others : I mean the fault of not acknowledging ; for an historian cannot read everything. But it must be said that mankind give little encouragement to candour on this point. Hallam, in his ' History of Literature,' states with his own usual instinct of honesty every case in which he depends upon others : Montucla does not. And what is the consequence ?—Montucla is trusted, and believed in, and cried up in the bulk ; while the smallest talker can lament that Hallam should be so unequal and apt to depend on others, without remembering to mention that Hallam himself gives the information. As to a universal history of any great subject being written entirely upon primary know-ledge, it is a thing of which the possibility is not yet proved by an example. Delambre attempted it with astronomy, and was removed by death before it was finished, to say nothing of the gaps he left.

Montucla was nothing of a bibliographer, and his descriptions of books in the first edition were insufficient.    The Abbé Rive fell foul of him, and as the phrase is, gave it him.    Montucla took it with great good humour, tried to mend, and, in his second edition, wished his critic had lived to see the *vernis de biblio-graphe* which he had given himself.

I have seen Montucla set down as an *esprit fort,* more than once : wrongly, I think.    When he mentions Barrow's address to the Almighty, he adds, ' On voit, au reste, par là, que Barrow étoit un pauvre philosophe ; car il croyait en l'immortalité de l'âme, et en une Divinité autre que la nature universelle.'    This is irony, not an expression of opinion.    In the book of mathe-matical recreations which Montucla constructed upon that of Ozanam, and Ozanam upon that of Van Etten, now best known in England by Hutton's similar treatment of Montucla, there is an amusing chapter on the quadrators.    Montucla refers to his own anonymous book of 1754 as a curious book published by Jombert. He seems to have been a little ashamed of writing about circle-squarers : what a slap on the face for an unborn Budgeteer !

Montucla says, speaking of France, that he finds three notions prevalent among the cyclometers : 1. that there is a large reward offered for success ; 2. that the longitude problem depends on that success ; 3. that the solution is the great end and object of geometry.    The same three notions are equally prevalent among the same class in England.    No reward has ever been offered by the government of either country.    The longitude problem in no way depends upon perfect solution : existing approximations are sufficient to a point of accuracy far beyond what can be wanted.    And geometry, content with what exists, has long passed on to other matters.    Sometimes a cyclometer persuades a skipper who has made land in the wrong place that the astronomers are in fault, for using a wrong measure of the circle ; and the skipper thinks it a very comfortable solution !    And this is the utmost that the problem ever has to do with longitude.

Antinewtonianismus.    By Cælestino Cominale, M.D.  Naples, 1754 and 1756, 2 vols. 4to.

The first volume upsets the theory of light ; the second vacuum, vis inertiæ, gravitation, and attraction.    I confess I never attempted these big Latin volumes, numbering 450 closely-printed quarto pages.    The man who slays Newton in a pamphlet is the man for me.    But I will lend them to anybody who will give security, himself in 500*l.*, and two sureties in 250*l.* each, that he will read them through, and give a full abstract; and I will not exact security for their return. I have never seen any mention of this book: it has a printer, but not a publisher, as happens with so many unrecorded books.

1755. The French Academy of Sciences came to the determination not to examine any more quadratures or kindred problems. This was the consequence, no doubt, of the publication of Montucla's book: the time was well chosen; for that book was a full justification of the resolution. The Royal Society followed the same course, I believe, a few years afterwards. When our Board of Longitude was in existence, most of its time was consumed in listening to schemes, many of which included the quadrature of the circle. It is certain that many quadrators have imagined the longitude problem to be connected with theirs : and no doubt the notion of a reward being offered by Government for a true quadrature is a result of the reward offered for the longitude. Let it also be noted that this longitude reward was not a premium upon excogitation of a mysterious difficulty. The legislature was made to know that the rational hopes of the problem were centred in the improvement of the lunar tables and the improvement of chronometers. To these objects alone, and by name, the offer was directed: several persons gained rewards for both; and the offer was finally repealed.

> Fundamentalis Figura Geometrica, primas tantum lineas circuli quadraturæ possibilitatis ostendens. By Niels Erichsen (Nicolaus Ericius), shipbuilder, of Copenhagen. Copenhagen, 1755, 12mo.

This was a gift from my oldest friend who was not a relative, Dr. Samuel Maitland of the 'Dark Ages.' He found it among his books, and could not imagine how he came by it : I could have told him. He once collected interpretations of the Apocalypse : and auction lots of such books often contain quadratures. The wonder is he never found more than one.

The quadrature is not worth notice. Erichsen is the only squarer I have met with who has distinctly asserted the particulars of that reward which has been so frequently thought to have been offered in England. He says that, in 1747, the Royal Society, on the 2nd of June, offered to give a large reward for the quadrature of the circle and a true explanation of magnetism, in addition to 30,000*l.* previously promised for the same. I need hardly say that the Royal Society had not 30,000*l.* at that time, and would not, if it had had such a sum, have spent it on the circle, nor on magnetic theory ; nor would it have coupled the two things.    On this book, see *Notes and Queries,* 1st S. xii. 306.    Perhaps Erichsen meant that the 30,000*l.* had been promised by the Government, and the addition by the Royal Society.

October 8, 1866.    I receive a letter from a cyclometer who understands that a reward is offered to any one who will square the circle, and that all competitors are to send their plans to me. The hoaxers have not yet failed out of the land.

> Theoria Philosophiæ Naturalis redacta ad unicam legem virium in natura existentium.    Editio *Veneta* prima.    By Roger Joseph Boscovich.    Venice, 1763, 4to.

The first edition is said to be of Vienna, 1758.    This is a celebrated work on the molecular theory of matter, grounded on the hypothesis of spheres of alternate attraction and repulsion. Boscovich was a Jesuit of varied pursuit.    During his measurement of a degree of the meridian, while on horseback or waiting for his observations, he composed a Latin poem of about five thousand verses on eclipses, with notes, which he dedicated to the Royal Society : 'De Solis et Lunæ defectibus,' London, Millar and Dodsley, 1760, 4to.

> Traité de paix entre Des Cartes et Newton, *précédé* des vies littéraires de ces deux chefs de la physique moderne. . . . By Aimé Henri Paulian.    Avignon, 1763, 12mo.

I have had these books for many a year without feeling the least desire to see how a lettered Jesuit would atone Descartes and Newton.    On looking at my two volumes, I find that one contains nothing but the literary life of Des Cartes ; the other nothing but the literary life of Newton.    The preface indicates more : and Watt mentions *three* volumes.    I dare say the first two contain all that is valuable.    On looking more attentively at the two volumes, I find them both readable and instructive ; the account of Newton is far above that of Voltaire, but not so popular.    But he should not have said that Newton's family came from Newton in Ireland.    Sir Rowland Hill gives fourteen *Newtons* in Ireland : twice the number of the cities that contended for the birth of Homer may now contend for the origin of Newton, on the word of Father Paulian.

> Philosophical Essays, in three parts. By R. Lovett, Lay Clerk of the Cathedral Church of Worcester. Worcester, 1766, 8vo.
>
> The Electrical Philosopher: containing a new system of physics founded upon the principle of an universal Plenum of elementary fire . . . By R. Lovett. Worcester, 1774, 8vo.

Mr. Lovett was one of those ether philosophers who bring in elastic fluid as an explanation by imposition of words, without deducing any one phenomenon from what we know of it. And yet he says that attraction has received no support from geometry; though geometry, applied to a particular law of attraction, had shown how to predict the motions of the bodies of the solar system. He, and many of his stamp, have not the least idea of the confirmation of a theory by accordance of deduced results with observation posterior to the theory.

> Lettres sur l'Atlantide de Platon, et sur l'ancien Histoire de l'Asie, pour servir de suite aux lettres sur l'origine des Sciences, adressées à M. de Voltaire, par M. Bailly. London and Paris, 1779, 8vo.

I might enter here all Bailly's histories of astronomy. The paradox which runs through them all more or less, is the doctrine that astronomy is of immense antiquity, coming from some forgotten source, probably the drowned island of Plato, peopled by a race whom Bailly makes, as has been said, to teach us everything except their existence and their name. These books, the first scientific histories which belong to readable literature, made a great impression by power of style : Delambre created a strong reaction, of injurious amount, in favour of history founded on contemporary documents, which early astronomy cannot furnish. These letters are addressed to Voltaire, and continue the discussion. There is one letter of Voltaire, being the fourth, dated Feb. 27, 1777, and signed ' le vieux malade de Ferney, V. puer centum annorum.' Then begin Bailly's letters, from January 16 to May 12, 1778. From some ambiguous expressions in the Preface, it would seem that these are fictitious letters, supposed to be addressed to Voltaire at their dates. Voltaire went to Paris February 10, 1778, and died there May 30. Nearly all this interval was his closing scene, and it is very unlikely that Bailly would have troubled him with these letters.

An inquiry into the cause of motion, or a general theory of physics. By S. Miller. London, 1781, 4to.

Newton all wrong : matter consists of two kinds of particles, one inert, the other elastic and capable of expanding themselves *ad infinitum*.

> Des Erreurs et de la Vérité, ou les hommes rappelés au principe universel de la science; ouvrage dans lequel, en faisant remarquer aux observateurs l'incertitude de leurs recherches, et leurs méprises continuelles, on leur indique la route qu'ils auroient dû suivre, pour acquérir l'évidence physique sur l'origine du bien et du mal, sur l'homme, sur la nature matérielle, et la nature sacrée; sur la base des gouvernements politiques, sur l'autorité des souverains, sur la justice civile et criminelle, sur les sciences, les langues, et les arts. Par un Ph. . . . Inc. . . . A Edimbourg. 1782. Two vols. 8vo.

This is the famous work of Louis Claude de Saint-Martin (1743–1803), for whose other works, vagaries included, the reader must look elsewhere : among other things, he was a translator of Jacob Behmen. The title promises much, and the writer has smart thoughts now and then; but the whole is the wearisome omniscience of the author's day and country, which no reader of our time can tolerate. Not that we dislike omniscience; but we have it of our own country, both home-made and imported; and fashions vary. But surely there can be but one omniscience? Must a man have but one wife? Nay, may not a man have a new wife while the old one is living? There was a famous instrumental professor forty years ago, who presented a friend to Madame ——. The friend started, and looked surprised; for, not many weeks before, he had been presented to another lady, with the same title, at Paris. The musician observed his surprise, and quietly said, ' Celle-ci est Madame —— de Londres.' In like manner we have a London omniscience now current, which would make any one start who only knew the old French article.

The book was printed at Lyon, but it was a trick of French authors to pretend to be afraid of prosecution : it made a book look wicked-like to have a feigned place of printing, and stimulated readers. A Government which had undergone Voltaire would never have drawn its sword upon quiet Saint-Martin. To make himself look still worse, he was only ph[ilosophe] Inc. . . , which is generally read *Inconnu*, but sometimes *Incrédule* :

most likely the ambiguity was intended. There is an awful paradox about the book, which explains, in part, its leaden sameness. It is all about *l'homme, l'homme, l'homme*, except as much as treats of *les hommes, les hommes, les hommes*; but not one single man is mentioned by name in its 500 pages. It reminds one of

> Water, water, everywhere,
> And not a drop to drink.

Not one opinion of any other man is referred to, in the way of agreement or of opposition. Not even a town is mentioned: there is nothing which brings a capital letter into the middle of a sentence, except, by the rarest accident, such a personification as *Justice*. A likely book to want an *Edimbourg* godfather!

Saint-Martin is great in mathematics. The number *four* essentially belongs to straight lines, and *nine* to curves. The object of a straight line is to perpetuate *ad infinitum* the production of a point from which it emanates. A circle ○ bounds the production of all its radii, tends to destroy them, and is in some sort their enemy. How is it possible that things so distinct should not be distinguished in their *number* as well as in their action? If this important observation had been made earlier, immense trouble would have been saved to the mathematicians, who would have been prevented from searching for a common measure to lines which have nothing in common. But, though all straight lines have the number *four*, it must not be supposed that they are all equal, for a line is the result of its law and its number; but though both are the same for all lines of a sort, they act differently, as to force, energy, and duration, in different individuals; which explains all differences of length, &c. I congratulate the reader who understands this; and I do not pity the one who does not.

Saint-Martin and his works are now as completely forgotten as if they had never been born, except so far as this, that some one may take up one of the works as of heretical character, and lay it down in disappointment, with the reflection that it is as dull as orthodoxy. For a person who was once in some vogue, it would be difficult to pick out a more fossil writer, from Aa to Zypœus, except,—though it is unusual for (,—) to represent an interval of more than a year—his unknown opponent. This opponent, in the very year of the 'Des Erreurs . . . .' published a book in two parts with the same fictitious place of printing;

Tableau Naturel des Rapports qui existent entre Dieu, l'Homme, et l'Univers. A Edimbourg, 1782, 8vo.

There is a motto from the *Des Erreurs* itself, ' Expliquer les choses par l'homme, et non l'homme par les choses. *Des Erreurs et de la Vérité*, par un PH. . . . ÎNC. . . ., p. 9.' This work is set down in various catalogues and biographies as written by the PH. . . . INC. . . . himself. But it is not usual for a writer to publish two works in the same year, one of which takes a motto from the other. And the second work is profuse in capitals and italics, and uses Hebrew learning: its style differs much from the first work. The first work sets out from man, and has nothing to do with God : the second is religious and raps the knuckles of the first as follows :—' Si nous voulons nous préserver de toutes les illusions, et surtout des amorces de l'orgueil par lesquelles l'homme est si souvent séduit, ne prenons jamais les hommes, mais toujours *Dieu* pour notre terme de comparaison.' The first uses *four* and *nine* in various ways, of which I have quoted one : the second says, ' Et ici se trouve déjà une explication des nombres *quatre* et *neuf*, qui ont peu embarrassé dans l'ouvrage déjà cité. L'homme s'est égaré en allant de *quatre* à *neuf* . . . .' The work cited is the *Erreurs*, &c., and the citation is in the motto, which is the text of the opposition sermon.

> Method to discover the difference of the earth's diameters ; proving its true ratio to be not less variable than as 45 is to 46, and shortest in its pole's axis 174 miles . . . likewise a method for fixing an universal standard for weights and measures. By Thomas Williams. London, 1788, 8vo.

Mr. Williams was a paradoxer in his day, and proposed what was, no doubt, laughed at by some. He proposed the sort of plan which the French—independently of course—carried into effect a few years after. He would have the 52nd degree of latitude divided into 100,000 parts and each part a geographical yard. The geographical tun was to be the cube of the geographical yard filled with sea-water taken some leagues from land. All multiples and subdivisions were to be decimal.

I was beginning to look up those who had made similar proposals, when a learned article on the proposal of a metrical system came under my eye in the *Times* of Sept. 15, 1863. The author cites Mouton, who would have the minute of a degree divided into 10,000 *virgulæ* ; James Cassini, whose foot was to be six thousandths of a minute; and Paucton, whose foot was the 400,000th of a degree. I have verified the first and third statements; surely the second ought to be the *six-thousandth*.

> An inquiry into the Copernican system . . . wherein it is proved, in the clearest manner, that the earth has only her diurnal motion . . . with an attempt to point out the only true way whereby mankind can receive any real benefit from the study of the heavenly bodies. By John Cunningham. London, 1789, 8vo.

The 'true way' appears to be the treatment of heaven and earth as emblematical of the Trinity.

> Cosmology. An inquiry into the cause of what is called gravitation or attraction, in which the motions of the heavenly bodies, and the preservation and operations of all nature, are deduced from an universal principle of efflux and reflux. By T. Vivian, vicar of Cornwood, Devon. Bath, 1792, 12mo.

Attraction, an influx of matter to the sun; centrifugal force, the solar rays; cohesion, the pressure of the atmosphere. The confusion about centrifugal *force*, so called, as demanding an external agent, is very common.

> The rights of MAN, being an answer to Mr. Burke's attack on the French Revolution. By Thomas Paine. In two parts. 1791–1792. 8vo. (Various editions.)
>
> A vindication of the rights of WOMAN, with strictures on political and moral subjects. By Mary Wollstonecraft. 1792. 8vo.
>
> A sketch of the rights of BOYS and GIRLS. By Launcelot Light, of Westminster School; and Lætitia Lookabout, of Queen's Square, Bloomsbury. [By the Rev. Samuel Parr, LL.D.] 1792. 8vo. (pp. 64).

When did we three meet before? The first work has sunk into oblivion : had it merited its title, it might have lived. It is what the French call a *pièce de circonstance*; it belongs in time to the French Revolution, and in matter to Burke's opinion of that movement. Those who only know its name think it was really an attempt to write a philosophical treatise on what we now call socialism. Silly government prosecutions gave it what it never could have got for itself.

Mary Wollstonecraft seldom has her name spelt right. I suppose the O! O! character she got made her Woolstonecraft.

Watt gives double insinuation, for his cross-reference sends us to Goodwin. No doubt the title of the book was an act of discipleship to Paine's 'Rights of Man'; but this title is very badly chosen. The book was marred by it, especially when the authoress and her husband assumed the right of dispensing with legal sanction until the approach of offspring brought them to a sense of their child's interest. Not a hint of such a claim is found in the book, which is mostly about female education. The right claimed for woman is to have the education of a rational human being, and not to be considered as nothing but woman throughout youthful training. The maxims of Mary Wollstonecraft are now, though not derived from her, largely followed in the education of girls, especially in home education: just as many of the political principles of Tom Paine, again not derived from him, are the guides of our actual legislation. I remember, forty years ago, an old lady who used to declare that she disliked girls from the age of sixteen to five-and-twenty. 'They are full,' said she, ' of *femalities.*' She spoke of their behaviour to women as well as to men. She would have been shocked to know that she was a follower of Mary Wollstonecraft, and had packed half her book into one sentence.

The third work is a satirical attack on Mary Wollstonecraft and Tom Paine. The details of the attack would convince any one that neither has anything which would now excite reprobation. It is utterly unworthy of Dr. Parr, and has quite disappeared from lists of his works, if it were ever there. That it was written by him I take to be evident, as follows. Nichols, who could not fail to know, says (Anecd., vol. ix. p. 120): 'This is a playful essay by a first-rate scholar, who is elsewhere noticed in this volume, but whose name I shall not bring forward on so trifling an occasion.' Who the scholar was is made obvious by Master Launcelot being made to talk of Bellendenus. Further, the same boy is made to say, 'Let Dr. Parr lay his hand upon his heart, if his conscience will let him, and ask himself how many thousands of waggon-loads of this article [birch] he has cruelly misapplied.' How could this apply to Parr, with his handful of private pupils, and no reputation for severity? Any one except himself would have called on the head-master of Westminster or Eton. I doubt whether the name of Parr could be connected with the rod by anything in print, except the above and an anecdote of his pupil, Tom Sheridan. The Doctor had dressed for a dinner visit, and was ready a quarter of an hour too soon to set off. 'Tom,' said he, 'I think I had better whip you now;

you are sure to do something while I am out.'—' I wish you would, sir!' said the boy; 'it would be a letter of licence for the whole evening.' The Doctor saw the force of the retort: my two tutelaries will see it by this time. They paid in advance; and I have given liberal interpretation to the order.

The following story of Dr. Parr was told me and others, about 1829, by the late Leonard Horner, who knew him intimately. Parr was staying in a house full of company, I think in the north of England. Some gentlemen from America were among the guests, and after dinner they disputed some of Parr's assertions or arguments. So the Doctor broke out with ' Do you know what country you come from ? You come from the place to which we used to send our thieves!' This made the host angry, and he gave Parr such a severe rebuke as sent him from the room in ill-humour. The rest walked on the lawn, amusing the Americans with sketches of the Doctor. There was a dark cloud overhead, and from that cloud presently came a voice which called *Tham* (Parr-lisp for *Sam*). The company were astonished for a moment, but thought the Doctor was calling his servant in the house, and that the apparent direction was an illusion arising out of inattention. But presently the sound was repeated, certainly from the cloud,

And nearer, clearer, deadlier than before.

There was now a little alarm: where could the Doctor have got to ? They ran to his bedroom, and there they discovered a sufficient rather than satisfactory explanation. The Doctor had taken his pipe into his bedroom, and had seated himself, in sulky mood, upon the higher bar of a large and deep old-fashioned grate with a high mantelshelf. Here he had tumbled backwards, and doubled himself up between the bars and the back of the grate. He was fixed tight, and when he called for help, he could only throw his voice up the chimney. The echo from the cloud was the warning which brought his friends to the rescue.

Days of political paradox were coming, at which we now stare. Cobbett said, about 1830, in earnest, that in the country every man who did not take off his hat to the clergyman was suspected, and ran a fair chance of having something brought against him. I heard this assertion canvassed, when it was made, in a party of elderly persons. The Radicals backed it, the old Tories rather denied it, but in a way which satisfied me they ought to have denied it less if they could not deny it more. But it must be said that the Governments stopped far short of what their partisans would have had them do. All who know Robert Robinson's very quiet assault on church-made festivals in his 'History and Mystery of Good Friday' (1777) will hear or remember with surprise that the *British Critic* pronounced it a direct, unprovoked, and malicious libel on the most sacred institutions of the national Church. It was reprinted again and again: in 1811 it was in a cheap form at 6s 6d. a hundred. When the Jacobin day came, the State was really in a fright: people thought twice before they published what would now be quite disregarded. I examined a quantity of letters addressed to George Dyer (Charles Lamb's G.D.) and what between the autographs of Thelwall, Hardy, Horne Tooke, and all the rebels, put together a packet which produced five guineas, or thereabouts, for the widow. Among them were the following verses, sent by the author—who would not put his name, even in a private letter, for fear of accidents—for consultation whether they could safely be sent to an editor: and they were *not* sent. The occasion was the public thanksgiving at St. Paul's for the naval victories, December 19, 1797.

> God bless me! what a thing!
> Have you heard that the King
>  Goes to St. Paul's?
> Good Lord! and when he's there,
> He'll roll his eyes in prayer,
> To make poor Johnny stare
>  At this fine thing.
>
> No doubt the plan is wise
> To blind poor Johnny's eyes
>  By this grand show;
> For should he once suppose
> That he's led by the nose,
> Down the whole fabric goes,
>  Church, lords, and king.
>
> As he shouts Duncan's praise,
> Mind how supplies they'll raise
>  In wondrous haste.
> For while upon the sea
> We gain one victory,
> John still a dupe will be
>  And taxes pay.
>
> Till from his little store
> Three-fourths or even more
>  Goes to the Crown.

> Ah, John ! you little think
> How fast we downward sink
> And touch the fatal brink
> At which we're slaves.

I would have indicted the author for not making his thirds and sevenths rhyme. As to the rhythm, it is not much better than what the French sang in the Calais theatre, when the Duke of Clarence took over Louis XVIII. in 1814.

> God save noble Clarence,
> Who brings our king to France ;
> God save Clarence !
> He maintains the glory
> Of the British navy.
> &c. &c.

Perhaps had this been published, the Government would have assailed it as a libel on the church service. They got into the way of defending themselves by making libels on the Church, of what were libels, if on anything, on the rulers of the State ; until the celebrated trials of Hone settled the point for ever, and established that juries will not convict for one offence, even though it have been committed, when they know the prosecution is directed at another offence and another intent.

The results of Hone's trials (William Hone, 1779-1842) are among the important constitutional victories of our century. He published parodies on the Creeds, the Lord's Prayer, the Catechism, &c., with intent to bring the Ministry into contempt : everybody knew that was his *purpose*. The Government indicted him for impious, profane, blasphemous intent, but not for seditious intent. They hoped to wear him out by proceeding day by day. December 18, 1817, they hid themselves under the Lord's Prayer, the Creed, and the Commandments ; December 19, under the Litany ; December 20, under the Athanasian Creed, an odd place for shelter when they could not find it in the previous places. Hone defended himself for six, seven, and eight hours on the several days : and the jury acquitted him in 15, 105, and 20 minutes. In the second trial the offence was laid both as profanity and as sedition, which seems to have made the jury hesitate. And they probably came to think that the second count was false pretence : but the length of their deliberation is a satisfactory addition to the value of the whole. In the first trial the Attorney-General (Shepherd) had the impudence to say that the libel had nothing of a political tendency about it, but was *avowedly* set off against the religion and worship of the Church of England. The whole is political in every sentence; neither more nor less political than the following, which is part of the parody on the Catechism. ' What is thy duty towards the Minister ? My duty towards the Minister is, to trust him as much as I can ; to honour him with all my words, with all my bows, with all my scrapes, and with all my cringes ; to flatter him ; to give him thanks ; to give up my whole soul to him ; to idolize his name, and obey his word, and serve him blindly all the days of his political life.' And the parody on the Creed begins, ' I believe in George, the Regent almighty, maker of new streets and Knights of the Bath.' This is what the Attorney-General said had nothing of a political tendency about it. But this was *on the first trial* : Hone was not known. The first day's trial was under Justice Abbott (afterwards C. J. Tenterden). It was perfectly understood, when Chief Justice Ellenborough appeared in Court on the second day, that he was very angry at the first result, and put his junior aside to try his own rougher dealing. But Hone tamed the lion. An eye-witness told me that when he implored of Hone not to detail his own father Bishop Law's views on the Athanasian Creed, which humble petition Hone kindly granted, he held by the desk for support. And the same when—which is not reported—the Attorney-General appealed to the Court for protection against a stinging attack which Hone made on the Bar : he *held on*, and said, ' Mr. Attorney, what *can* I do ! ' I was a boy of twelve years old, but so strong was the feeling of exultation at the verdicts that boys at school were not prohibited from seeing the parodies, which would have been held at any other time quite unfit to meet their eyes. I was not able to comprehend all about the Lord Chief Justice until I read and heard again in after years. In the meantime, Joe Miller had given me the story of the leopard which was sent home on board a ship of war, and was in two days made as docile as a cat by the sailors. ' You have got that fellow well under,' said an officer. ' Lord bless your honour ! ' said Jack, ' if the Emperor of Marocky would send us a cock rhinoceros, we'd bring him to his bearings in no time ! ' When I came to the subject again, it pleased me to entertain the question whether, if the Emperor had sent a cock rhinoceros to preside on the third day in the King's Bench, Hone would have mastered *him* : I forget how I settled it. There grew up a story that Hone caused Lord Ellenborough's death, but this could not have been true. Lord Ellenborough resigned his seat in a few months, and died just a year after the trials; but sixty-eight years may have had more to do with it than his defeat.

A large subscription was raised for Hone, headed by the Duke of Bedford for 105*l.* Many of the leading ante-ministerialists joined: but there were many of the other side who avowed their disapprobation of the false pretence. Many could not venture their names. In the list I find: A member of the House of Lords, an enemy to persecution, and especially to religious persecution employed for political purposes—No parodist, but an enemy to persecution—A juryman on the third day's trial— Ellen Borough —My name would ruin me—Oh! minions of Pitt—Oil for the Hone—The Ghosts of Jeffries and Sir William Roy [Ghosts of Jeffries in abundance]—A conscientious Jury and a conscientious Attorney, 1*l.* 6*s.* 8*d.*—To Mr. Hone, for defending in his own person the freedom of the press, attacked for a political object, under the old pretence of supporting Religion—A cut at corruption —An Earldom for myself and a translation for my brother—One who disapproves of parodies, but abhors persecution—From a schoolboy who wishes Mr. Hone to have a very grand subscription —' For delicacy's sake forbear,' and ' Felix trembled '—' I will go myself to-morrow '—Judge Jeffries' works rebound in calf by Law —Keep us from Law, and from the Shepherd's paw—I must not give you my name, but God bless you!—As much like Judge Jeffries as the present times will permit—May Jeffries' fame and Jeffries' fate on every modern Jeffries wait—No parodist, but an admirer of the man who has proved the fallacy of the Lawyer's Law, that when a man is his own advocate he has a fool for his client—A Mussulman who thinks it would not be an impious libel to parody the Koran—May the suspenders of the Habeas Corpus Act be speedily suspended—Three times twelve for thrice-tried Hone, who cleared the cases himself alone, and won three heats by twelve to one, 1*l.* 16*s.*—A conscientious attorney, 1*l.* 6*s.* 8*d.*—Rev. T. B. Morris, rector of Shelfanger, who disapproves of the parodies, but abhors the making an affected zeal for religion the pretext for political persecution—A Lawyer opposed in principle to Law—For the Hone that set the razor that shaved the rats—Rev. Dr. Samuel Parr, who most seriously disapproves of all parodies upon the hallowed language of Scripture and the contents of the Prayer-book, but acquits Mr. Hone of intentional impiety, admires his talents. and fortitude, and applauds the good sense and integrity of his juries—Religion without hypocrisy, and Law without partiality—O Law! O Law! O Law!

These are specimens of a great many allusive mottoes. The subscription was very large, and would have bought a handsome annuity, but Hone employed it in the bookselling trade, and did not thrive. His 'Everyday Book' and his 'Apocryphal New Testament' are useful books. On an annuity he would have thriven as an antiquarian writer and collector. It is well that the attack upon the right to ridicule Ministers roused a dormant power which was equal to the occasion. Hone declared, on his honour, that he had never addressed a meeting in his life, nor spoken a word before more than twelve persons. Had he—which however could not then be done—employed counsel, and had a *guilty defence* made for him, he would very likely have been convicted, and the work would have been left to be done by another. No question that the parodies disgusted all who reverenced Christianity, and who could not separate the serious and the ludicrous, and prevent their existence in combination.

My extracts, &c., are from the nineteenth, seventeenth, and sixteenth editions of the three trials, which seem to have been contemporaneous (all in 1818) as they are made up into one book, with additional title over all, and the motto 'Thrice the brindled cat hath mew'd.' They are published by Hone himself, who I should have said was a publisher as well as was to be. And though the trials only ended Dec. 20, 1817, the preface attached to this common title is dated Jan. 23, 1818.

The spirit which was roused against the false dealing of the Government, i.e. the pretence of prosecuting for impiety when all the world knew the real offence was, if anything, sedition—was not got up at the moment : there had been previous exhibitions of it. For example, in the spring of 1818 Mr. Russell, a little printer in Birmingham, was indicted for publishing the Political Litany on which Hone was afterwards tried. He took his witnesses to the summer Warwick assizes, and was told that the indictment had been removed by certiorari into the King's Bench. He had notice of trial for the spring assizes at Warwick: he took his witnesses there, and the trial was postponed by the Crown. He then had notice for the summer assizes at Warwick ; and so on. The policy seems to have been to wear out the obnoxious parties, either by delays or by heaping on trials. The Government was odious, and knew it could *not* get verdicts against ridicule, and *could* get verdicts against impiety. No difficulty was found in convicting the sellers of Paine's works, and the like. When Hone was held to bail it was seen that a crisis was at hand. All parties in politics furnished him with parodies in proof of religious persons having made instruments of them. The parodies by Addison and Luther were contributed by a Tory lawyer, who was afterwards a judge.

Hone had published, in 1817, tracts of purely political ridicule: 'official account of the noble lord's bite,' 'trial of the dog for biting the noble lord,' &c. These were not touched. After the trials, it is manifest that Hone was to be unassailed, do what he might. 'The Political House that Jack built,' in 1819; 'The Man in the Moon,' 1820; 'The Queen's Matrimonial Ladder,' 'Non mi ricordo,' 'The R—l fowls,' 1820; 'The Political Showman at home,' with plates by G. Cruickshank, 1821 [he did all the plates]; 'The Spirit of Despotism,' 1821—would have been legitimate marks for prosecution in previous years. The biting caricature of several of these works are remembered to this day. 'The Spirit of Despotism' was a tract of 1795, of which a few copies had been privately circulated with great secrecy. Hone reprinted it, and prefixed the following address to 'Robert Stewart, *alias* Lord Castlereagh'—'It appears to me that if, unhappily, your counsels are allowed much longer to prevail in the Brunswick Cabinet, they will bring on a crisis, in which the king may be dethroned or the people enslaved. Experience has shown that the people will not be enslaved—the alternative is the affair of your employers.' Hone might say this without notice.

In 1819 Mr. Murray published Lord Byron's 'Don Juan,' and Hone followed it with 'Don John, or Don Juan unmasked,' a little account of what the publisher to the Admiralty was allowed to issue without prosecution. The parody on the Commandments was a case very much in point: and Hone makes a stinging allusion to the use of the '*unutterable Name*, with a profane levity unsurpassed by any other two lines in the English language.' The lines are

'Tis strange—the Hebrew noun which means 'I am,'
The English always use to govern d——n.

Hone ends with: 'Lord Byron's dedication of "Don Juan" to Lord Castlereagh was suppressed by Mr. Murray from delicacy to Ministers. *Q.* Why did not Mr. Murray suppress Lord Byron's *parody* on the Ten Commandments? *A.* Because it contains nothing in ridicule of Ministers, and therefore nothing that *they* could suppose would lead to the displeasure of Almighty God.'

The little matters on which I have dwelt will never appear in history from their political importance, except in a few words of result. As a mode of thought, silly evasions of all kinds belong to such a work as the present. Ignorance, which seats itself in the chair of knowledge, is a mother of revolutions in politics, and of unread pamphlets in circle-squaring. From 1815 to 1830 the question of revolution or no revolution lurked in all our English discussions. The high classes must govern; the high classes shall not govern; and thereupon issue was to be joined. In 1828–1833 the question came to issue; and it was, Revolution with or without civil war; choose. The choice was wisely made; and the Reform Bill started a new system so well dovetailed into the old that the joinings are hardly visible. And now, in 1867, the thing is repeated with a marked subsidence of symptoms; and the party which has taken the place of the extinct Tories is carrying through Parliament a wider extension of the franchise than their opponents would have ventured. Napoleon used to say that a decided nose was a sign of power: on which it has been remarked that he had good reason to say so before the play was done. And so had our country; it was saved from a religious war, and from a civil war, by the power of that nose over its colleagues.

> The Commentaries of Proclus. Translated by Thomas Taylor. London, 1792, 2 vols. 4to.

The reputation of 'the Platonist' begins to grow, and will continue to grow. The most authentic account is in the *Penny Cyclopædia,* written by one of the few persons who knew him well, and one of the fewer who possess all his works. At page lvi. of the Introduction is Taylor's notion of the way to find the circumference. It is not geometrical, for it proceeds on the motion of a point: the words ' on account of the simplicity of the impulsive motion, such a line must be either straight or circular' will suffice to show how Platonic it is. Taylor certainly professed a kind of heathenism. D'Israeli said, ' Mr. T. Taylor, the Platonic philosopher and the modern Plethon, consonant to that philosophy, professes polytheism.' Taylor printed this in large type, in a page by itself after the dedication, without any disavowal. I have seen the following, Greek and translation both, in his handwriting :—' Πας ἀγαθος ἠ ἀγαθος ἐθνικος· και πας χριστιανος ἠ χριστιανος κακος. Every good man, so far as he is a good man, is a heathen ; and every Christian, so far as he is a Christian, is a bad man.' Whether Taylor had in his head the Christian of the New Testament, or whether he drew from those members of the ' religious world ' who make manifest the religious flesh and the religious devil, cannot be decided by us, and perhaps was not known to himself. If a heathen, he was a virtuous one.

(1795.) This is the date of a very remarkable paradox. The religious world—to use a name claimed by a doctrinal sect— had long set its face against amusing literature, and all works of imagination. Bunyan, Milton, and a few others were irresistible; but a long face was pulled at every attempt to produce something readable for poor people and *poor children*. In 1795, a benevolent association began to circulate the works of a lady who had been herself a dramatist, and had nourished a pleasant vein of satire in the society of Garrick and his friends; all which is carefully suppressed in some biographies. Hannah More's *Cheap Repository Tracts*, which were bought by millions of copies, destroyed the vicious publications with which the hawkers deluged the country, by the simple process of furnishing the hawkers with something more saleable.

*Dramatic fiction*, in which the *characters* are drawn by themselves, was, at the middle of the last century, the monopoly of writers who required indecorum, such as Fielding and Smollett. All, or nearly all, which could be permitted to the young, was dry narrative, written by people who could not make their personages *talk character*; they all spoke alike. The author of the *Rambler* is ridiculed, because his young ladies talk Johnsonese; but the satirists forget that all the presentable novel-writers were equally incompetent; even the author of 'Zeluco' (1789) is the strongest possible case in point.

Dr. Moore, the father of the hero of Corunna, with good narrative power, some sly humour, and much observation of character, would have been, in our day, a writer of the *Peacock* family. Nevertheless, to one who is accustomed to our style of things, it is comic to read the dialogue of a jealous husband, a suspected wife, a faithless maid-servant, a tool of a nurse, a wrong-headed pomposity of a priest, and a sensible physician, all talking Dr. Moore through their masks. Certainly an Irish soldier does say *by Jasus*, and a cockney footman *this here* and *that there*; and this and the like is all the painting of characters which is effected out of the mouths of the bearers by a narrator of great power. I suspect that some novelists repressed their power under a rule that a narrative should narrate, and that the dramatic should be confined to the drama.

I make no exception in favour of Miss Burney; though she was the forerunner of a new era. Suppose a country in which dress is always of one colour; suppose an importer who brings in cargoes of blue stuff, red stuff, green stuff, &c., and exhibits dresses of these several colours, that person is the similitude of Miss

Burney. It would be a delightful change from a universal dull brown, to see one person all red, another all blue, &c. ; but the real inventor of pleasant dress would be the one who could mix his colours and keep down the bright and gaudy. Miss Burney's introduction was so charming, by contrast, that she nailed such men as Johnson, Burke, Garrick, &c., to her books. But when a person who has read them with keen pleasure in boyhood, as I did, comes back to them after a long period, during which he has made acquaintance with the great novelists of our century, three-quarters of the pleasure is replaced by wonder that he had not seen he was at a puppet-show, not at a drama. Take some *labelled* characters out of our humourists, let them be put together into one piece, to speak only as labelled : let there be a Dominie with nothing but ' Prodigious ! ' a Dick Swiveller with nothing but adapted quotations ; a Dr. Folliott with nothing but sneers at Lord Brougham ; and the whole will pack up into one of Miss Burney's novels.

Maria Edgeworth, Sydney Owenson (Lady Morgan), Jane Austen, Walter Scott, &c., are all of our century ; as are, I believe, all the Minerva Press novels, as they were called, which show some of the power in question. Perhaps dramatic talent found its best encouragement in the drama itself. But I cannot ascertain that any such power was directed at the multitude, whether educated or uneducated, with natural mixture of character, under the restraints of decorum, until the use of it by two religious writers of the school called ' evangelical,' Hannah More and Rowland Hill. The *Village Dialogues*, though not equal to the *Repository Tracts*, are in many parts an approach, and perhaps a copy ; there is frequently humorous satire, in that most effective form, self-display. They were published in 1800, and, partly at least, by the Religious Tract Society, the lineal successor of the *Repository* association, though knowing nothing about its predecessor. I think it right to add that Rowland Hill here mentioned is not the regenerator of the Post Office. Some do not distinguish accurately ; I have heard of more than one who took me to have had a logical controversy with a diplomatist who died some years before I was born.

A few years ago, an attempt was made by myself and others to collect some information about the *Cheap Repository* (see *Notes and Queries*, 3rd Series, vi. 241, 290, 353 ; *Christian Observer*, Dec. 1864, pp. 944–49). It appeared that after the Religious Tract Society had existed more than fifty years, a friend presented it with a copy of the original prospectus of the *Reposi-*

*tory*, a thing the existence of which was not known. In this prospectus it is announced that from the plan ' will be carefully excluded whatever is enthusiastic, absurd, or superstitious.' The ' evangelical' party had, from the foundation of the Religious Tract Society, regretted that the *Repository Tracts* ' did not contain a fuller statement of the great evangelical principles;' while in the prospectus it is also stated that ' no cause of any particular party is intended to be served by it, but general Christianity will be promoted upon practical principles.' This explains what has often been noticed, that the tracts contain a mild form of the ' evangelical' doctrine, free from that more fervid dogmatism which appears in the *Village Dialogues*; and such as H. More's friend, Bishop Porteus—a great promoter of the scheme—might approve. The Religious Tract Society (in 1863) republished some of H. More's tracts, with alterations, additions, and omissions *ad libitum*. This is an improper way of dealing with the works of the dead; especially when the reprints are of popular works. A small type addition to the preface contains : ' Some alterations and abridgments have been made to adapt them to the present times and the aim of the Religious Tract Society.' I think every publicity ought to be given to the existence of such a practice; and I reprint what I said on the subject in *Notes and Queries*.

Alterations in works which the Society republishes are a necessary part of their plan, though such notes as they should judge to be corrective would be the best way of proceeding. But the fact of alteration should be very distinctly announced on the title of the work itself, not left to a little bit of small type at the end of the preface, in the place where trade advertisements, or directions to the binder, are often found. And the places in which alteration has been made should be pointed out, either by marks of omission, when omission is the alteration, or by putting the altered sentences in brackets, when change has been made. May any one alter the works of the dead at his own discretion ? We all know that readers in general will take each sentence to be that of the author whose name is on the title; so that a correcting republisher *makes use of his author's name to teach his own variation*. The tortuous logic of ' the trade,' which is content when ' the world' is satisfied, is not easily answered, any more than an eel is easily caught; but the Religious Tract Society may be *convinced* [in the old sense] in a sentence. On which course would they feel most safe in giving their account to the God of truth ? ' In your own conscience, now ? '

I have tracked out a good many of the variations made by the Religious Tract Society in the recently published volume of *Repository Tracts.* Most of them are doctrinal insertions or amplifications, to the matter of which Hannah More would not have objected—all that can be brought against them is the want of notice.   But I have found two which the respect I have for the Religious Tract Society, in spite of much difference on various points, must not prevent my designating as paltry.   In the story of Mary Wood, a kind-hearted clergyman converses with the poor girl who has ruined herself by lying.   In the original, he 'assisted her in the great work of repentance ;' in the reprint it is to be shown in some detail how he did this.   He is to begin by pointing out that 'the heart is deceitful above all things and desperately wicked.'   Now the clergyman's name is *Heartwell*: so to prevent his name from contradicting his doctrine, he is actually cut down to *Harwell*.   Hannah More meant this good man for one of those described in Acts xv. 8, 9, and his name was appropriate.

Again, Mr. Flatterwell, in persuasion of Parley the porter to let him into the castle, declares that the worst he will do is to 'play an innocent game of cards just to keep you awake, or sing a cheerful song with the maids.'   Oh fie ! Miss Hannah More ! and you a single lady too, and a contemporary of the virtuous Bowdler !   Though Flatterwell be an allegory of the devil, this is really too indecorous, even for him.   Out with the three last words ! and out it is.

The Society cuts a poor figure before a literary tribunal. Nothing was wanted except an admission that the remarks made by me were unanswerable, and this was immediately furnished by the Secretary (*N.* and *Q.* 3 S. vi. 290).   In a reply of which six parts out of seven are a very amplified statement that the Society did not intend to reprint *all* Hannah More's tracts, the remaining seventh is as follows :—

I am not careful [perhaps this should be *careful not*] to notice Professor De Morgan's objections to the changes in 'Mary Wood' or 'Parley the Porter,' but would merely reiterate that the tracts were neither designed nor announced to be 'reprints' of the originals [design is only known to the designers ; as to announcement, the title is ' 'Tis all for the best, The Shepherd of Salisbury Plain, and other narratives, by Mrs. Hannah More ']; and much less [this must be *careful not*; further removed from answer than *not careful*] can I occupy your space by a treatise on the Professor's question : ' May any one alter the works of the dead at his own discretion ? '

To which I say—Thanks for help !

I predict that Hannah More's *Cheap Repository Tracts* will somewhat resemble the *Pilgrim's Progress* in their fate. Written for the cottage, and long remaining in their original position, they will become classical works of their kind. Most assuredly this will happen if my assertion cannot be upset, namely—That they contain the first specimens of fiction addressed to the world at large, and widely circulated, in which dramatic—as distinguished from puppet—power is shown, and without indecorum.

According to some statements I have seen, but which I have not verified, other publishing bodies, such as the Christian Knowledge Society, have taken the same liberty with the names of the dead as the Religious Tract Society. If it be so, the impropriety is the work of the smaller spirits, who have not been sufficiently overlooked. There must be an overwhelming majority in the higher councils to feel that, whenever *altered* works are published, *the fact of alteration should be made as prominent as the name of the author*. Everything short of this is suppression of truth, and will ultimately destroy the credit of the Society. Equally necessary is it that the alterations should be noted. When it comes to be known that the author before him is altered, he knows not where nor how nor by whom, the lowest reader will lose his interest.

The principles of Algebra. By William Frend. London, 1796, 8vo. Second Part, 1799.

This Algebra, says Dr. Peacock, shows ' great distrust of the results of algebraical science which were in existence at the time when it was written.' Truly it does; for, as Dr. Peacock had shown by full citation, it makes war of extermination upon all that distinguishes algebra from arithmetic. Robert Simson and Baron Maseres were Mr. Frend's predecessors in this opinion.

The genuine respect which I entertained for my father-in-law did not prevent my canvassing with perfect freedom his anti-algebraical and anti-Newtonian opinions, in a long obituary memoir read at the Astronomical Society in February 1842, which was written by me. It was copied into the *Athenæum* of March 19. It must be said that if the manner in which algebra *was* presented to the learner had been true algebra, he would have been right: and if he had confined himself to protesting against the imposition of attraction as a fundamental part of the existence of matter, he would have been in unity with a great many, including Newton himself. I wish he had preferred

amendment to rejection when he was a college tutor : he wrote
and spoke English with a clearness which is seldom equalled.

His anti-Newtonian discussions are confined to the preliminary
chapters of his ‘Evening Amusements,’ a series of astronomical
lessons in nineteen volumes, following the moon through a period
of the golden numbers.

There is a mistake about him which can never be destroyed.
It is constantly said that, at his celebrated trial in 1792, for
sedition and opposition to the Liturgy, &c., he was *expelled* the
University.  He was *banished*.  People cannot see the difference;
but it made all the difference to Mr. Frend.  He held his fellow-
ship and its profits till his marriage in 1808, and was a member
of the University and of its Senate till his death in 1841, as any
Cambridge Calendar up to 1841 will show.  That they would have
expelled him if they could, is perfectly true ; and there is a funny
story—also perfectly true—about their first proceedings being
under a statute which would have given the power, had it not been
discovered during the proceedings that the statute did not exist.
It had come so near to existence as to be entered into the Vice-
Chancellor’s book for his signature, which it wanted, as was not
seen till Mr. Frend exposed it : in fact, the statute had never
actually passed.

There is an absurd mistake in Gunning’s ‘Reminiscences of
Cambridge.’  In quoting a passage of Mr. Frend’s pamphlet,
which was very obnoxious to the existing Government, it is
printed that the poor market-women complained that they were
to be *scotched* a quarter of their wages by taxation; and attention
is called to the word by its being three times printed in italics.
In the pamphlet it is ‘sconced’; that very common old word for
fined or mulcted.

Lord Lyndhurst, who has [1863] just passed away under a load
of years and honours, was Mr. Frend’s private pupil at Cambridge.
At the time of the celebrated trial, he and two others amused
themselves, and vented the feeling which was very strong among
the undergraduates, by chalking the walls of Cambridge with
‘ Frend for ever!’  While thus engaged in what, using the term
legally, we are probably to call his first publication, he and his
friends were surprised by the proctors.  Flight and chase followed
of course : Copley and one of the others, Serjeant Rough,
escaped ; the third, whose name I forget, but who afterwards, I
have been told, was a bishop,[1] being lame, was captured and
impositioned.  Looking at the Cambridge Calendar to verify the

---

[1] Herbert Marsh, afterwards Bishop of Peterborough, a relation of my father, (Ed.)

fact that Copley was an undergraduate at the time, I find that there are but two other men in the list of honours of his year whose names are now widely remembered. And they were both celebrated schoolmasters ; Butler of Harrow, and Tate of Richmond.

But Mr. Frend had another noted pupil. I once had a conversation with a very remarkable man, who was generally called ' Place, the tailor,' but who was politician, political economist, &c., &c. He sat in the room above his shop—he was then a thriving master tailor at Charing Cross—surrounded by books enough for nine, to shame a proverb. The blue books alone, cut up into strips, would have measured Great Britain for oh-no-we-never-mention-'ems, the Highlands included. I cannot find a biography of this worthy and able man. I happened to mention William Frend, and he said, ' Ah! my old master, as I always call him. Many and many a time, and year after year, did he come in every now and then to give me instruction, while I was sitting on the board, working for my living, you know.'

Place, who really was a sound economist, is joined with Cobbett, because they were together at one time, and because he was, in 1800, &c., a great Radical. But for Cobbett he had a great contempt. He told me the following story. He and others were advising with Cobbett about the defence he was to make on a trial for seditious libel which was coming on. Said Place, ' You must put in the letters you have received from Ministers, members of the Commons from the Speaker downwards, &c., about your Register, and their wish to have subjects noted. You must then ask the jury whether a person so addressed must be considered as a common sower of sedition, &c. You will be acquitted; nay, if your intention should get about, very likely they will manage to stop proceedings.' Cobbett was too much disturbed to listen ; he walked about the room ejaculating ' D—— the prison !' and the like. He had not the sense to follow the advice, and was convicted.

Cobbett, to go on with the chain, was a political acrobat, ready for any kind of posture. A friend of mine gave me several times an account of a mission to him. A Tory member—those who know the old Tory world may look for his initials in initials of two consecutive words of ' Pay his money with interest '—who was, of course, a political opponent, thought Cobbett had been hardly used, and determined to subscribe handsomely towards the expenses he was incurring as a candidate. My friend was commissioned to hand over the money—a bag of sovereigns, that notes might not be traced. He went into Cobbett's committee-room,

told the patriot his errand, and put the money on the table.
'And to whom, sir, am I indebted?' said Cobbett. 'The donor,'
was the answer, 'is Mr. Andrew Theophilus Smith,' or some such
unlikely pair of baptismals. 'Ah!' said Cobbett, 'I have known M
A. T. S. a long time! he was always a true friend of his countr<sub>y</sub>.

To return to Place. He is a noted instance of the advantage
of our jury system, which never asks a man's politics, &c. The
late King of Hanover, when Duke of Cumberland, being unpopular,
was brought under unjust suspicions by the suicide of his valet:
he must have seduced the wife and murdered the husband. The
charges were as absurd as those brought against the Englishman
in the Frenchman's attempt at satirical verses upon him :—

> The Englishman is a very bad man ;
> He drink the beer and he steal the can :
> He kiss the wife and he beat the man ;
> And the Englishman is a very G—— d——.

The charges were revived in a much later day, and the defence
might have given some trouble. But Place, who had been the
foreman at the inquest, came forward, and settled the question in
a few lines. Everyone knew that the old Radical was quite free
of all disposition to suppress truth from wish to curry favour with
royalty.

John Speed, the author of the English History (1632) which
Bishop Nicolson calls the best chronicle extant, was a man, like
Place, of no education but what he gave himself. The bishop
says he would have done better if he had had better training:
but what, he adds, could have been expected from a tailor! This
Speed was, as well as Place. But he was released from manual
labour by Sir Fulk Grevil, who enabled him to study.

I have elsewhere noticed that those who oppose the mysteries
of algebra do not ridicule them ; this I want the cyclometers to
do. Of the three who wrote against the great point, the negative
quantity, and the uses of 0 which are connected with it, only
one could fire a squib. That Robert Simson should do such a
thing will be judged impossible by all who admit tradition. I
do not vouch for the following ; I give it as a proof of the
impression which prevailed about him :—

He used to sit at his open window on the ground floor, as deep
in geometry as a Robert Simson ought to be. Here he would be
accosted by beggars, to whom he generally gave a trifle , he
roused himself to hear a few words of the story, made his dona-
tion, and instantly dropped down into his depths. Some wags

one day stopped a mendicant who was on his way to the window, with 'Now, my man, do as we tell you, and you will get something from that gentleman, and a shilling from us besides. You will go and say you are in distress, he will ask you who you are, and you will say you are Robert Simson, son of John Simson of Kirktonhill.' The man did as he was told ; Simson quietly gave him a coin, and dropped off. The wags watched a little, and saw him rouse himself again, and exclaim 'Robert Simson, son of John Simson of Kirktonhill! why, that is myself. That man must be an impostor.' Lord Brougham tells the same story, with some difference of details.

Baron Maseres was, as a writer, dry ; those who know his writings will feel that he seldom could have taken in a joke or issued a pun. Maseres was the fourth wrangler of 1752, and first Chancellor's medallist (or highest in classics); his second was Porteus (afterwards Bishop of London). Waring came five years after him : he could not get Maseres through the second page of his first work on algebra ; a negative quantity stood like a lion in the way. In 1758 he published his 'Dissertation on the Use of the Negative Sign,' 4to. There are some who care little about + and −, who would give it house-room for the sake of the four words 'Printed by Samuel Richardson.'

Maseres speaks as follows: 'A single quantity can never be marked with either of those signs, or considered as either affirmative or negative ; for if any single quantity, as $b$, is marked either with the sign + or with the sign − without assigning some other quantity, as $a$, to which it is to be added, or from which it is to be subtracted, the mark will have no meaning or signification : thus if it be said that the square of −5, or the product of −5 into −5, is equal to +25, such an assertion must either signify no more than that 5 times 5 is equal to 25 without any regard to the signs, or it must be mere nonsense and unintelligible jargon. I speak according to the foregoing definition, by which the affirmativeness or negativeness of any quantity implies a relation to another quantity of the same kind to which it is added, or from which it is subtracted ; for it may perhaps be very clear and intelligible to those who have formed to themselves some other idea of affirmative and negative quantities different from that above defined.'

Nothing can be more correct, or more identically logical: +5 and −5, standing alone, are jargon if +5 and −5 are to be understood as without reference to another quantity. But those who have 'formed to themselves some other idea' see meaning

enough.  The great difficulty of the opponents of algebra lay in
want of power or will to see extension of terms.  Maseres is right
when he implies that extension, accompanied by its refusal,
makes jargon.  One of my paradoxers was present at a meeting
of the Royal Society (in 1864, I think) and asked permission to make some remarks upon a paper.  He rambled into
other things, and, naming me, said that I had written a
book in which two sides of a triangle are pronounced *equal* to
the third.  So they are, in the sense in which the word is used
in complete algebra; in which $A + B = C$ makes A, B, C, three
sides of a triangle, and declares that going over A and B, one after
the other, is equivalent, in change of place, to going over C at
once.  My critic, who might, if he pleased, have objected to
extension, insisted upon reading me in unextended meaning.

On the other hand, it must be said that those who wrote on
the other idea wrote very obscurely about it, and justified Des
Cartes (*De Methodo*) when he said: ' Algebram vero, ut solet
doceri, animadverti certis regulis et numerandi formulis ita esse
contentam, ut videatur potius ars quædam confusa, cujus usu
ingenium quodam modo turbatur et obscuratur, quam scientia
qua excolatur et perspicacius reddatur.'  Maseres wrote this
sentence on the title of his own copy of his own work, now before
me ; he would have made it his motto if he had found it earlier.

There is, I believe, in Cobbett's ' Annual Register,' an account
of an interview between Maseres and Cobbett when in prison.

The conversation of Maseres was lively, and full of serious anecdote : but only one attempt at humorous satire is recorded of
him ; it is an instructive one.  He was born in 1731 (Dec. 15),
and his father was a refugee.  French was the language of the
house, with the pronunciation of the time of Louis XIV.  He
lived until 1824 (May 19), and saw the race of refugees who
were driven out by the first Revolution.  Their pronunciation
differed greatly from his own ; and he used to amuse himself by
mimicking them.  Those who heard him and them had the two
schools of pronunciation before them at once; a thing which
seldom happens.  It might even yet be worth while to examine
the Canadian pronunciation.

Maseres went as Attorney-General to Quebec ; and was appointed Cursitor Baron of our Exchequer in 1773.  There is a
curious story about his mission to Canada, which I have heard as
good tradition, but have never seen in print.  The reader shall
have it as cheap as I ; and I confess I rather believe it.  Maseres
was inveterately honest ; he could not, at the bar, bear to see his

own client victorious, when he knew his cause was a bad one. On a certain occasion he was in a cause which he knew would go against him if a certain case were quoted. Neither the judge nor the opposite counsel seemed to remember this case, and Maseres could not help dropping an allusion which brought it out. His business as a barrister fell off, of course. Some time after, Mr. Pitt (Chatham) wanted a lawyer to send to Canada on a private mission, and wanted a *very honest man*. Some one mentioned Maseres, and told the above story : Pitt saw that he had got the man he wanted. The mission was satisfactorily performed, and Maseres remained as Attorney-General.

The 'Doctrine of Life Annuities' (4to. 726 pages, 1783) is a strange paradox. Its size, the heavy dissertations on the national debt, and the depth of algebra supposed known, put it out of the question as an elementary work, and it is unfitted for the higher student by its elaborate attempt at elementary character, shown in its rejection of forms derived from chances in favour of *the average*, and its exhibition of the separate values of the years of an annuity, as arithmetical illustrations. It is a climax of unsaleability, unreadability, and inutility. For intrinsic nullity of interest, and dilution of little matter with much ink, I can compare this book to nothing but that of Claude de St. Martin, elsewhere mentioned, or the lectures ' On the Nature and Properties of Logarithms,' by James Little, Dublin, 1830, 8vo. (254 heavy pages of many words and few symbols), a wonderful weight of weariness.

The stock of this work on annuities, very little diminished, was given by the author to William Frend, who paid warehouse room for it until about 1835, when he consulted me as to its disposal. As no publisher could be found who would take it as a gift, for any purpose of sale, it was consigned, all but a few copies, to a buyer of waste paper.

Baron Maseres's republications are well known : the *Scriptores Logarithmici* is a set of valuable reprints, mixed with much which might better have entered into another collection. It is not so well known that there is a volume of optical reprints, *Scriptores Optici*, London, 1823, 4to, edited for the veteran of ninety-two by Mr. Babbage at twenty-nine. This excellent volume contains James Gregory, Des Cartes, Halley, Barrow, and the optical writings of Huyghens, the *Principia* of the undulatory theory. It also contains, by the sort of whim in which such men as Maseres, myself, and some others are apt to indulge, a reprint of 'The great and new Art of weighing

Vanity,' by M. Patrick Mathers, Arch-Bedel to the University of St. Andrews, Glasgow, 1672. Professor Sinclair, of Glasgow, a good man at clearing mines of the water which they did not want, and furnishing cities with the water which they did want, seems to have written absurdly about hydrostatics, and to have attacked a certain Sanders, M.A. So Sanders, assisted by James Gregory, published a heavy bit of jocosity about him. This story of the authorship rested on a note made in his copy by Robert Gray, M.D. ; but it has since been fully confirmed by a letter of James Gregory to Collins, in the Macclesfield Correspondence. 'There is one Master Sinclair, who did write the *Ars Magna et Nova,* a pitiful ignorant fellow, who hath lately written horrid nonsense in the hydrostatics, and hath abused a master in the University, one Mr. Sanders, in print. This Mr. Sanders . . . is resolved to cause the Bedel of the University to write against him. . . . We resolve to make excellent sport with him.'

On this I make two remarks : First, I have learnt from experience that old notes, made in books by their possessors, are statements of high authority : they are almost always confirmed. I do not receive them without hesitation ; but I believe that of all the statements about books which rest on one authority, there is a larger percentage of truth in the written word than in the printed word. Secondly, I mourn to think that when the New Zealander picks up his old copy of this book, and reads it by the associations of his own day, he may, in spite of the many assurances I have received that my *Athenæum* Budget was amusing, feel me to be as heavy as I feel James Gregory and Sanders. But he will see that I knew what was coming, which Gregory did not.

It was left for William Frend to prove that an impugner of algebra could attempt ridicule. He was, in 1803, editor of a periodical *The Gentleman's Monthly Miscellany,* which lasted a few months. To this, among other things, he contributed the following, in burlesque of the use made of 0, to which he objected. The imitation of Rabelais, a writer in whom he delighted, is good : to those who have never dipped, it may give such a notion as they would not easily get elsewhere. The point of the satire is not so good. But in truth it is not easy to make pungent scoffs upon what is common sense to all mankind. Who can laugh with effect at six times nothing is nothing, as false or unintelligible ? In an article intended for that undistinguishing know-0 the 'general reader,' there would have been no force of satire, if *division* by 0 had been separated from multiplication by the same.

I have followed the above by another squib, by the same author, on the English language. The satire is covertly aimed at theological phraseology; and any one who watches this subject will see that it is a very just observation that the Greek words are not boiled enough.

PANTAGRUEL'S DECISION *of the* QUESTION *about* NOTHING.

PANTAGRUEL determined to have a snug afternoon with Epistemon and Panurge. Dinner was ordered to be set in a small parlour, and a particular batch of Hermitage with some choice Burgundy to be drawn from a remote corner of the cellar upon the occasion. By way of lunch, about an hour before dinner, Pantagruel was composing his stomach with German sausages, reindeer's tongues, oysters, brawn, and half a dozen different sorts of English beer just come into fashion, when a most thundering knocking was heard at the great gate, and from the noise they expected it to announce the arrival at least of the First Consul, or king Gargantua. Panurge was sent to reconnoitre, and after a quarter of an hour's absence, returned with the news that the University of Pontemaca was waiting his highness's leisure in the great hall, to propound a question which had turned the brains of thirty-nine students, and had flung twenty-seven more into a high fever. With all my heart, says Pantagruel, and swallowed down three quarts of Burton ale; but remember, it wants but an hour of dinner time, and the question must be asked in as few words as possible; for I cannot deprive myself of the pleasure I expected to enjoy in the company of my good friends for a set of mad-headed masters. I wish brother John was here to settle these matters with the black gentry.

Having said or rather growled this, he proceeded to the hall of ceremony, and mounted his throne; Epistemon and Panurge standing on each side, but two steps below him. Then advanced to the throne the three beadles of the University of Pontemaca with their silver staves on their shoulders, and velvet caps on their heads, and they were followed by three times three doctors, and thrice three times three masters of art; for everything was done in Pontemaca by the number three, and on this account the address was written on parchment, one foot in breadth, and thrice three times thrice three feet in length. The beadles struck the ground with their heads and their staves three times in approaching the throne; the doctors struck the ground with their heads thrice three times, and the masters did the same thrice each time, beating the ground with their heads thrice three times. This was the accustomed form of approaching the throne, time out of mind, and it was said to be emblematic of the usual prostration of science to the throne of greatness.

The mathematical professor, after having spit, and hawked, and cleared his throat, and blown his nose on a handkerchief lent to him, for he had forgotten to bring his own, began to read the address. In this he was assisted by three masters of arts, one of whom, with a silver pen, pointed out the stops; the second with a small stick rapped his knuckles when he was to raise or lower his voice; and a third pulled his hair behind when he was to look Pantagruel in the face. Pantagruel began to chafe like a lion: he turned first on one side, then on the other: he listened and groaned, and groaned and listened, and was in the utmost cogitabundity of cogitation. His countenance began to brighten, when, at the end of an hour, the reader stammered out these words:

'It has therefore been most clearly proved, that as all matter may be divided into parts infinitely smaller than the infinitely smallest part of the infinitesimal of nothing, so nothing has all the properties of something, and may become, by just and lawful right, susceptible of addition, subtraction, multiplication, division, squaring, and cubing: that it is to all intents and purposes as good as anything that has been, is, or can be taught in the nine universities of the land, and to deprive it of its rights is a most cruel innovation and usurpation, tending to destroy all just subordination in the world, making all universities superfluous, levelling vice-chancellors, doctors, and proctors, masters, bachelors, and scholars, to the mean and contemptible state of butchers and tallow-chandlers, bricklayers and chimney-sweepers, who, if it were not for these learned mysteries, might think that they knew as much as their betters. Every one then, who has the good of science at heart, must pray for the interference of his highness to put a stop to all the disputes about nothing, and by his decision to convince all gainsayers that the science of nothing is taught in the best manner in the universities; to the great edification and improvement of all the youth in the land.'

Here Pantagruel whispered in the ear of Panurge, who nodded to Epistemon, and they two left the assembly, and did not return for an hour, till the orator had finished his task. The three beadles had thrice struck the ground with their heads and staves, the doctors had finished their compliments, and the masters were making their twenty-seven prostrations. Epistemon and Panurge went up to Pantagruel, whom they found fast asleep and snoring; nor could he be roused but by as many tugs as there had been bowings from the corps of learning. At last he opened his eyes, gave a good stretch, made half a dozen yawns, and called for a stoup of wine. I thank you, my masters, says he; so sound a nap I have not had since I came from the island of Priestfolly. Have you dined, my masters? They answered the question by as many bows as at entrance; but his highness left them to the care of Panurge, and retired to the little parlour with Epistemon, where they burst into a fit of laughter, declaring that this learned Baragouin about nothing was just as intelligible as the lawyer's

Galimathias. Panurge conducted the learned body into a large saloon, and each in his way hearing a clattering of plates and glasses, congratulated himself on his approaching good cheer. There they were left by Panurge, who took his chair by Pantagruel just as the soup was removed, but he made up for the want of that part of his dinner by a pint of Champagne. The learning of the university had whetted their appetites ; what they each ate it is needless to recite ; good wine, good stories, and hearty laughs went round, and three hours elapsed before one soul of them recollected the hungry students of Pontemaca.

Epistemon reminded them of the business in hand, and orders were given for a fresh dozen of hermitage to be put upon table, and the royal attendants to get ready. As soon as the dozen bottles were emptied, Pantagruel rose from table, the royal trumpets sounded, and he was accompanied by the great officers of his court into the large dining hall, where was a table with forty-two covers. Pantagruel sat at the head, Epistemon at the bottom, and Panurge in the middle, opposite an immense silver tureen, which would hold fifty gallons of soup. The wise men of Pontemaca then took their seats according to seniority. Every countenance glistened with delight ; the music struck up ; the dishes were uncovered. Panurge had enough to do to handle the immense silver ladle : Pantagruel and Epistemon had no time for eating, they were fully employed in carving. The bill of fare announced the names of a hundred different dishes. From Panurge's ladle came into the soup plate as much as he took every time out of the tureen ; and as it was the rule of the court that every one should appear to eat, as long as he sat at table, there was the clattering of nine and thirty spoons against the silver soup-plates for a quarter of an hour. They were then removed, and knives and forks were in motion for half an hour. Glasses were continually handed round in the mean time, and then everything was removed, except the great tureen of soup. The second course was now served up, in dispatching which half an hour was consumed ; and at the conclusion the wise men of Pontemaca had just as much in their stomachs as Pantagruel in his head from their address : for nothing was cooked up for them in every possible shape that Panurge could devise.

Wine-glasses, large decanters, fruit dishes, and plates were now set on. Pantagruel and Epistemon alternately gave bumper toasts : the University of Pontemaca, the eye of the world, the mother of taste and good sense and universal learning, the patroness of utility, and the second only to Pantagruel in wisdom and virtue (for these were her titles), was drank standing with thrice three times three, and huzzas and clatterings of glasses ; but to such wine the wise men of Pontemaca had not been accustomed ; and though Pantagruel did not suffer one to rise from table till the eighty-first glass had been emptied, not even the weakest headed master of arts felt his head in the least indisposed. The decanters indeed were often removed, but they were brought back replenished, filled always with nothing.

Silence was now proclaimed, and in a trice Panurge leaped into the large silver tureen. Thence he made his bows to Pantagruel and the whole company, and commenced an oration of signs, which lasted an hour and a half, and in which he went over all the matter contained in the Pontemacan address ; and though the wise men looked very serious during the whole time, Pantagruel himself and his whole court could not help indulging in repeated bursts of laughter. It was universally acknowledged that he excelled himself, and that the arguments by which he beat the English masters of arts at Paris were nothing to the exquisite selection of attitudes which he this day assumed. The greatest shouts of applause were excited when he was running thrice round the tureen on its rim, with his left hand holding his nose, and the other exercising itself nine and thirty times on his back. In this attitude he concluded with his back to the professor of mathematics ; and at the instant he gave his last flap, by a sudden jump, and turning heels over head in the air, he presented himself face to face to the professor, and standing on his left leg, with his left hand holding his nose, he presented to him, in a white satin bag, Pantagruel's royal decree. Then advancing his right leg, he fixed it on the professor's head, and after three turns, in which he clapped his sides with both hands thrice three times, down he leaped, and Pantagruel, Epistemon, and himself took their leaves of the wise men of Pontemaca.

The wise men now retired, and by royal orders were accompanied by a guard, and according to the etiquette of the court, no one having a royal order could stop at any public house till it was delivered. The procession arrived at Pontemaca at nine o'clock the next morning, and the sound of bells from every church and college announced their arrival. The congregation was assembled ; the royal decree was saluted in the same manner as if his highness had been there in person ; and after the proper ceremonies had been performed, the satin bag was opened exactly at twelve o'clock. A finely emblazoned roll was drawn forth, and the public orator read to the gaping assembly the following words :

' They who can make something out of nothing shall have nothing to eat at the court of—PANTAGRUEL.'

### ORIGIN of the ENGLISH LANGUAGE, related by a SWEDE.

SOME months ago in a party in Holland, consisting of natives of various countries, the merit of their respective languages became a topic of conversation. A Swede, who had been a great traveller, and could converse in most of the modern languages of Europe, laughed very heartily at an Englishman, who had ventured to speak in praise of the tongue of his dear country. I never had any trouble, says he, in learning English. To my very great surprise, the moment I sat foot on shore at Gravesend, I found out, that I could understand, with very little trouble, every word that was said. It was a mere jargon, made up of German, French, and Italian, with now and then a word from the Spanish, Latin, or Greek. I had only to bring my mouth to their mode of speaking, which was done with ease in less than a week, and I was every where taken for a true-born Englishman; a privilege by the way of no small importance in a country, where each man, God knows why, thinks his foggy island superior to any other part of the world : and though his door is never free from some dun or other coming for a tax, and if he steps out of it he is sure to be knocked down or to have his pocket picked, yet he has the insolence to think every foreigner a miserable slave, and his country the seat of every thing wretched. They may talk of liberty as they please, but Spain or Turkey for my money : barring the bowstring and the inquisition, they are the most comfortable countries under heaven, and you need not be afraid of either, if you do not talk of religion and politics. I do not see much difference too in this respect in England, for when I was there, one of their most eminent men for learning was put in prison for a couple of years, and got his death for translating one of Æsop's fables into English, which every child in Spain and Turkey is taught, as soon as he comes out of his leading strings. Here all the company unanimously cried out against the Swede, that it was impossible : for in England, the land of liberty, the only thing its worst enemies could say against it, was, that they paid for their liberty a much greater price than it was worth.—Every man there had a fair trial according to laws, which every body could understand ;· and the judges were cool, patient, discerning men, who never took the part of the crown against the prisoner, but gave him every assistance possible for his defence.

The Swede was borne down, but not convinced; and he seemed determined to spit out all his venom. Well, says he, at any rate you will not deny that the English have not got a language of their own, and that they came by it in a very odd way. Of this at least I am certain, for the whole history was related to me by a witch in Lapland, whilst I was bargaining for a wind. Here the company were all in unison again for the story.

In antient times, said the old hag, the English occupied a spot in Tartary, where they lived sulkily by themselves, unknowing and unknown. By a great convulsion that took place in China, the inhabitants of that and the adjoining parts of Tartary were driven from their seats, and after various wanderings took up their abode in Germany. During this time no body could understand the English, for they did not talk, but hissed like so many snakes. The poor people felt uneasy under this circumstance, and in one of their parliaments, or rather hissing meetings, it was determined to seek for a remedy : and an embassy was sent to some of our sisterhood then living on Mount Hecla. They were put to a nonplus, and summoned the Devil to their

K

relief. To him the English presented their petitions, and explained their sad case; and he, upon certain conditions, promised to befriend them, and to give them a language. The poor Devil was little aware of what he had promised; but he is, as all the world knows, a man of too much honour to break his word. Up and down the world then he went in quest of this new language: visited all the universities, and all the schools, and all the courts of law, and all the play-houses, and all the prisons; never was poor devil so fagged. It would have made your heart bleed to see him. Thrice did he go round the earth in every parallel of latitude; and at last, wearied and jaded out, back came he to Hecla in despair, and would have thrown himself into the volcano, if he had been made of combustible materials. Luckily at that time our sisters were engaged in settling the balance of Europe; and whilst they were looking over projects, and counter-projects, and ultimatums, and post ultimatums, the poor Devil, unable to assist them, was groaning in a corner and ruminating over his sad condition.

On a sudden, a hellish joy overspread his countenance; up he jumped, and, like Archimedes of old, ran like a madman amongst the throng, turning over tables, and papers, and witches, roaring out for a full hour together nothing else but 'tis found, 'tis found! Away were sent the sisterhood in every direction, some to traverse all corners of the earth, and others to prepare a larger caldron than had ever yet been set upon Hecla. The affairs of Europe were at a stand: its balance was thrown aside; prime ministers and ambassadors were every where in the utmost confusion; and, by the way, they have never been able to find the balance since that time, and all the fine speeches upon the subject, with which your newspapers are every now and then filled, are all mere hocus-pocus and rhodomontade. However, the caldron was soon set on, and the air was darkened by witches riding on broomsticks, bringing a couple of folios under each arm, and across each shoulder. I remember the time exactly: it was just as the council of Nice had broken up, so that they got books and papers there dog cheap; but it was a bad thing for the poor English, as these were the worst materials that entered into the caldron. Besides, as the Devil wanted some amusement, and had not seen an account of the transactions of this famous council, he had all the books brought from it laid before him, and split his sides almost with laughing, whilst he was reading the speeches and decrees of so many of his old friends and acquaintance. All this while the witches were depositing their loads in the great caldron. There were books from the Dalai Lama, and from China: there were books from the Hindoos, and tallies from the Caffres: there were paintings from Mexico, and rocks of hieroglyphics from Egypt: the last country supplied besides the swathings of two thousand mummies, and four-fifths of the famed library of Alexandria. Bubble! bubble! toil and trouble! never was a day of more labour and anxiety; and if our good master had but flung in the Greek books at the proper time, they would have made a complete job of it. He was a little too impatient: as the caldron frothed up, he skimmed it off with a great ladle, and filled some thousands of our wind-bags with the froth, which the English with great joy carried back to their own country. These bags were sent to every district: the chiefs first took their fill, and then the common people; hence they now speak a language which no foreigner can understand, unless he has learned half a dozen other languages; and the poor people, not one in ten, understand a third part of what is said to them. The hissing, however, they have not entirely got rid of, and every seven years, when the Devil, according to agreement, pays them a visit, they entertain him at their common halls and county meetings with their original language.

The good natured old hag told me several other circumstances, relative to this curious transaction, which, as there is an Englishman in company, it will be prudent to pass over in silence: but I cannot help mentioning one thing which she told me as a very great secret. You know, says she to me, that the English have more religions among them than any other nation in Europe, and that there is more teaching and sermonizing with them than in any other country. The fact is this; it matters not who gets up to teach them, the hard words of the Greek were not sufficiently boiled, and whenever they get into a sentence, the poor people's brains are turned, and they know no more what the preacher is talking about, than if he harangued them in Arabic. Take my word for it if you please; but if not, when you get to England, desire the bettermost sort of people that you are acquainted with to read to you an act of parliament, which of course is written in the clearest and plainest stile in which any thing can be written, and you will find that not one in ten will be able to make tolerable sense of it. The language would have been an excellent language, if it had not been for the council of Nice, and the words had been well boiled.

Here the company burst out into a fit of laughter. The Englishman got up and shook hands with the Swede: *si non è vero*, said he, *è ben trovato*. But, however I may laugh at it here, I would not advise you to tell this story on the other side of the water. So here's a bumper to Old England for ever, and God save the king.'

---

The accounts given of extraordinary children and adolescents frequently defy credence. I will give two well-attested instances. The celebrated mathematician, Alexis Claude Clairault (now Clairaut) was certainly born in May, 1713. His treatise on curves of double curvature (printed in 1731) received the approbation of the Academy of Sciences, August 23, 1729. Fontenelle, in his certificate of this, calls the author sixteen years of age, and does not strive to exaggerate the wonder, as he might have done, by reminding his readers that this work, of original and sustained mathematical investigation, must have been coming from the pen at the ages of fourteen and fifteen. The truth was, as attested by De Molières, Clairaut had given public proofs of his power at twelve years old. His age being thus publicly certified, all doubt is removed : say he had been—though great wonder would still have been left—twenty-one instead of sixteen, his appearance, and the remembrances of his friends, schoolfellows, &c., would have made it utterly hopeless to knock off five years of that age while he was on view in Paris as a young lion. De Molières, who examined the work officially for the *Garde des Sceaux*, is transported beyond the bounds of official gravity, and says that it 'ne mérite pas seulement d'être imprimé, mais d'être admiré comme un prodige d'imagination, de conception, et de capacité.'

That Blaise Pascal was born in June, 1623, is perfectly well established and uncontested. That he wrote his conic sections at the age of sixteen might be difficult to establish, though tolerably well attested, if it were not for one circumstance, for the book was not published. The celebrated theorem, *Pascal's hexagram*, makes all the rest come very easy. Now Curabelle, in a work published in 1644, sneers at Desargues, whom he quotes, for having, in 1642, deferred a discussion until *cette grande proposition nommée la Pascale verra le jour*. That is, by the time Pascal was nineteen, the *hexagram* was circulating under a name derived from the author. The common story about Pascal, given by his sister, is an absurdity which no doubt has prejudiced many against tales of early proficiency. He is made, when quite a boy, to invent geometry *in the order of Euclid's propositions*: as if that order were natural sequence of investigation. The hexagram at ten years old would be a hundred times less unlikely.

The instances named are painfully astonishing : I give one which has fallen out of sight, because it will preserve an imperfect biography. John Wilson is Wilson of that Ilk, that is, of *Wilson's Theorem*. It is this : If $p$ be a prime number, the product of all the numbers up to $p-1$, increased by 1, is divisible without remainder by $p$. All mathematicians know this as Wilson's theorem, but few know who Wilson was. He was born August 6, 1741, at the Howe in Applethwaite, and he was heir to a small estate at Troutbeck in Westmoreland. He was sent to Peterhouse, at Cambridge, and, while an undergraduate was considered stronger in algebra than any one in the University, except Professor Waring, one of the most powerful algebraists of the century.[1]  He was the senior wrangler of 1761, and was then for some time a private tutor.  When Paley, then in his third year, determined to make a push for the senior wranglership, which he got, Wilson was recommended to him as a tutor.  Both were ardent in their work, except that sometimes Paley, when he came for his lesson, would find *gone a fishing* written on his tutor's outer door: which was insult added to injury, for Paley was very fond of fishing.  Wilson soon left Cambridge, and went to the bar.  He practised on the northern circuit with great success; and, one day, while passing his vacation on his little property at Troutbeck, he received information, to his great surprise, that Lord Thurlow, with whom he had no acquaintance, had recommended him to be a Judge of the Court of Common Pleas.  He died, Oct. 18, 1793, with a very high reputation as a lawyer and a Judge.  These facts are partly from Meadley's 'Life of Paley,' no doubt from Paley himself, partly from the *Gentleman's Magazine*, and from an epitaph written by Bishop Watson.  Wilson did not publish anything: the theorem by which he has cut his name in the theory of numbers was communicated to Waring, by whom it was published.  He married, in 1788, a daughter of Serjeant Adair, and left issue.  *Had a family*, many will say: but a man and his wife are a family, even without children.  An actuary may be allowed to be accurate in this matter, of which I was reminded by what an actuary wrote of another actuary.  William Morgan, in the life of his uncle Dr. Richard Price, says that the Doctor and his wife were 'never blessed with an addition to their family.'  I never met with such accuracy elsewhere.  Of William Morgan I add that my surname and pursuits have sometimes, to my credit be it said, made a confusion between him and me.  Dates are nothing to the mistaken; the last three years of Morgan's life were the first three years of my actuary-life (1830–33).  The mistake was to my advantage as well as to my credit.  I owe to it the acquaintance of one of the noblest of the human race, I mean Elizabeth Fry, who came to me for advice about a philanthropic design, which involved life questions, under a general impression that some Morgan had attended to such things.[2]

---

[1] He wrote, in 1760, a tract in defence of Waring, a point of whose algebra had been assailed by a Dr. Powell.  Waring wrote another tract of the same date.

[2] Mrs. Fry certainly believed that the writer was the old actuary of the Equitable, when she first consulted him upon the benevolent Assurance project; but we were introduced to her by our old and dear friend Lady Noel Byron, by whom she had

A treatise on the sublime science of heliography, satisfactorily demonstrating our great orb of light, the sun, to be absolutely no other than a body of ice! Overturning all the received systems of the universe hitherto extant; proving the celebrated and indefatigable Sir Isaac Newton, in his theory of the solar system, to be as far distant from the truth, as any of the heathen authors of Greece or Rome. By Charles Palmer, Gent. London, 1798, 8vo.

Mr. Palmer burned some tobacco with a burning glass, saw that a lens of ice would do as well, and then says—

'If we admit that the sun could be removed, and a terrestrial body of ice placed in its stead, it would produce the same effect. The sun is a crystaline body receiving the radience of God, and operates on this earth in a similar manner as the light of the sun does when applied to a convex mirror or glass.'

Nov. 10, 1801. The Rev. Thomas Cormouls, minister of Tettenhall, addressed a letter to Sir Wm. Herschel, from which I extract the following :—

Here it may be asked, then, how came the doctrines of Newton to solve all astronomic Phenomina, and all problems concerning the same, both *a parte ante* and *a parte post*. It is answered that he certainly wrought the principles he made use of into strickt analogy with the real Phenomina of the heavens, and that the rules and results arizing from them agree with them and resolve accurately all questions concerning them. Though they are not fact and true, or nature, but analogous to it, in the manner of the artificial numbers of logarithms, sines, &c. A very important question arises here, Did Newton mean to impose upon the world? By no means: he received and used the doctrines reddy formed; he did a little extend and contract his principles when wanted, and commit a few oversights of consequences. But when he was very much advanced in life, he suspected the fundamental nullity of them : but I have from a certain anecdote strong ground to believe that he knew it before his decease, and intended to have retracted his error. But, however, somebody did deceive, if not wilfully, neglently at least. That was a man to whom the world has great obligations too. It was no less a philosopher than Galileo.

That Newton wanted to retract before his death, is a notion not uncommon among paradoxers. Nevertheless, there is no

been long known and venerated, and who referred her to Mr. De Morgan for advice. An unusual degree of confidence in, and appreciation of each other, arose on their first meeting between the two, who had so much that was externally different, and so much that was essentially alike, in their natures.—(Ed.)

retraction in the third edition of the 'Principia,' published when Newton was eighty-four years old! The moral of the above is, that a gentleman who prefers instructing William Herschel to learning how to spell, may find a proper niche in a proper place, for warning to others. It seems that gravitation is not truth, but only the logarithm of it.

The mathematical and philosophical works of the Right Rev. John Wilkins . . . In two volumes. London, 1802, 8vo.

This work, or at least part of the edition—all for aught I know —is printed on wood ; that is, on paper made from wood-pulp. It has a rough surface, and when held before a candle is of very unequal transparency. There is in it a reprint of the works on the earth and moon. The discourse on the possibility of going to the moon, in this and the edition of 1640, is incorporated : but from the account in the life prefixed, and a mention by D'Israeli, I should suppose that it had originally a separate title-page, and some circulation as a separate tract. Wilkins treats this subject half seriously, half jocosely; he has evidently not quite made up his mind. He is clear that 'arts are not yet come to their solstice,' and that posterity will bring hidden things to light. As to the difficulty of carrying food, he thinks, scoffing Puritan that he is, the Papists may be trained to fast the voyage, or may find the bread of their Eucharist ' serve well enough for their *viaticum*.' He also puts the case that the story of Domingo Gonsales may be realized, namely, that wild geese find their way to the moon. It will be remembered—to use the usual substitute for, It has been forgotten—that the posthumous work of Bishop Francis Godwin of Llandaff was published in 1638, the very year of Wilkins's first edition, in time for him to mention it at the end. Godwin makes Domingo Gonsales get to the moon in a chariot drawn by wild geese, and, as old books would say, discourses fully on that head. It is not a little amusing that Wilkins should have been seriously accused of plagiarizing Godwin, Wilkins writing in earnest, or nearly so, and Godwin writing fiction. It may serve to show philosophers how very near pure speculation comes to fable. From the sublime to the ridiculous there is but a step: which is the sublime, and which the ridiculous, every one must settle for himself. With me, good fiction is the sublime, and bad speculation the ridiculous. The number of bishops in my list is small. I might, had I possessed the book, have opened the list of quadrators with an Archbishop of Canterbury, or at least with a divine who was not wholly not archbishop. Thomas Bradwardine (Bragvardinus, Bragadinus) was elected in 1348; the Pope put in another, who died unconsecrated; and Bradwardine was again elected in 1349, and lived five weeks longer, dying, I suppose, unconfirmed and unconsecrated. Leland says he held the see a year, *unus tantum annulus*, which seems to be a confusion: the whole business, from the first election, took about a year. He squared the circle, and his performance was printed at Paris in 1494. I have never seen it, nor any work of the author, except a tract on proportion.

As Bradwardine's works are very scarce indeed, I give two titles from one of the Libri catalogues.

'ARITHMETIC. BRAUARDINI (Thomæ) Arithmetica speculativa revisa et correcta a Petro Sanchez Ciruelo Aragonesi, black letter, *elegant woodcut title-page*, VERY RARE, *folio. Parisiis, per Thomam Anguelast (pro Olivier Senant), s.a. circa* 1510.

'This book, by Thomas Bradwardine, Archbishop of Canterbury, must be exceedingly scarce as it has escaped the notice of Professor De Morgan, who, in his *Arithmetical Books*, speaks of a treatise of the same author on proportions, printed at Vienna in 1515, but does not mention the present work.

'Bradwardine (Archbp. T.). Brauardini (Thomæ) Geometria speculativa, cum Tractatu de Quadratura Circuli bene revisa a Petro Sanchez Ciruelo, SCARCE, *folio. Parisiis, J. Petit,* 1511.

'In this work we find the *polygones étoilés*, see Chasles (*Aperçu,* pp. 480, 487, 521, 523, &c.) on the merit of the discoveries of this English mathematician, who was Archbishop of Canterbury in the XIVth Century (tempore Edward III. A.D. 1349); and who applied geometry to theology. M. Chasles says that the present work of Bradwardine contains "Une théorie nouvelle qui doit faire honneur au XIVe Siècle."'

The titles do not make it quite sure that Bradwardine is the quadrator; it may be Peter Sanchez after all.

Nouvelle théorie des parallèles. Par Adolphe Kircher [so signed at the end of the appendix]. Paris, 1803, 8vo.

An alleged emendation of Legendre. The author refers to attempts by Hoffman, 1801, by Hauff, 1799, and to a work of Karsten, or at least a theory of Karsten, contained in 'Tentamen novæ parallelarum theoriæ notione situs fundatæ; auctore G. C.

Schwal, Stuttgardæ, 1801, en 8 volumes.' Surely this is a misprint; *eight* volumes on the theory of parallels? If there be such a work, I trust I and it may never meet, though ever so far produced.

> Soluzione . . . della quadratura del Circolo. By ·Gaetano Rossi. London, 1804, 8vo.

The three remarkable points of this book are, that the household of the Prince of Wales took ten copies, Signora Grassini sixteen, and that the circumference is $3\frac{1}{5}$ diameters. That is, the appetite of Grassini for quadrature exceeded that of the whole household (*loggia*) of the Prince of Wales in the ratio in which the semi-circumference exceeds the diameter. And these are the first two in the list of subscribers. Did the author see this theorem?

> Britain independent of commerce; or proofs, deduced from an investigation into the true causes of the wealth of nations, that our riches, prosperity, and power are derived from sources inherent in ourselves, and would not be affected, even though our commerce were annihilated. By Wm. Spence. 4th edition, 1808, 8vo.

A patriotic paradox, being in alleviation of the Commerce panic which the measures of Napoleon I.—who *felt* our Commerce, while Mr. Spence only *saw* it—had awakened. In this very month (August, 1866), the Pres. Brit. Assoc. has applied a similar salve to the coal panic; it is fit that science, which rubbed the sore, should find a plaster. We ought to have an iron panic and a timber panic; and a solemn embassy to the Americans, to beg them not to whittle, would be desirable. There was a gold panic beginning, before the new fields were discovered. For myself, I am the unknown and unpitied victim of a chronic gutta-percha panic: I never could get on without it; to me, gutta percha and Rowland Hill are the great discoveries of our day; and not unconnected either, gutta percha being to the submarine post what Rowland Hill is to the superterrene. I should be sorry to lose cow-choke—I gave up trying to spell it many years ago—but if gutta percha go, I go too. I think, that perhaps when, five hundred years hence, the people say to the Brit. Assoc. (if it then exist) ' Pray, gentlemen, is it not time for the coal to be exhausted?' they will be answered out of Molière (who will certainly then exist): *Cela était autrefois ainsi, mais nous avons changé tout cela.* A great many people

think that if the coal be used up, it will be announced some
unexpected morning by all the yards being shut up and written
notice outside, 'Coal all gone!' just like the 'Please, ma'am,
there ain't no more sugar,' with which the maid servant damps
her mistress just at breakfast-time. But these persons should
be informed that there is every reason to think that there will
be time, as the city gentleman said, to *venienti* the *occurrite
morbo*.

> An appeal to the republic of letters in behalf of injured science,
> from the opinions and proceedings of some modern authors of
> elements of geometry. By George Douglas. Edinburgh,
> 1810, 8vo.

Mr. Douglas was the author of a very good set of mathematical
tables, and of other works. He criticizes Simson, Playfair, and
others,—sometimes, I think, very justly. There is a curious
phrase, which occurs more than once. When he wants to say
that something or other was done before Simson or another was
born, he says 'before he existed, at least as an author.' He
seems to reserve the possibility of Simson's *pre-existence*, but at
the same time to assume that he never wrote anything in his
previous state. Tell me that Simson pre-existed in any other
way than as editor of some pre-existent Euclid? Tell Apella!

1810. In this year Jean Wood, Professor of Mathematics in
the University of Virginia (Richmond), addressed a printed
circular to 'Dr. Herschel, Astronomer, Greenwich Observatory.'
No mistake was more common than the natural one of imagining
that the *Private Astronomer* of the king was the *Astronomer
Royal*. The letter was on the difference of velocities of the two
sides of the earth, arising from the composition of the rotation
and the orbital motion. The *paradox* is a fair one, and
deserving of investigation; but, perhaps it would not be easy to
deduce from it tides, trade-winds, aerolithes, &c., as Mr. Wood
thought he had done in a work from which he gives an extract,
and which he describes as published. The composition of rota-
tions, &c., is not for the world at large: the paradox of the
non-rotation of the moon about her axis is an instance. How
many persons know that when a wheel rolls on the ground, the
lowest point is moving upwards, the highest point forwards, and
the intermediate points in all degrees of betwixt and between?
This is too short an explanation, with some good difficulties.

The Elements of Geometry. In 2 vols. [By the Rev. J. Dobson, B.D.] Cambridge, 1815. 4to.

Of this unpunctuating paradoxer I shall give an account in his own way: he would not stop for any one; why should I stop for him? It is worth while to try how unpunctuated sentences will read.

The reverend J Dobson BD late fellow of saint Johns college Cambridge was rector of Brandesburton in Yorkshire he was seventh wrangler in 1798 and died in 1847 he was of that sort of eccentricity which permits account of his private life if we may not rather say that in such cases private life becomes public there is a tradition that he was called Death Dobson on account of his head and aspect of countenance being not very unlike the ordinary pictures of a human skull his mode of life is reported to have been very singular whenever he visited Cambridge he was never known to go twice to the same inn he never would sleep at the rectory with another person in the house some ancient charwoman used to attend to the house but never slept in it he has been known in the time of coach travelling to have deferred his return to Yorkshire on account of his disinclination to travel with a lady in the coach he continued his mathematical studies until his death and till his executors sold the type all his tracts to the number of five were kept in type at the university press none of these tracts had any stops except full stops at the end of paragraphs only neither had they capitals except one at the beginning of a paragraph so that a full stop was generally followed by some white as there is not a single proper name in the whole of the book I have I am not able to say whether he would have used capitals before proper names I have inserted them as usual for which I hope his spirit will forgive me if I be wrong he also published the elements of geometry in two volumes quarto Cambridge 1815 this book had also no stops except when a comma was wanted between letters as in the straight lines AB, BC I should also say that though the title is unpunctuated in the author's part it seems the publishers would not stand it in their imprint this imprint is punctuated as usual and Deighton and Sons to prove the completeness of their allegiance have managed that comma semicolon colon and period shall all appear in it why could they not have contrived interrogation and exclamation this is a good precedent to establish the separate right of the publisher over the imprint it is said that only twenty of the tracts were printed and very few indeed of the book on geometry it is doubtful whether any were sold there is a copy of the geometry in the university library at Cambridge and I have one myself the matter of the geometry differs entirely from Euclid and is so fearfully prolix that I am sure no mortal except the author ever read it the man went on without stops and without stop save for a period at the end of a paragraph this is the unpunctuated account of the unpunctuating geometer *suum cuique tribuito* Mrs Thrale would have been amused at a Dobson who managed to come to a full stop without either of the three warnings.

I do not find any difficulty in reading Dobson's geometry; and I have read more of it to try reading without stops than I should have done had it been printed in the usual way. Those who dip into the middle of my paragraph may be surprised for a moment to see that ' on account of his disinclination to travel with a lady in the coach he continued his mathematical studies until his death and [further, of course] until his executors sold the type.' But a person reading straight through would hardly take it so. I should add that, in order to give a fair trial, I did not compose as I wrote, but copied the words of the correspondent who gave me the facts, so far as they went.

*Philosophia Sacra, or the principles of natural Philosophy. Extracted from Divine Revelation. By the Rev. Samuel Pike. Edited by the Rev. Samuel Kittle. Edinburgh, 1815, 8vo.*

This is a work of modified Hutchinsonianism, which I have seen cited by several. Though rather dark on the subject, it seems not to contradict the motion of the earth, or the doctrine of gravitation, Mr. Kittle gives a list of some Hutchinsonians, —as Bishop Horne ; Dr. Stukeley ; the Rev. W. Jones, author of ' Physiological Disquisitions;' Mr. Spearman, author of ' Letters on the Septuagint ' and editor of Hutchinson ; Mr. Barker, author of ' Reflexions on Learning'; Dr. Catcott, author of a work on the creation, &c. ; Dr. Robertson, author of a ' Treatise on the Hebrew Language;' Dr. Holloway, author of ' Originals, Physical and Theological;' Dr. Walter Hodges, author of a work on *Elohim* ; Lord President Forbes (ob. 1747).

The Rev. William Jones, above mentioned, (1726–1800), the friend and biographer of Bishop Horne, and his stout defender, is best known as William Jones of Nayland, who (1757) published the ' Catholic Doctrine of the Trinity;' he was also strong for the Hutchinsonian physical trinity of fire, light, and spirit. This well-known work was generally recommended, as the defence of the orthodox system, to those who could not go into the learning of the subject. There is now a work more suited to our time : 'The Rock of Ages,' by the Rev. E. H. Bickersteth, now published by the Religious Tract Society, without date, answered by the Rev. Dr. Sadler, in a work (1859) entitled *Gloria Patri*, in which, says Mr. Bickersteth, 'the author has not even attempted to grapple with my main propositions.' I have read largely on the controversy, and I think I know what this means. Moreover, when I see the note 'There are two other passages to which Unitarians sometimes refer, but the deduction they draw from them is, in each case, refuted by the context'— I think I see why the two texts are not named. Nevertheless, the author is a little more disposed to yield to criticism than his foregoers ; he does not insist on texts and readings which the greatest editors have rejected. And he writes with courtesy, both direct and oblique, towards his antagonists ; which, on his side of this subject, is like letting in fresh air. So that I suspect the two books will together make a tolerably good introduction to the subject for those who cannot go deep. Mr. Bickersteth's book is well arranged and indexed, which is a point of superiority to Jones of Nayland. There is a point which I should gravely recommend to writers on the orthodox side. The Unitarians in England have frequently contended that the method of proving the divinity of Jesus Christ from the New Testament would equally prove the divinity of Moses. I have not fallen in the way of any orthodox answers specially directed at the repeated tracts written by Unitarians in proof of their assertion. If there be any, they should be more known ; if there be none, some should be written. Which ever side may be right, the treatment of this point would be indeed coming to close quarters. The heterodox assertion was first supported, it is said, by John Bidle or Biddle (1615–1662) of Magdalen College, Oxford, the earliest of the English Unitarian writers, previously known by a translation of part of Virgil and part of Juvenal. But I cannot find that he wrote on it. It is the subject of ' αἱρεσεων ἀναστασις, or a new way of deciding old controversies. By Basanistes. Third edition, enlarged,' London, 1815, 8vo. It is the appendix to the amusing, 'Six more letters to Granville Sharp, Esq., . . . By Gregory Blunt, Esq.' London, 8vo., 1803. This much I can confidently say, that the study of these tracts would prevent orthodox writers from some curious slips, which are slips obvious to all sides of opinion. The lower defenders of orthodoxy frequently vex the spirits of the higher ones.

Since writing the above I have procured Dr. Sadler's answer.

I thought I knew what the challenger meant when he said the respondent had not grappled with his main propositions. I should say that he is clung on to from beginning to end. But perhaps Mr. B. has his own meaning of logical terms, such as *proposition*: he certainly has his own meaning of *cumulative*. He says his evidence is cumulative; not a catena, the strength of which is in its weakest part, but distinct and independent lines, each of which corroborates the other. This is the very opposite of *cumulative*: it is *distributive*. When different arguments are each necessary to a conclusion, the evidence is *cumulative*; when any one will do, even though they strengthen each other, it is *distributive*. The word *cumulative* is a synonym of the law word *constructive*; a whole which will do made out of parts which separately will not. Lord Strafford opens his defence with the use of both words: 'They have invented a kind of *accumulated* or *constructive* evidence; by which many actions, either totally innocent in themselves, or criminal in a much inferior degree, shall, when united, *amount to* treason.' The conclusion is, that Mr. B. is a Cambridge man; the Oxford men do not confuse the elementary terms of logic. O dear old Cambridge! when the New Zealander comes let him find among the relics of your later sons some proof of attention to the elementary laws of thought. A little-go of logic, please!

Mr. B., though apparently not a Hutchinsonian, has a nibble at a physical Trinity. 'If, as we gaze on the sun shining in the firmament, we see any faint adumbration of the doctrine of the Trinity in the fontal orb, the light ever generated, and the heat proceeding from the sun and its beams—threefold and yet one, the sun, its light, and its heat,—that luminous globe, and the radiance ever flowing from it, are both evident to the eye; but the vital warmth is felt, not seen, and is only manifested in the life it transfuses through creation. The proof of its real existence is self-demonstrating.'

We shall see how Revilo[1] illustrates orthodoxy by mathematics. It was my duty to have found one of the many illustrations from physics; but perhaps I should have forgotten it if this instance had not come in my way. It is very bad physics. The sun, apart from its light, evident to the eye! Heat more self-demonstrating than light, because *felt*! Heat only manifested by the life it diffuses! Light implied not necessary to life! But the theology is worse than Sabellianism. To adumbrate—i.e. make

---

[1] The name assumed by a writer who professed to give a mathematical explanation of the Trinity, see farther on.—(Ed.)

a picture of—the orthodox doctrine, the sun must be heavenly body, the light heavenly body, the heat heavenly body: and yet, not three heavenly bodies, but one heavenly body. The truth is, that this illustration and many others most strikingly illustrate the Trinity of fundamental doctrine held by the Unitarians, in all its differences from the Trinity of persons held by the Orthodox. Be right which may, the right or wrong of the Unitarians shines out in the comparison. Dr. Sadler confirms me—by which I mean that I wrote the above before I saw what he says—in the following words : ' The sun is one object with two *properties*, and these properties have a parallel not in the second and third persons of the Trinity, but in the attributes of Deity.' The letting light alone, as self-evident, and making heat self-demonstrating, because felt—i.e. perceptible now and then—has the character of the Irishman's astronomy :—

> Long life to the moon, for a dear noble cratur,
> Which serves us for lamplight all night in the dark,
> While the sun only shines in the day, which, by natur,
> Wants no light at all, as ye all may remark.

*Sir Richard Phillips* (born 1768) was conspicuous in 1793, when he was sentenced to a year's imprisonment for selling Paine's 'Rights of Man ;' and again when, in 1807, he was knighted as Sheriff of London. As a bookseller, he was able to enforce his astronomical opinions in more ways than others. For instance, in James Mitchell's 'Dictionary of the Mathematical and Physical Sciences,' 1823, 12mo., which, though he was not technically a publisher, was printed for him—a book I should recommend to the collector of works of reference—there is a temperate description of his doctrines, which one may almost swear was one of his conditions previous to undertaking the work. Phillips himself was not only an anti-Newtonian, but carried to a fearful excess the notion that statesmen and Newtonians were in league to deceive the world. He saw this plot in Mrs. Airy's pension, and in Mrs. Somerville's. In 1836, he did me the honour to attempt my conversion. In his first letter he says :—

Sir Richard Phillips has an inveterate abhorrence of all the pretended wisdom of philosophy derived from the monks and doctors of the middle ages, and not less of those of higher name who merely sought to make the monkish philosophy more plausible, or so to disguise it as to mystify the mob of small thinkers.

So little did his writings show any knowledge of antiquity, that I strongly suspect, if required to name one of the monkish doctors, he would have answered—Aristotle. These schoolmen, and the 'philosophical trinity of gravitating force, projectile force, and void space,' were the bogies of his life.

I think he began to publish speculations in the *Monthly Magazine* (of which he was editor) in July 1817 : these he republished separately in 1818. In the Preface, perhaps judging the feelings of others by his own, he says that he 'fully expects to be vilified, reviled, and anathematized, for many years to come.' Poor man! he was let alone. He appeals with confidence to the 'impartial decision of posterity;' but posterity does not appoint a hearing for one per cent. of the appeals which are made; and it is much to be feared that an article in such a work of reference as this will furnish nearly all her materials fifty years hence. The following, addressed to M. Arago, in 1835, will give posterity as good a notion as she will probably need :—

Even the present year has afforded EVER-MEMORABLE examples, paralleled only by that of the Romish Conclave which persecuted Galileo. Policy has adopted that maxim of Machiavel which teaches that it is *more prudent* to *reward* partisans than to *persecute* opponents. Hence, a bigotted party had influence enough with the late short-lived administration [I think he is wrong as to the administration] of Wellington, Peel, &c., to confer munificent royal pensions on three writers whose sole distinction was their advocacy of the Newtonian philosophy. A Cambridge professor last year published an elaborate volume in illustration of *Gravitation*, and on him has been conferred a pension of 300*l*. per annum. A lady has written a light popular view of the Newtonian Dogmas, and she has been complimented by a pension of 200*l*. per annum. And another writer, who has recently published a volume to prove that the only true philosophy is that of Moses, has been endowed with a pension of 200*l*. per annum. Neither of them were needy persons, and the political and ecclesiastical bearing of the whole was indicated by another pension of 300*l*. bestowed on a political writer, the advocate of all abuses and prejudices. Whether the conduct of the Romish Conclave was more base for visiting with legal penalties the promulgation of the doctrines that the Earth turns on its axis and revolves around the Sun; or that of the British Court, for its craft in conferring pensions on the opponents of the plain corollary, that all the motions on the Earth are 'part and parcel' of these great motions, and those again and all like them consecutive displays of still greater motions in equality of action and reaction, is A QUESTION which must be reserved for the casuists of other generations. . . I cannot expect that on a sudden you and your friends will come to my conclusion, that the present philosophy of the Schools and Universities of Europe, based on faith in witchcraft, magic, &c., is a system of execrable nonsense, *by which quacks live on the faith of fools*; but I desire a free and fair examination of my Aphorisms, and if a few are admitted to be true, merely as courteous concessions to arithmetic, my purpose will be effected, for men will thus be led to think; and if they think, then the fabric of false assumptions, and degrading superstitions will soon tumble in ruins.

This for posterity. For the present time I ground the fame of Sir R. Phillips on his having squared the circle without knowing it, or intending to do it. In the *Protest* presently noted he discovered that 'the force taken as 1 is equal to the sum of all its fractions . . . . thus $1 = \frac{1}{4} + \frac{1}{9} + \frac{1}{16} + \frac{1}{25}$, &c., carried to infinity.' This the mathematician instantly sees is equivalent to the theorem that the circumference of any circle is double of the diagonal of the cube on its diameter.

I have examined the following works of Sir R. Phillips, and heard of many others :—

> Essays on the proximate mechanical causes of the general phenomena of the Universe, 1818, 12mo.
>
> Protest against the prevailing principles of natural philosophy, with the development of a common sense system (no date, 8vo. pp. 16).
>
> Four dialogues between an Oxford Tutor and a disciple of the common-sense philosophy, relative to the proximate causes of material phenomena. 8vo. 1824.
>
> A century of original aphorisms on the proximate causes of the phenomena of nature, 1835, 12mo.

Sir Richard Phillips had four valuable qualities; honesty, zeal, ability, and courage. He applied them all to teaching matters about which he knew nothing; and gained himself an uncomfortable life and a ridiculous memory.

> Astronomy made plain; or only way the true perpendicular distance of the Sun, Moon, or Stars, from this earth, can be obtained. By Wm. Wood. Chatham, 1819, 12mo.

If this theory be true, it will follow, of course, that this earth is the only one God made, and that it does not whirl round the sun, but *vice versâ*, the sun round it.

> Historic doubts relative to Napoleon Buonaparte. London, 1819, 8vo.

This tract has since been acknowledged by Archbishop Whately and reprinted. It is certainly a paradox : but differs from most of those in my list as being a joke, and a satire upon the reasoning of those who cannot receive narrative, no matter what the evidence, which is to them utterly improbable à *priori*. But had it been serious earnest, it would not have been so absurd as many of those which I have brought forward. The next on the list is not a joke.

The idea of the satire is not new. Dr. King, in the dispute on the genuineness of Phalaris, proved with humour that Bentley did not write his own dissertation. An attempt has lately been made, for the honour of Moses, to prove, without humour, that Bishop Colenso did not write his own book. This is intolerable : anybody who tries to use such a weapon without banter, plenty and good, and of form suited to the subject, should get the drubbing which the poor man got in the Oriental tale for striking the dervishes with the wrong hand.

The excellent and distinguished author of this tract has ceased to live. I call him the Paley of our day : with more learning, and more purpose than his predecessor ; but perhaps they might have changed places if they had changed centuries. The clever satire above named is not the only work which he published without his name. The following was attributed to him, I believe rightly : 'Considerations on the Law of Libel, as relating to Publications on the subject of Religion, by John Search.' London, 1833, 8vo. This tract excited little attention : for those who should have answered, could not. Moreover, it wanted a prosecution to call attention to it : the fear of calling such attention may have prevented prosecutions. Those who have read it will have seen why.

The theological review elsewhere mentioned attributes the pamphlet of John Search on blasphemous libel to Lord Brougham. This is quite absurd : the writer states points of law on credence where the judge must have spoken with authority. Besides which, a hundred points of style are decisive between the two. I think any one who knows Whately's writings will soon arrive at my conclusion. Lord Brougham himself informs me that he has no knowledge whatever of the pamphlet.

It is stated in *Notes and Queries* (3 S. xi. 511) that Search was answered by the Bishop of Ferns as S.N., with a rejoinder by Blanco White. These circumstances increase the probability that Whately was written against and for.

> Voltaire Chrétien ; preuves tirées de ses ouvrages. Paris, 1820, 12mo.

If Voltaire have not succeeded in proving himself a strong theist and a strong anti-revelationist, who is to succeed in proving himself one thing or the other in any matter whatsoever? By occasional confusion between theism and Christianity; by taking advantage of the formal phrases of adhesion to the Roman Church, which very often occur, and are often the happiest bits of irony in an ironical production; by citations of his morality, which is decidedly Christian, though often attributed to Brahmins; and so on—the author makes a fair case for his paradox, in the eyes of those who know no more than he tells them. If he had said that Voltaire was a better Christian than himself knew of, towards all mankind except men of letters, I for one should have agreed with him.

*Christian!* the word has degenerated into a synonym of *man*, in what are called Christian countries. So we have the parrot who ' swore for all the world like a Christian,' and the two dogs who ' hated each other just like Christians.' When the Irish duellist of the last century, whose name may be spared in consideration of its historic fame and the worthy people who bear it, was (June 12, 1786) about to take the consequence of his last brutal murder, the rope broke, and the criminal got up, and exclaimed, ' By —— Mr. Sheriff, you ought to be ashamed of yourself! this rope is not strong enough to hang a dog, far less a Christian!' But such things as this are far from the worst depravations. As to a word so defiled by usage, it is well to know that there is a way of escape from it, without renouncing the New Testament. I suppose any one may assume for himself what I have sometimes heard contended for, that no New Testament word is to be used in religion in any sense except that of the New Testament. This granted, the question is settled. The word *Christian,* which occurs three times, is never recognised as anything but a term of contempt from those without the pale to those within. Thus, Herod Agrippa, who was deep in Jewish literature, and a correspondent of Josephus, says to Paul, (Acts xxvi. 28) ' Almost thou persuadest me to be (what I and other followers of the state religion depise under the name) a Christian.' Again, (Acts xi. 26) ' The disciples (as they called *themselves*) were called (by the surrounding heathens) Christians first in Antioch.' Thirdly, (1 Peter iv. 16) ' Let none of you suffer as a *murderer.* . . . But if as a *Christian* (as the heathen call it by whom the suffering comes), let him not be ashamed.' That is to say, no *disciple* ever called *himself* a *Christian,* or applied the name, as from himself, to another disciple, from one end of the New Testament to the other; and no disciple need apply that name to himself in our day, if he dislike the associations with which the conduct of Christians has clothed it.

> Address of M. Hoene Wronski to the British Board of Longitude, upon the actual state of the mathematics, their reform, and upon the new celestial mechanics, giving the definitive solution of the problem of longitude. London, 1820, 8vo.

M. Wronski was the author of seven quartos on mathematics, showing very great power of generalization. He was also deep in the transcendental philosophy, and had the Absolute at his fingers' ends. All this knowledge was rendered useless by a persuasion that he had greatly advanced beyond the whole world, with many hints that the Absolute would not be forthcoming, unless prepaid. He was a man of the widest extremes. At one time he desired people to see all possible mathematics in

$$Fx = A_0\Omega_0 + A_1\Omega_1 + A_2\Omega_2 + A_3\Omega_3 + \&c.$$

which he did not explain, though there is meaning to it in the quartos. At another time he was proposing the general solution of the fifth degree by help of 625 independent equations of one form and 125 of another. The first separate memoir from any Transactions that I ever possessed was given to me when at Cambridge; the refutation (1819) of this asserted solution, presented to the Academy of Lisbon by Evangelista Torriano. I cannot say I read it. The tract above is an attack on modern mathematicians in general, and on the Board of Longitude, and Dr. Young.

1820. In this year died Dr. Isaac Milner, President of Queens' College, Cambridge, one of the class of rational paradoxers. Under this name I include all who, in private life, and in matters which concern themselves, take their own course, and suit their own notions, no matter what other people may think of them. These men will put things to uses they were never intended for, to the great distress and disgust of their gregarious friends. I am one of the class, and I could write a little book of cases in which I have incurred absolute reproach for not 'doing as other people do.' I will name two of my atrocities: I took one of those butter-dishes which have for a top a dome with holes in it, which is turned inward, out of reach of accident, when not in use. Turning the dome inwards, I filled the dish with water, and put a sponge in the dome: the holes let it fill with water, and I had a penwiper, always moist, and worth its price five times over. 'Why! what do you mean? It was made to hold butter. You are always at some queer thing or other!' I bought a leaden comb, intended to dye the hair, it being supposed that the application of lead will have this effect. I did not try: but I divided the comb into two, separating the part of closed prongs from the other; and thus I had two ruling machines. The lead marks paper, and by drawing the end of one of the machines along a ruler, I could rule twenty lines at a time, quite fit to write on. I thought I should have killed a friend to whom I explained it: he could not for the life of him understand how leaden *lines* on paper would dye the hair.

But Dr. Milner went beyond me. He wanted a seat suited to his shape, and he defied opinion to a fearful point. He spread a thick block of putty over a wooden chair and sat in it until it had taken a ceroplast copy of the proper seat. This he gave to a carpenter to be imitated in wood. One of the few now living who knew him—my friend, General Perronet Thompson— answers for the wood, which was shown him by Milner himself; but he does not vouch for the material being putty, which was in the story told me at Cambridge; William Frend also remembered it. Perhaps the Doctor took off his great seal in green wax, like the Crown; but some soft material he certainly adopted; and very comfortable he found the wooden copy.

The same gentleman vouches for Milner's lamp: but this had visible *science* in it; the vulgar see no science in the construction of the chair. A hollow semi-cylinder, but not with a circular curve, revolved on pivots. The curve was calculated on the law that, whatever quantity of oil might be in the lamp, the position of equilibrium just brought the oil up to the edge of the cylinder, at which a bit of wick was placed. As the wick exhausted the oil, the cylinder slowly revolved about the pivots so as to keep the oil always touching the wick.

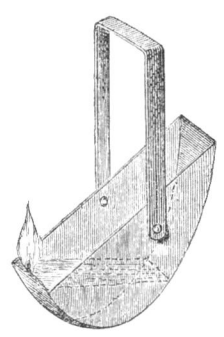

Great discoveries are always laughed at: but it is very often not the laugh of incredulity; it is a mode of distorting the sense of inferiority into a sense of superiority, or a mimicry of superiority interposed between the laugher and his feeling of inferiority. Two persons in conversation agreed that it was often a nuisance not to be able to lay hands on a bit of paper to mark the place in a book, every bit of paper on the table was sure to contain something not to be spared. I very quietly said that I always had a stock of bookmarkers ready cut, with a proper place for them : my readers owe many of my anecdotes to this absurd practice. My two colloquials burst into a fit of laughter ; about what? Incredulity was out of the question ; and there could be nothing foolish in my taking measures to avoid what they knew was an inconvenience. I was in this matter obviously their superior, and so they laughed at me. Much more candid was the Royal Duke of the last century, who was noted for slow ideas. 'The rain comes into my mouth,' said he, while riding. 'Had not your Royal Highness better shut your mouth?' said the equerry. The Prince did so, and ought, by rule, to have laughed heartily at his adviser ; instead of this, he said quietly, ' It doesn't come in now.'

> De Attentionis mensura causisque primariis. By J. F. Herbart. Kœnigsberg, 1822, 4to.

This celebrated philosopher maintained that mathematics ought to be applied to psychology, in a separate tract, published also in 1822 : the one above seems, therefore, to be his challenge on the subject. It is on *attention*, and I think it will hardly support Herbart's thesis. As a specimen of his formula, let $t$ be the time elapsed since the consideration began, $\beta$ the whole perceptive intensity of the individual, $\phi$ the whole of his mental force, and $z$ the force given to a notion by attention during the time $t$. Then,

$$z = \phi(1 - \varepsilon^{-\beta t})$$

Now for a test. There is a *jactura*, $v$, the meaning of which I do not comprehend. If there be anything in it, my mathematical readers ought to interpret it from the formula

$$v = \frac{\pi \phi \beta}{1 - \beta} \varepsilon^{-\beta t} + C \varepsilon^{-t}:$$

and to this task I leave them, wishing them better luck than mine. The time may come when other manifestations of mind, besides *belief*, shall be submitted to calculation : at that time, should it arrive, a final decision may be passed upon Herbart.

The theory of the Whizgig considered ; in as much as it mechanically exemplifies the three working properties of nature ; which
are now set forth under the guise of this toy, for children of all
ages.   London, 1822, 12mo. (pp. 24, B. McMillan, Bow Street,
Covent Garden.)

The toy called the *whizgig* will be remembered by many.   The
writer is a follower of Jacob Behmen, William Law, Richard
Clarke, and Eugenius Philalethes.   Jacob Behmen first announced the three working properties of nature, which Newton
stole, as described in the *Gentleman's Magazine*, July, 1782,
p. 329.   These laws are illustrated in the whizgig.   There is the
harsh astringent, attractive compression ; the bitter compunction,
repulsive expansion ; and the stinging anguish, duplex motion.
The author hints that he has written other works, to which he
gives no clue.   I have heard that Behmen was pillaged by Newton, and Swedenborg by Laplace, and Pythagoras by Copernicus,
and Epicurus by Dalton, &c.   I do not think this mention will
revive Behmen ; but it may the whizgig, a very pretty toy, and
philosophical withal, for few of those who used it could explain it.

A Grammar of infinite forms ; or the mathematical elements of
ancient philosophy and mythology.   By Wm. Howison.   Edinburgh, 1823, 8vo.

A curious combination of geometry and mythology.   Perseus,
for instance, is treated under the head, ' the evolution of diminishing hyperbolic branches.'

The Mythological Astronomy of the Ancients ; part the second :
or the key of Urania, the wards of which will unlock all the
mysteries of antiquity.   Norwich, 1823, 12mo.
A Companion to the Mythological Astronomy, &c., containing
remarks on recent publications. . . Norwich, 1824, 12mo.
A new Theory of the Earth and of planetary motion ; in which it
is demonstrated that the Sun is vicegerent of his own system.
Norwich, 1825, 12mo.
The analyzation of the writings of the Jews, so far as they are
found to have any connection with the sublime science of
astronomy.   [This is pp. 97–180 of some other work, being all
I have seen.]

These works are all by Sampson Arnold Mackey, for whom see
*Notes and Queries*, 1st S. viii. 468, 565, ix. 89, 179.   Had it

not been for actual quotations given by one correspondent only (1st S. viii. 565), that journal would have handed him down as a man of some real learning. An extraordinary man he certainly was: it is not one illiterate shoemaker in a thousand who could work upon such a singular mass of Sanscrit and Greek words, without showing evidence of being able to read a line in any language but his own, or to spell that correctly. He was an uneducated Godfrey Higgins. A few extracts will put this in a strong light: one for history of science, one for astronomy, and one for philology:—

'Sir Isaac Newton was of opinion that " the atmosphere of the earth was the sensory of God ; by which he was enabled to see quite round the earth :" which proves that Sir Isaac had no idea that God could see through the earth.

Sir Richard [Phillips] has given the most rational explanation of the cause of the earth's elliptical orbit that I have ever seen in print. It is because the earth presents its watery hemisphere to the sun at one time and that of solid land the other; but why has he made his Oxonian astonished at the coincidence ? It is what I taught in my attic twelve years before.

Again, admitting that the Eloim were powerful and intelligent beings that managed these things, we would accuse *them* of being the authors of all the sufferings of Chrisna. And as they and the constellation of Leo were below the horizon, and consequently cut off from the end of the zodiac, there were but eleven constellations of the zodiac to be seen ; the three at the end were wanted, but those three would be accused of bringing Chrisna into the troubles which at last ended in his death. All this would be expressed in the Eastern language by saying that Chrisna was persecuted by those Judoth Ishcarioth ! ! ! ! ! [the five notes of exclamation are the author's]. But the astronomy of those distant ages, when the sun was at the south pole in winter, would leave five of those Decans cut off from our view, in the latitude of twenty-eight degrees; hence Chrisna died of wounds from five Decans, but the whole five may be included in Judoth Ishcarioth! for the phrase means *the men that are wanted at the extreme parts*. Ishcarioth is a compound of *ish*, a man, and *carat* wanted or taken away, and *oth* the plural termination, more ancient than *im*. . . '

I might show at length how Michael is the sun, and the D'-ev-'l, in French Di-ob-al, also 'L-evi-ath-an—the *evi* being the radical part both of d*evi*l and l*evi*athan—is the Nile, which the sun dried up for Moses to pass : a battle celebrated by Jude. Also how *Moses*, the same name as *Muses*, is from *mesha*, drawn out of the water, 'and hence we called our land which is saved from the water by the name of *marsh*.' But it will be of more use to collect the character of S. A. M. from such correspondents of

*Notes and Queries* as have written after superficial examination.
Great astronomical and philological attainments; much ability
and learning; had evidently read and studied deeply; remarkable for the originality of his views upon the very abstruse
subject of mythological astronomy, in which he exhibited great
sagacity.   Certainly his views were *original*; but their sagacity,
if it be allowable to copy his own mode of etymologizing, is of an
*ori-gin-ale* cast, resembling that of a person who puts to his
mouth liquors both distilled and fermented.

> Principles of the Kantesian, or transcendental philosophy.   By
> Thomas Wirgman.   London, 1824, 8vo.

Mr. Wirgman's mind was somewhat attuned to psychology;
but he was cracky and vagarious.   He had been a fashionable
jeweller in St. James's Street, no doubt the son or grandson of
Wirgman at 'the well-known toy-shop in St. James's Street,'
where Sam Johnson smartened himself with silver buckles.
(Boswell, æt. 69).   He would not have the ridiculous large ones
in fashion; and he would give no more than a guinea a pair;
such, says Boswell, in Italics, were the *principles* of the business:
and I think this may be the first place in which the philosophical word was brought down from heaven to mix with men.
However this may be, *my* Wirgman sold snuff-boxes, among
other things, and fifty years ago a fashionable snuff-boxer would
be under inducement, if not positively obliged, to have a stock
with very objectionable pictures.   So it happened that Wirgman
—by reason of a trifle too much candour—came under the notice
of the *Suppression* Society, and ran considerable risk.   Mr.
Brougham was his counsel; and managed to get him acquitted.
Years and years after this, when Mr. Brougham was deep in the
formation of the London University (now University College),
Mr. Wirgman called on him.   'What now?' said Mr. B. with his
most sarcastic look—a very perfect thing of its kind—'you're
in a scrape again, I suppose!' 'No! indeed!' said W., 'my present
object is to ask your interest for the chair of *Moral Philosophy*
in the new University!'   He had taken up Kant!

Mr. Wirgman, an itinerant paradoxer, called on me in 1831: he
came to convert me.   'I assure you,' said he, 'I am nothing but
an old brute of a jeweller;' and his eye and manner were of the
extreme of jocosity, as good in their way, as the satire of his
former counsel.   I mention him as one of that class who go away
quite satisfied that they have wrought conviction.   'Now,' said he,

'I'll make it clear to you! Suppose a number of gold-fishes in a glass bowl—you understand? Well! I come with my cigar, and go puff, puff, puff, over the bowl, until there is a little cloud of smoke: now, tell me, what will the gold-fishes say to that?' 'I should imagine,' said I, 'that they would not know what to make of it.' 'By Jove! you're a Kantian;' said he, and with this and the like, he left me, vowing that it was delightful to talk to so intelligent a person. The greatest compliment Wirgman ever received was from James Mill, who used to say he did not *understand* Kant. That such a man as Mill should think this worth saying is a feather in the cap of the jocose jeweller.

Some of my readers will stare at my supposing that Boswell may have been the first down-bringer of the word *principles* into common life; the best answer will be a prior instance of the word as true vernacular; it has never happened to me to notice one. Many words have very common uses which are not old. Take the following from Nichols (*Anecd.* ix. 263): 'Lord Thurlow presents his best respects to Mr. and Mrs. Thicknesse, and assures them that he knows of no cause to complain of any part of Mr. Thicknesse's carriage; least of all the circumstance of sending the head to Ormond Street.' Surely Mr. T. had lent Lord T. a satisfactory carriage with a moveable head, and the above is a polite answer to inquiries. Not a bit of it! *carriage* is here *conduct*, and the *head* is a *bust*. The vehicles of the rich, at the time, were coaches, chariots, chaises, &c., never carriages, which were rather *carts*. Gibbon has the word for baggage-waggons. In Jane Austen's novels the word carriage is established.

*John Walsh*, of Cork (1786–1847).—This discoverer has had the honour of a biography from Prof. Boole, who, at my request, collected information about him on the scene of his labours. It is in the *Philosophical Magazine* for November, 1851, and will, I hope, be transferred to some biographical collection where it may find a larger class of readers, It is the best biography of a single hero of the kind that I know. Mr. Walsh introduced himself to me, as he did to many others, in the anterowlandian days of the Post-office; his unpaid letters were double, treble, &c. They contained his pamphlets, and cost their weight in silver: all have the name of the author, and all are in octavo or in quarto letter-form: most are in four pages, and all dated from Cork. I have the following by me:—

The Geometric Base. 1825.—The theory of plane angles. 1827.—
Three Letters to Dr. Francis Sadleir. 1838.—The invention of
polar geometry. By Irelandus. 1839.—The theory of partial
functions. Letter to Lord Brougham. 1839.—On the invention
of polar geometry. 1839.—Letter to the Editor of the Edinburgh Review. 1840.—Irish Manufacture. A new method of
tangents. 1841.—The normal diameter in curves. 1843.—
Letter to Sir R. Peel. 1845.—[Hints that Government should
compel the introduction of Walsh's Geometry into Universities.]
—Solution of Equations of the higher orders. 1845.

Besides these, there is a ' Metalogia,' and I know not how many
others.

Mr. Boole, who has taken the moral and social features of
Walsh's delusions from the commiserating point of view, which
makes ridicule out of place, has been obliged to treat Walsh as
Scott's Alan Fairford treated his client Peter Peebles ; namely,
keep the scarecrow out of court while his case was argued. My
plan requires me to bring him in : and when he comes in at the
door, pity and sympathy fly out at the window. Let the reader
remember that he was not an ignoramus in mathematics : he
might have won his spurs if he could have first served as an
esquire. Though so illiterate that even in Ireland he never
picked up anything more Latin than *Irelandus*, he was a very
pretty mathematician spoiled in the making by intense self-
opinion.

This is part of a private letter to me at the back of a page of
print : I had never addressed a word to him :—

' There are no limits in mathematics, and those that assert there are,
are infinite ruffians, ignorant, lying blackguards. There is no differential calculus, no Taylor's theorem, no calculus of variations, &c.
in mathematics. There is no quackery whatever in mathematics ; no
$\frac{0}{0}$ equal to anything. What sheer ignorant blackguardism that!
In mechanics the parallelogram of forces is quackery, and is dangerous ; for nothing is at rest, or in uniform, or in rectilinear motion, in
the universe. Variable motion is an essential property of matter.
Laplace's demonstration of the parallelogram of forces is a begging of
the question ; and the attempts of them all to show that the difference
of twenty minutes between the sidereal and actual revolution of the
earth round the sun arises from the tugging of the Sun and Moon at
the pot-belly of the earth, without being sure even that the earth has
a pot-belly at all, is perfect quackery. The said difference arising
from and demonstrating the revolution of the Sun itself round some
distant centre.'

In the letter to Lord Brougham we read as follows :—

' I ask the Royal Society of London, I ask the Saxon crew of that crazy hulk, where is the dogma of their philosophic god now ? . . . When the Royal Society of London, and the Academy of Sciences of Paris, shall have read this memorandum, how will they appear ? Like two cur dogs in the paws of the noblest beast of the forest . . . Just as this note was going to press, a volume lately published by you was put into my hands, wherein you attempt to defend the fluxions and Principia of Newton.    Man ! what are you about ?    You come forward now with your special pleading, and fraught with national prejudice, to defend, like the philosopher Grassi, the persecutor of Galileo, principles and reasoning which, unless you are actually insane, or an ignorant quack in mathematics, you know are mathematically false. What a moral lesson this for the students of the University of London from its head !    Man ! demonstrate corollary 3, in this note, by the lying dogma of Newton, or turn your thoughts to something you understand.'

                                        ' WALSH IRELANDUS.'

Mr. Walsh—honour to his memory—once had the consideration to save me postage by addressing a pamphlet under cover to a Member of Parliament, with an explanatory letter.    In that letter he gives a candid opinion of himself :—

(1838.)   ' Mr. Walsh takes leave to send the enclosed corrected copy to Mr. Hutton as one of the Council of the University of London, and to save postage for the Professor of Mathematics there.    He will find in it geometry more deep and subtle, and at the same time more simple and elegant, than it was ever contemplated human genius could invent.'

He then proceeds to set forth that a certain ' tomfoolery lemma,' with its ' tomfoolery ' superstructure, ' never had existence outside the shallow brains of its inventor,' Euclid.    He then proceeds thus :—

' The same spirit that animated those philosophers who sent Galileo to the Inquisition animates all the philosophers of the present day without exception.    If anything can free them from the yoke of error, it is the [Walsh] problem of double tangence.    But free them it will, how deeply soever they may be sunk into mental slavery—and God knows that is deeply enough ; and they bear it with an admirable grace ; for none bear slavery with a better grace than tyrants.    The lads must adopt my theory . . . It will be a sad reverse for all our great professors to be compelled to become schoolboys in their gray years.    But the sore scratch is to be compelled, as they had before been compelled one thousand years ago, to have recourse to Ireland for instruction.'

The following ' Impromptu ' is no doubt by Walsh himself : he was more of a poet than of an astronomer :—

> ' Through ages unfriended,
> With sophistry blended,
> Deep science in Chaos had slept ;
> Its limits were fettered,
> Its voters unlettered,
> Its students in movements but crept.
> Till, despite of great foes,
> Great WALSH first arose,
> And with logical might did unravel
> Those mazes of knowledge,
> Ne'er known in a college,
> Though sought for with unceasing travail.
> With cheers we now hail him,
> May success never fail him,
> In Polar Geometrical mining ;
> Till his foes be as tamed
> As his works are far-famed
> For true philosophic refining.'

Walsh's system is, that all mathematics and physics are wrong : there is hardly one proposition in Euclid which is demonstrated. His example ought to warn all who rely on their own evidence to their own success. He was not, properly speaking, insane ; he only spoke his mind more freely than many others of his class. The poor fellow died in the Cork union, during the famine. He had lived a happy life, contemplating his own perfections, like Brahma on the lotos-leaf.

The year 1825 brings me to about the middle of my *Athenæum* list : that is, so far as mere number of names mentioned is concerned. Freedom of opinion, beyond a doubt, is gaining ground, for good or for evil, according to what the speaker happens to think : admission of authority is no longer made in the old way. If we take soul-cure and body-cure, divinity and medicine, it is manifest that a change has come over us. Time was when it was enough that dose or dogma should be certified by ' Il a été ordonné, Monsieur, il a été ordonné,' as the apothecary said when he wanted to operate upon poor de Porceaugnac. Very much changed : but whether for good or for evil does not now matter ; the question is, whether contempt of *demonstration* such as our paradoxers show has augmented with the rejection of *dogmatic authority*. It ought to be just the other way : for the worship of reason is the system on which, if we trust them, the deniers of guidance ground their plan of life. The following attempt at an experiment on this point is the best which I can make; and, so far as I know, the first that ever was made.

Say that my list of paradoxers divides in 1825: this of itself proves nothing, because so many of the earlier books are lost, or not likely to be come at. It would be a fearful rate of increase which would make the number of paradoxes since 1825 equal to the whole number before that date. Let us turn now to another collection of mine, arithmetical books, of which I have published a list. The two collections are similarly circumstanced as to new and old books; the paradoxes had no care given to the collection of either; the arithmetical books equal care to both. The list of arithmetical books, published in 1847, divides at 1735; the paradoxes, up to 1863, divide at 1825. If we take the process which is most against the distinction, and allow every year from 1847 to 1863 to add a year to 1735, we should say that the arithmetical writers divide at 1751. This rough process may serve, with sufficient certainty, to show that the proportion of paradoxes to books of sober demonstration is on the increase; and probably, quite as much as the proportion of heterodoxes to books of orthodox adherence. So that divinity and medicine may say to geometry, Don't *you* sneer: if rationalism, homœopathy, and their congeners are on the rise among us, your enemies are increasing quite as fast. But geometry replies—Dear friends, content yourselves with the rational inference that the rise of heterodoxy within your pales is not conclusive against you, taken alone; for it rises at the same time within mine. Store within your garners the precious argument that you are not proved wrong by increase of dissent; because there is increase of dissent against exact science. But do not therefore *even* yourselves to me: remember that you, Dame Divinity, have inflicted every kind of penalty, from the stake to the stocks, in aid of your reasoning; remember that you, Mother Medicine, have, not many years ago applied to Parliament for increase of forcible hindrance of antipharmacopœal drenches, pills, and powders. Who ever heard of my asking the legislature to fine blundering circle-squarers? Remember that the D in dogma is the D in decay; but the D in demonstration is the D in durability.

I have known a medical man—a young one—who was seriously of opinion that the country ought to be divided into medical parishes, with a practitioner appointed to each, and a penalty for calling in any but the incumbent curer. How should people know how to choose? The hair-dressers once petitioned Parliament for an act to compel people to wear wigs. My own opinion is of the opposite extreme, as in the following letter (*Examiner*, April 5, 1856); which, to my surprise, I saw reprinted in a medical journal, as a plan not absolutely to be rejected. I am perfectly satisfied that it would greatly promote true medical orthodoxy, the predominance of well educated thinkers, and the development of their desirable differences.

SIR. The Medical Bill and the medical question generally is one on which experience would teach, if people would be taught.

The great soul question took three hundred years to settle : the little body question might be settled in thirty years, if the decisions in the former question were studied.

Time was when the State believed, as honestly as ever it believed anything, that it *might*, *could*, and *should* find out true doctrine for the poor ignorant community; to which, like a worthy honest state, it added *would*. Accordingly, by the assistance of a Church, which undertook the physic, the surgery, and the pharmacy of sound doctrine all by itself, it sent forth its legally qualified teachers into every parish, and woe to the man who called in any other. They burnt that man, they whipped him, they imprisoned him, they did everything but what was Christian to him, all for his soul's health and the amendment of his excesses.

But men would not submit. To the argument that the State was a father to the ignorant, they replied that it was at best the ignorant father of an ignorant son, and that a blind man could find his way into a ditch without another blind man to help him. And when the State said—But here we have the Church, which knows all about it, the ignorant community declared that it had a right to judge that question, and that it would judge it. It also said that the Church was never one thing long, and that it progressed, on the whole, rather more slowly than the ignorant community.

The end of it was, in this country, that every one who chose taught all who chose to let him teach, on condition only of an open and true registration. The State was allowed to patronise one particular Church, so that no one need trouble himself to choose a pastor from the mere necessity of choosing. But every church is allowed its colleges, its studies, its diplomas ; and every man is allowed his choice. There is no proof that our souls are worse off than in the sixteenth century; and, judging by fruits, there is much reason to hope they are better off.

Now the little body question is a perfect parallel to the great soul question in all its circumstances. The only things in which the parallel fails are the following: Every one who believes in a future state sees that the soul question is incomparably more important than the body question, and every one can try the body question by experiment to a larger extent than the soul question. The proverb, which always has a spark of truth at the bottom, says that every man of forty is either a fool or a physician; but did even the proverb maker ever dare to say that every man is at any age either a fool or a fit teacher of religion?

Common sense points out the following settlement of the medical question: and to this it will come sooner or later.

Let every man who chooses—subject to one common law of manslaughter for all the *crass* cases—doctor the bodies of all who choose to trust him, and recover payment according to agreement in the courts of law. Provided always that every person practising should be registered at a moderate fee in a register to be republished every six months.

Let the register give the name, address, and asserted qualification of each candidate—as licentiate, or doctor, or what not, of this or that college, hall, university, &c., home or foreign. Let it be competent to any man to describe himself as qualified by study in public schools without a diploma, or by private study, or even by intuition or divine inspiration, if he please. But whatever he holds his qualification to be, that let him declare. Let all qualification which of its own nature admits of proof be proved, as by the diploma or certificate, &c., leaving things which cannot be proved, as asserted private study, intuition, inspiration, &c., to work their own way.

Let it be highly penal to assert to the patient any qualification which is not in the register, and let the register be sold very cheap. Let the registrar give each registered practitioner a copy of the register in his own case; let any patient have power to demand a sight of this copy; and let no money for attendance be recoverable in any case in which there has been false representation.

Let any party in any suit have a right to produce what medical testimony he pleases. Let the medical witness produce his register, and let his evidence be for the jury, as is that of an engineer or a practitioner of any art which is not attested by diplomas.

Let any man who practises without venturing to put his name on the register be liable to fine and imprisonment.

The consequence would be that, as now, anybody who pleases might practise ; for the medical world is well aware that there is no power of preventing what they call quacks from practising. But very different from what is now, every man who practises would be obliged to tell the whole world what his claim is, and would run a great risk if he dared to tell his patient in private anything different from what he had told the whole world.

The consequence would be that a real education in anatomy, physiology, chemistry, surgery, and what is known of the thing called medicine, would acquire more importance than it now has.

It is curious to see how completely the medical man of the nineteenth century squares with the priest of the sixteenth century. The clergy of all sects are now better divines and better men than they ever were. They have lost Bacon's reproach that they took a smaller measure of things than any other educated men ; and the physicians are now in this particular the rear-guard of the learned world ; though it may be true that the rear in our day is further on in the march than the van of Bacon's day. Nor will they ever recover the lost position until medicine is as free as religion.

To this it must come. To this the public, which will decide for itself, has determined it shall come. To this the public has, in fact, brought it, but on a plan which it is not desirable to make permanent. We will be as free to take care of our bodies as of our souls and of our goods. This is the profession of all who sign as I do, and the practice of most of those who would not like the name HETEROPATH.'

> The motion of the Sun in the Ecliptic, proved to be uniform in a circular orbit . . . with preliminary observations on the fallacy of the Solar System. By Bartholomew Prescott, 1825, 8vo.

The author had published, in 1803, a 'Defence of the Divine System,' which I never saw ; also, 'On the inverted scheme of Copernicus.' The above work is clever in its satire.

> Manifesto of the Christian Evidence Society, established Nov. 12, 1824. Twenty-four plain questions to honest men.

These are two broadsides of August and November, 1826, signed by Robert Taylor, A.B., Orator of the Christian Evidence

Society. This gentleman was a clergyman, and was convicted of blasphemy in 1827, for which he suffered imprisonment, and got the name of the *Devil's Chaplain*. The following are quotations : —

'For the book of Revelation, there was no original Greek at all, but *Erasmus* wrote it himself in Switzerland, in the year 1516. Bishop Marsh, vol. i. p. 320.'—'Is not God the author of your reason ? Can he then be the author of anything which is contrary to your reason ? If reason be a sufficient guide, why should God give you any other ? if it be not a sufficient guide, why has he given you *that* ? '

I remember a votary of the Society being asked to substitute for *reason* ' the right leg,' and for *guide* ' support,' and to answer the two last questions : he said there must be a quibble, but he did not see what. It is pleasant to reflect that the *argumentum à carcere* is obsolete. One great defect of it was that it did not go far enough : there should have been laws against subscriptions for blasphemers, against dealing at their shops, and against rich widows marrying them.

Had I taken in theology, I must have entered books against Christianity. I mention the above, and Paine's ' Age of Reason,' simply because they are the only English modern works that ever came in my way without my asking for them. The three parts of the ' Age of Reason ' were published in Paris 1793, Paris 1795, and New York 1807. Carlile's edition is of London, 1818, 8vo. It must be republished when the time comes, to show what stuff governments and clergy were afraid of at the beginning of this century. I should never have seen the book, if it had not been prohibited : a bookseller put it under my nose with a fearful look round him ; and I could do no less, in common curiosity, than buy a work which had been so complimented by church and state. And when I had read it, I said in my mind to church and state,—Confound you ! you have taken me in worse than any reviewer I ever met with. I forget what I gave for the book, but I ought to have been able to claim compensation somewhere.

Cabbala Algebraica. Auctore Gul. Lud. Christmann. Stuttgard, 1827, 4to.

Eighty closely printed pages of an attempt to solve equations of every degree, which has a process called by the author *cabbala*. An anonymous correspondent spells *cabbala* as follows, χαββαλλ, and makes 666 out of its letters. This gentleman has sent me, since my Budget commenced, a little heap of satirical communications, each having a 666 or two; for instance, alluding to my remarks on the spelling of *chemistry*, he finds the fated number in χιμεια. With these are challenges to explain them, and hints about the end of the world. All these letters have different fantastic seals; one of them with the legend 'keep your temper,'—another bearing 'bank token five pence.' The only signature is a triangle with a little circle in it, which I interpret to mean that the writer confesses himself to be the round man stuck in the three-cornered hole, to be explained as in Sydney Smith's joke.

There is a kind of Cabbala Alphabetica which the investigators of the numerals in words would do well to take up: it is the formation of sentences which contain all the letters of the alphabet, and each only once. No one has done it with *v* and *j* treated as consonants; but you and I can do it. Dr. Whewell and I amused ourselves, some years ago, with attempts. He could not make sense, though he joined words: he gave me

Phiz, styx, wrong, buck, flame, quid.

I gave him the following, which he agreed was 'admirable sense:' I certainly think the words would never have come together except in this way:—

I, quartz pyx, who fling muck beds.

I long thought that no human being could say this under any circumstances. At last I happened to be reading a religious writer—as he thought himself—who threw aspersions on his opponents thick and threefold. Heyday! came into my head, this fellow flings muck beds; he must be a quartz pyx. And then I remembered that a pyx is a sacred vessel, and quartz is a hard stone, as hard as the heart of a religious foe-curser. So that the line is the motto of the ferocious sectarian, who turns his religious vessels into mud-holders, for the benefit of those who will not see what he sees.

I can find no circumstances for the following, which I received from another:—

Fritz! quick! land! hew gypsum box.

From other quarters I have the following:—

Dumpy quiz! whirl back fogs next.

This might be said in time of haze to the queer little figure in

the Dutch weather-toy, which comes out or goes in with the
change in the atmosphere.   Again,

Export my fund! Quiz black whigs.

This Squire Western might have said, who was always afraid of
the whigs sending the sinking-fund over to Hanover.   But the
following is the best : it is good advice to a young man, very well
expressed under the circumstances :—

Get nymph ; quiz sad brow ; fix luck.

Which in more sober English would be, Marry; be cheerful;
watch your business.   There is more edification, more religion in
this than in all the 666-interpretations put together.

Such things would make excellent writing copies, for they
secure attention to every letter; *v* and *j* might be placed at the
end.

> The Celtic Druids. By Godfrey Higgins, Esq. of Skellow Grange,
> near Doncaster.   London, 1827, 4to.
> Anacalypsis, or an attempt to draw aside the veil of the Saitic
> Isis : or an inquiry into the origin of languages, nations,. and
> religions.   By Godfrey Higgins, &c. . . . . . . London, 1836,
> 2 vols. 4to.

The first work had an additional preface and a new index in
1829.   Possibly, in future time, will be found bound up with
copies of the second work two sheets which Mr. Higgins circulated among his friends in 1831 : the first a ' Recapitulation,' the
second ' Book vi. ch. 1.'

The system of these works is that—

The Buddhists of Upper India (of whom the Phenician Canaanite,
Melchizedek, was a priest), who built the Pyramids, Stonehenge
Carnac, &c. will be shown to have founded all the ancient mythologies
of the world, which, however varied and corrupted in recent times,
were originally one, and that one founded on principles sublime,
beautiful, and true.

These works contain an immense quantity of learning, very
honestly put together.   I presume the enormous number of facts,
and the goodness of the index, to be the reasons why the *Ana-
calypsis* found a permanent place in the *old* reading-room of the
British Museum, even before the change which greatly increased
the number of books left free to the reader in that room.

Mr. Higgins, whom I knew well in the last six years of his life,
and respected as a good, learned, and (in his own way) *pious* man,

was thoroughly and completely the man of a system. He had that sort of mental connection with his theory that made his statements of his authorities trustworthy: for, besides perfect integrity, he had no bias towards alteration of facts: he saw his system in the way the fact was presented to him by his authority, be that what it might.

He was very sure of a fact which he got from any of his authorities: nothing could shake him. Imagine a conversation between him and an Indian officer who had paid long attention to Hindoo antiquities and their remains: a third person was present, *ego qui scribo*. *G. H.* 'You know that in the temples of I-forget-who the Ceres is always sculptured precisely as in Greece.' *Col.*——, 'I really do not remember it, and I have seen most of these temples.' *G. H.* 'It is so, I assure you, especially at I-forget-where.' *Col.*——, 'Well, I am sure! I was encamped for six weeks at the gate of that very temple, and, except a little shooting, had nothing to do but to examine its details, which I did, day after day, and I found nothing of the kind.' It was of no use at all.

Godfrey Higgins began life by exposing and conquering, at the expense of two years of his studies, some shocking abuses which existed in the York Lunatic Asylum. This was a proceeding which called much attention to the treatment of the insane, and produced much good effect. He was very resolute and energetic. The magistracy of his time had scruples about using the severity of law to people of such station as well-to-do farmers, &c.: they would allow a great deal of resistance, and endeavour to mollify the rebels into obedience. A young farmer flatly refused to pay under an order of affiliation made upon him by Godfrey Higgins. He was duly warned; and persisted: he shortly found himself in gaol. He went there sure to conquer the Justice, and the first thing he did was to demand to see his lawyer. He was told, to his horror, that as soon as he had been cropped and prison-dressed, he might see as many lawyers as he pleased, to be looked at, laughed at, and advised that there was but one way out of the scrape. Higgins was, in his speculations, a regular counterpart of Bailly; but the celebrated Mayor of Paris had not his nerve. It is impossible to say, if their characters had been changed, whether the unfortunate crisis in which Bailly was not equal to the occasion would have led to very different results if Higgins had been in his place: but assuredly constitutional liberty would have had one chance more. There are two works of his by which he was known, apart from his paradoxes.

First, ' An apology for the life and character of the celebrated prophet of Arabia, called Mohamed, or the Illustrious.' London, 8vo. 1829. The reader will look at this writing of our English Buddhist with suspicious eye, but he will not be able to avoid confessing that the Arabian prophet has some reparation to demand at the hands of Christians. Next, ' Horæ Sabbaticæ; or an attempt to correct certain superstitions and vulgar errors respecting the Sabbath. Second edition, with a large appendix.' London, 12mo. 1833. This book was very heterodox at the time, but it has furnished material for some of the clergy of our day.

I never could quite make out whether Godfrey Higgins took that system which he traced to the Buddhists to have a Divine origin, or to be the result of good men's meditations. Himself a strong theist, and believer in a future state, one would suppose that he would refer a *universal* religion, spread in different forms over the whole earth from one source, directly to the universal Parent. And this I suspect he did, whether he knew it or not. The external evidence is balanced. In his preface he says—

' I cannot help smiling when I consider that the priests have objected to admit my former book, " the Celtic Druids," into libraries, because it was antichristian; and it has been attacked by Deists, because it was superfluously religious. The learned Deist, the Rev. R. Taylor [already mentioned], has designated me as *the religious* Mr. Higgins.'

The time will come when some profound historian of literature will make himself much clearer on the point than I am.

> The triumphal Chariot of Friction : or a familiar elucidation of the origin of magnetic attraction, &c. &c. By William Pope. London, 1829, 4to.

Part of this work is on a dipping-needle of the author's construction. It must have been under the impression that a book of naval magnetism was proposed, that a great many officers, the Royal Naval Club, &c. lent their names to the subscription list. How must they have been surprised to find, right opposite to the list of subscribers, the plate presenting ' the three emphatic letters, J. A. O.' And how much more when they saw it set forth that if a square be inscribed in a circle, a circle within that, then a square again, &c., it is impossible to have more than fourteen circles, let the first circle be as large as you please. From this the seven attributes of God are unfolded; and further, that all matter was *moral*, until Lucifer *churned* it into *physical* ' as far as the third circle in Deity': this Lucifer, called Leviathan in Job, being thus the moving cause of chaos. I shall say no more, except that the friction of the air is the cause of magnetism.

> Remarks on the Architecture, Sculpture, and Zodiac of Palmyra; with a Key to the Inscriptions. By B. Prescot. London, 1830, 8vo.

Mr. Prescot gives the signs of the zodiac a Hebrew origin.

> Epitome de mathématiques. Par F. Jacotot, Avocat. 3ième édition. Paris, 1830, 8vo. (pp. 18).
> Méthode Jacotot. Choix de propositions mathématiques. Par P. Y. de Séprés. 2nde édition. Paris, 1830, 8vo. (pp. 82).

Of Jacotot's method, which had some vogue in Paris, the principle was *Tout est dans tout,* and the process *Apprendre quelque chose, et à y rapporter tout le reste.* The first tract has a proposition in conic sections and its preliminaries: the second has twenty exercises, of which the first is finding the greatest common measure of two numbers, and the last is the motion of a point on a surface, acted on by given forces. This is topped up with the problem of sound in a tube, and a slice of Laplace's theory of the tides. All to be studied until known by heart, and all the rest will come, or at least join on easily when it comes. There is much truth in the assertion that new knowledge hooks on easily to a little of the old, thoroughly mastered. The day is coming when it will be found out that crammed erudition, got up for examinations, does not cast out any hooks for more.

> Lettre à MM. les Membres de l'Académie Royale des Sciences, contenant un développement de la réfutation du système de la gravitation universelle, qui leur a été présentée le 30 août, 1830. Par Félix Passot. Paris, 1830, 8vo.

Works of this sort are less common in France than in England. In France there is only the Academy of Sciences to go to: in England there is a reading public out of the Royal Society, &c.

About 1830 was published, in the *Library of Useful Knowledge,* the tract on *Probability,* the joint work of the late Sir John Lubbock and Mr. Drinkwater (Bethune). It is one of the best elementary openings of the subject. A binder put my name on the outside (the work was anonymous) and the consequence was that nothing could drive out of people's heads that it was

written by me.  I do not know how many denials I have made,
from a passage in one of my own works to a letter in the *Times* :
and I am not sure that I have succeeded in establishing the
truth, even now.  I accordingly note the fact once more.  But
as a book has no right here unless it contain a paradox—or thing
counter to general opinion or practice—I will produce two small
ones.  Sir John Lubbock, with whom lay the executive arrangement, had a strong objection to the last word in ' Theory of
Probabilities,' he maintained that the singular *probability,* should
be used ; and I hold him quite right.

The second case was this : My friend Sir J. L., with a large
cluster of intellectual qualities, and another of social qualities,
had one point of character which I will not call bad and cannot
call good ; he never used a slang expression.  To such a length
did he carry his dislike, that he could not bear *head* and *tail,*
even in a work on games of chance : so he used *obverse* and *reverse.*
I stared when I first saw this : but, to my delight, I found that
the force of circumstances beat him at last.  He was obliged to
take an example from the race-course, and the name of one of
the horses was *Bessy Bedlam* ! And he did not put her down as
*Elizabeth Bethlehem,* but forced himself to follow the jockeys.

[Almanach Romain sur la Loterie Royale de France, ou les
    Étrennes nécessaires aux Actionnaires et Receveurs de la dite
    Loterie.  Par M. Menut de St.-Mesmin.  Paris, 1830.  12mo.

This book contains all the drawings of the French lottery (two
or three, each month) from 1758 to 1830.  It is intended for
those who thought they could predict the future drawings from
the past : and various sets of *sympathetic* numbers are given to
help them.  The principle is, that anything which has not
happened for a long time must be soon to come.  At *rouge et
noir,* for example, when the red has won five times running,
sagacious gamblers stake on the black, for they think the turn
which must come at last is nearer than it was.  So it is : but
observation would have shown that if a large number of those
cases had been registered which show a run of five for the red,
the next game would just as often have made the run into six
as have turned in favour of the black.  But the gambling
reasoner is incorrigible : if he would but take to squaring the
circle, what a load of misery would be saved.  A writer of 1823,
who appeared to be thoroughly acquainted with the gambling of
Paris and London, says that the gamesters by profession are
haunted by a secret foreboding of their future destruction, and

seem as if they said to the banker at the table, as the gladiators said to the emperor, *Morituri te salutant.*

In the French lottery, five numbers out of ninety were drawn at a time. Any person, in any part of the country, might stake any sum upon any event he pleased, as that 27 should be drawn; that 42 and 81 should be drawn; that 42 and 81 should be drawn, and 42 first; and so on up to a *quine déterminé*, if he chose, which is betting on five given numbers in a given order. Thus, in July, 1821, one of the drawings was

$$8 \quad 46 \quad 16 \quad 64 \quad 13.$$

A gambler had actually predicted the five numbers (but not their order), and won 131,350 francs on a trifling stake. M. Menut seems to insinuate that the hint what numbers to choose was given at his own office. Another won 20,852 francs on the quaterne 8, 16, 46, 64, in this very drawing. These gains, of course, were widely advertised: of the multitudes who lost nothing was said. The enormous number of those who played is proved to all who have studied chances arithmetically by the numbers of simple quaternes which were gained: in 1822, fourteen; in 1823, six; in 1824, sixteen; in 1825, nine, &c.

The paradoxes of what is called chance, or hazard, might themselves make a small volume. All the world understands that there is a long run, a general average; but great part of the world is surprised that this general average should be computed and predicted. There are many remarkable cases of verification; and one of them relates to the quadrature of the circle. I give some account of this and another. Throw a penny time after time until *head* arrives, which it will do before long: let this be called a *set*. Accordingly, H is the smallest set, TH the next smallest, then TTH, &c. For abbreviation, let a set in which seven *tails* occur before *head* turns up be $T^7H$. In an immense number of trials of sets, about half will be H; about a quarter TH; about an eighth, $T^2H$. Buffon tried 2,048 sets; and several have followed him. It will tend to illustrate the principle if I give all the results; namely, that many trials will with moral certainty show an approach—and the greater the greater the number of trials—to that average which sober reasoning predicts. In the first column is the most likely number of the theory: the next column gives Buffon's result; the three next are results obtained from trial by correspondents of mine. In each case the number of trials is 2,048.

| H | . | 1,024 | . | 1,061 | . | 1,048 | . | 1,017 | . | 1,039 |
|---|---|---|---|---|---|---|---|---|---|---|
| TH | . | 512 | . | 494 | . | 507 | . | 547 | . | 480 |
| T²H | . | 256 | . | 232 | . | 248 | . | 235 | . | 267 |
| T³H | . | 128 | . | 137 | . | 99 | . | 118 | . | 126 |
| T⁴H | . | 64 | . | 56 | . | 71 | . | 72 | . | 67 |
| T⁵H | . | 32 | . | 29 | . | 38 | . | 32 | . | 33 |
| T⁶H | . | 16 | . | 25 | . | 17 | . | 10 | . | 19 |
| T⁷H | . | 8 | . | 8 | . | 9 | . | 9 | . | 10 |
| T⁸H | . | 4 | . | 6 | . | 5 | . | 3 | . | 3 |
| T⁹H | . | 2 | . | | . | 3 | . | 2 | . | 4 |
| T¹⁰H | . | 1 | . | | . | 1 | . | 1 | | |
| T¹¹H | | | | | | 0 | . | 1 | | |
| T¹²H | | | | | | 0 | . | 0 | | |
| T¹³H | | 1 | . | | | 1 | . | 0 | | |
| T¹⁴H | | | | | | 0 | . | 0 | | |
| T¹⁵H | | | | | | 1 | . | 1 | | |
| &c. | | | | | | 0 | . | 0 | | |

$$\overline{\quad\quad} \quad \overline{\quad\quad} \quad \overline{\quad\quad} \quad \overline{\quad\quad} \quad \overline{\quad\quad}$$

2,048  .  2,048  .  2,048  .  2,048  .  2,048

In very many trials, then, we may depend upon something like the predicted average. Conversely, from many trials we may form a guess at what the average will be. Thus, in Buffon's experiment the 2,048 first throws of the sets gave *head* in 1,061 cases: we have a right to infer that in the long run something like 1,061 out of 2,048 is the proportion of heads, even before we know the reasons for the equality of chance, which tell us that 1,024 out of 2,048 is the real truth. I now come to the way in which such considerations have led to a mode in which mere pitch-and-toss has given a more accurate approach to the quadrature of the circle than has been reached by some of my paradoxers. What would my friend [1] in No. 14 have said to this? The method is as follows: Suppose a planked floor of the usual kind, with thin visible seams between the planks. Let there be a thin straight rod, or wire, not so long as the breadth of the plank. This rod, being tossed up at hazard, will either fall quite clear of the seams, or will lay across one seam. Now Buffon, and after him Laplace, proved the following: That in the long run the fraction of the whole number of trials in which a seam is intersected will be the fraction which twice the length of the rod is of the circumference of the circle having the breadth of a plank for its diameter. In 1855 Mr. *Ambrose* Smith, of Aberdeen, made 3,204 trials with a rod three-fifths of the distance between the planks: there were 1,213 clear intersections, and 11 contacts on which it was difficult to decide. Divide these

[1] See p. 172. This article was a supplement to No. 14 in the *Athenæum* Budget.

contacts equally, and we have 1,218½ to 3,204 for the ratio of 6 to 5$\pi$, presuming that the greatness of the number of trials gives something near to the final average, or result in the long run: this gives $\pi = 3 \cdot 1553$. If all the 11 contacts had been treated as intersections, the result would have been $\pi = 3 \cdot 1412$, exceedingly near. A pupil of mine made 600 trials with a rod of the length between the seams, and got $\pi = 3 \cdot 137$.

This method will hardly be believed until it has been repeated so often that 'there never could have been any doubt about it.'

The first experiment strongly illustrates a truth of the theory, well confirmed by practice: whatever can happen will happen if we make trials enough. Who would undertake to throw tail eight times running? Nevertheless, in the 8,192 sets tail 8 times running occurred 17 times; 9 times running, 9 times; 10 times running, twice; 11 times and 13 times, each once; and 15 times, twice.]

1830. The celebrated interminable fraction $3 \cdot 14159 \ldots$, which the mathematician calls $\pi$, is the ratio of the circumference to the diameter. But it is thousands of things besides. It is constantly turning up in mathematics: and if arithmetic and algebra had been studied without geometry, $\pi$ must have come in somehow, though at what stage or under what name must have depended upon the casualties of algebraical invention. This will readily be seen when it is stated that $\pi$ is nothing but four times the series

$$1 - \tfrac{1}{3} + \tfrac{1}{5} - \tfrac{1}{7} + \tfrac{1}{9} - \tfrac{1}{11} + \ldots \ldots$$

*ad infinitum.* It would be wonderful if so simple a series had but one kind of occurrence. As it is, our trigonometry being founded on the circle, $\pi$ first appears as the ratio stated. If, for instance, a deep study of probable fluctuation from the average had preceded geometry, $\pi$ might have emerged as a number perfectly indispensable in such problems as—What is the chance of the number of aces lying between a million $+ x$ and a million $- x$, when six million of throws are made with a die? I have not gone into any detail of all those cases in which the paradoxer finds out, by his unassisted acumen, that results of mathematical investigation *cannot be*: in fact, this discovery is only an accompaniment, though a necessary one, of his paradoxical statement of that which *must be*. Logicians are beginning to see that the notion of *horse* is inseparably connected with that of *non-horse*: that the first without the second would be no notion at all. And it is clear that the positive affirmation of that which contradicts mathematical demonstration cannot but be accompanied by a declaration, mostly overtly made, that demonstration is false.   If the mathematician were interested in punishing this indiscretion, he could make his denier ridiculous by inventing asserted results which would completely take him in.

More than thirty years ago I had a friend, now long gone,.who was a mathematician, but not of the higher branches : he was, *inter alia*, thoroughly up in all that relates to mortality, life assurance, &c.   One day, explaining to him how it should be ascertained what the chance is of the survivors of a large number of persons now alive lying between given limits of number at the end of a certain time, I came, of course, upon the introduction of $\pi$, which I could only describe as the ratio of the circumference of a circle to its diameter.   ' Oh, my dear friend ! that must be a delusion ; what can the circle have to do with the numbers alive at the end of a given time ? '—' I cannot demonstrate it to you ; but it is demonstrated.'—' Oh ! stuff ! I think you can prove anything with your differential calculus : figment, depend upon it.'   I said no more ; but, a few days afterwards, I went to him and very gravely told him that I had discovered the law of human mortality in the Carlisle Table, of which he thought very highly.   I told him that the law was involved in this circumstance.   Take the table of expectation of life, choose any age, take its expectation and make the nearest integer a new age, do the same with that, and so on ; begin at what age you like, you are sure to end at the place where the age past is equal, or most nearly equal, to the expectation to come.   ' You don't mean that this always happens ? '—' Try it.'   He did try, again and again ; and found it as I said.   ' This is, indeed, a curious thing ; this *is* a discovery.'   I might have sent him about trumpeting the law of life : but I contented myself with informing him thàt the same thing would happen with any table whatsoever in which the first column goes up and the second goes down ; and that if a proficient in the higher mathematics chose to palm a figment upon him, he could do without the circle : *à corsaire, corsaire et demi*, the French proverb says.   'Oh !' it was remarked, 'I see, this was Milne !'   It was *not* Milne : I remember well showing the formula to him some time afterwards.   He raised no difficulty about $\pi$ ; he knew the forms of Laplace's results, and he was much interested.   Besides, Milne never said stuff ! and figment !   And he would not have been taken in : he would have quietly tried it with the Northampton and all the other tables, and would have got at the truth.

The first book of Euclid's Elements. With alterations and familiar notes. Being an attempt to get rid of axioms altogether; and to establish the theory of parallel lines, without the introduction of any principle not common to other parts of the elements. By a member of the University of Cambridge. Third edition. In usum serenissimæ filiolæ. London, 1830.

The author was Lieut.-Col. (now General) Perronet Thompson, the author of the ' Catechism on the Corn Laws.' I reviewed the fourth edition—which had the name of ' Geometry without Axioms,' 1833—in the quarterly *Journal of Education* for January, 1834. Col. Thompson, who then was a contributor to— if not editor of—the *Westminster Review*, replied in an article. the authorship of which could not be mistaken.

Some more attempts upon the problem, by the same author, will be found in the sequel. They are all of acute and legitimate speculation ; but they do not conquer the difficulty in the manner demanded by the conditions of the problem. The paradox of parallels does not contribute much to my pages : its cases are to be found for the most part in geometrical systems, or in notes to them. Most of them consist in the proposal of additional postulates; some are attempts to do without any new postulate. Gen. Perronet Thompson, whose paradoxes are always constructed on much study of previous writers, has collected in the work above-named, a budget of attempts, the heads of which are in the *Penny* and *English Cyclopædias*, at ' Parallels.' He has given thirty instances, selected from what he had found.

Lagrange, in one of the later years of his life, imagined that he had overcome the difficulty. He went so far as to write a paper, which he took with him to the Institute, and began to read it. But in the first paragraph something struck him which he had not observed : he muttered *Il faut que j'y songe encore*, and put the paper in his pocket.

---

The following paragraph appeared in the *Morning Post*, May 4, 1831 :—

' We understand that although, owing to circumstances with which the public are not concerned, Mr. Goulburn declined becoming a candidate for University honours, that his scientific attainments are far from inconsiderable. He is well known to be the author of an essay in the Philosophical Transactions on the accurate rectification of a circular arc, and of an investigation of the equation of a lunar caustic—a problem likely to become of great use in nautical astronomy.'

This hoax—which would probably have succeeded with any journal—was palmed upon the *Morning Post*, which supported Mr. Goulburn, by some Cambridge wags who supported Mr. Lubbock, the other candidate for the University of Cambridge. Putting on the usual concealment, I may say that I always suspected Dr-nkw-t-r B-th-n- of having a share in the matter. The skill of the hoax lies in avoiding the words 'quadrature of the circle,' which all know, and speaking of 'the accurate rectification of a circular arc,' which all do not know for its synonyme. The *Morning Post* next day gave a reproof to hoaxers in general, without referring to any particular case. It must be added, that although there are *caustics* in mathematics, there is no *lunar* caustic.

So far as Mr. Goulburn was concerned, the above was poetic justice. He was the minister who, in the old time, told a deputation from the Astronomical Society that the Government 'did not care twopence for all the science in the country.' There may be some still alive who remember this : I heard it from more than one of those who were present, and are now gone. Matters are much changed. I was thirty years in office at the Astronomical Society ; and, to my certain knowledge, every Government of that period, Whig and Tory, showed itself ready to help with influence when wanted, and with money whenever there was an answer for the House of Commons. The following correction subsequently appeared. Referring to the hoax about Mr. Goulburn, Messrs. C. H. and Thompson Cooper have corrected an error, by stating that the election which gave rise to the hoax was that in which Messrs. Goulburn and Yates Peel defeated Lord Palmerston and Mr. Cavendish. They add that Mr. Gunning, the well-known Esquire Bedell of the University, attributed the hoax to the late Rev. R. Sheepshanks, to whom, they state, are also attributed certain clever fictitious biographies—of public men, as I understand it—which were palmed upon the editor of the *Cambridge Chronicle*, who never suspected their genuineness to the day of his death. Being in most confidential intercourse with Mr. Sheepshanks, both at the time and all the rest of his life (twenty-five years), and never having heard him allude to any such things—which were not in his line, though he had satirical power of quite another kind—I feel satisfied he had nothing to do with them. I may add that others, his nearest friends, and also members of his family, never heard him allude to these hoaxes as their author, and disbelieve his authorship as much as I do myself. I say this not as imputing any blame to the true author, such hoaxes being fair election jokes in all time, but merely to put the saddle off the wrong horse, and to give one more instance of the insecurity of imputed authorship. Had Mr. Sheepshanks ever told me that he had perpetrated the hoax, I should have had no hesitation in giving it to him. I consider all clever election squibs, free from bitterness and personal imputation, as giving the multitude good channels for the vent of feelings which but for them would certainly find bad ones.

[But I now suspect that Mr. Babbage had some hand in the hoax. He gives it in his ' Passages, &c.' and is evidently writing from memory, for he gives the wrong year. But he has given the paragraph, though not accurately, yet with such a recollection of the points as brings suspicion of the authorship upon him, perhaps in conjunction with D. B. Both were on Cavendish's committee. Mr. Babbage adds, that ' late one evening a cab drove up in hot haste to the office of the *Morning Post,* delivered the copy as coming from Mr. Goulburn's committee, and at the same time ordered fifty extra copies of the *Post* to be sent next morning to their committee-room. I think the man—the only one I ever heard of—who knew all about the cab and the extra copies must have known more.]

*Demonville.*—A Frenchman's Christian name is his own secret, unless there be two of the surname. M. Demonville is a very good instance of the difference between a French and English discoverer. In England there is a public to listen to discoveries in mathematical subjects made without mathematics: a public which will hear, and wonder, and think it possible that the pretensions of the discoverer have some foundation. The unnoticed man may possibly be right: and the old country-town reputation which I once heard of, attaching to a man who ' had written a book about the signs of the zodiac which all the philosophers in London could not answer,' is fame as far as it goes. Accordingly, we have plenty of discoverers who, even in astronomy, pronounce the learned in error because of mathematics. In France, beyond the sphere of influence of the Academy of Sciences, there is no one to cast a thought upon the matter: all who take the least interest repose entire faith in the Institute. Hence the French discoverer turns all his thoughts to the Institute, and looks for his only hearing in that quarter. He therefore throws no slur upon the means of knowledge, but would say, with M. Demonville—'A l'égard de M. Poisson, j'envie loyalement la millième partie de ses connaissances mathématiques, pour prouver mon systême d'astronomie aux plus incrédules.' This system is that the only bodies of our system are the earth, the sun, and the moon; all the others being illusions, caused by reflexion of the sun and moon from the ice of the polar regions. In mathematics, addition and subtraction are for men; multiplication and division, which are in truth creation and destruction, are prerogatives of Deity. But *nothing* multiplied by *nothing* is *one*. M. Demonville obtained an introduction to William the Fourth, who desired the opinion of the Royal Society upon his system: the answer was very brief. The King was quite right; so was the Society: the fault lay with those who advised His Majesty on a matter they knew nothing about. The writings of M. Demonville in my possession are as follows. The dates—which were only on covers torn off in binding—were about 1831-34 :—

'Petit cours d'astronomie' followed by ' Sur l'unité mathématique.' —Principes de la physique de la création implicitement admis dans la notice sur le tonnerre par M. Arago.—Question de longitude sur mer.—Vrai systême du monde (pp. 92). Same title, four pages, small type. Same title, four pages, addressed to the British Association. Same title, four pages, addressed to M. Mathieu. Same title, four pages, on M. Bouvard's report.—Résumé de la physique de la création; troisième partie du vrai système du monde.

> The quadrature of the circle discovered, by Arthur Parsey, author of the 'art of miniature painting.' Submitted to the consideration of the Royal Society, on whose protection the author humbly throws himself. London, 1832, 8vo.

Mr. Parsey was an artist, who also made himself conspicuous by a new view of perspective. Seeing that the sides of a tower, for instance, would appear to meet in a point if the tower were high enough, he thought that these sides ought to slope to one another in the picture. On this theory he published a small work, of which I have not the title, with a Grecian temple in the frontispiece, stated, if I remember rightly, to be the first picture which had ever been drawn in true perspective. Of course the building looked very Egyptian, with its sloping sides. The answer to his notion is easy enough. What is called the picture is not the picture from which the mind takes its perception; that picture is on the retina. The *intermediate* picture, as it may be called—the human artist's work—is itself seen perspectively. If the tower were so high that the sides, though parallel, appeared to meet in a point, the picture must also be so high that the *picture-sides*, though parallel, would appear to meet in a point. I never saw this answer given, though I have seen and heard the remarks of artists on Mr. Parsey's work. I am inclined to think it is commonly supposed that the artist's picture is the representation which comes before the mind: this is not true; we might as well say the same of the object itself. In July 1831, reading an article on squaring the circle, and finding that there was a difficulty, he set to work, got a light denied to all the mathematicians in—some would say through—a crack, and advertised in the *Times* that he had done the trick. He then prepared this work, in which, those who read it will see how, he showed that 3·14159......should be 3·0625. He might have found out his error by *stepping* a draughtsman's circle with the compasses.

Perspective has not had many paradoxes. The only other one I remember is that of a writer on perspective, whose name I forget, and whose four pages I do not possess. He circulated remarks on my notes on the subject, published in the *Athenæum*, in which he denies that the stereographic projection is a case of perspective, the reason being that the whole hemisphere makes too large a picture for the eye conveniently to grasp at once. That is to say, it is no perspective because there is too much perspective.

> Principles of Geometry familiarly illustrated. By the Rev. W. Ritchie, LL.D. London, 1833, 12mo.
> A new Exposition of the system of Euclid's Elements, being an attempt to establish his work on a different basis. By Alfred Day, LL.D. London, 1839, 12mo.

These works belong to a small class which have the peculiarity of insisting that in the general propositions of geometry a proposition gives its converse: that 'Every B is A' follows from 'Every A is B.' Dr. Ritchie says, 'If it be proved that the equality of two of the angles of a triangle depends *essentially* upon the equality of the opposite sides, it follows that the equality of the opposite sides depends *essentially* on the equality of the angles.' Dr. Day puts it as follows:—

'That the converses of Euclid, so called, where no particular limitation is specified or implied in the leading proposition, more than in the converse, must be necessarily true; for as by the nature of the reasoning the leading proposition must be universally true, should the converse not be so, it cannot be so universally, but has at least all the exceptions conveyed in the leading proposition, and the case is therefore unadapted to geometric reasoning ; or, what is the same thing, by the very nature of geometric reasoning, the particular exceptions to the extended converse must be identical with some one or other of the cases under the universal affirmative proposition with which we set forth, which is absurd.'

On this I cannot help transferring to my reader the words of the Pacha when he orders the bastinado,—May it do you good ! A rational study of logic is much wanted to show many mathematicians, of all degrees of proficiency, that there is nothing in the *reasoning* of mathematics which differs from other reasoning. Dr. Day repeated his argument in ' A Treatise on Proportion,' London, 1840, 8vo. Dr. Ritchie was a very clear-headed man. He published, in 1818, a work on arithmetic, with rational explanations. This was too early for such an improvement, and nearly the whole of this excellent work was sold as waste paper. His elementary introduction to the Differential Calculus was drawn up while he was learning the subject late in life. Books of this sort are often very effective on points of difficulty.

Letter to the Royal Astronomical Society in refutation of Mistaken Notions held in common, by the Society, and by all the Newtonian philosophers. By Capt. Forman, R.N. Shepton-Mallet, 1833, 8vo.

Capt. Forman wrote against the whole system of gravitation, and got no notice. He then wrote to Lord Brougham, Sir J. Herschel, and others I suppose, desiring them to procure notice of his books in the reviews : this not being acceded to, he wrote (in print) to Lord John Russell to complain of their ' dishonest' conduct. He then sent a manuscript letter to the Astronomical Society, inviting controversy : he was answered by a recommendation to study dynamics. The above pamphlet was the consequence, in which, calling the Council of the Society ' craven dunghill cocks,' he set them right about their doctrines. From all I can learn, the life of a worthy man and a creditable officer was completely embittered by his want of power to see that no person is bound in reason to enter into controversy with every one who chooses to invite him to the field. This mistake is not peculiar to philosophers, whether of orthodoxy or paradoxy ; a majority of educated persons imply, by their modes of proceeding, that no one has a right to any opinion which he is not prepared to defend against all comers.

David and Goliath, or an attempt to prove that the Newtonian system of Astronomy is directly opposed to the Scriptures. By Wm. Lauder, Sen., Mere, Wilts. Mere, 1833, 12mo.

Newton is Goliath; Mr. Lauder is David. David took five pebbles; Mr. Lauder takes five arguments. He expects opposition; for Paul and Jesus both met with it.

Mr. Lauder, in his comparison, seems to put himself in the divinely inspired class. This would not be a fair inference in every case; but we know not what to think when we remember that a tolerable number of cyclometers have attributed their knowledge to direct revelation. The works of this class are very scarce; I can only mention one or two from Montucla. Alphonso Cano de Molina, in the last century, upset all Euclid, and squared the circle upon the ruins; he found a follower, Janson, who translated him from Spanish into Latin. He declared that he believed in Euclid, until God, who humbles the proud, taught him better. One Paul Yvon, called from his estate de la Leu, a merchant at Rochelle, supported by his book-keeper, M. Pujos, and a Scotchman, John Dunbar, solved the problem by divine grace, in a manner which was to convert all Jews, Infidels, &c. There seem to have been editions of his work in 1619 and 1628, and a controversial 'Examen' in 1630, by Robert Sara. There was a noted discussion, in which Mydorge, Hardy, and others took part against de la Leu. I cannot find this name either in Lipenius or Murhard, and I should not have known the dates if it had not been for one of the keenest bibliographers of any time, my friend Prince Balthasar Boncompagni, who is trying to find copies of the works, and has managed to find copies of the titles. In 1750, Henry Sullamar, an Englishman, squared the circle by the number of the Beast: he published a pamphlet every two or three years; but I cannot find any mention of him in English works. In France, in 1753, M. de Causans, of the Guards, cut a circular piece of turf, squared it, and deduced original sin and the Trinity. He found out that the circle was equal to the square in which it is inscribed; and he offered a reward for detection of any error, and actually deposited 10,000 francs as earnest of 300,000. But the courts would not allow any one to recover.

1834. In this year Sir John Herschel set up his telescope at Feldhausen, Cape of Good Hope. He did much for astronomy, but not much for the Budget of Paradoxes. He gives me, however, the following story. He showed a resident a remarkable blood-red star, and some little time after he heard of a sermon preached in those parts in which it was asserted that the statements of the Bible must be true, for that Sir J. H. had seen in his telescope 'the very place where wicked people go.'

But red is not always the colour. Sir J. Herschel has in his possession a letter written to his father, Sir W. H., dated April 3, 1787, and signed ' Eliza Cumyns,' begging to know if any of the stars be *indigo* in colour, ' because, if there be, I think it may be deemed a strong conjectural illustration of the expression, so often used by our Saviour in the Holy Gospels, that "the disobedient shall be cast into outer darkness ; " for as the Almighty Being can doubtless confine any of his creatures, whether corporeal or spiritual, to what part of his creation He pleases, if therefore any of the stars (which are beyond all doubt so many suns to other systems) be of so dark a colour as that above mentioned, they may be calculated to give the most insufferable heat to those dolorous systems dependent upon them (and to reprobate spirits placed there), without one ray of cheerful light ; and may therefore be the scenes of future punishments.' This letter is addressed to Dr. Heirschel at Slow. Some have placed the infernal regions inside the earth, but others have filled this internal cavity—for cavity they will have—with refulgent light, and made it the abode of the blessed. It is difficult to build without knowing the number to be provided for. A friend of mine heard the following (part) dialogue between two strong Scotch Calvinists : ' Noo! hoo manny d'ye thank there are of the alact on the arth at this moment?—Eh ! mabbee a doozen— Hoot ! mon ! nae so mony as thot ! '

1834. From 1769 to 1834 the *Nautical Almanac* was published on a plan which gradually fell behind what was wanted. In 1834 the new series began, under a new superintendent (Lieut. W. S. Stratford). There had been a long scientific controversy, which would not be generally intelligible. To set some of the points before the reader, I reprint a cutting which I have by me. It is from the Nautical *Magazine*, but I did hear that some had an idea that it was in the Nautical *Almanac* itself. It certainly was not, and I feel satisfied the Lords of the Admiralty would not have permitted the insertion ; they are never in advance of their age. The Almanac for 1834 was published in July 1833.

THE NEW NAUTICAL ALMANAC.—Extract from the 'Primum Mobile,' and 'Milky Way Gazette.' Communicated by AEROLITH.

A meeting of the different bodies composing the Solar System was this day held at the Dragon's Tail, for the purpose of taking into consideration the alterations and amendments introduced into the New Nautical Almanac. The honourable luminaries had been individually summoned by fast-sailing comets, and there was a remarkably full attendance. Among the visitors we *observed* several nebulæ, and almost all the stars whose proper motions would admit of their being present.

The SUN was unanimously called to the focus. The small planets took the oaths, and their places, after a short discussion, in which it was decided that the places should be those of the Almanac itself, with leave reserved to move for corrections.

Petitions were presented from $a$ and $\delta$ Ursæ Minoris, complaining of being put on daily duty, and praying for an increase of salary.—Laid on the plane of the ecliptic.

The trustees of the eccentricity[1] and inclination funds reported a balance of ·00001 in the former, and a deficit of $0''·009$ in the latter. This announcement caused considerable surprise, and a committee was moved for, to ascertain which of the bodies had more or less than his share. After some discussion, in which the small planets offered to consent to a reduction, if necessary, the motion was carried.

The FOCAL BODY then rose to address the meeting. He remarked that the subject on which they were assembled was one of great importance to the routes and revolutions of the heavenly bodies. For himself, though a private arrangement between two of his honourable neighbours (here he looked hard at the Earth and Venus) had prevented his hitherto paying that close attention to the predictions of the Nautical Almanac which he declared he always had wished to do; yet he felt consoled by knowing that the conductors of that work had every disposition to take his peculiar circumstances into consideration. He declared that he had never passed the wires of a transit without deeply feeling his inability to adapt himself to the present state of his theory; a feeling which he was afraid had sometimes caused a slight tremor in his limb. Before he sat down, he expressed a hope that honourable luminaries would refrain as much as possible from eclipsing each other, or causing mutual perturba-

---

[1] See Sir J. Herschel's *Astronomy*, p. 369.

tions. Indeed, he should be very sorry to see any interruption of the harmony of the spheres. (Applause.)

The several articles of the New Nautical Almanac were then read over without any comment; only we observed that Saturn shook his ring at every novelty, and Jupiter gave his belt a hitch, and winked at the satellites at page 21 of each month.

The MOON rose, to propose a resolution. No one, he said, would be surprised at his bringing this matter forward in the way he did, when it was considered in how complete and satisfactory a manner his motions were now represented. He must own he had trembled when the Lords of the Admiralty dissolved the Board of Longitude, but his tranquillity was more than reestablished by the adoption of the new system. He did not know but that any little assistance he could give in Nautical Astronomy was becoming of less and less value every day, owing to the improvement of chronometers. But there was one thing, of which nothing could deprive him—he meant the regulation of the tides. And, perhaps, when his attention was not occupied by more than the latter, he should be able to introduce a little more regularity into the phenomena. (Here the honourable luminary gave a sort of modest libration, which convulsed the meeting with laughter.) They might laugh at his natural infirmity if they pleased, but he could assure them it arose only from the necessity he was under, when young, of watching the motions of his worthy primary. He then moved a resolution highly laudatory of the alterations which appeared in the New Nautical Almanac.

The EARTH rose, to second the motion. His honourable satellite had fully expressed his opinions on the subject. He joined his honourable friend in the focus in wishing to pay every attention to the Nautical Almanac, but, really, when so important an alteration had taken place in his magnetic pole[1] (hear) and there might, for aught he knew, be a successful attempt to reach his pole of rotation, he thought he could not answer for the preservation of the precession in its present state. (Here the hon. luminary, scratching his side, exclaimed, as he sat down, ' More steam-boats—confound 'em!')

An honourable satellite (whose name we could not learn) proposed that the resolution should be immediately despatched, corrected for refraction, when he was called to order by the Focal Body, who reminded him that it was contrary to the moving

---

[1] Captain Ross had just stuck a bit of brass there.

orders of the system to take cognizance of what passed inside the atmosphere of any planet.

SATURN and PALLAS rose together. (Cries of ' New member !' and the former gave way.) The latter, in a long and eloquent speech, praised the liberality with which he and his colleagues had at length been relieved from astronomical disqualifications. He thought that it was contrary to the spirit of the laws of gravitation to exclude any planet from office on account of the eccentricity or inclination of his orbit. Honourable luminaries need not talk of the want of convergency of his series. What had they to do with any private arrangements between him and the general equations of the system ? (Murmurs from the opposition.) So long as he obeyed the laws of motion, to which he had that day taken a solemn oath, he would ask, were old planets, which were now so well known that nobody trusted them, to . . . .

The FOCAL BODY said he was sorry to break the continuity of the proceedings, but he thought that remarks upon character, with a negative sign, would introduce differences of too high an order. The honourable luminary must eliminate the expression which he had brought out, in finite terms, and use smaller inequalities in future. (Hear, hear.)

PALLAS explained, that he was far from meaning to reflect upon the orbital character of any planet present. He only meant to protest against being judged by any laws but those of gravitation, and the differential calculus: he thought it most unjust that astronomers should prevent the small planets from being observed, and then reproach them with the imperfections of the tables, which were the result of their own narrow-minded policy. (Cheers.)

SATURN thought that, as an old planet, he had not been treated with due respect. (Hear, from his satellites.) He had long foretold the wreck of the system from the friends of innovation. Why, he might ask, were his satellites to be excluded, when small planets, trumpery comets, which could not keep their mean distances (cries of oh ! oh !), double stars, with graphical approximations, and such obscure riff-raff of the heavens (great uproar) found room enough. So help him Arithmetic, nothing could come of it, but a stoppage of all revolution. His hon. friend in the focus might smile, for he would be a gainer by such an event; but as for him (Saturn), he had something to lose, and hon. luminaries well knew that, whatever they might think *under* an atmosphere, *above* it continual revolution was the only way of preventing perpetual anarchy.  As to the hon. luminary who had risen before him, he was not surprised at his remarks, for he had invariably observed that he and his colleagues allowed themselves *too much latitude.*  The stability of the system required that they should be brought down, and he, for one, would exert all his powers of attraction to accomplish that end.  If other bodies would cordially unite with him, particularly his noble friend next him, than whom no luminary possessed greater weight—

JUPITER rose to order.  He conceived his noble friend had no right to allude to him in that manner, and was much surprised at his proposal, considering the matters which remained in dispute between them.  In the present state of affairs, he would take care never to be in conjunction with his hon. neighbour one moment longer than he could help.  (Cries of ' Order, order, no long inequalities,' during which he sat down.)

SATURN proceeded to say, that he did not know till then that a planet with a ring could affront one who had only a belt, by proposing mutual co-operation.  He would now come to the subject under discussion.  He should think meanly of his hon. colleagues if they consented to bestow their approbation upon a mere astronomical production.  Had they forgotten that they once were considered the arbiters of fate, and the prognosticators of man's destiny?  What had lost them that proud position?  Was it not the infernal march of intellect, which, after having turned the earth topsy-turvy, was now disturbing the very universe.  For himself (others might do as they pleased), but he stuck to the venerable Partridge, and the Stationers' Company, and trusted that they would outlive infidels and anarchists, whether of Astronomical or Diffusion of Knowledge Societies.  (Cries of oh ! oh !)

MARS said he had been told, for he must confess he had not seen the work, that the places of the planets were given for Sundays.  This, he must be allowed to say, was an indecorum he had not expected ; and he was convinced the Lords of the Admiralty had given no orders to that effect.  He hoped this point would be considered in the measure which had been introduced in another place, and that some one would move that the prohibition against travelling on Sundays extend to the heavenly as well as earthly bodies.

Several of the stars here declared, that they had been much annoyed by being observed on Sunday evenings, during the hours of divine service.

The room was then cleared for a division, but we are unable to state what took place. Several comets-at-arms were sent for, and we heard rumours of a personal collision having taken place between two luminaries in opposition. We were afterwards told, that the resolution was carried by a majority, and the luminaries elongated at 2 h. 15 m. 33,41 s. sidereal time.

*‌.‌* It is reported, but we hope without foundation, that Saturn, and several other discontented planets, have accepted an invitation from Sirius to join his system, on the most liberal appointments. We believe the report to have originated in nothing more than the discovery of the annual parallax of Sirius from the orbit of Saturn ; but we may safely assure our readers that no steps have as yet been taken to open any communication.

We are also happy to state, that there is no truth in the rumour of the laws of gravitation being about to be repealed. We have traced this report, and find it originated with a gentleman living near Bath (Captain Forman, R.N.), whose name we forbear to mention.

A great excitement has been observed among the nebulæ, visible to the earth's southern hemisphere, particularly among those which have not yet been discovered from thence. We are at a loss to conjecture the cause, but we shall not fail to report to our readers the news of any movement which may take place. (Sir J. Herschel's visit. He could just see this before he went out.)

> A Treatise on the Divine System of the Universe, by Captain Woodley, R.N., and as demonstrated by his Universal Time-piece, and universal method of determining a ship's longitude by the apparent true place of the moon ; with an introduction refuting the solar system of Copernicus, the Newtonian philosophy, and mathematics. 1834. 8vo.
> Description of the Universal Time-piece. (4 pp. 12mo.)

I think this divine system was published several years before, and was republished with an introduction in 1834. Capt. Woodley was very sure that the earth does not move : he pointed out to me, in a conversation I had with him, something—I forget what—in the motion of the Great Bear, visible to any eye, which could not possibly be if the earth moved. He was exceedingly ignorant, as the following quotation from his account of the usual opinion will show :—

The north pole of the Earth's axis deserts, they say, the north star or pole of the Heavens, at the rate of 1° in 71¾ years . . . The fact is, nothing can be more certain than that the Stars have not changed their latitudes or declinations *one degree* in the last 71¾ years.

This is a strong specimen of a class of men by whom all accessible persons who have made any name in science are hunted. It is a pity that they cannot be admitted into scientific societies, and allowed fairly to state their cases, and stand quiet cross-examination, being kept in their answers very close to the questions, and the answers written down. I am perfectly satisfied that if one meeting in the year were devoted to the hearing of those who chose to come forward on such conditions, much good would be done. But I strongly suspect few would come forward at first, and none in a little while : and I have had some experience of the method I recommend, privately tried. Capt. Woodley was proposed, a little after 1834, as a Fellow of the Astronomical Society ; and, not caring whether he moved the sun or the earth, or both—I could not have stood *neither*—I signed the proposal. I always had a sneaking kindness for paradoxers, such a one, perhaps, as Petit André had for his *lambs*, as he called them. There was so little feeling against his opinions, that he only failed by a fraction of a ball. Had I myself voted, he would have been elected ; but being engaged in conversation, and not having heard the slightest objection to him, I did not think it worth while to cross the room for the purpose. I regretted this at the time, but had I known how ignorant he was I should not have supported him. Probably those who voted against him knew more of his books than I then did.

I remember no other instance of exclusion from a scientific society on the ground of opinion, even if this be one ; of which it may be that ignorance had more to do with it than paradoxy. Mr. Frend, a strong anti-Newtonian, was a Fellow of the Astronomical Society, and for some years in the Council. Lieut. Kerigan was elected to the Royal Society at a time when his proposers must have known that his immediate object was to put F.R.S. on the title-page of a work against the tides. To give all I know, I may add that the editor of some very ignorant bombast about the ' forehead of the solar sky,' who did not know the difference between *Bailly* and *Baily*, received hints which induced him to withdraw his proposal for election into the Astronomical Society. But this was an act of kindness ; for if he had seen Mr. Baily in the chair, with his head on, he might have been political historian enough to faint away.

De la formation des Corps. Par Paul Laurent. Nancy, 1834, 8vo.

Atoms, and ether, and ovules or eggs, which are planets, and their eggs, which are satellites. These speculators can create worlds, in which they cannot be refuted ; but none of them dare attack the problem of a grain of wheat, and its passage from a seed to a plant, bearing scores of seeds like what it was itself.

An account of the Rev. John Flamsteed, the First Astronomer-Royal . . . By Francis Baily, Esq. London, 1835, 4to. Supplement, London, 1837, 4to.

My friend Francis Baily was a paradoxer : he brought forward things counter to universal opinion. That Newton was impeccable in every point was the national creed ; and failings of temper and conduct would have been utterly disbelieved, if the paradox had not come supported by very unusual evidence. Anybody who impeached Newton on existing evidence might as well have been squaring the circle, for any attention he would have got. About this book I will tell a story. It was published by the Admiralty for distribution ; and the distribution was entrusted to Mr. Baily. On the eve of its appearance, rumours of its extraordinary revelations got about, and persons of influence applied to the Admiralty for copies. The Lords were in a difficulty : but on looking at the list they saw names, as they thought, which were so obscure that they had a right to assume Mr. Baily had included persons who had no claim to such a compliment as presentation from the Admiralty. The Secretary requested Mr. Baily to call upon him. ' Mr. Baily, my Lords are inclined to think that some of the persons in this list are perhaps not of that note which would justify their Lordships in presenting this work.'—' To whom does your observation apply, Mr. Secretary ? '—' Well, now, let us examine the list ; let me see ; now,—now,—now,—come !—here's Gauss—*who's Gauss?*'—' Gauss, Mr. Secretary, is the oldest mathematician now living, and is generally thought to be the greatest.'—' O-o-oh ! Well, Mr. Baily, we will see about it, and I will write you a letter.' The letter expressed their Lordships' perfect satisfaction with the list.

There was a controversy about the revelations made in this work ; but as the eccentric anomalies took no part in it, there is nothing for my purpose. The following valentine from Mrs. Flamsteed, which I found among Baily's papers, illustrates some of the points :—

'3 Astronomers' Row, Paradise: February 14, 1836.

'Dear Sir,—I suppose you hardly expected to receive a letter from me, dated from this place; but the truth is, a gentleman from our street was appointed guardian angel to the American Treaty, in which there is some astronomical question about boundaries. He has got leave to go back to fetch some instruments which he left behind, and I take this opportunity of making your acquaintance. That America has become a wonderful place since I was down among you; you have no idea how grand the fire at New York looked up here. Poor dear Mr. Flamsteed does not know I am writing a letter to a gentleman on Valentine's day; he is walked out with Sir Isaac Newton (they are pretty good friends now, though they do squabble a little sometimes) and Sir William Herschel, to see a new nebula. Sir Isaac says he can't make out at all how it is managed; and I am sure I cannot help him. I never bothered my head about those things down below, and I don't intend to begin here.

I have just received the news of your having written a book about my poor dear man. It's a chance that I heard it at all; for the truth is, the scientific gentlemen are somehow or other become so wicked, and go so little to church, that very few of them are considered fit company for this place. If it had not been for Dr. Brinkley, who came here of course, I should not have heard about it. He seems a nice man, but is not yet used to our ways. As to Mr. Halley, he is of course not here; which is lucky for him, for Mr. Flamsteed swore the moment he caught him in a place where there are no magistrates, he would make a sacrifice of him to heavenly truth. It was very generous in Mr. F. not appearing against Sir Isaac when he came up, for I am told that if he had, Sir Isaac would not have been allowed to come in at all. I should have been sorry for that, for he is a companionable man enough, only holds his head rather higher than he should do. I met him the other pay walking with Mr. Whiston, and disputing about the deluge. "Well, Mrs. Flamsteed," says he, "does old Poke-the-Stars understand gravitation yet?" Now you must know that is rather a sore point with poor dear Mr. Flamsteed. He says that Sir Isaac is as crochetty about the moon as ever; and as to what some people say about what has been done since his time, he says he should like to see somebody who knows something about it of himself. For it is very singular that none of the people who have carried on Sir Isaac's notions have been allowed to come here.

I hope you have not forgotten to tell how badly Sir Isaac used Mr. Flamsteed about that book. I have never quite forgiven him; as for Mr. Flamsteed, he says that as long as he does not come for observations, he does not care about it, and that he will never trust him with any papers again as long as he lives. I shall never forget what a rage he came home in when Sir Isaac had called him a puppy. He struck the stairs all the way up with his crutch, and said puppy at every step, and all the evening, as soon as ever a star appeared in the telescope, he called it puppy. I could not think what was the matter, and when I asked, he only called me puppy.

I shall be very glad to see you if you come our way, Pray keep up some appearances, and go to church a little. St. Peter is always uncommonly civil to astronomers, and indeed to all scientific persons, and never bothers them with many questions. If they can make anything out of a case, he is sure to let them in. Indeed, he says, it is perfectly out of the question expecting a mathematician to be as religious as an apostle, but that it is as much as his place is worth to let in the greater number of those who come. So try if you cannot manage it, for I am very curious to know whether you found all the letters. I remain, dear sir, your faithful servant,

MARGARET FLAMSTEED.

Francis Baily, Esq.

P.S. Mr. Flamsteed has come in, and says he left Sir Isaac riding cockhorse upon the nebula, and poring over it as if it were a book. He has brought in his old acquaintance Ozanam, who says that it was always his maxim on earth, that "il appartient aux docteurs de Sorbonne de disputer, au Pape de prononcer, et au mathématicien d'aller en Paradis en ligne perpendiculaire."'

The Secretary of the Admiralty was completely extinguished. I can recall but two instances of demolition as complete, though no doubt there are many others. The first is in

Simon Stevin and M. Dumortier. Nieuport, 1845, 12mo.

M. Dumortier was a member of the Academy of Brussels : there was a discussion, I believe, about a national Pantheon for Belgium. The name of Stevinus suggested itself as naturally as that of Newton to an Englishman ; probably no Belgian is better known to foreigners as illustrious in science. Stevinus is great in the *Mécanique Analytique* of Lagrange ; Stevinus is great in the *Tristram Shandy* of Sterne. M. Dumortier, who believed that not one Belgian in a thousand knew Stevinus, and who confesses with ironical shame that he was not the odd man, protested against placing the statue of an obscure man in the Pantheon, to give foreigners the notion that Belgium could show nothing greater. The work above named is a slashing retort : any one who knows the history of science ever so little may imagine what a dressing was given, by mere extract from foreign writers. The tract is a letter signed J. du Fan, but this is a pseudonym of Mr. Van de Weyer. The Academician says Stevinus was a man. who was not without merit for the time at which he lived: Sir! is the answer, he was as much before his own time as you are behind yours. How came a man who had never heard of Stevinus to be a member of the Brussels Academy?

The second story was told me by Mr. Crabb Robinson, who was long connected with the *Times*, and intimately acquainted with Mr. W***. When W*** was an undergraduate at Cambridge, taking a walk, he came to a stile, on which sat a bumpkin who did not make way for him: the gown in that day looked down on the town. 'Why do you not make way for a gentleman?' —'Eh?'—'Yes, why do you not move? You deserve a good hiding, and you shall get it if you don't take care?' The bumpkin raised his muscular figure on its feet, patted his menacer on the head, and said, very quietly,—'Young man! I'm Cribb.' W*** seized the great pugilist's hand, and shook it warmly, got him to his own rooms in college, collected some friends, and had a symposium which lasted until the large end of the small hours.

God's Creation of the Universe as it is, in support of the Scriptures. By Mr. Finleyson. Sixth Edition, 1835, 8vo.

This writer, by his own account, succeeded in delivering the famous Lieut. Richard Brothers from the lunatic asylum, and tending him, not as a keeper but as a disciple, till he died. Brothers was, by his own account, the nephew of the Almighty, and Finleyson ought to have been the nephew of Brothers. For Napoleon came to him in a vision, with a broken sword and an arrow in his side, beseeching help: Finleyson pulled out the arrow, but refused to give a new sword; whereby poor Napoleon, though he got off with life, lost the battle of Waterloo. This story was written to the Duke of Wellington, ending with 'I pulled out the arrow, but left the broken sword. Your Grace can supply the rest, and what followed is amply recorded in history.' The book contains a long account of applications to Government to do three things: to pay 2,000*l.* for care taken of Brothers, to pay 10,000*l.* for discovery of the longitude, and to prohibit the teaching of the Newtonian system, which makes God a liar. The successive administrations were threatened that they would have to turn out if they refused, which, it is remarked, came to pass in every case. I have heard of a joke of Lord Macaulay, that the House of Commons must be the Beast of the Revelations, since 658 members, with the officers necessary for the action of the House, make 666. Macaulay read most things,

and the greater part of the rest : so that he might be suspected
of having appropriated as a joke one of Finleyson's serious points
—'I wrote Earl Grey upon the 13th of July, 1831, informing
him that his Reform Bill could not be carried, as it reduced the
members below the present amount of 658, which, with the
eight principal clerks or officers of the House, make the number
666.' But a witness has informed me that Macaulay's joke was
made in his hearing a great many years before the Reform Bill
was proposed ; in fact, when both were students at Cambridge.
Earl Grey was, according to Finleyson, a descendant of Uriah
the Hittite. For a specimen of Lieut. Brothers, this book would
be worth picking up. Perhaps a specimen of the Lieutenant's
poetry may be acceptable : Brothers *loquitur*, remember :—

> Jerusalem ! Jerusalem ! shall be built again !
>   More rich, more grand than ever ;
> And through it shall Jordan flow ! (!)
>   My people's favourite river.
> There I'll erect a splendid throne,
>   And build on the wasted place ;
> To fulfil my ancient covenant
>   To King David and his race.
> \*       \*       \*       \*       \*       \*
> Euphrates' stream shall flow with ships,
>   And also my wedded Nile ;
> And on my coast shall cities rise,
>   Each one distant but a mile.
> \*       \*       \*       \*       \*       \*
> My friends the Russians on the north
>   With Persees and Arabs round,
> Do show the limits of my land,
>   Here ! Here ! then I mark the ground.

Among the paradoxers are some of the theologians who in
their own organs of the press venture to criticise science. These
may hold their ground when they confine themselves to the
geology of long past periods and to general cosmogony : for it is
the tug of Greek against Greek ; and both sides deal much in
what is grand when called *hypothesis*, petty when called *supposition*. And very often they are not conspicuous when they
venture upon things within knowledge ; wrong, but not quite
wrong enough for a Budget of Paradoxes. One case, however,
is destined to live, as an instance of a school which finds writers,
editors, and readers. The double stars have been seen from the
seventeenth century, and diligently observed by many from the

time of Wm. Herschel, who first devoted continuous attention to
them.   The year 1836 was that of a remarkable triumph of
astronomical prediction.   The theory of gravitation had been
applied to the motion of binary stars about each other, in elliptic
orbits, and in that year the two stars of γ Virginis, as had been
predicted should happen within a few years of that time—for
years are small quantities in such long revolutions—the two
stars came to their nearest: in fact, they appeared to be one as
much with the telescope as without it.   This remarkable turn-
ing-point of the history of a long and widely-known branch of
astronomy was followed by an article in the *Church of England
Quarterly Review* for April 1837, written against the Useful
Knowledge Society.   The notion that there are any such things
as double stars is (p. 460) implied to be imposture or delusion,
as in the following extract.   I suspect that I myself am the
*Sidrophel,* and that my companion to the maps of the stars,
written for the Society and published in 1836, is the work to
which the writer refers :—

We have forgotten the name of that Sidrophel who lately discovered
that the fixed stars were not single stars, but appear in the heavens,
like soles at Billingsgate, in pairs ; while a second astronomer, under
the influence of that competition in trade which the political economists
tell us is so advantageous to the public, professes to show us, through
his superior telescope, that the apparently single stars are really three.
Before such wondrous mandarins of science, how continually must
*homunculi* like ourselves keep in the background, lest we come between
the wind and their nobility.

If the *homunculus* who wrote this be still above ground, how
devoutly must he hope he may be able to keep in the background !   But the chief blame falls on the editor. . The title of
the article is—

The new school of superficial pantology ; a speech intended to be
delivered before a defunct Mechanics' Institute.   By Swallow Swift,
late M.P. for the Borough of Cockney-Cloud, Witsbury : reprinted
Balloon Island, Bubble year, month *Ventose.*   Long live Charlatan !

As a rule, orthodox theologians should avoid humour, a weapon
which all history shows to be very difficult to employ in favour
of establishment, and which, nine times out of ten, leaves its
wielder fighting on the side of heterodoxy.   Theological argument, when not enlivened by bigotry, is seldom worse than
narcotic : but theological fun, when not covert heresy, is almost
always sialagogue.   The article in question is a craze, which no
editor should have admitted, except after severe inspection by

qualified persons. The author of this wit committed a mistake which occurs now and then in old satire, the confusion between himself and the party aimed at. He ought to be reviewing this fictitious book, but every now and then the article becomes the book itself; not by quotation, but by the writer forgetting that *he* is not Mr. Swallow Swift, but his reviewer. In fact, he and Mr. S. Swift had each had a dose of the *Devil's Elixir.* A novel so called, published about forty years ago, proceeds upon a legend of this kind. If two parties both drink of the elixir, their identities get curiously intermingled ; each turns up in the character of the other throughout the three volumes, without having his ideas clear as to whether he be himself or the other. There is a similar confusion in the answer made to the famous *Epistolæ Obscurorum Virorum* : it is headed *Lamentationes Obscurorum Virorum.* This is not a retort of the writer, throwing back the imputation : the obscure men who had been satirized are themselves made, by name, to wince under the disapprobation which the Pope had expressed at the satire upon themselves.

Of course the book here reviewed is a transparent forgery. But I do not know how often it may have happened that the book, in the journals which always put a title at the head, may have been written after the review. About the year 1830 a friend showed me the proof of an article of his on the malt tax, for the next number of the *Edinburgh Review.* Nothing was wanting except the title of the book reviewed ; I asked what it was. He sat down, and wrote as follows at the head, ' The Maltster's Guide (pp. 124),' and said that would do as well as anything.

But I myself, it will be remarked, have employed such humour as I can command ' in favour of establishment.' What it is worth I am not to judge ; as usual in such cases, those who are of my cabal pronounce it good, but cyclometers and other paradoxers either call it very poor, or commend it as sheer buffoonery. Be it one or the other, I observe that all the effective ridicule is, in this subject, on the side of establishment. This is partly due to the difficulty of quizzing plain and sober demonstration ; but so much, if not more, to the ignorance of the paradoxers. For that which cannot be *ridiculed,* can be *turned into ridicule* by those who know how. But by the time a person is deep enough in *negative* quantities, and *impossible* quantities, to be able to satirise them, he is caught, and being inclined to become a *user,* shrinks from being an *abuser.* Imagine a person with a gift of ridicule, and knowledge enough, trying his hand on the junction of the assertions which he will find in various books of algebra.  First, that a negative quantity has no logarithm ; secondly, that a negative quantity has no square root ; thirdly, that the first non-existent is to the second as the circumference of a circle to its diameter.  One great reason of the allowance of such unsound modes of expression is the confidence felt by the writers that $\sqrt{-1}$ and $\log(-1)$ will make their way, however inaccurately described. I heartily wish that the cyclometers had knowledge enough to attack the weak points of algebraical diction : they would soon work a beneficial change.

> Recueil de ma vie, mes ouvrages et mes pensées.  Par Thomas Ignace Marie Forster.  Brussels, 1836, 12mo.

Mr. Forster, an Englishman settled at Bruges, was an observer in many subjects, but especially in meteorology.  He communicated to the Astronomical Society, in 1848, the information that, in the registers kept by his grandfather, his father, and himself, beginning in 1767, new moon on Saturday was followed, nineteen times out of twenty, by twenty days of rain and wind.  This statement being published in the *Athenœum*, a cluster of correspondents averred that the belief is common among seamen, in all parts of the world, and among landsmen too.  Some one quoted a distich—

> ' Saturday's moon and Sunday's full
> Never were fine and never *wull*.'

Another brought forward—

> ' If a Saturday's moon
> Comes once in seven years it comes too soon.'

Mr. Forster did not say he was aware of the proverbial character of the phenomenon.  He was a very eccentric man.  He treated his dogs as friends, and buried them with ceremony.  He quarrelled with the *curé* of his parish, who remarked that he could not take his dogs to heaven with him.  I will go nowhere, said he, where I cannot take my dog.  He was a sincere Catholic : but there is a point beyond which even churches have no influence.

The following is some account of the announcement of 1849. The *Athenœum* (Feb. 17), giving an account of the meeting of the Astronomical Society in December, 1858, says :

' Dr. Forster of Bruges, who is well known as a meteorologist, made a communication at which our readers will stare : he declares that by journals of the weather kept by his grandfather, father, and himself, ever since 1767, to the present time, *whenever the new moon has fallen on a Saturday, the following twenty days have been wet and windy*, in nineteen cases out of twenty. In spite of our friend Zadkiel and the others who declare that we would smother every truth that does not happen to agree with us, we are glad to see that the Society had the sense to publish this communication, coming, as it does, from a veteran observer, and one whose love of truth is undoubted. It must be that the fact is so set down in the journals, because Dr. Forster says it: and whether it be only a fact of the journals, or one of the heavens, can soon be tried. The new moon *of* March next, falls on *Saturday* the 24th, at 2 in the afternoon. We shall certainly look out.'

The following appeared in the number of March 31 :—

' The first *Saturday Moon* since Dr. Forster's announcement came off a week ago. We had previously received a number of letters from different correspondents—all to the effect that the notion of new moon on Saturday bringing wet weather is one of widely extended currency. One correspondent (who gives his name) states that he has constantly heard it at sea, and among the farmers and peasantry in Scotland, Ireland, and the North of England. He proceeds thus : " Since 1826, nineteen years of the time I have spent in a seafaring life. I have constantly observed, though unable to account for, the phenomenon. I have also heard the stormy qualities of a Saturday's moon remarked by American, French, and Spanish seamen ; and, still more distant, a Chinese pilot, who was once doing duty on board my vessel seemed to be perfectly cognizant of the fact." So that it seems we have, in giving currency to what we only knew as a very curious communication from an earnest meteorologist, been repeating what is common enough among sailors and farmers. Another correspondent affirms that the thing is most devoutly believed in by seamen ; who would as soon sail on a Friday as be in the Channel after a Saturday moon.— After a tolerable course of dry weather, there was some snow, accompanied by wind on Saturday last, here in London ; there were also heavy louring clouds. Sunday was cloudy and cold, with a little rain ; Monday was louring; Tuesday unsettled; Wednesday quite over-clouded, with rain in the morning. The present occasion shows only a general change of weather, with a tendency towards rain. If Dr. Forster's theory be true, it is decidedly one of the minor instances, as far as London weather is concerned.—It will take a good deal of evidence to make us believe in the omen of a Saturday Moon. But, as we have said of the Poughkeepsie Seer, the thing is very curious whether true or false. Whence comes this universal proverb—and a hundred others—while the meteorological observer cannot, when he puts down a long series of results, detect any weather cycles at all? One of our correspondents wrote us something of a lecture for en-couraging, he said, the notion that *names* could influence the weather. He mistakes the question. If there be any weather cycles depending on the moon, it is possible that one of them may be so related to the week cycle of seven days, as to show recurrences which are of the kind

stated, or any other.  For example, we know that if the new moon of
March fall on a Saturday in this year, it will most probably fall on
a Saturday nineteen years hence.  This is not connected with the
spelling of Saturday—but with the connexion between the motions of
the sun and moon.  Nothing but the Moon can settle the question—
and we are willing to wait on her for further information.  If the
adage be true, then the philosopher has missed what lies before his
eyes; if false, then the world can be led by the nose in spite of the
eyes.  Both these things happen sometimes; and we are willing to
take whichever of the two solutions is borne out by future facts.  In
the mean time, we announce the next Saturday Moon for the 18th
of August.'

How many coincidences are required to establish a law of
connexion ?  It depends on the way in which the mind views the
matter in question.  Many of the paradoxers are quite set up by
a very few instances.  I will now tell a story about myself, and
then ask them a question.

So far as instances can prove a law, the following is proved : no
failure has occurred.  Let a clergyman be known to me, whether
by personal acquaintance or correspondence, or by being frequently
brought before me by those with whom I am connected in private
life : that clergyman does not, except in few cases, become a
bishop ; but, *if* he become a bishop, he is sure, first or last, to
become an arch-bishop.  This has happened in every case.  As
follows :—

1.  My last schoolmaster, a former Fellow of Oriel, was a very
intimate college friend of Richard Whately, a younger man.
Struck by his friend's talents, he used to talk of him perpetually,
and predict his future eminence.  Before I was sixteen, and
before Whately had even given his Bampton Lectures, I was very
familiar with his name, and some of his sayings.  I need not say
that he became Archbishop of Dublin.

2.  When I was a child, a first cousin of John Bird Sumner
married a sister of my mother.  I cannot remember the time
when I first heard his name, but it was made very familiar to me.
In time he became Bishop of Chester, and then, Archbishop of
Canterbury.  My reader may say that Dr. C. R. Sumner, Bishop
of Winchester, has just as good a claim : but it is not so : those
connected with me had more knowledge of Dr. J. B. Sumner ;
and said nothing, or next to nothing, of the other.  Rumour says
that the Bishop of Winchester has *declined* an Archbishopric : if
so, my rule is a rule of gradations.

3.  Thomas Musgrave, Fellow of Trinity College, Cambridge,
was *Dean* of the college when I was an undergraduate : this

brought me into connexion with him, he giving impositions for not going to chapel, I writing them out according. We had also friendly intercourse in after life; I forgiving, he probably forgetting. Honest Tom Musgrave, as he used to be called, became Bishop of Hereford, and Archbishop of York.

4. About the time when I went to Cambridge, I heard a great deal about Mr. C. T. Longley, of Christchurch, from a cousin of my own of the same college, long since deceased, who spoke of him much, and most affectionately. Dr. Longley passed from Durham to York, and thence to Canterbury. I cannot quite make out the two Archbishoprics ; I do not remember any other private channel through which the name came to me : perhaps Dr. Longley, having two strings to his bow, would have been one Archbishop if I had never heard of him.

5. When Dr. Wm. Thomson was appointed to the see of Gloucester in 1861, he and I had been correspondents on the subject of logic—on which we had both written—for about fourteen years. On his elevation I wrote to him, giving the preceding instances, and informing him that he would certainly be an Archbishop. The case was a strong one, and the law acted rapidly ; for Dr. Thomson's elevation to the see of York took place in 1862.

Here are five cases ; and there is no opposing instance. I have searched the almanacs since 1828, and can find no instance of a Bishop not finally Archbishop of whom I had known through private sources, direct or indirect. Now what do my paradoxers say ? Is this a pre-established harmony, or a chain of coincidences ? And how many instances will it require to establish a law ?

> Some account of the great astronomical discoveries lately made by Sir John Herschel at the Cape of Good Hope. Second Edition. London, 12mo. 1836.

This is a curious hoax, evidently written by a person versed in astronomy and clever at introducing probable circumstances and undesigned coincidences. It first appeared in a newspaper. It makes Sir J. Herschel discover men, animals, &c. in the moon, of which much detail is given. There seems to have been a French edition, the original, and English editions in America, whence the work came into Britain : but whether the French was published in America or at Paris I do not know. There is no doubt that it was produced in the United States, by M. Nicollet, an astronomer, once of Paris, and a fugitive of some kind. About him I have heard two stories. First, that he fled to America with funds not his own, and that this book was a mere device to raise the wind. Secondly, that he was a *protégé* of Laplace, and of the Polignac party, and also an outspoken man. That after the revolution he was so obnoxious to the republican party that he judged it prudent to quit France; which he did in debt, leaving money for his creditors, but not enough, with M. Bouvard. In America he connected himself with an assurance office. The moon-story was written, and sent to France, chiefly with the intention of entrapping M. Arago, Nicollet's especial foe, into the belief of it. And those who narrate this version of the story wind up by saying that M. Arago *was* entrapped, and circulated the wonders through Paris, until a letter from Nicollet to M. Bouvard explained the hoax. I have no personal knowledge of either story: but as the poor man had to endure the first, it is but right that the second should be told with it.

The Weather Almanac for the Year 1838. By P. Murphy, Esq. M.N.S.

By M.N.S. is meant *member of no society*. This almanac bears on the title-page two recommendations. The *Morning Post* calls it one of the most important-if-true publications of our generation. The *Times* says: ' If the basis of his theory prove sound, and its principles be sanctioned by a more extended experience, it is not too much to say that the importance of the discovery is equal to that of the longitude.' Cautious journalist ! Three times that of the longitude would have been too little to say. That the landsman might predict the weather of all the year, at its beginning, Jack would cheerfully give up astronomical longitude—*the* problem—altogether, and fall back on chronometers with the older Ls, lead, latitude, and look-out, applied to dead-reckoning. Mr. Murphy attempted to give the weather day by day: thus the first seven days of March bore Changeable ; Rain ; Rain ; Rain-*wind* ; Changeable ; Fair ; Changeable. To aim at such precision as to put a fair day between two changeable ones by weather theory was going very near the wind and weather too. Murphy opened the year with cold and frost; and the weather did the same. But Murphy, opposite to Saturday, January 20, put down ' Fair, Probable lowest degree of winter temperature.' When this Saturday came, it was not merely the probably coldest of 1838, but certainly the coldest of many consecutive years. Without knowing anything of Murphy, I felt it prudent to cover my nose with my glove as I walked the street at eight in the morning. The fortune of the Almanac was made. Nobody waited to see whether the future would dement the prophecy: the shop was beset in a manner which brought the police to keep order; and it was said that the Almanac for 1838 was a gain of 5,000*l.* to the owners. It very soon appeared that this was only a lucky hit: the weather-prophet had a modified reputation for a few years; and is now no more heard of. A work of his will presently appear in the list.

> Letter from Alexandria on the evidence of the practical application of the quadrature of the circle in the great pyramids of Gizeh. By H. C. Agnew, Esq. London, 1838, 4to.

Mr. Agnew detects proportions which he thinks were suggested by those of the circumference and diameter of a circle.

> The creed of St. Athanasius proved by a mathematical parallel. Before you censure, condemn, or approve; read, examine, and understand. E. B. Revilo. London, 1839, 8vo.

This author really believed himself, and was in earnest. He is not the only person who has written nonsense by confounding the mathematical infinite (of quantity) with what speculators now more correctly express by the unlimited, the unconditioned, or the absolute. This tract is worth preserving, as the extreme case of a particular kind. The following is a specimen. Infinity being represented by $\infty$, as usual, and $f$, $s$, $g$, being finite integers, the three Persons are denoted by $\infty^f$, $(m\,\infty)^s$, $\infty^g$, the finite fraction $m$ representing human nature, as opposed to $\infty$. The clauses of the Creed are then given with their mathematical parallels. I extract a couple:—

But the Godhead of the Father, of the Son, and of the Holy Ghost, is all one: the glory equal, the Majesty co-eternal.

It has been shown that $\infty^f$, $\infty^g$, and $(m\,\infty)^s$, together, are but $\infty$, and that each is $\infty$, and any magnitude in existence represented by $\infty$ always was and always will be: for it cannot be made, or destroyed, and yet exists.

Equal to the Father, as touching his Godhead: and inferior to the Father, as touching his Manhood.

$(m\,\infty)^s$ is equal to $\infty^f$ as touching $\infty$, but inferior to $\infty^f$ as touching $m$: because $m$ is not infinite.

I might have passed this over, as beneath even my present subject, but for the way in which I became acquainted with it. A bookseller, *not the publisher*, handed it to me over his counter: one who had published mathematical works. He said, with an

air of important communication, Have you seen *this*, Sir! In
reply, I recommended him to show it to my friend Mr. ——, for
whom he had published mathematics.  Educated men, used to
books, and to the converse of learned men, look with mysterious
wonder on such productions as this: for which reason I have
made a quotation which many will judge had better have been
omitted.  But it would have been an imposition on the public if
I were, omitting this and some other uses of the Bible and
Common Prayer, to pretend that I had given a true picture of
my school.

[Since the publication of the above, it has  been stated that the
author is Mr. Oliver Byrne, the author of the *Dual Arithmetic*
mentioned further on : E. B. Revilo seems to be obviously a
reversal.]

> Old and new logic contrasted : being an attempt to elucidate, for
> ordinary comprehension, how Lord Bacon delivered the human
> mind from its 2,000 years' enslavement under Aristotle.  By
> Justin Brenan.  London, 1839, 12mo.

Logic, though the other exact science, has not had the sort of
assailants who have clustered about Mathematics.  There is a
sect which disputes the utility of logic, but there are no special
points, like the quadrature of the circle, which excite dispute
among those who admit other things.  The old story about
Aristotle having one logic to trammel us, and Bacon another to
set us free,—always laughed at by those who really knew either
Aristotle or Bacon,—now begins to be understood by a large
section of the educated world.  The author of this tract connects
the old logic with the indecencies of the classical writers, and the
new with moral purity : he appeals to women, who, 'when they
see plainly the demoralizing tendency of syllogistic logic, they
will, no doubt, exert their powerful influence against it, and
support the Baconian method.'  This is the only work against
logic which I can introduce, but it is a rare one, I mean in
contents.  I quote the author's idea of a syllogism :—

> The basis of this system is the syllogism.  This is a form of couch-
> ing the substance of your argument or investigation into one short
> line or sentence—then corroborating or supporting it in another, and
> drawing your conclusion or proof in a third.

On this definition he gives an example, as follows : ' Every sin
deserves death,' the substance of the ' argument or investigation.'
Then comes, ' Every unlawful wish is a sin,' which ' corroborates
or supports ' the preceding : and, lastly, ' therefore every unlaw-

ful wish deserves death,' which is the 'conclusion or proof.' We learn, also, that 'sometimes the first is called the premises (*sic*), and sometimes the first premiss;' as also that 'the first is sometimes called the proposition, or subject, or affirmative, and the next the predicate, and sometimes the middle term.' To which is added, with a mark of exclamation at the end, 'but, in analyzing the syllogism, there is a middle term, and a predicate too, in each of the lines!' It is clear that Aristotle never enslaved this mind.

I have said that logic has no paradoxers, but I was speaking of old time. This science has slept until our own day: Hamilton says there has been 'no progress made in the *general* development of the syllogism since the time of Aristotle; and in regard to the few *partial* improvements, the professed historians seem altogether ignorant.' But in our time, the paradoxer, the opponent of common opinion, has appeared in this field. I do not refer to Prof. Boole, who is not a *paradoxer*, but a *discoverer*: his system could neither oppose nor support common opinion, for its grounds were not within the conception of any one. I speak especially of two others, who fought like cat and dog: one was dogmatical, the other categorical. The first was Hamilton himself—Sir William Hamilton of Edinburgh, the metaphysician, not Sir William *Rowan* Hamilton of Dublin, the mathematician, a combination of peculiar genius with unprecedented learning, erudite in all he could want except mathematics, for which he had no turn, and in which he had not even a schoolboy's knowledge, thanks to the Oxford of his younger day. The other was the author of this work, so fully described in Hamilton's writings that there is no occasion to describe him here. I shall try to say a few words in common language about the paradoxers.

Hamilton's great paradox was the *quantification of the predicate*; a fearful phrase, easily explained. We all know that when we say 'Men are animals,' a form wholly unquantified in phrase, we speak of *all* men, but not of all animals: it is *some or all*, some may be all for aught the proposition says. This some-may-be-all-for-aught-we-say, or *not-none*, is the logician's *some*. One would suppose that 'all men are some animals,' would have been the logical phrase in all time: but the predicate never was quantified. The few who alluded to the possibility of such a thing found reasons for not adopting it over and above the great reason, that Aristotle did not adopt it. For Aristotle never ruled in physics or metaphysics *in the old time* with near so much of

absolute sway as he has ruled in logic *down to our own time.*
The logicians knew that in the proposition ' all men are animals '
the ' animal ' is not *universal,* but *particular*: yet no one dared
to say that *all* men are *some* animals, and to invent the phrase,
' *some* animals are *all* men ' until Hamilton leaped the ditch,
and not only completed a system of enunciation, but applied it to
syllogism.

My own case is as peculiar as his: I have proposed to introduce mathematical *thought* into logic to an extent which makes
the old stagers cry

> St. Aristotle! what wild notions!
> Serve a *ne exeat regno* on him!

Hard upon twenty years ago, a friend and opponent, who stands
high in these matters, and who is not nearly such a sectary of
Aristotle and establishment as most, wrote to me as follows:—
' It is said that next to the man who forms the taste of a nation,
the greatest genius is the man who corrupts it. I mean therefore
no disrespect, but very much the reverse, when I say that I
have hitherto always considered you as a great logical heresiarch.'
Coleridge says he thinks that it was Sir Joshua Reynolds who
made the remark: which, to copy a bull I once heard, I cannot
deny, because I was not there when he said it. My friend did
not call me to repentance and reconciliation with the church:
I think he had a guess that I was a reprobate sinner. My
offences at that time were but small: I went on spinning syllogism systems, all alien from the common logic, until I had six,
the initial letters of which, put together, from the names I gave
before I saw what they would make, bar all repentance by the
words

> RUE NOT!

leaving to the followers of the old school the comfortable option
of placing the letters thus:

> TRUE? NO!

It should however be stated that the question is not about
absolute truth or falsehood. No one denies that anything I call
an inference is an inference: they say that my alterations are
*extra-logical* ; that they are *material,* not *formal* ; and that logic
is a *formal* science.

The distinction between material and formal is easily made,
where the usual perversions are not required. A *form* is an
empty machine, such as ' Every X is Y;' it may be supplied
with *matter,* as in ' Every *man* is *animal.*' The logicians will

not see that their *formal* proposition, ' Every X is Y,' is material
in three points, the degree of assertion, the quantity of the
proposition, and the copula.   The purely formal proposition is
' There is the probability *a* that X stands in the relation L to
Y.'  The time will come when it will be regretted that logic
went without paradoxers for two thousand years : and when much
that has been said on the distinction of form and matter will
breed jokes.

I give one instance of one mood of each of the systems, in the
order of the letters first written above.

*Relative.*—In this system the formal relation is taken, that is,
the copula may be any whatever.   As a material instance, in
which the *relations* are those of consanguinity (of men understood), take the following : X is the brother of Y ; X is not the
uncle of Z ; therefore, Z is not the child of Y.   The discussion of
relation, and of the objections to the extension, is in the *Cambridge Transactions*, vol. x, part 2 ; a crabbed conglomerate.

*Undecided.*—In this system one premise, and want of power
over another, infer want of power over a conclusion.   As ' Some
men are not capable of tracing consequences ; we cannot be sure
that there are beings responsible for consequences who are incapable of tracing consequences ; therefore, we cannot be sure
that all men are responsible for the consequences of their actions.'

*Exemplar.*—This, long after it suggested itself to me as a
means of correcting a defect in Hamilton's system, I saw to be
the very system of Aristotle himself, though his followers have
drifted into another.   It makes its subject and predicate examples, thus : Any one man is an animal ; any one animal is a
mortal ; therefore, any one man is a mortal.

*Numerical.*—Suppose 100 Ys to exist : then if 70 Xs be Ys,
and 40 Zs be Ys, it follows that 10 Xs (at least) are Zs.   Hamilton, whose mind could not generalize on symbols, saw that the
word *most* would come under this system, and admitted, as valid,
such a syllogism as ' most Ys are Xs ; most Ys are Zs ; therefore,
some Xs are Zs.'

*Onymatic.*—This is the ordinary system much enlarged in
propositional forms.   It is fully discussed in my *Syllabus of
Logic.*

*Transposed.*—In this syllogism the quantity in one premise is
transposed into the other.   As, some Xs are not Ys ; for every X
there is a Y which is Z ; therefore, some Zs are not Xs.

Sir William Hamilton of Edinburgh was one of the best

friends and allies I ever had. When I first began to publish speculation on this subject, he introduced me to the logical world as having plagiarized from him. This drew their attention: a mathematician might have written about logic under forms which had something of mathematical look long enough before the Aristotelians would have troubled themselves with him: as was done by John Bernoulli, James Bernoulli, Lambert, and Gergonne; who, when our discussion began, were not known even to omnilegent Hamilton. He retracted his accusation of *wilful* theft in a manly way when he found it untenable; but on this point he wavered a little, and was convinced to the last that I had taken his principle unconsciously. He thought I had done the same with Ploucquet and Lambert. It was his pet notion that I did not understand the commonest principles of logic, that I did not always know the difference between the middle term of a syllogism and its conclusion. It went against his grain to imagine that a mathematician could be a logician. So long as he took me to be riding my own hobby, he laughed consumedly: but when he thought he could make out that I was mounted behind Ploucquet or Lambert, the current ran thus:—'It would indeed have been little short of a miracle had he, ignorant even of the common principles of logic, been able of himself to rise to generalization so lofty and so accurate as are supposed in the peculiar doctrines of both the rival logicians, Lambert and Ploucquet—how useless soever these may in practice prove to be.' All this has been sufficiently discussed elsewhere: 'but, masters, remember that I am an ass.'

I know that I never saw Lambert's work until after all Hamilton supposed me to have taken was written: he himself, who read almost everything, knew nothing about it until after I did. I cannot prove what I say about my knowledge of Lambert: but the means of doing it may turn up. For, by the casual turning up of an old letter, I *have* found the means of clearing myself as to Ploucquet. Hamilton assumed that (unconsciously) I took from Ploucquet the notion of a logical notation in which the symbol of the conclusion is seen in the joint symbols of the premises. For example, in my own fashion I write down ( . ) ( . ), two symbols of premises. By these symbols I see that there is a valid conclusion, and that it may be written in symbol by striking out the two middle parentheses, which gives ( . . ) and reading the two negative dots as an affirmative. And so I see in ( . ) ( . ) that ( ) is the conclusion. This, in full, is the perception that 'all are either Xs or Ys' and 'all are either Ys or Zs' necessitates 'some Xs are Zs.' Now in Ploucquet's book of 1763, is found,

'Deleatur in præmissis medius; id quod restat indicat conclusionem.' In the paper in which I explain my symbols—which are altogether different from Ploucquet's—there is found 'Erase the symbols of the middle term ; the remaining symbols show the inference.' There is very great likeness: and I would have excused Hamilton for his notion if he had fairly given reference to the part of the book in which his quotation was found. For I had shown in my *Formal Logic* what part of Ploucquet's book I had used : and a fair disputant would either have strengthened his point by showing that I had been at his part of the book, or allowed me the advantage of it being apparent that I had not given evidence of having seen that part of the book. My good friend, though an honest man, was sometimes unwilling to allow due advantage to controversial opponents.

But to my point. The only work of Ploucquet I ever saw was lent me by my friend Dr. Logan, with whom I have often corresponded on logic, &c. I chanced (in 1865) to turn up the letter which he sent me (Sept. 12, 1847) *with the book.* Part of it runs thus :—' I congratulate you on your success in your logical researches [that is, in asking for the book, I had described some results]. Since the reading of your first paper I have been satisfied as to the possibility of inventing a logical notation in which the rationale of the inference is contained in the symbol, though I never attempted to verify it [what I communicated, then, satisfied the writer that I had done and communicated what he, from my previous paper, suspected to be practicable]. I send you Ploucquet's dissertation. . . .'

It now being manifest that I cannot be souring grapes which have been taken from me, I will say what I never said in print before. There is not the slightest merit in making the symbols of the premises yield that of the conclusion by erasure : *the thing must do itself in every system which symbolises quantities.* For in every syllogism (except the inverted *Bramantip* of the Aristotelians) the conclusion is manifest in this way without symbols. This *Bramantip* destroys system in the Aristotelian lot : and circumstances which I have pointed out destroy it in Hamilton's own collection. But in that enlargement of the reputed Aristotelian system which I have called *onymatic*, and in that correction of Hamilton's system which I have called *exemplar*, the rule of erasure is universal, and may be seen without symbols.

Our first controversy was in 1846. In 1847, in my *Formal Logic*, I gave him back a little satire for satire, just to show, as I stated, that I could employ ridicule if I pleased. He was so offended with the appendix in which this was contained, that he would not accept the copy of the book I sent him, but returned it.    Copies of controversial works, sent from opponent to opponent, are not *presents*, in the usual sense : it was a marked success to make him angry enough to forget this.    It had some effect however : during the rest of his life I wished to avoid provocation ; for I could not feel sure that excitement might not produce consequences.    I allowed his slashing account of me in the *Discussions* to pass unanswered : and before that, when he proposed to open a controversy in the *Athenæum* upon my second Cambridge paper, I merely deferred the dispute until the next edition of my *Formal Logic*.    I cannot expect the account in the *Discussions* to amuse an unconcerned reader as much as it amused myself : but for a cut-and-thrust, might-and-main, tooth-and-nail, hammer-and-tongs assault, I can particularly recommend it.    I never knew, until I read it, how much I should enjoy a thundering onslaught on myself, done with racy insolence by a master hand, to whom my good genius had whispered *Ita feri ut se sentiat emori*.    Since that time I have, as the Irishman said, become ' dry moulded for want of a bating.'    Some of my paradoxers have done their best : but theirs is mere twopenny—' small swipes,' as Peter Peebles said.    Brandy for heroes !    I hope a reviewer or two will have mercy on me, and will give me as good discipline as Strafford would have given to Hampden and his set : ' much beholden,' said he, ' should they be to any one that should thoroughly take pains with them in that kind '—meaning *objective* flagellation.    And I shall be the same to any one who will serve me so—but in a literary and periodical sense : my corporeal cuticle is as thin as my neighbours'.

Sir W. H. was suffering under local paralysis before our controversy commenced : and though his mind was quite unaffected, a retort of as downright a character as the attack might have produced serious effect upon a person who had shown himself sensible of ridicule.    Had a second attack of his disorder followed an answer from me, I should have been held to have caused it : though, looking at Hamilton's genial love of combat, I strongly suspected that a retort in kind

> Would cheer his heart, and warm his blood,
> And make him fight, and do him good.

But I could not venture to risk it.    So all I did, in reply to the article in the *Discussions*, was to write to him the following note : which. as illustrating an etiquette of controversy, I insert.

' I beg to acknowledge and thank you for . . . It is necessary that I should say a word on my retention of this work, with reference to your return of the copy of my ' Formal Logic,' which I presented to you on its publication : a return made on the ground of your disapproval of the account of our controversy which that work contained. According to my view of the subject, any one whose dealing with the author of a book is specially attacked in it, has a right to expect from the author that part of the book in which the attack is made, together with so much of the remaining part as is fairly context. And I hold that the acceptance by the party assailed of such work or part of a work does not imply any amount of approval of the contents, or of want of disapproval. On this principle (though I am not prepared to add the word *alone*) I forwarded to you the whole of my work on " Formal Logic " and my second Cambridge Memoir. And on this principle I should have held you wanting in due regard to my literary rights if you had not forwarded to me your asterisked pages, with all else that was necessary to a full understanding of their scope and meaning, so far as the contents of the book would furnish it. For the remaining portion, which it would be a hundred pities to separate from the pages in which I am directly concerned, I am your debtor on another principle ; and shall be glad to remain so if you will allow me to make a feint of balancing the account by the offer of two small works on subjects as little connected with our discussion as the " Epistolæ Obscurorum Virorum," or the Lutheran dispute. I trust that by accepting my " Opuscula " you will enable me to avoid the use of the knife, and leave me to cut you up with the pen as occasion shall serve, I remain, &c. (April 21, 1852).'

I received polite thanks, but not a word about the body of the letter : my argument, I suppose, was admitted.

I find among my miscellaneous papers the following *jeu d'esprit*, or *jeu de bêtise*, whichever the reader pleases—I care not—intended, before I saw ground for abstaining, to have, as the phrase is, come in somehow. I think I could manage to bring anything into anything : certainly into a Budget of Paradoxes. Sir W. H. rather piqued himself upon some caniculars, or doggrel verses, which he had put together *in memoriam [technicam]* of the way in which A E I O are used in logic : he added U, Y, for the addition of *meet*, &c. to the system. I took the liberty of concocting some counter-doggrel, just to show that a mathematician may have architectonic power as well as a metaphysician.

## DOGGREL.

### By Sir W. Hamilton.

A it affirms of *this, these, all,*
    Whilst E denies of *any* ;
I it affirms (whilst O denies)
    Of some (or few, or many).

Thus A affirms, as E denies,
    And definitely either ;
Thus I affirms, as O denies,
    And definitely neither.

A half, left semidefinite,
    Is worthy of its score ;
U, then, affirms, as Y denies,
    This, neither less nor more.

Indefinito-definites,
    I, UI, YO, last we come ;
And this affirms, as that denies
    Of *more, most* (*half, plus, some*).


## COUNTER DOGGREL.

### By Prof. De Morgan.
### (1847.)

Great A affirms of all ;
    Sir William does so too :
When the subject is ' my suspicion,'
    And the predicate ' must be true.'

Great E denies of all ;
    Sir William of all but one :
When he speaks about this present time,
    And of those who in logic have done.

Great I takes up but *some* ;
    Sir William ! my dear soul !
Why then in all your writings,
    Does ' Great I ' fill[1] the whole !

---

[1] A very truculently unjust assertion : for Sir W. was as great a setter up of some as he was a puller down of others. His writings are a congeries of praises and blames, both *cruel smart*, as they say in the States. But the combined instigation of

Great O says some are not;
　　Sir William's readers catch,
That some (modern) Athens is not without
　　An Aristotle to match.

' A half, left semi-definite,
　　Is worthy of its score :'
This looked very much like balderdash,
　　And neither less nor more.

It puzzled me like anything;
　　In fact, it puzzled me worse :
Isn't schoolman's logic hard enough,
　　Without being in Sibyl's verse ?

At last, thinks I, 'tis German ;
　　And I'll try it with some beer !
The landlord asked what bothered me so,
　　And at once he made it clear.

It's *half-and-half*, the gentleman means;
　　Don't you see he talks of *score ?*
That's the bit of a memorandum
　　That we chalk behind the door.

*Semi-definite* 's outlandish ;
　　But I see, in half a squint,
That he speaks of the lubbers who call for a quart,
　　When they can't manage more than a pint.

Now I'll read it into English,
　　And then you'll answer me this :
If it isn't good logic all the world round,
　　I should like to know what is ?

When you call for a pot of half-and-half,
　　If you're lost to sense of shame,
You may leave it *semi-definite*,
　　But you pay for it all just the same.

*　　　　*　　　　*　　　　*

I am unspeakably comforted when I look over the above in
remembering that the question is not whether it be Pindaric or

prose, rhyme, and retort would send Aristides himself to Tartarus, if it were not
pretty certain that Minos would grant a *stet processus* under the circumstances. The
first two verses are exaggerations standing on a basis of truth. The fourth verse is
quite true : Sir W. H. was an Edinburgh Aristotle, with the differences of ancient and
modern Athens well marked, especially the *perfervidum ingenium Scotorum.*

P

Horatian, but whether the copy be as good as the original.    And
I say it is : and will take no denial.

Long live—long will live—the glad memory of William
Hamilton, Good, Learned, Acute, and Disputatious!    He fought
upon principle : the motto of his book is—

> Truth, like a torch, the more it's shook it shines.

There is something in this; but metaphors, like puddings,
quarrels, rivers, and arguments, always have two sides to them.
For instance,

> Truth, like a torch, the more it's shook it shines ;
> But those who want to use it, hold it steady.
> They shake the flame who like a glare to gaze at,
> They keep it still who want a light to see by.

Theory of Parallels. The proof of Euclid's axiom looked for in
   the properties of the Equiangular Spiral. By Lieut.-Col. G. Perronet Thompson. The same, second edition, revised and corrected. The same, third edition, shortened, and freed from
   dependence on the theory of limits. The same, fourth edition,
   ditto, ditto. All London, 1840, 8vo.

To explain these editions it should be noted that General
Thompson rapidly modified his notions, and republished his tracts
accordingly.

Vestiges of the Natural History of Creation. London, 1840, 12mo.

This is the first edition of this celebrated work. Its form is a
case of the theory : the book is an undeniable duodecimo, but the
size of its paper gives it the look of not the smallest of octavos.
Does not this illustrate the law of development, the gradation of
families, the transference of species, and so on? If so, I claim
the discovery of this esoteric testimony of the book to its own
contents ; I defy any one to point out the reviewer who has
mentioned it. The work itself is described by its author as 'the
first attempt to connect the natural sciences into a history of
creation.' The attempt was commenced, and has been carried
on, both with marked talent, and will be continued. Great
advantage will result : at the worst we are but in the alchemy of
some new chemistry, or the astrology of some new astronomy.
Perhaps it would be as well not to be too sure on the matter,
until we have an antidote to possible consequences as ex-

hibited under another theory, on which it is as reasonable to speculate as on that of the 'Vestiges.' I met long ago with a splendid player on the guitar, who assured me, and was confirmed by his friends, that he *never practised*, except in thought, and did not possess an instrument: he kept his fingers acting in his mind, until they got their habits; and thus he learnt the most difficult novelties of execution. Now what if this should be a minor segment of a higher law? What if, by constantly thinking of ourselves as descended from primæval monkeys, we should, —if this be true—actually *get our tails again?* What if the first man who was detected with such an appendage should be obliged to confess himself the author of the 'Vestiges'—a person yet unknown—who would naturally get the start of his species by having had the earliest habit of thinking on the matter? I confess I never hear a man of note talk fluently about it without a curious glance at his proportions, to see whether there may be ground to conjecture that he may have more of 'mortal coil' than others, in anaxyridical concealment. I do not feel sure that even a paternal love for his theory would induce him, in the case I am supposing, to exhibit himself at the British Association,

> With a hole behind which his tail peeped through.

The first sentence of this book (1840) is a cast of the log, which shows our rate of progress. 'It is familiar knowledge that the earth which we inhabit is a globe of somewhat less than 8,000 miles in diameter, being one of a series of eleven which revolve at different distances around the sun. The *eleven!* Not to mention the Iscariot which Le Verrier and Adams calculated into existence, there is more than a septuagint of *new* planetoids.

> The Constitution and Rules of the Ancient and Universal 'Benefit Society' established by Jesus Christ, exhibited, and its advantages and claims maintained, against all Modern and merely Human Institutions of the kind: A Letter very respectfully addressed to the Rev. James Everett, and occasioned by certain remarks made by him, in a speech to the Members of the 'Wesleyan Centenary Institute' Benefit Society. Dated York, Dec. 7, 1840. By Thomas Smith. 12mo. (pp. 8.)

The Wesleyan minister addressed had advocated provision against old age, &c.: the writer declares all *private* provision unchristian. After decent maintenance and relief of family claims of indigence, he holds that all the rest is to go to the 'Benefit Society,' of which he draws up the rules, in technical form, with chapters of 'Officers,' 'Contributors,' &c., from the Acts of the Apostles, &c., and some of the early Fathers. He holds that a Christian may not 'make a *private* provision against the contingencies of the future:' and that the great 'Benefit Society' is the divinely-ordained recipient of all the surplus of his income; capital, beyond what is necessary for business, he is to have none. A real good speculator shuts his eyes by instinct, when opening them would not serve the purpose: he has the vizor of the Irish fairy tale, which fell of itself over the eyes of the wearer the moment he turned them upon the enchanted light which would have destroyed him if he had caught sight of it. 'Whiles it remained, was it not thine own? and after it was sold, was it (the purchase-money) not in thine own power?' would have been awkward to quote, and accordingly nothing is stated except the well-known result, which is rule 3, cap. 5, 'Prevention of Abuses.' By putting his principles together, the author can be made, logically, to mean that the successors of the apostles should put to death all contributors who are detected in not paying their full premiums.

I have known one or two cases in which policy-holders have surrendered their policies through having arrived at a conviction that direct provision is unlawful. So far as I could make it out, these parties did not think it unlawful to lay by out of income, except when this was done in a manner which involved calculation of death-chances. It is singular they did not see that the entrance of chance of death was the entrance of the very principle of the benefit society described in the Acts of the Apostles. The family of the one who died young received more in proportion to *premiums* paid than the family of the one who died old. Every one who understands life assurance sees that—*bonus* apart—the difference between an assurance office and a savings bank consists in the adoption, *pro tanto*, of the principle of community of goods. In the original constitution of the oldest assurance office, the *Amicable Society*, the plan with which they started was nothing but this: persons of all ages under forty-five paid one common premium, and the proceeds were divided among the representatives of those who died within the year.

[I omitted from its proper place a manuscript quadrature (3·1416 exactly) addressed to an eminent mathematician, dated in 1842 from the debtors' ward of a country gaol. The unfortunate speculator says, 'I have laboured many years to find the precise ratio.' I have heard of several cases in which squaring the circle has produced an inability to square accounts. I remind those who feel a kind of inspiration to employ native genius upon difficulties, without gradual progression from elements, that the call is one which becomes stronger and stronger, and may lead, as it has led, to abandonment of the duties of life, and all the consequences.]

1842. Provisional Prospectus of the Double Acting Rotary Engine Company. Also Mechanic's Magazine, March 26, 1842.

Perpetual motion by a drum with one vertical half in mercury, the other in a vacuum: the drum, I suppose, working round for ever to find an easy position. Steam to be superseded: steam and electricity convulsions of nature never intended by Providence for the use of man. The price of the present engines, as old iron, will buy new engines that will work without fuel and at no expense. Guaranteed by the Count de Predaval, the discoverer. I was to have been a Director, but my name got no further than ink, and not so far as official notification of the honour, partly owing to my having communicated to the *Mechanic's Magazine* information privately given to me, which gave premature publicity, and knocked up the plan.

An Exposition of the Nature, Force, Action, and other properties of Gravitation on the Planets. London, 1842, 12mo.
An Investigation of the principles of the Rules for determining the Measures of the Areas and Circumferences of Circular Plane Surfaces . . . London, 1844, 8vo.

These are anonymous; but the author (whom I believe to be Mr. Denison, presently noted) is described as author of a new system of mathematics, and also of mechanics. He had need have both, for he shows that the line which has a square equal to a given circle, has a cube equal to the sphere on the same diameter: that is, in old mathematics, the diameter is to the circumference as 9 to 16! Again, admitting that the velocities of planets in circular orbits are inversely as the square roots of their distances, that is, admitting Kepler's law, he manages to prove that gravitation is inversely as the square *root* of the distance: and suspects magnetism of doing the difference between this and Newton's law. Magnetism and electricity are, in physics, the member of parliament and the cabman—at every man's bidding, as Henry Warburton said.

The above is an outrageous quadrature. In the preceding year, 1841, was published what I suppose at first to be a Maori quadrature, by Maccook. But I get it from a cutting out of some French periodical, and I incline to think that it must be by a Mr. M'Cook. He maks $\pi$ to be $2 + 2\sqrt{(8\sqrt{2} - 11)}$.

> Refutation of a Pamphlet written by the Rev. John Mackey, R.C.P., entitled 'A method of making a cube double of a cube, founded on the principles of elementary geometry,' wherein his principles are proved erroneous, and the required solution not yet obtained. By Robert Murphy. Mallow, 1824, 12mo.

This refutation was the production of an Irish boy of eighteen years old, self-educated in mathematics, the son of a shoemaker at Mallow. He died in 1843, leaving a name which is well known among mathematicians. His works on the theory of equations and on electricity, and his papers in the *Cambridge Transactions*, are all of high genius. The only account of him which I know of is that which I wrote for the *Supplement* of the *Penny Cyclopædia*. He was thrown by his talents into a good income at Cambridge, with no social training except penury, and very little intellectual training except mathematics. He fell into dissipation, and his scientific career was almost arrested: but he had great good in him, to my knowledge. A sentence in a letter from the late Dean Peacock to me—giving some advice about the means of serving Murphy—sets out the old case: 'Murphy is a man whose *special* education is in advance of his *general*; and such men are almost always difficult subjects to manage.' This article having been omitted in its proper place, I put it at 1843, the date of Murphy's death.

> The Invisible Universe disclosed; or, the real Plan and Government of the Universe. By Henry Coleman Johnson, Esq. London, 1843, 8vo.

The book opens abruptly with—

"First demonstration. Concerning the centre: showing that, because the centre is an innermost point at an equal distance between two extreme points of a right line, and from every two relative and opposite intermediate points, it is composed of the two extreme internal points of each half of the line; each extreme internal point attracting towards itself all parts of that half to which it belongs . . ."

Of course the circle is squared : and the circumference is $3\frac{1}{2\frac{1}{1}}$ diameters.

Combination of the Zodiacal and Cometical Systems. Printed for the London Society, Exeter Hall. Price Sixpence. (n.d. 1843.)

What this London Society was, or the 'combination,' did not appear. There was a remarkable comet in 1843, the tail of which was at first confounded with what is called the *zodiacal light*. This nicely-printed little tract, evidently got up with less care for expense than is usual in such works, brings together all the announcements of the astronomers, and adds a short head and tail piece, which I shall quote entire. As the announcements are very ordinary astronomy, the reader will be able to detect, if detection be possible, what. is the meaning and force of the ' Combination of the Zodiacal and Cometical Systems ' :—

"*Premonition*. It has pleased the AUTHOR OF CREATION to cause (to His *human and reasoning* Creatures of this generation, by a '*combined*' appearance in His *Zodiacal* and *Cometical* systems) a '*warning Crisis*' of universal concernment to this our GLOBE. It is this ' *Crisis* ' that has so generally 'ROUSED ' at this moment the '*nations throughout the Earth*' that no equal interest has ever before been excited by MAN ; unless it be in that caused by the ' PAGAN-TEMPLE IN ROME,' which is recorded by the elder Pliny, '*Nat. Hist.*' i. 23. iii. 3. HARDOUIN."

After the accounts given by the unperceiving astronomers, comes what follows :—-

"Such has been (*hitherto*) the only object discerned by the ' *Wise of this World*,' in this *twofold union* of the ' *Zodiacal* ' and ' *Cometical* ' systems : yet it is nevertheless a most ' *Thrilling Warning*,' to *all* the inhabitants of this precarious and transitory EARTH. We have no authorized intimation, or reasonable prospective contemplation, of ' *current time* ' beyond a year 1860, of the present century ; or rather, except ' *the interval which may now remain from the present year* 1843, *to a year* 1860 ' (ἡμερας EΞHKONTA—'*threescore or sixty days* '—' *I have appointed each* " DAY " *for a* " YEAR,*" ' *Ezek.* iv. 6) : and we know, from our ' *common experience*,' how speedily such a measure of time will pass away.

No words can be ' *more explicit* ' than these of OUR BLESSED LORD : viz. ' THIS GOSPEL *of the Kingdom shall be preached in* ALL *the* EARTH, *for a Witness to* ALL NATIONS ; AND THEN, *shall the* END COME.' The ' *next* 18 *years* ' must therefore supply the interval of the ' *special Episcopal forerunners*.'

(Matt. xxiv. 14.)

See the ' JEWISH INTELLIGENCER ' of the present month (*April*)

p. 153, for the 'Debates in Parliament,' respecting the BISHOP OF JERUSALEM, viz. Dr. Bowring, Mr. Hume, Sir R. Inglis, Sir R. Peel, Viscount Palmerston."

I have quoted this at length, to show the awful threats which were published at a time of some little excitement about the phenomenon, under the name of the London Society. The assumption of a corporate appearance is a very unfair trick : and there are junctures at which harm might be done by it.

> Wealth the name and number of the Beast, 666, in the Book of Revelation. [By John Taylor.] London, 1844, 8vo.

Whether Junius or the Beast be the more difficult to identify, must be referred to Mr. Taylor, the only person who has attempted both. His cogent argument on the political secret is not unworthily matched in his treatment of the theological riddle. He sees the solution in $\varepsilon\dot{\upsilon}\pi o \rho i a$, which occurs in the Acts of the Apostles as the word for wealth in one of its most disgusting forms, and makes 666 in the most straightforward way. This explanation has as good a chance as any other. The work contains a general attempt at explanation of the Apocalypse, and some history of opinion on the subject. It has not the prolixity which is so common a fault of apocalyptic commentators.

> A practical Treatise on Eclipses . . . with remarks on the anomalies of the present Theory of the Tides. By T. Kerigan, F.R.S. 1844, 8vo.

Containing also a refutation of the theory of the tides, and afterwards increased by a supplement, 'Additional facts and arguments against the theory of the tides,' in answer to a short notice in the Athenæum journal. Mr. Kerigan was a lieutenant in the Navy: he obtained admission to the Royal Society just before the publication of his book.

> A new theory of Gravitation. By Joseph Denison, Esq. London, 1844, 12mo.
> Commentaries on the Principia. By the author of ' A new theory of Gravitation.' London, 1846, 8vo.

Honour to the speculator who can be put in his proper place by one sentence, be that place where it may.

' But we have shown that the velocities are inversely as the square roots of the mean distances from the sun ; wherefore, by equality of ratios, the forces of the sun's gravitation upon them are also inversely as the square roots of their distances from the sun.'

In the years 1818 and 1845 the full moon fell on Easter Day, having been particularly directed to fall before it in the act for the change of style, and in the English missals and prayer-books of all time : perhaps it would be more correct to say that Easter Day was directed to fall after the full moon ; ' but the principle is the same.' No explanation was given in 1818, but Easter was kept by the tables, in defiance of the rule, and of several protests. A chronological panic was beginning in December 1844, which was stopped by the *Times* newspaper printing extracts from an article of mine in the *Companion to the Almanac* for 1845, which had then just appeared. No one had guessed the true reason, which is that the thing called the moon in the Gregorian Calendar is not the moon of the heavens, but a fictitious imitation put wrong on purpose, as will presently appear, partly to keep Easter out of the way of the Jews' Passover, partly for convenience of calculation. The apparent error happens but rarely ; and all the work will perhaps have to be gone over next time. I now give two bits of paradox.

Some theologians were angry at this explanation. A review called the *Christian Observer* (of which Christianity I do not know) got up a crushing article against me. I did not look at it, feeling sure that an article on such a subject which appeared on January 1, 1845, against a publication made in December 1844, must be a second-hand job. But some years afterwards (Sept. 10, 1850), the reviews, &c. having been just placed at the disposal of readers in the *old* reading-room of the Museum, I made a tour of inspection, came upon my critic on his perch, and took a look at him. I was very glad to remember this, for, though expecting only second-hand, yet even of this there is good and bad ; and I expected to find some hints in the good second-hand of a respectable clerical publication. I read on, therefore, attentively, but not long : I soon came to the information that some additions to Delambre's statement of the rule for finding Easter, belonging to distant years, had been made by Sir Harris Nicolas ! Now as I myself furnished my friend Sir H. N. with Delambre's digest of Clavius's rule, which I translated out of algebra into common language for the purpose, I was pretty sure this was the ignorant reading of a person to whom Sir H. N. was the highest *arithmetical* authority on the subject. A person pretending to chronology, without being able to distinguish the historical points—so clearly as they stand out—in which Sir H. N. speaks with authority, from the arithmetical points of pure reckoning on which he does not pretend to do more than directly repeat others, must be as fit to talk about the construction of Easter Tables as the Spanish are to talk French. I need hardly say that the additions for distant years are as much from Clavius as the rest: my reviewer was not deep enough in his subject to know that Clavius made and published, from his rules, the full table up to A.D. 5000, for all the moveable feasts of every year! I gave only a glance at the rest: I found I was either knave or fool, with a leaning to the second opinion; and I came away satisfied that my critic was either ignoramus or novice, with a leaning to the first. I afterwards found an ambiguity of expression in Sir H. N.'s account—whether his or mine I could not tell—which might mislead a novice or content an ignoramus, but would have been properly read or further inquired into by a competent person.

The second case is this. Shortly after the publication of my article, a gentleman called at my house, and, finding I was not at home, sent up his card—with a stylish west-end club on it—to my wife, begging for a few words on pressing business. With many well-expressed apologies, he stated that he had been alarmed by hearing that Prof. De M. had an intention of altering Easter next year. Mrs. De M. kept her countenance, and assured him that I had no such intention, and further, that she greatly doubted my having the power to do it. Was she quite sure? his authority was very good: fresh assurances given. He was greatly relieved, for he had some horses training for after Easter, which would not be ready to run if it were altered the wrong way. A doubt comes over him: would Mrs. De M., in the event of her being mistaken, give him the very earliest information? Promise given; profusion of thanks; more apologies; and departure.

Now, candid reader!—or uncandid either!—which most deserves to be laughed at? A public instructor, who undertakes to settle for the world whether a reader of Clavius, the constructor of the Gregorian Calendar, is fool or knave, upon information derived from a compiler—in this matter—of his own day; or a gentleman of horse and dog associations who, misapprehending something which he heard about a current topic, infers that the reader of Clavius had the ear of the Government on a proposed alteration. I suppose the querist had heard some one say, perhaps, that the day ought to be set right, and some one else remark that I might be consulted, as the only person who had discussed the matter from the original source of the Calendar.

To give a better chance of the explanation being at once produced, next time the real full moon and Easter Day shall fall together, I insert here a summary which was printed in the Irish Prayer-book of the Ecclesiastical Society. If the amusement given by paradoxers should prevent a useless discussion some years hence, I and the paradoxers shall have done a little good between us—at any rate, I have done my best to keep the heavy weight afloat by tying bladders to it. I think the next occurrence will be in 1875.

## EASTER DAY.

In the years 1818 and 1845, Easter Day, as given by the *rules in* 24 Geo. II. cap. 23. (known as the act for the *change of style*) contradicted the *precept* given in the preliminary explanations. The precept is as follows :—

' *Easter Day*, on which the rest ' of the moveable feasts ' depend, is always the First Sunday after the Full Moon, which happens upon or next after the Twenty-first Day of *March* ; and if the Full Moon happens upon a Sunday, *Easter Day* is the Sunday after.'

But in 1818 and 1845, the full moon fell on a Sunday, and yet the rules gave *that same Sunday* for Easter Day. Much discussion was produced by this circumstance in 1818 : but a repetition of it in 1845 was nearly altogether prevented by a timely[1] reference to the intention of those who conducted the Gregorian reformation of the Calendar. Nevertheless, seeing that the apparent error of the Calendar is due to the precept in the Act of Parliament, which is both erroneous and insufficient, and that the difficulty will recur so often as Easter Day falls on the day of full moon, it may be advisable to select from the two articles cited in the note such of their conclusions and rules, without proof or controversy, as will enable the reader to understand the main points of the Easter question, and, should he desire it, to calculate for himself the Easter of the old or new style, for any given year.

1. In the very earliest age of Christianity, a controversy arose as to the mode of keeping Easter, some desiring to perpetuate the *Passover*, others to keep the *festival of the Resurrection*. The first afterwards obtained the name of *Quartadecimans*, from their Easter being always kept on the *fourteenth day* of the moon (Exod. xii. 18, Levit. xxiii. 5.). But though it is unquestionable that a Judaizing party existed, it is also likely that many dissented on chronological grounds. It is clear that no *perfect* anniversary can take place, except when the fourteenth

---

[1] In the *Companion to the Almanac* for 1845 is a paper by Professor De Morgan, ' On the Ecclesiastical Calendar,' the statements of which, so far as concerns the Gregorian Calendar, are taken direct from the work of Clavius, the principal agent in the arrangement of the reformed reckoning. This was followed, in the *Companion to the Almanac* for 1846, by a second paper, by the same author, headed 'On the Earliest Printed Almanacs,' much of which is written in direct supplement to the former article.

of the moon, and with it the passover, falls on a Friday. Suppose, for instance, it falls on a Tuesday: one of three things must be done. Either (which seems never to have been proposed) the crucifixion and resurrection must be celebrated on Tuesday and Sunday, with a wrong interval; or the former on Tuesday, the latter on Thursday, abandoning the first day of the week; or the former on Friday, and the latter on Sunday, abandoning the paschal commemoration of the crucifixion.

The last mode has been, as every one knows, finally adopted. The disputes of the first three centuries did not turn on any *calendar* questions. The Easter question was merely the symbol of the struggle between what we may call the Jewish and Gentile sects of Christians : and it nearly divided the Christian world, the Easterns, for the most part, being *Quartadecimans*. It is very important to note that there is no recorded dispute about a method of predicting the new moon, that is, no general dispute leading to formation of sects : there may have been difficulties, and discussions about them. The Metonic cycle, presently mentioned, must have been used by many, perhaps most, churches.

2.   The question came before the Nicene Council (A.D. 325) not as an astronomical, but as a doctrinal, question : it was, in fact, this, Shall the *passover* [1] be treated as a part of Christianity ? The Council resolved this question in the negative, and the only information on its premises and conclusion, or either, which comes from itself, is contained in the following sentence of the synodical epistle, which epistle is preserved by Socrates and Theodoret. ' We also send you the good news concerning the unanimous consent of all in reference to the celebration of the most solemn feast of Easter, for this difference also has been made up by the assistance of your prayers : so that all the brethren in the East, who formerly celebrated this festival *at the same time as the Jews*, will in future conform *to the Romans and to us*, and to all who have of old observed *our manner* of celebrating Easter.' This is all that can be found on the subject : none of the stories about the Council ordaining the astronomical mode of finding Easter, and introducing the Metonic cycle into ecclesiastical reckoning, have any contemporary evidence : the canons which purport to be those of the Nicene Council do not contain a word about Easter ; and this is evidence, whether we suppose those canons to be genuine or spurious.

3.   The astronomical dispute about a lunar cycle for the prediction of Easter either commenced, or became prominent, by the extinction of greater ones, soon after the time of the Nicene Council. Pope Innocent I. met with difficulty in 414. S. Leo, in 454, ordained that Easter of 455 should be April 24; which is right. It is useless to

---

[1] It may be necessary to remind some English readers that in Latin and its derived European languages, what we call Easter is called the passover (*pascha*). The Quartadecimans had the *name* on their side : a possession which often is, in this world, nine points of the law.

record details of these disputes in a summary : the result was, that in the year 463, Pope Hilarius employed Victorinus of Aquitaine to correct the Calendar, and Victorinus formed a rule which lasted until the sixteenth century. He combined the Metonic cycle and the solar cycle, presently described. But this cycle bears the name of Dionysius Exiguus, a Scythian settled at Rome, about A.D. 530, who adapted it to his new yearly reckoning, when he abandoned the æra of Diocletian as a commencement, and constructed that which is now in common use.

4. With Dionysius, if not before, terminated all difference as to the mode of keeping Easter which is of historical note : the increasing defects of the Easter Cycle produced in time the remonstrance of persons versed in astronomy, among whom may be mentioned Roger Bacon, Sacrobosco, Cardinal Cusa, Regiomontanus, &c. From the middle of the sixth to that of the sixteenth century, one rule was observed.

5. The mode of applying astronomy to chronology has always involved these two principles. First, the actual position of the heavenly body is not the object of consideration, but what astronomers call its *mean place*, which may be described thus. Let a fictitious sun or moon move in the heavens, in such manner as to revolve among the fixed stars at an average rate, avoiding the alternate accelerations and retardations which take place in every planetary motion. Thus the fictitious (say *mean*) sun and moon are always very near to the real sun and moon. The ordinary clocks show time by the mean, not the real, sun : and it was always laid down that Easter depends on the opposition (or full moon) of the mean sun and moon, not of the real ones. Thus we see that, were the Calendar ever so correct as to the *mean* moon, it would be occasionally false as to the *true* one : if, for instance, the opposition of the mean sun and moon took place at one second before midnight, and that of the real bodies only two seconds afterwards, the calendar day of full moon would be one day before that of the common almanacs. Here is a way in which the discussions of 1818 and 1845 might have arisen : the British legislature has defined *the moon* as the regulator of the paschal calendar. But this was only a part of the mistake.

6. Secondly, in the absence of perfectly accurate knowledge of the solar and lunar motion (and for convenience, even if such knowledge existed), cycles are, and always have been, taken, which serve to represent those motions nearly. The famous Metonic cycle, which is introduced into ecclesiastical chronology under the name of the cycle of the golden numbers, is a period of 19 Julian[1] years. This period, in the old Calendar, was taken to contain exactly 235 *lunations*, or intervals between new moons, of the mean moon. Now the state of the case is this :—

[1] The Julian year is a year of the Julian Calendar, in which there is leap year *every* fourth year. Its average length is therefore 365 days and a quarter.

19 average Julian years make 6939 days 18 hours.

235 average lunations make 6939 days 16 hours 31 minutes.

So that successive cycles of golden numbers, supposing the first to start right, amount to making the new moons fall too late, gradually, so that the mean moon *of this cycle* gains 1 hour 29 minutes in 19 years upon the mean moon of the heavens, or about a day in 300 years. When the Calendar was reformed, the calendar new moons were four days in advance of the mean moon of the heavens: so that, for instance, calendar full moon on the 18th usually meant real full moon on the 14th.

7. If the difference above had not existed, the moon of the heavens (the mean moon at least), would have returned permanently to the same days of the month in 19 years; with an occasional slip arising from the unequal distribution of the leap years, of which a period contains sometimes five and sometimes four. As a general rule, the days of new and full moon in any one year would have been also the days of new and full moon of a year having 19 more units in its date. Again, if there had been no leap years, the days of the month would have returned to the same days of the week every seven years. The introduction of occasional 29ths of February disturbs this, and makes the permanent return of month days to week days occur only after 28 years. If all had been true, the lapse of 28 times 19, or 532 years, would have restored the year in every point: that is, A.D. 1, for instance, and A.D. 533, would have had the same almanac in every matter relating to week days, month days, sun, and moon (mean sun and moon at least). And on the supposition of its truth, the old system of Dionysius was framed. Its errors are, first, that the moments of mean new moon advance too much by 1h. 29m. in 19 average Julian years; secondly, that the average Julian year of $365\frac{1}{4}$ days is too long by 11m. 10s.

8. The Council of Trent, moved by the representations made on the state of the Calendar, referred the consideration of it to the Pope. In 1577, Gregory XIII. submitted to the Roman Catholic Princes and Universities a plan presented to him by the representatives of Aloysius Lilius, then deceased. This plan being approved of, the Pope nomi nated a commission to consider its details, the working member of which was the Jesuit Clavius. A short work was prepared by Clavius, descriptive of the new Calendar: this was published[1] in 1582, with the Pope's bull (dated February 24, 1581) prefixed. A larger work was prepared by Clavius, containing fuller explanation, and entitled 'Romani Calendarii a Gregorio XIII. Pontifice Maximo restituti Explicatio.' This was published at Rome in 1603, and again in the collection of the works of Clavius in 1612.

---

[1] The title of this work, which is the authority on all points of the new Calendar, is 'Kalendarium Gregorianum Perpetuum. Cum Privilegio Summi Pontificis Et Aliorum Principum. Romæ, Ex Officina Dominici Basæ. MDLXXXII. Cum Licentia Superiorum' (quarto, pp. 60).

9. The following extracts from Clavius settle the question of the meaning of the term *moon*, as used in the Calendar :—

' Who, except a few who think they are very sharp-sighted in this matter, is so blind as not to see that the 14th of the moon and the full moon are not the same things in the Church of God ? . . . Although the Church, in finding the new moon, and from it the 14th day, *uses neither the true nor the mean motion of the moon*, but measures only according to the order of a cycle, it is nevertheless undeniable that the mean full moons found from astronomical tables are of the greatest use in determining the cycle which is to be preferred . . . the new moons of which cycle, in order to the due celebration of Easter, should be so arranged that the 14th days of those moons, reckoning from the day of new moon *inclusive*, should not fall two or more days before the mean full moon, but only one day, or else on the very day itself, or not long after. And even thus far the Church need not take very great pains . . . for it is sufficient that all should reckon by the 14th day of the moon in the cycle, even though sometimes it *should be more than one day before or after* the mean full moon . . . We have taken pains that in our cycle the new moons should *follow* the real new moons, so that the 14th of the moon should fall either the day before the mean full moon, or on that day, or not long after ; and this was done on purpose, for if the new moon of the cycle fell on the same day as the mean new moon of the astronomers, it might chance that we should celebrate Easter on the same day as the Jews or the Quartadeciman heretics, which would be absurd, or else before them, which would be still more absurd.'

From this it appears that Clavius continued the Calendar of his predecessors in the choice of the *fourteenth* day of the moon. Our legislature lays down the day of the *full moon* : and this mistake appears to be rather English than Protestant ; for it occurs in missals published in the reign of Queen Mary. The calendar lunation being $29\frac{1}{2}$ days, the middle day is the *fifteenth* day, and this is and was reckoned as the day of the full moon. There is every right to presume that the original passover was a feast of the *real full moon* : but it is most probable that the moons were then reckoned, not from the astronomical conjunction with the sun, which nobody sees except at an eclipse, but from the day of *first visibility* of the new moon. In fine climates this would be the day or two days after conjunction ; and the fourteenth day from that of first visibility inclusive, would very often be the day of full moon. The following is then the proper correction of the precept in the Act of Parliament :—

Easter Day, on which the rest depend, is always the First Sunday after the *fourteenth day* of the *calendar* moon which happens upon or next after the Twenty-first day of March, *according to the rules laid down for the construction of the Calendar* ; and if the *fourteenth day* happens upon a Sunday, Easter Day is the Sunday after.

10. Further, it appears that Clavius valued the celebration of the

festival after the Jews, &c., more than astronomical correctness. He gives comparison tables which would startle a believer in the astronomical intention of his Calendar : they are to show that a calendar in which the moon is always made a day older than by him, *represents the heavens better than he has done, or meant to do.* But it must be observed that this diminution of the real moon's age has a tendency to make the English explanation often practically accordant with the Calendar. For the fourteenth day of Clavius *is* generally the fifteenth day of the mean moon of the heavens, and therefore most often that of the real moon. But for this, 1818 and 1845 would not have been the only instances of our day in which the English precept would have contradicted the Calendar.

11. In the construction of the Calendar, Clavius adopted the ancient cycle of 532 years, but, we may say, without ever allowing it to run out. At certain periods, a shift is made from one part of the cycle into another. This is done whenever what should be Julian leap year is made a common year, as in 1700, 1800, 1900, 2100, &c. It is also done at certain times to correct the error of 1h. 19m., before referred to, in each cycle of golden numbers : Clavius, to meet his view of the amount of that error, put forward the moon's age a day 8 times in 2,500 years. As we cannot enter at full length into the explanation, we must content ourselves with giving a set of rules, independent of tables, by which the reader may find Easter for himself in any year, either by the old Calendar or the new. Any one who has much occasion to find Easters and moveable feasts should procure Francœur's [1] tables.

12. *Rule for determining Easter Day of the Gregorian Calendar in any year of the new style.* To the several parts of the rule are annexed, by way of example, the results for the year 1849.

I. Add 1 to the given year. (1850).

II. Take the quotient of the given year divided by 4, neglecting the remainder. (462).

III. Take 16 from the centurial figures of the given year, if it can be done, and take the remainder. (2).

IV. Take the quotient of III. divided by 4, neglecting the remainder. (0).

V. From the sum of I., II., and IV., substract III. (2310).

VI. Find the remainder of V. divided by 7. (0).

VII. Subtract VI. from 7; this is the number of the dominical letter $\begin{smallmatrix} A\,B\,C\,D\,E\,F\,G \\ 1\,2\,3\,4\,5\,6\,7 \end{smallmatrix}$. (7; dominical letter G).

VIII. Divide I. by 19, the remainder (or 19, if no remainder) is the *golden number.* (7).

---

[1] 'Manuels-Roret. Théorie du Calendrier et collection de tous les Calendriers des Années passées et futures. . . .Par L. B. Francœur, . . .Paris, à la librairie encyclopédique de Roret, rue Hautefeuille, 10 bis. 1842.' (12mo.) In this valuable manual, the 35 possible almanacs are given at length, with such preliminary tables as will enable any one to find, by mere inspection, which almanac he is to choose for any year, whether of old or new style. [1866. I may now refer to my own 'Book of Almanacs,' for the same purpose].

IX. From the centurial figures of the year subtract 17, divide by 25, and keep the quotient. (0).

X. Subtract IX. and 15 from the centurial figures, divide by 3, and keep the quotient. (1).

XI. To VIII. add ten times the next less number, divide by 30, and keep the remainder. (7).

XII. To XI. add X. and IV., and take away III., throwing out thirties, if any. If this give 24, change it into 25. If 25, change it into 26, whenever the golden number is greater than 11. If 0, change it into 30. Thus we have the epact, or age of the *Calendar* moon at the beginning of the year. (6).

| | |
|---|---|
| *When the Epact is 23, or less.* | *When the Epact is greater than 23.* |
| XIII. Subtract XII., the epact, from 45. (39). | XIII. Subtract XII., the epact, from 75. |
| XIV. Subtract the epact from 27, divide by 7, and keep the remainder, or 7, if there be no remainder. (7). | XIV. Subtract the epact from 57, divide by 7, and keep the remainder, or 7, if there be no remainder. |

XV. To XIII. add VII., the dominical number, (and 7 besides, if XIV. be greater than VII.,) and subtract XIV., the result is the day of March, or if more than 31, subtract 31, and the result is the day of April, on which Easter Sunday falls. (39; Easter Day is April 8).

In the following examples, the several results leading to the final conclusion are tabulated.

| Given year | 1592 | 1637 | 1723 | 1853 | 2018 | 4686 |
|---|---|---|---|---|---|---|
| I. | 1593 | 1638 | 1724 | 1854 | 2019 | 4687 |
| II. | 398 | 409 | 430 | 463 | 504 | 1171 |
| III. | — | 0 | 1 | 2 | 4 | 30 |
| IV. | — | 0 | 0 | 0 | 1 | 7 |
| V. | 1991 | 2047 | 2153 | 2315 | 2520 | 5835 |
| VI. | 3 | 3 | 4 | 5 | 0 | 4 |
| VII. | 4 | 4 | 3 | 2 | 7 | 3 |
| VIII. | 16 | 4 | 14 | 11 | 5 | 13 |
| IX. | — | — | 0 | 0 | 0 | 1 |
| X. | 0 | 0 | 0 | 1 | 1 | 10 |
| XI. | 16 | 4 | 24 | 21 | 15 | 13 |
| XII. | 16 | 4 | 23 | 20 | 13 | 0 say 30 |
| XIII. | 29 | 41 | 22 | 25 | 32 | 45 |
| XIV. | 4 | 2 | 4 | 7 | 7 | 6 |
| XV. | 29 | 43 | 28 | 27 | 32 | 49 |
| Easter Day. | Mar. 29 | Apr. 12 | Mar. 28 | Mar. 27 | Apr. 1 | Apr. 18 |

13. *Rule for determining Easter Day of the Antegregorian Calendar in any year of the old style.* To the several parts of the rule are annexed, by way of example, the results for the year 1287. The steps are numbered to correspond with the steps of the Gregorian rule, so that it can be seen what augmentations the latter requires.

I. Set down the given year. (1287).

II. Take the quotient of the given year divided by 4, neglecting the remainder (321).

V. Take 4 more than the sum of I. and II. (1612).

VI. Find the remainder of V. divided by 7. (2).

VII. Subtract VI. from 7; this is the number of the dominical letter $\frac{A\,B\,C\,D\,E\,F\,G}{1\,2\,3\,4\,5\,6\,7}$.

(5; dominical letter E).

Q

VIII. Divide one more than the given year by 19, the remainder (or 19 if no remainder) is the golden number. (15).

XII. Divide 3 less than 11 times VIII. by 30 ; the remainder (or 30 if there be no remainder) is the epact. (12).

| When the Epact is 23, or less. | When the Epact is greater than 23. |
|---|---|
| XIII. Subtract XII., the epact, from 45. (33). | XIII. Subtract XII., the epact, from 75. |
| XIV. Subtract the epact, from 27, divide by 7, and keep the remainder, or 7, if there be no remainder. (1). | XIV. Subtract the epact from 57, divide by 7, and keep the remainder, or 7, if there be no remainder. |

XV. To XIII. add VII., the dominical number, (and 7 besides if XIV. be greater than VII.,) and subtract XIV., the result is the day of March, or if more than 31, subtract 31, and the result is the day of April, on which Easter Sunday (old style) falls. (37 ; Easter Day is April 6).

These rules completely represent the old and new Calendars, so far as Easter is concerned. For further explanation we must refer to the articles cited at the commencement.

The annexed is the table of new and full moons of the Gregorian Calendar, cleared of the errors made for the purpose of preventing Easter from coinciding with the Jewish Passover.

The second table (page 228) contains *epacts*, or ages of the moon at the beginning of the year : thus in 1913, the epact is 22 : in 1868 it is 6. This table goes from 1850 to 1999 : should the New Zealander not have arrived by that time, and should the churches of England and Rome then survive, the epact table may be continued from their liturgy-books. The way of using the table is as follows : Take the epact of the required year, and find it in the first or last column of the first table, in line with it are seen the calendar days of new and full moon. Thus, when the epact is 17, the new and full moons of March fall on the 13th and 28th. The result is, for the most part, correct : but in a minority of cases there is an error of a day. When this happens, the error is almost always a fraction of a day much less than twelve hours. Thus, when the table gives full moon on the 27th, and the real truth is the 28th, we may be sure it is early on the 28th. For example, the year 1867. The epact is 25, and we find in the table :

| | J. | F. | M. | Ap. | M. | Ju. | Jl. | Au. | S. | O. | N. | D. |
|---|---|---|---|---|---|---|---|---|---|---|---|---|
| New . . . | 5+ | 4 | 5+ | 4 | 3+ | 2 | 1,31 | 29 | 28— | 27 | 26 | 25 |
| Full . . . | 20 | 19—20 | | 19—18 | | 17 | 16 | 15 | 13— | 13 | 11+ | 11 |

When the truth is the day after + is written after the date; when the day before, —. Thus, the new moon of March is on the 6th ; the full moon of April is on the 18th.

# Table of New and Full Moon.

| | Jan. | Feb. | Mar. | Apr. | May | June | July | Aug. | Sep. | Oct. | Nov. | Dec. | |
|---|---|---|---|---|---|---|---|---|---|---|---|---|---|
| 1 | 29<br>14 | 27<br>13 | 29<br>14 | 27<br>13 | 27<br>12 | 25<br>11 | 25<br>10 | 23<br>9 | 22<br>7 | 21<br>7 | 20<br>5 | 19<br>5 | 1 |
| 2 | 28<br>13 | 26<br>12 | 28<br>13 | 26<br>12 | 26<br>11 | 24<br>10 | 24<br>9 | 22<br>8 | 21<br>6 | 20<br>6 | 19<br>4 | 18<br>4 | 2 |
| 3 | 27<br>12 | 25<br>11 | 27<br>12 | 25<br>11 | 25<br>10 | 23<br>9 | 23<br>8 | 21<br>7 | 20<br>5 | 19<br>5 | 18<br>3 | 17<br>3 | 3 |
| 4 | 26<br>11 | 24<br>10 | 26<br>11 | 24<br>10 | 24<br>9 | 22<br>8 | 22<br>7 | 20<br>6 | 19<br>4 | 18<br>4 | 17<br>2 | 16<br>2,31 | 4 |
| 5 | 25<br>10 | 23<br>9 | 25<br>10 | 23<br>9 | 23<br>8 | 21<br>7 | 21<br>6 | 19<br>5 | 18<br>3 | 17<br>3 | 16<br>1 | 15<br>1,30 | 5 |
| 6 | 24<br>9 | 22<br>8 | 24<br>9 | 22<br>8 | 22<br>7 | 20<br>6 | 20<br>5 | 18<br>4 | 17<br>2 | 16<br>2,31 | 15<br>30 | 14<br>29 | 6 |
| 7 | 23<br>8 | 21<br>7 | 23<br>8 | 21<br>7 | 21<br>6 | 19<br>5 | 19<br>4 | 17<br>3. | 16<br>1 | 15<br>1,30 | 14<br>29 | 13<br>28 | 7 |
| 8 | 22<br>7 | 20<br>6 | 22<br>7 | 20<br>6 | 20<br>5 | 18<br>4 | 18<br>3 | 16<br>2,31 | 15<br>30 | 14<br>29 | 13<br>28 | 12<br>27 | 8 |
| 9 | 21<br>6 | 19<br>5 | 21<br>6 | 19<br>5 | 19<br>4 | 17<br>3 | 17<br>2 | 15<br>1,30 | 14<br>29 | 13<br>28 | 12<br>27 | 11<br>26 | 9 |
| 10 | 20<br>5 | 18<br>4 | 20<br>5 | 18<br>4 | 18<br>3 | 16<br>2 | 16<br>1,31 | 1!<br>29 | 13<br>28 | 12<br>27 | 11<br>26 | 10<br>25 | 10 |
| 11 | 19<br>4 | 17<br>3 | 19<br>4 | 17<br>3 | 17<br>2 | 15<br>1,30 | 15<br>30 | 13<br>28 | 12<br>27 | 11<br>26 | 10<br>25 | 9<br>24 | 11 |
| 12 | 18<br>3 | 16<br>2 | 18<br>3 | 16<br>2 | 16<br>1,31 | 14<br>29 | 14<br>29 | 12<br>27 | 11<br>26 | 10<br>25 | 9<br>24 | 8<br>23 | 12 |
| 13 | 17<br>2 | 15<br>1 | 17<br>2 | 15<br>1,30 | 15<br>30 | 13<br>28 | 13<br>28 | 11<br>26 | 10<br>25 | 9<br>24 | 8<br>23 | 7<br>22 | 13 |
| 14 | 16<br>1,31 | 14<br>— | 16<br>1,31 | 14<br>29 | 14<br>29 | 12<br>27 | 12<br>27 | 10<br>25 | 9<br>24 | 8<br>23 | 7<br>22 | 6<br>21 | 14 |
| 15 | 15<br>30 | 13<br>28 | 15<br>30 | 13<br>28 | 13<br>28 | 11<br>26 | 11<br>26 | 9<br>24 | 8<br>23 | 7<br>22 | 6<br>21 | 5<br>20 | 15 |
| 16 | 14<br>29 | 12<br>27 | 14<br>29 | 12<br>27 | 12<br>27 | 10<br>25 | 10<br>25 | 8<br>23 | 7<br>22 | 6<br>21 | 5<br>20 | 4<br>19 | 16 |
| 17 | 13<br>28 | 11<br>26 | 13<br>28 | 11<br>26 | 11<br>26 | 9<br>24 | 9<br>24 | 7<br>22 | 6<br>21 | 5<br>20 | 4<br>19 | 3<br>18 | 17 |
| 18 | 12<br>27 | 10<br>25 | 12<br>27 | 10<br>25 | 10<br>25 | 8<br>23 | 8<br>23 | 6<br>21 | 5<br>20 | 4<br>19 | 3<br>18 | 2<br>17 | 18 |
| 19 | 11<br>26 | 9<br>24 | 11<br>26 | 9<br>24 | 9<br>24 | 7<br>22 | 7<br>22 | 5<br>20 | 4<br>19 | 3<br>18 | 2<br>17 | 1,31<br>16 | 19 |
| 20 | 10<br>25 | 8<br>23 | 10<br>25 | 8<br>23 | 8<br>23 | 6<br>21 | 6<br>21 | 4<br>19 | 3<br>18 | 2<br>17 | 1,31<br>16 | 30<br>15 | 20 |
| 21 | 9<br>24 | 7<br>22 | 9<br>24 | 7<br>22 | 7<br>22 | 5<br>20 | 5<br>20 | 3<br>18 | 2<br>17 | 1,31<br>16 | 29<br>15 | 29<br>14 | 21 |
| 22 | 8<br>23 | 6<br>21 | 8<br>23 | 6<br>21 | 6<br>21 | 4<br>19 | 4<br>19 | 2<br>17 | 1,30<br>16 | 30<br>15 | 28<br>14 | 28<br>13 | 22 |
| 23 | 7<br>22 | 5<br>20 | 7<br>22 | 5<br>20 | 5<br>20 | 3<br>18 | 3<br>18 | 1,31<br>16 | 29<br>15 | 29<br>14 | 27<br>13 | 27<br>12 | 23 |
| 24 | 6<br>21 | 5<br>19 | 6<br>21 | 5<br>19 | 4<br>19 | 2<br>17 | 2<br>17 | 1,30<br>15 | 29<br>14 | 28<br>13 | 27<br>12 | 26<br>11 | 24 |
| 25 | 5<br>20 | 4<br>19 | 5<br>20 | 4<br>19 | 3<br>18 | 2<br>17 | 1,31<br>16 | 29<br>15 | 28<br>13 | 27<br>13 | 26<br>11 | 25<br>11 | 25 |
| 26 | 4<br>19 | 3<br>18 | 4<br>19 | 3<br>18 | 2<br>17 | 1,30<br>16 | 30<br>15 | 28<br>14 | 27<br>12 | 26<br>12 | 25<br>10 | 24<br>10 | 26 |
| 27 | 3<br>18 | 2<br>17 | 3<br>18 | 2<br>17 | 1,31<br>16 | 29<br>15 | 29<br>14 | 27<br>13 | 26<br>11 | 25<br>11 | 24<br>9 | 23<br>9 | 27 |
| 28 | 2<br>17 | 1<br>16 | 2<br>17 | 1,30<br>16 | 30<br>15 | 28<br>14 | 28<br>13 | 26<br>12 | 25<br>10 | 24<br>10 | 23<br>8 | 22<br>8 | 28 |
| 29 | 1,31<br>16 | —<br>15 | 1,31<br>16 | 29<br>15 | 29<br>14 | 27<br>13 | 27<br>12 | 25<br>11 | 24<br>9 | 23<br>9 | 22<br>7 | 21<br>7 | 29 |
| 30 | 30<br>15 | 28<br>14 | 30<br>15 | 28<br>14 | 28<br>13 | 26<br>12 | 26<br>11 | 24<br>10 | 23<br>8 | 22<br>8 | 21<br>6 | 20<br>6 | 30 |
| | Jan. | Feb. | Mar. | Apr. | May | June | July | Aug. | Sep. | Oct. | Nov. | Dec. | |

|     | 0 | 1 | 2 | 3 | 4 | 5 | 6 | 7 | 8 | 9 |
|-----|---|---|---|---|---|---|---|---|---|---|
| 185 | 17 | 28 | 19 | 20 | 2 | 12 | 23 | 4 | 15 | 26 |
| 186 | 7 | 18 | 30 | 11 | 22 | 3 | 14 | 25 | 6 | 17 |
| 187 | 28 | 9 | 20 | 1 | 12 | 23 | 4 | 15 | 26 | 7 |
| 188 | 18 | 30 | 11 | 22 | 3 | 14 | 25 | 6 | 17 | 28 |
| 189 | 9 | 21 | 1 | 12 | 23 | 4 | 15 | 26 | 7 | 18 |
| 190 | 29 | 10 | 21 | 2 | 13 | 24 | 5 | 16 | 27 | 8 |
| 191 | 19 | 30 | 11 | 22 | 3 | 14 | 26 | 6 | 17 | 29 |
| 192 | 10 | 21 | 2 | 13 | 24 | 5 | 16 | 27 | 8 | 19 |
| 193 | 30 | 11 | 22 | 3 | 14 | 26 | 6 | 17 | 29 | 10 |
| 194 | 21 | 2 | 13 | 24 | 5 | 16 | 27 | 8 | 19 | 30 |
| 195 | 11 | 22 | 3 | 14 | 26 | 6 | 17 | 29 | 10 | 21 |
| 196 | 2 | 13 | 24 | 5 | 16 | 27 | 8 | 19 | 30 | 11 |
| 197 | 22 | 3 | 14 | 26 | 6 | 17 | 29 | 10 | 21 | 2 |
| 198 | 13 | 24 | 5 | 16 | 27 | 8 | 19 | 30 | 11 | 22 |
| 199 | 3 | 14 | 26 | 6 | 17 | 29 | 10 | 21 | 2 | 13 |

I now introduce a small paradox of my own : and as I am not able to prove it, I am compelled to declare that any one who shall dissent must be either very foolish or very dishonest, and will make me quite uncomfortable about the state of his soul. This being settled once for all, I proceed to say that the necessity of arriving at the truth about the assertions that the Nicene Council laid down astronomical tests led me to look at Fathers, Church histories, &c. to an extent which I never dreamed of before. One conclusion which I arrived at was, that the Nicene Fathers had a knack of sticking to the question which many later councils could not acquire. In our own day, it is not permitted to Convocation seriously to discuss any one of the points which are bearing so hard upon their resources of defence—the cursing clauses of the Athanasian Creed, for example. And it may be collected that the prohibition arises partly from fear that there is no saying where a beginning, if allowed, would end. There seems to be a suspicion that debate, once let loose, would play up old Trent with the liturgy, and bring the whole book to book. But if any one will examine the real Nicene Creed, without the augmentation, he will admire the way in which the framers stuck to the point, and settled what they had to decide, according to their view of it. With such a presumption of good sense in their favour, it becomes easier to believe in any claim which may be made on their behalf to tact or sagacity in settling any other matter. And I strongly suspect such a claim may be made for them on the Easter question.

I collect from many little indications, both before and after the Council, that the division of the Christian world into Judaical and Gentile, though not giving rise to a sectarian distinction expressed by names, was of far greater force and meaning than historians prominently admit. I took *note* of many indications of this, but not *notes*, as it was not to my purpose. If it were so, we must admire the discretion of the Council. The Easter question was the fighting ground of the struggle: the Eastern or Judaical Christians, with some varieties of usage and meaning, would have the Passover itself to be the great feast, but taken in a Christian sense ; the Western or Gentile Christians, would have the commemoration of the Resurrection, connected with the Passover only by chronology. To shift the Passover in time, under its name, *Pascha*, without allusion to any of the force of the change, was gently cutting away the ground from under the feet of the Conservatives. And it was done in a very quiet way : no allusion to the precise character of the change; no hint that the question was about two different festivals: ' all the brethren in the East, who formerly celebrated this festival at the same time as the Jews, will in future conform to the Romans and to us.' The Judaizers meant to be keeping the Passover *as* a Christian feast : they are gently assumed to be keeping, *not* the Passover, *but* a Christian feast; and a doctrinal decision is quietly, but efficiently, announced under the form of a chronological ordinance. Had the Council issued theses of doctrine, and excommunicated all dissentients, the rupture of the East and West would have taken place earlier by centuries than it did. The only place in which I ever saw any part of my paradox advanced, was in an article in the *Examiner* newspaper, towards the end of 1866, after the above was written.

A story about Christopher Clavius, the workman of the new Calendar. I chanced to pick up ' Albertus Pighius Campensis de æquinoctiorum solsticiorumque inventione . . . . Ejusdem de ratione Paschalis celebrationis, De que Restitutione ecclesiastici Kalendarii,' Paris, 1520, folio. On the title-page were decayed words followed by ' . . hristophor . . C . . ii, 1556 (or 8),' the last blank not entirely erased by time, but showing the lower halves of an *l* and of an *a*, and rather too much room for a *v*. It looked very like *E Libris Christophori Clavii* 1556. By the courtesy of some members of the Jesuit body in London, I procured a tracing of the signature of Clavius from Rome, and the shapes of the letters, and the modes of junction and disjunction, put the matter beyond question. Even the extra space was explained; he wrote himself Clauius. Now in 1556, Clavius was nineteen years old: it thus appears probable that the framer of the Gregorian Calendar was selected, not merely as a learned astronomer, but as one who had attended to the calendar, and to works on its reformation, from early youth. When on the subject I found reason to think that Clavius had really read this work, and taken from it a phrase or two and a notion or two. Observe the advantage of writing the baptismal name at full length.

> The discovery of a general resolution of all superior finite equations, of every numerical both algebraick and transcendent form. By A. P. Vogel, mathematician at Leipzick. Leipzick and London, 1845, 8vo.

This work is written in the English of a German who has not mastered the idiom: but it is always intelligible. It professes to solve equations of every degree ' in a more extent sense, and till to every degree of exactness.' The general solution of equations of *all* degrees is a vexed question, which cannot have the mysterious interest of the circle problem, and is of a comparatively modern date. Mr. Vogel announces a forthcoming treatise in which are resolved the ' last impossibilities of pure mathematics.'

> Elective Polarity the Universal Agent. By Frances Barbara Burton, authoress of ' Astronomy familiarized,' ' Physical Astronomy,' &c. London, 1845, 8vo.

The title gives a notion of the theory. The first sentence states, that 12,500 years ago $a$ Lyræ was the pole-star, and attributes the immense magnitude of the now fossil animals to a star of such ' polaric intensity as Vega pouring its magnetic streams through our planet.' Miss Burton was a lady of property, and of very respectable acquirements, especially in Hebrew; she was eccentric in all things.

1867.—Miss Burton is revived by the writer of a book on meteorology which makes use of the planets: she is one of his leading minds.

In the year 1845 the old *Mathematical Society* was merged in the Astronomical Society. The circle-squarers, &c., thrive more in England than in any other country : there are most weeds where there is the largest crop. Speculation, though not encouraged by our Government so much as by those of the Continent, has had, not indeed such forcing, but much wider diffusion : few tanks, but many rivulets. On this point I quote from the preface to the reprint of the work of Ramchundra, which I superintended for the late Court of Directors of the East India Company.—

'That sound judgment which gives men well to know what is best for them, as well as that faculty of invention which leads to development of resources and to the increase of wealth and comfort, are both materially advanced, perhaps cannot rapidly be advanced without, a great taste for pure speculation among the general mass of the people, down to the lowest of those who can read and write. England is a marked example. Many persons will be surprised at this assertion. They imagine that our country is the great instance of the refusal of all *unpractical* knowledge in favour of what is *useful*. I affirm, on the contrary, that there is no country in Europe in which there has been so wide a diffusion of speculation, theory, or what other unpractical word the reader pleases. In our country, the scientific *society* is always formed and maintained by the people ; in every other, the scientific *academy*—most aptly named—has been the creation of the government, of which it has never ceased to be the nursling. In all the parts of England in which manufacturing pursuits have given the artisan some command of time, the cultivation of mathematics and other speculative studies has been, as is well known, a very frequent occupation. In no other country has the weaver at his loom bent over the *Principia* of Newton ; in no other country has the man of weekly wages maintained his own scientific periodical. With us, since the beginning of the last century, scores upon scores—perhaps hundreds, for I am far from knowing all—of annuals have run, some their ten years, some their half-century, some their century and a half, containing questions to be answered, from which many of our examiners in the Universities have culled materials for the academical contests. And these questions have always been answered, and in cases without number by the lower order of purchasers, the mechanics, the weavers, and the printers' workmen. I cannot here digress to point out the manner in which the concentration of manufactures, and the general diffusion of education, have affected the state of things ; I speak of the time during which the present system took its rise, and of the circumstances under which many of its most effective promoters were trained. In all this there is nothing which stands out, like the state-nourished academy, with its few great names and brilliant single achievements. This country has differed from all others in the wide diffusion of the disposition to speculate, which disposition has found its place among the ordinary habits of life, moderate in its action, healthy in its amount.'

Among the most remarkable proofs of the diffusion of speculation was the Mathematical Society, which flourished from 1717 to 1845.　Its habitat was Spitalfields, and I think most of its existence was passed in Crispin Street,　It was originally a plain society, belonging to the studious artisan.　The members met for discussion once a week ; and I believe I am correct in saying that each man had his pipe, his pot, and his problem.　One of their old rules was that, ' If any member shall so far forget himself and the respect due to the Society as in the warmth of debate to threaten or offer personal violence to any other member, he shall be liable to immediate expulsion, or to pay such fine as the majority of the members present shall decide.'　But their great rule, printed large on the back of the title page of their last book of regulations, was ' By the constitution of the Society, it is the duty of every member, if he be asked any mathematical or philosophical question by another member, to instruct him in the plainest and easiest manner he is able.'　We shall presently see that, in old time, the rule had a more homely form.

I have been told that De Moivre was a member of this Society. This I cannot verify: circumstances render it unlikely ; even though the French refugees clustered in Spitalfields ; many of them were of the Society, which there is some reason to think was founded by them.　But Dollond, Thomas Simpson, Saunderson, Crossley, and others of known name, were certainly members.　The Society gradually declined, and in 1845 was reduced to nineteen members.　An arrangement was made by which sixteen of these members, who were not already in the Astronomical Society became Fellows without contribution, all the books and other property of the old Society being transferred to the new one.　I was one of the committee which made the preliminary inquiries, and the reason of the decline was soon manifest.　The only question which could arise was whether the members of the society of working men—for this repute still continued—were of that class of educated men who could associate with the Fellows of the Astronomical Society on terms agreeable to all parties.　We found that the artisan element had been extinct for many years ; there was not a man but might, as to education, manners, and position, have become a Fellow in the usual way.　The fact was that life in Spitalfields had become harder : and the weaver could only live from hand to mouth, and not up to the brain.　The material of the old Society no longer existed.

In 1798, experimental lectures were given, a small charge for admission being taken at the door : by this hangs a tale—and a song. Many years ago, I found among papers of a deceased friend, who certainly never had anything to do with the Society, and who passed all his life far from London, a song, headed ' Song sung at a Mathematical Society in London, at a dinner given to Mr. Fletcher, a solicitor, who had defended the Society gratis.' Mr. Williams, the Assistant Secretary of the Astronomical Society, formerly Secretary of the Mathematical Society, remembered that the Society had had a solicitor named Fletcher among the members. Some years elapsed before it struck me that my old friend Benjamin Gompertz, who had long been a member, might have some recollection of the matter. The following is an extract of a letter from him (July 9, 1861) :—

As to the Mathematical Society, of which I was a member when only 18 years of age, [Mr. G. was born in 1779], having been, contrary to the rules, elected under the age of 21. How I came to be a member of that Society—and continued so until it joined the Astronomical Society, and was then the President—was : I happened to pass a bookseller's small shop, of second-hand books, kept by a poor taylor, but a good mathematician, John Griffiths. I was very pleased to meet a mathematician, and I asked him if he would give me some lessons ; and his reply was that I was more capable to teach him, but he belonged to a society of mathematicians, and he would introduce me. I accepted the offer, and I was elected, and had many scholars then to teach, as one of the rules was, if a member asked for information, and applied to any one who could give it, he was obliged to give it, or fine one penny. Though I might say much with respect to the Society which would be interesting, I will for the present reply only to your question. I well knew Mr. Fletcher, who was a very clever and very scientific person. He did, as solicitor, defend an action brought by an informer against the Society—I think for 5,000l.—for giving lectures to the public in philosophical subjects [i.e. for unlicensed public exhibition with money taken at the doors]. I think the price for admission was one shilling, and we used to have, if I rightly recollect, from two to three hundred visitors. Mr. Fletcher was successful in his defence, and we got out of our trouble. There was a collection made to reward his services, but he did not accept of any reward : and I think we gave him a dinner, as you state, and enjoyed ourselves ; no doubt with astronomical songs and other songs ; but my recollection does not enable me to say if the astronomical song was a drinking song. I think the anxiety caused by that action was the cause of some of the members' death. [They had, no doubt, broken the law in ignorance ; and by the sum named, the informer must have been present, and sued for a penalty on every shilling he could prove to have been taken].

I by no means guarantee that the whole song I proceed to give is what was sung at the dinner : I suspect, by the completeness of the chain, that augmentations have been made. My deceased friend was just the man to add some verses, or the addition may have been made before it came into his hands, or since his decease, for the scraps containing the verses passed through several hands before they came into mine. We may, however, be pretty sure that the original is substantially contained in what is given, and that the character is therefore preserved. I have had myself to repair damages every now and then, in the way of conjectural restoration of defects caused by ill-usage.

### THE ASTRONOMER'S DRINKING-SONG.

' Whoe'er would search the starry sky,
    Its secrets to divine, sir,
Should take his glass—I mean, should try
    A glass or two of wine, sir !
True virtue lies in golden mean,
    And man must wet his clay, sir ;
Join these two maxims, and 'tis seen
    He should drink his bottle a day, sir !

Old Archimedes, reverend sage !
    By trump of fame renowned, sir,
Deep problems solved in every page,
    And the sphere's curved surface found, sir :
Himself he would have far outshone,
    And borne a wider sway, sir,
Had he our modern secret known,
    And drank his bottle a day, sir !

When Ptolemy, now long ago,
    Believed the earth stood still, sir,
He never would have blundered so,
    Had he but drunk his fill, sir :
He'd then have felt[1] it circulate,
    And would have learnt to say, sir,
The true way to investigate
    Is to drink your bottle a day, sir !

Copernicus, that learned wight,
    The glory of his nation,
With draughts of wine refreshed his sight,
    And saw the earth's rotation ;

---

[1] Dr. Whewell, when I communicated this song to him, started the opinion, which I had before him, that this was a very good idea, of which too little was made.

Each planet then its orb described,
    The moon got under way, sir;
These truths from nature he imbibed
    For he drank his bottle a day, sir!

The noble[1] Tycho placed the stars,
    Each in its due location;
He lost his nose[2] by spite of Mars,
    But *that* was no privation:
Had he but lost his mouth, I grant
    He would have felt dismay, sir,
Bless you! *he* knew what he should want
    To drink his bottle a day, sir!

Cold water makes no lucky hits;
    On mysteries the head runs:
Small drink let Kepler time his wits
    On the regular polyhedrons:
He took to wine, and it changed the chime,
    His genius swept away, sir,
Through area varying[3] as the time
    At the rate of a bottle a day, sir!

Poor Galileo, forced to rat
    Before the Inquisition,
*E pur si muove* was the pat
    He gave them in addition:
He meant, whate'er you think you prove,
    The earth must go its way, sirs;
Spite of your teeth I'll make it move,
    For I'll drink my bottle a day, sirs!

Great Newton, who was never beat
    Whatever fools may think, sir;
Though sometimes he forgot to eat,
    He never forgot to drink, sir:
Descartes[4] took nought but lemonade,
    To conquer him was play, sir;
The first advance that Newton made
    Was to drink his bottle a day, sir!

[1] The common epithet of rank: *nobilis Tycho*, as he was a nobleman.  The writer had been at history.

[2] He lost it in a duel, with Manderupius Pasbergius. A contemporary, T. B. Laurus, insinuates that they fought to settle which was the best mathematician! This seems odd, but it must be remembered they fought in the dark, 'in tenebris densis'; and it is a nice problem to shave off a nose in the dark, without any other harm.

[3] Referring to Kepler's celebrated law of planetary motion.  He had previously wasted his time on analogies between the planetary orbits and the polyhedrons.

[4] As great a lie as ever was told: but in 1800 a compliment to Newton without a fling at Descartes would have been held a lopsided structure.

> D'Alembert, Euler, and Clairaut,
>     Though they increased our store, sir,
> Much further had been seen to go
>     Had they tippled a little more, sir !
> Lagrange gets mellow with Laplace,
>     And both are wont to say, sir,
> The *philosophe* who's not an ass
>     Will drink his bottle a day, sir !
>
> Astronomers ! what can avail
>     Those who calumniate us ;
> Experiment can never fail
>     With such an apparatus :
> Let him who'd have his merits known
>     Remember what I say, sir ;
> Fair science shines on him alone
>     Who drinks his bottle a day, sir !
>
> How light we reck of those who mock
>     By this we'll make to appear, sir,
> We'll dine by the sidereal[1] clock
>     For one more bottle a year, sir :
> But choose which pendulum you will,
>     You'll never make your way, sir,
> Unless you drink—and drink your fill,—
>     At least a bottle a day, sir ! '

Old times are changed, old manners gone!
There is a new Mathematical Society, and I am, at this present writing (1866), its first President. We are very high in the newest developements, and bid fair to take a place among the scientific establishments. Benjamin Gompertz, who was President of the old Society when it expired, was the link between the old and new body : he was a member of *ours* at his death. But not a drop of liquor is seen at our meetings, except a decanter of water : all our heavy is a fermentation of symbols ; and we do not draw it mild. There is no penny fine for reticence or occult science ; and as to a song ! not the ghost of a chance.

1826. The time may have come when the original documents connected with the discovery of Neptune may be worth revising. The following are extracts from the *Athenæum* of October 3 and October 17 :—

---

[1] The *sidereal* day is about four minutes short of the solar ; there are 366 sidereal days in the year.

## LE VERRIER'S PLANET.

We have received, at the last moment before making up for press, the following letter from Sir John Herschel, in reference to the matter referred to in the communication from Mr. Hind given below :—

Collingwood, Oct. 1.

' In my address to the British Association assembled at Southampton, on the occasion of my resigning the chair to Sir R. Murchison, I stated, among the remarkable astronomical events of the last twelvemonth, that it had added a new planet to our list,—adding, " it has done more, —it has given us the probable prospect of the discovery of another. We see it as Columbus saw America from the shores of Spain. Its movements have been felt, trembling along the far-reaching line of our analysis, with a certainty hardly inferior to that of ocular demonstration."— These expressions are not reported in any of the papers which profess to give an account of the proceedings, but I appeal to all present whether they were not used.

Give me leave to state my reasons for this confidence ; and, in so doing, to call attention to some facts which deserve to be put on record in the history of this noble discovery. On July 12, 1842, the late illustrious astronomer, Bessel, honoured me with a visit at my present residence. On the evening of that day, conversing on the great work of the planetary reductions undertaken by the Astronomer Royal—then in progress, and since published,[1]—M. Bessel remarked that the motions of Uranus, as he had satisfied himself by careful examination of the recorded observations, could not be accounted for by the perturbations of the known planets ; and that the deviations far exceeded any possible limits of error of observation. In reply to the question, Whether the deviations in question might not be due to the action of an unknown planet?—he stated that he considered it highly probable that such was the case,—being systematic, and such as might be produced by an exterior planet. I then inquired whether he had attempted, from the indications afforded by these perturbations, to discover the position of the unknown body,—in order that " a hue and cry" might be raised for it. From his reply, the words of which I do not call to mind, I collected that he had not then gone into that inquiry ; but proposed to do so, having now completed certain works which had occupied too much of his time. And, accordingly, in a letter which I received from him after his return to Königsberg, dated November 14, 1842, he says,—" In reference to our conversation at Collingwood, I *announce* to you (*melde* ich Ihnen) that Uranus is not forgotten." Doubtless, therefore, among his papers will be found some researches on the subject.

[1] The expense of this magnificent work was defrayed by Government grants, obtained, at the instance of the British Association, in 1833.

The remarkable calculations of M. Le Verrier—which have pointed out, as now appears, nearly the true situation of the new planet, by resolving the inverse problem of the perturbations—if uncorroborated by repetition of the numerical calculations by another hand, or by independent investigation from another quarter, would hardly justify so strong an assurance as that conveyed by my expressions above alluded to. But it was known to me, at that time, (I will take the liberty to cite the Astronomer Royal as my authority) that a similar investigation had been independently entered into, and a conclusion as to the situation of the new planet very nearly coincident with M. Le Verrier's arrived at (in entire ignorance of his conclusions), by a young Cambridge mathematician, Mr. Adams;—who will, I hope, pardon this mention of his name (the matter being one of great historical moment), —and who will, doubtless, in his own good time and manner, place his calculations before the public.

J. F. W. Herschel.'

## Discovery of Le Verrier's Planet.

Mr. Hind announces to the *Times* that he has received a letter from Dr. Brünnow, of the Royal Observatory at Berlin, giving the very important information that Le Verrier's planet was found by M. Galle, on the night of September 23. 'In announcing this grand discovery,' he says, ' I think it better to copy Dr. Brünnow's letter.'

Berlin, Sept. 25.

' My dear Sir,—M. Le Verrier's planet was discovered here the 23rd of September, by M. Galle. It is a star of the 8th magnitude, but with a diameter of two or three seconds. Here are its places :—

|          h. m. s.        |    R. A.      |    Declination.   |
|--------------------------|---------------|-------------------|
| Sept. 23, 12  0 14·6 M.T. | 328° 19′ 16·0″ | —13° 24   8·2″   |
| Sept. 24,  8 54 40·9 M.T. | 328° 18′ 14·3″ | —13° 24′ 29·7″   |

The planet is now retrograde, its motion amounting daily to four seconds of time.

Yours most respectfully,          BRÜNNOW.'

' This discovery,' Mr. Hind says, ' may be justly considered one of the greatest triumphs of theoretical Astronomy;' and he adds, in a postscript, that the planet was observed at Mr. Bishop's Observatory, in the Regent's Park, on Wednesday night, notwithstanding the moonlight and hazy sky. ' It appears bright,' he says, ' and with a power of 320 I can see the disc. The following position is the result of instrumental comparisons with 33 Aquarii :—

Sept. 30, at 8h. 16m. 21s. Greenwich mean time--
  Right ascension of planet   .  .   21h. 52m. 47·15s.
  South declination .   .   .   .   13° 27' 20''.'

## THE NEW PLANET.

Cambridge Observatory, Oct. 15.

The allusion made by Sir John Herschel, in his letter contained in the *Athenæum* of October 3, to the theoretical researches of Mr. Adams, respecting the newly-discovered planet, has induced me to request that you would make the following communication public. It is right that I should first say that I have Mr. Adams's permission to make the statements that follow, so far as they relate to his labours. I do not propose to enter into a detail of the steps by which Mr. Adams was led, by his spontaneous and independent researches, to a conclusion that a planet must exist more distant than Uranus. The matter is of too great historical moment not to receive a more formal record than it would be proper to give it here. My immediate object is to show, while the attention of the scientific public is more particularly directed to the subject, that, with respect to this remarkable discovery, English astronomers may lay claim to some merit.

Mr. Adams formed the resolution of trying, by calculation, to account for the anomalies in the motion of Uranus on the hypothesis of a more distant planet, when he was an undergraduate in this University, and when his exertions for the academical distinction, which he obtained in January 1843, left him no time for pursuing the research. In the course of that year, he arrived at an approximation to the position of the supposed planet; which, however, he did not consider to be worthy of confidence, on account of his not having employed a sufficient number of observations of Uranus. Accordingly, he requested my intervention to obtain for him the early Greenwich observations, then in course of reduction;—which the Astronomer Royal immediately supplied, in the kindest possible manner. This was in February, 1844. In September, 1845, Mr. Adams communicated to me values which he had obtained for the heliocentric longitude, excentricity of orbit, longitude of perihelion, and mass, of an assumed exterior planet,—deduced entirely from unaccounted-for perturbations of Uranus. The same results, somewhat corrected, he communicated, in October, to the Astronomer Royal. M. Le Verrier, in an investigation which was published in June of 1846, assigned very nearly the same heliocentric longitude for the probable position of the planet as Mr. Adams had arrived at, but gave no results respecting its mass and the form of its orbit. The coincidence as to position from two entirely independent investigations naturally inspired confidence; and the Astronomer Royal shortly after suggested the employing of the Northumberland telescope of this Observatory in a systematic search after the hypothetical planet; re-commending, at the same time, a definite plan of operations. I undertook to make the search,—and commenced observing on July 29. The observations were directed, in the first instance, to the part of the heavens which theory had pointed out as the most probable place of the planet; in selecting which I was guided by a paper drawn up for me by Mr. Adams. Not having hour XXI. of the Berlin star-maps—of the publication of which I was not aware—I had to proceed on the principle of comparison of observations made at intervals. On July 30, I went over a zone 9' broad, in such a manner as to include all stars to the eleventh magnitude. On August 4, I took a broader zone,— and recorded a place of the planet. My next observations were on August 12; when I met with a star of the eighth magnitude in the zone which I had gone over on July 30,—and which did not then contain this star. Of course, this was the planet;—the place of which was, thus, recorded a second time in four days of observing. A comparison of the observations of July 30 and August 12 would, according to the principle of search which I employed, have shown me the planet. I did not make the comparison till after the detection of it at Berlin— partly because I had an impression that a much more extensive search was required to give any probability of discovery—and partly from the press of other occupation. The planet, however, was *secured*, and two positions of it recorded six weeks earlier here than in any other observatory,—and in a systematic search expressly undertaken for that purpose. I give now the positions of the planet on August 4 and August 12.

Greenwich mean time.

| | | | | |
|---|---|---|---|---|
| Aug. 4, 13h. 36m. 25s. | . | . | R.A. | 21h. 58m. 14·70s. |
| | | | N.P.D. | 102° 57' 32·2" |
| Aug. 12, 13h. 3m. 26s. | . | . | R.A. | 21h. 57m. 26·13s. |
| | | | N.P.D. | 103° 2' 0·2" |

From these places compared with recent observations Mr. Adams has obtained the following results:—

| | |
|---|---|
| Distance of the planet from the sun . . . . . | 30·05 |
| Inclination of the orbit . . . . . . . | 1° 45' |
| Longitude of the descending node . . . . . | 309° 43' |
| Heliocentric longitude, Aug. 4 . . . . . | 326° 39' |

The present distance from the sun is, therefore, thirty times the earth's mean distance;—which is somewhat less than the theory had indicated. The other elements of the orbit cannot be approximated to till the observations shall have been continued for a longer period.

The part taken by Mr. Adams in the theoretical search after this planet will, perhaps, be considered to justify the suggesting of a name. With his consent, I mention *Oceanus* as one which may possibly receive the votes of astronomers.—I have authority to state that Mr. Adams's investigations will, in a short time, be published in detail.

J. CHALLIS.'

## ASTRONOMICAL POLICE REPORT.

" An ill-looking kind of body, who declined to give any name, was brought before the Academy of Sciences, charged with having assaulted a gentleman of the name of Uranus in the public highway. The prosecutor was a youngish looking person, wrapped up in two or three great coats; and looked chillier than anything imaginable, except the prisoner,—whose teeth absolutely shook, all the time.

Policeman Le Verrier stated that he saw the prosecutor walking along the pavement, — and sometimes turning sideways, and sometimes running up to the railings and jerking about in a strange way. Calculated that somebody must be pulling his coat, or otherwise assaulting him. It was so dark that he could not see; but thought, if he watched the direction in which the next odd move was made, he might find out something. When the time came, he set Brünnow, a constable in another division of the same force, to watch where he told him; and Brünnow caught the prisoner lurking about in the very spot,—trying to look as if he was minding his own business. Had suspected for a long time that somebody was lurking about in the neighbourhood. Brünnow was then called, and deposed to his catching the prisoner as described.

*M. Arago.*—Was the prosecutor sober?

*Le Verrier.*—Lord, yes, your worship; no man who had a drop in him ever looks so cold as he did.

*M. Arago.*—Did you see the assault?

*Le Verrier.*—I can't say I did; but I told Brünnow exactly how he'd be crouched down,—just as he was.

*M. Arago* (*to Brünnow*).—Did *you* see the assault?

*Brünnow.*—No, your worship; but I caught the prisoner.

*M. Arago.*—How do you know there was any assault at all?

*Le Verrier.*—I reckoned it could'nt be otherwise, when I saw the prosecutor making those odd turns on the pavement.

*M. Arago.*—You reckon and you calculate! Why, you'll tell me, next, that you policemen may sit at home and find out all that's going on in the streets by arithmetic. Did you ever bring a case of this kind before me till now?

*Le Verrier.*—Why, you see, your worship, the police are growing cleverer and cleverer every day. We can't help it:—it grows upon us.

*M. Arago.*—You're getting too clever for me. What does the prosecutor know about the matter?

The prosecutor said, all he knew was that he was pulled behind by somebody several times. On being further examined, he said that he had seen the prisoner often, but did not know his name, nor how he got his living; but had understood he was called Neptune. He himself had paid rates and taxes a good many years now. Had a family of six,—two of whom got their own living.

The prisoner, being called on for his defence, said that it was a quarrel. He had pushed the prosecutor—and the prosecutor had pushed him. They had known each other a long time, and were always quarrelling;—he did not know why. It was their nature, he supposed. He further said, that the prosecutor had given a false account of himself;—that he went about under different names. Sometimes he was called Uranus, sometimes Herschel, and sometimes Georgium Sidus; and he had no character for regularity in the neighbourhood. Indeed, he was sometimes not to be seen for a long time at once.

The prosecutor, on being asked, admitted, after a little hesitation, that he had pushed and pulled the prisoner too. In the altercation which followed, it was found very difficult to make out which began :—and the worthy magistrate' seemed to think they must have begun together.

*M. Arago.*—Prisoner, have you any family?

The prisoner declined answering that question at present. He said he thought the police might as well reckon it out whether he had or not.

*M. Arago* said he didn't much differ from that opinion.—He ᴊhen addressed both prosecutor and prisoner; and told them that if they couldn't settle their differences without quarrelling in the streets, he should certainly commit them both next time. In the meantime, he called upon both to enter into their own recognizances ; and directed the police to have an eye upon both, —observing that the prisoner would be likely to want it a long time, and the prosecutor would be not a hair the worse for it."

This squib was written by a person who was among the astronomers : and it illustrates the fact that Le Verrier had sole possession of the field until Mr. Challis's letter appeared. Sir John Herschel's previous communication should have paved the way : but the wonder of the discovery drove it out of many heads. There is an excellent account of the whole matter in Professor

Grant's 'History of Physical Astronomy.' The squib scandalized some grave people, who wrote severe admonitions to the editor. There are formalists who spend much time in writing propriety to journals, to which they serve as foolometers. In a letter to the *Athenæum*, speaking of the way in which people hawk fine terms for common things, I said that these people ought to have a new translation of the Bible, which should contain the verse 'gentleman and lady, created He them.' The editor was handsomely fired and brimstoned!

> A new theory of the tides: in which the errors of the usual theory are demonstrated; and proof shewn that the full moon is not the cause of a concomitant spring tide, but actually the cause of the neaps . . . By Comm$^r$. Debenham, R.N. London, 1846, 8vo.

The author replied to a criticism in the *Athenæum*, and I remember how, in a very few words, he showed that he had read nothing on the subject. The reviewer spoke of the forces of the planets (*i.e.* the Sun and Moon) on the Ocean, on which the author remarks, 'But N.B. the Sun is no planet, Mr. Critic.' Had he read any of the actual investigations on the usual theory, he would have known that to this day the sun and moon continue to be called *planets*—though the phrase is disappearing—in speaking of the tides; the sense, of course, being the old one, wandering bodies.

A large class of the paradoxers, when they meet with something which taken in their sense is absurd, do not take the trouble to find out the intended meaning, but walk off with the words laden with their own first construction. Such men are hardly fit to walk the streets without an interpreter. I was startled for a moment, at the time when a recent happy—and more recently happier—marriage occupied the public thoughts, by seeing in a haberdasher's window, in staring large letters, an unpunctuated sentence which read itself to me as 'Princess Alexandra! collar and cuff!' It immediately occurred to me that had I been any one of some scores out of my paradoxers, I should, no doubt, have proceeded to raise the mob against the unscrupulous person who dared to hint to a young bride such maleficent—or at least immellificent—conduct towards her new lord. But, as it was, certain material contexts in the shop window suggested a less savage explanation. A paradoxer should not stop at reading the advertisements of Newton or Laplace: he should learn to look at the stock of goods.

I think I must have an eye for double readings, when presented : though I never guess riddles.   On the day on which I
first walked into the *Panizzi* reading room—as it ought to be
called—at the Museum, I began my circuit of the wall-shelves
at the ladies' end: and perfectly coincided in the propriety of
the Bibles and theological works being placed there.   But the
very first book I looked on the back of had, in flaming gold
letters, the following inscription—'Blast the Antinomians!'   If
a line had been drawn below the first word, Dr. Blast's history
of the Antinomians would not have been so fearfully misinter-
preted.   It seems that neither the binder nor the arranger of the
room had caught my reading.   The book was removed before the
catalogue of books of reference was printed.

> Two systems of astronomy : first, the Newtonian system, showing
> the rise and progress thereof, with a short historical account;
> the general theory with a variety of remarks thereon : second,
> the system in accordance with the Holy Scriptures, showing
> the rise and progress from Enoch, the seventh from Adam, the
> prophets, Moses, and others, in the first Testament ; our Lord
> Jesus Christ, and his apostles, in the new or second Testament ;
> Reeve and Muggleton, in the third and last Testament; with
> a variety of remarks thereon.   By Isaac Frost.   London, 1846,
> 4to.

A very handsomely printed volume, with beautiful plates.
Many readers who have heard of Muggletonians have never had
any distinct idea of Lodowick Muggleton, the inspired tailor,
(1608–1698) who about 1650 received his commission from
heaven, wrote a Testament, founded a sect, and descended to
posterity.   Of Reeve less is usually said ; according to Mr. Frost,
he and Muggleton are the two 'witnesses.'   I shall content
myself with one specimen of Mr. Frost's science:

"I was once invited to hear read over 'Guthrie on Astronomy,' and
when the reading was concluded I was asked my opinion thereon;
when I said, 'Doctor, it appears to me that Sir I. Newton has only
given two proofs in support of his theory of the earth revolving round
the sun : all the rest is assertion without any proofs.'—'What are
they?' inquired the Doctor.—'Well,' I said, 'they are, first, the
power of attraction to keep the earth to the sun; the second is the
power of repulsion, by virtue of the centrifugal motion of the earth :
all the rest appears to me assertion without proof.'   The Doctor considered a short time, and then said, 'It certainly did appear so.'   I
said, 'Sir Isaac has certainly obtained the credit of completing the
system, but really he has only half done his work.'—'How is that,'

inquired my friend the Doctor. My reply was this: 'You will observe his system shows the earth traverses round the sun on an inclined plane ; the consequence is, there are four powers required to make his system complete :

1st. The power of *attraction*.
2ndly. The power of *repulsion*.
3rdly. The power of *ascending* the inclined plane.
4thly. The power of *descending* the inclined plane.

You will thus easily see the *four* powers required, and Newton has only accounted for *two* ; the work is therefore only half done.' Upon due reflection the Doctor said, 'It certainly was necessary to have these *four* points cleared up before the system could be said to be complete.' "

I have no doubt that Mr. Frost, and many others on my list, have really encountered doctors who could be puzzled by such stuff as this, or nearly as bad, among the votaries of existing systems, and have been encouraged thereby to print their objections. But justice requires me to say that from the words 'power of repulsion by virtue of the centrifugal motion of the earth,' Mr. Frost may be suspected of having something more like a notion of the much-mistaken term ' centrifugal force ' than many paradoxers of greater fame. The Muggletonian sect is not altogether friendless : over and above this handsome volume, the works of Reeve and Muggleton were printed, in 1832, in three quarto volumes. See *Notes and Queries*, 1st Series, v. 80 ; 3rd Series, iii, 303.

[The system laid down by Mr. Frost, though intended to be substantially that of Lodowick - Muggleton, is not so vagarious. It is worthy of note how very different have been the fates of two contemporary paradoxers, Muggleton and George Fox. They were friends and associates, and commenced their careers about the same time, 1647–1650. The followers of Fox have made their sect an institution, and deserve to be called the pioneers of philanthropy. But though there must still be Muggletonians, since expensive books are published by men who take the name, no sect of that name is known to the world. Nevertheless, Fox and Muggleton are men of one type, developed by the same circumstances : it is for those who investigate such men to point out why their teachings have had fates so different. Macaulay says it was because Fox found followers of more sense than himself. True enough : but why did Fox find such followers and not Muggleton ? The two were equally crazy, to all appearance : and the difference required must be sought in the doctrines themselves.

Fox was not a *rational* man : but the success of his sect and doctrines entitles him to a letter of alteration of the phrase which I am surprised has not become current.   When Conduitt, the husband of Newton's half-niece, wrote a circular to Newton's friends, just after his death, inviting them to bear their parts in a proper biography, he said, ' As Sir I. Newton was a *national* man, I think every one ought to contribute to a work intended to do him justice.'   Here is the very phrase which is often wanted to signify that celebrity which puts its mark, good or bad, on the national history, in a manner which cannot be asserted of many notorious or famous historical characters.   Thus George Fox and Newton are both *national* men.   Dr. Roget's *Thesaurus* gives more than fifty synonyms—*colleagues* would be the better word—of ' *celebrated*,' any one of which might be applied, either in prose or poetry, to Newton or to his works, no one of which comes near to the meaning which Conduitt's adjective immediately suggests.

The truth is, that we are too *monarchical* to be *national*. We have the Queen's army, the Queen's navy, the Queen's highway, the Queen's English, &c. ; nothing is national except the *debt*.   That this remark is not new is an addition to its force ; it has hardly been repeated since it was first made.   It is some excuse that *nation* is not vernacular English : the *country* is our word, and *country man* is appropriated.]

> Astronomical Aphorisms, or Theory of Nature ; founded on the immutable basis of Meteoric Action.   By P. Murphy, Esq. London, 1847, 12mo.

This is by the framer of the Weather Almanac, who appeals to that work as corroborative of his theory of planetary temperature, years after all the world knew by experience that this meteorological theory was just as good as the others.

> The conspiracy of the Bullionists as it affects the present system of the money laws.   By Caleb Quotem.   Birmingham, 1847, 8vo. (pp. 16).

This pamphlet is one of a class of which I know very little, in which the effects of the laws relating to this or that political bone of contention are imputed to deliberate conspiracy of one class to rob another of what the one knew ought to belong to the other. The success of such writers in believing what they have a bias to believe, would, if they knew themselves, make them think it equally likely that the inculpated classes might really believe what it is *their* interest to believe. The idea of a *guilty* understanding existing among fundholders, or landholders, or any holders, all the country over, and never detected except by bouncing pamphleteers, is a theory which should have been left for Cobbett to propose, and for Apella to believe.

[*August*, 1866. A pamphlet shows how to pay the National Debt. Advance paper to railways, &c., receivable in payment of taxes. The railways pay interest and principal in money, with which you pay your national debt, and redeem your notes. Twenty-five years of interest redeems the notes, and then the principal pays the debt. Notes to be kept up to value by penalties.]

The Reasoner. No. 45. Edited by G. J. Holyoake. Price 2*d*.
Is there sufficient proof of the existence of God? 8vo. 1847.

This acorn of the holy oak was forwarded to me with a manuscript note, signed by the editor, on the part of the 'London Society of Theological Utilitarians,' who say 'they trust you may be induced to give this momentous subject your consideration.' The supposition that a middle-aged person, known as a student of thought on more subjects than one, had that particular subject yet to begin, is a specimen of what I will call the *assumption-trick* of controversy, a habit which pervades all sides of all subjects. The tract is a proof of the good policy of letting opinions find their level, without any assistance from the Court of Queen's Bench. Twenty years earlier the thesis would have been positive, ' There is sufficient proof of the non-existence of God,' and bitter in its tone. As it stands, we have a moderate and respectful treatment—wrong only in making the opponent argue absurdly, as usually happens when one side invents the other—of a question in which a great many Christians have agreed with the atheist: that question being—Can the existence of God be proved independently of revelation? Many very religious persons answer this question in the negative, as well as Mr. Holyoake. And, this point being settled, all who agree in the negative separate into those who can endure scepticism, and those who cannot: the second class find their way to Christianity. This very number of ' The Reasoner ' announces the secession of one of its correspondents, and his adoption of the Christian faith. This would not have happened twenty years before : nor, had it happened, would it have been respectfully announced.

There are people who are very unfortunate in the expression of their meaning. Mr. Holyoake, in the name of the 'London Society,' &c., forwarded a pamphlet on the existence of God, and said that the Society trusted I 'may be induced to give' the subject my 'consideration.' How could I know the Society was one person, who supposed I *had* arrived at a conclusion, and wanted a '*guiding word*'? But so it seems it was : Mr. Holyoake, in the *English Leader* of October 15, 1864, and in a private letter to me, writes as follows :—

" The gentleman who was the author of the argument, and who asked me to send it to Mr. De Morgan, never assumed that that gentleman had ' that particular subject to begin '—on the contrary, he supposed that one whom we all knew to be eminent as a thinker *had* come to a conclusion upon it, and would perhaps vouchsafe a guiding word to one who was, as yet, seeking the solution of the Great Problem of Theology. I told my friend that ' Mr. De Morgan was doubtless preoccupied, and that he must be content to wait. On some day of courtesy and leisure he might have the kindness to write.' Nor was I wrong—the answer appears in your pages at the lapse of seventeen years."

I suppose Mr. Holyoake's way of putting his request was the *stylus curiæ* of the Society. A worthy Quaker who was sued for debt in the King's Bench was horrified to find himself charged in the declaration with detaining his creditor's money by force and arms, contrary to the peace of our Lord the King, &c. It's only the *stylus curiæ*, said a friend : I don't know *curiæ*, said the Quaker, but he shouldn't style us peace-breakers.

The notion that the *non*-existence of God can be *proved*, has died out under the light of discussion : had the only lights shone from the pulpit and the prison, so great a step would never have been made. The question now is as above. The dictum that Christianity is 'part and parcel of the law of the land' is also abrogated : at the same time, and the coincidence is not an accident, it is becoming somewhat nearer the truth that the law of the land is part and parcel of Christianity. It must also be noticed that *Christianity* was part and parcel of the articles of *war*; and so was *duelling*. Any officer speaking against religion was to be cashiered ; and any officer receiving an affront without, in the last resort, attempting to kill his opponent, was also to be cashiered. Though somewhat of a book-hunter, I have never been able to ascertain the date of the collected remonstrances of the prelates in the House of Lords against this overt inculcation of murder, under the soft name of *satisfaction* : it is neither in Watt, nor in Lowndes, nor in any edition of Brunet ; and there is no copy in the British Museum. Was the collected edition really published ?

[The publication of the above in the *Athenæum* has not produced reference to a single copy. The collected edition seems to be doubted. I have even met one or two persons who doubt the fact of the Bishops having remonstrated at all : but their doubt was founded on an absurd supposition, namely, that it was *no business of theirs* ; that it was not the business of the prelates of the Church in union with the State to remonstrate against the Crown commanding murder! Some say that the edition was published, but under an irrelevant title, which prevented people from knowing what it was about. Such things have happened : for example, arranged extracts from Wellington's general orders, which would have attracted attention, fell dead under the title of ' Principles of War.' It is surmised that the book I am looking for also contains the protests of the Reverend bench against other things besides the Thou-shalt-do-murder of the Articles (of war), and is called ' First Elements of Religion' or some similar title. Time clears up all things.]

With the general run of the philosophical atheists of the last century the notion of a God was an hypothesis. There was left an admitted possibility that the vague somewhat which went by more names than one, might be personal, intelligent, and superintendent. In the works of Laplace, who is sometimes called an atheist from his writings, there is nothing from which such an inference can be drawn : unless indeed a Reverend Fellow of the Royal Society may be held to be the fool who said in his heart, &c. &c., if his contributions to the *Philosophical Transactions* go no higher than *nature*. The following anecdote is well known in Paris, but has never been printed entire. Laplace once went in form to present some edition of his ' Système du Monde' to the First Consul, or Emperor. Napoleon, whom some wags had told that this book contained no mention of the name of God, and who was fond of putting embarrassing questions, received it with—' M. Laplace, they tell me you have written this large book on the system of the universe, and have never even mentioned its Creator.' Laplace, who, though the most supple of politicians, was as stiff as a martyr on every point of his philosophy or religion (*ex. gr.* even under Charles X. he never concealed his dislike of the priests), drew himself up, and answered bluntly, ' Je n'avais pas besoin de cette hypothèse-là.' Napoleon, greatly amused, told this reply to Lagrange, who exclaimed, ' Ah ! c'est une belle hypothèse ; ça explique beaucoup de choses.'

It is commonly said that the last words of Laplace were ' Ce que nous connaissons est peu de chose ; ce que nous ignorons est immense.' This looks like a parody on Newton's pebbles : the following is the true account ; it comes to me through one remove from Poisson. After the publication (in 1825) of the fifth volume of the *Mécanique Céleste*, Laplace became gradually weaker, and with it musing and abstracted. He thought much on the great problems of existence, and often muttered to himself *Qu'est ce que c'est que tout cela !* After many alternations, he appeared at last so permanently prostrated that his family applied to his favorite pupil, M. Poisson, to try to get a word from him. Poisson paid a visit, and after a few words of salutation, said ' J'ai une bonne nouvelle à vous annoncer : on a reçu au Bureau des Longitudes une lettre d'Allemagne annonçant que M. Bessel a vérifié par l'observation vos découvertes théoriques sur les satellites de Jupiter.' Laplace opened his eyes and answered with deep gravity, ' *L'homme ne poursuit que des chimères.*' He never spoke again. His death took place March 5, 1827.

The language used by the two great geometers illustrates what I have said : a supreme and guiding intelligence—apart from a blind rule called *nature of things*—was an *hypothesis.* The absolute denial of such a ruling power was not in the plan of the higher philosophers : it was left for the smaller fry. A round assertion of the non-existence of anything which stands in the way is the refuge of a certain class of minds : but it succeeds only with things subjective ; the objective offers resistance. A philosopher of the appropriative class tried it upon the constable who appropriated *him* : I deny your existence, said he ; Come along, all the same, said the unpsychological policeman.

Euler was a believer in God, downright and straightforward. The following story is told by Thiébault, in his *Souvenirs de vingt ans de séjour à Berlin,* published in his old age, about 1804. This volume was fully received as trustworthy ; and Marshal Mollendorff told the Duc de Bassano in 1807 that it was the most veracious of books written by the most honest of men. Thiébault says that he has no personal knowledge of the truth of the story, but·that it was believed throughout the whole of the north of Europe. Diderot paid a visit to the Russian Court at the invitation of the Empress. He conversed very freely, and gave the younger members of the Court circle a good deal of lively atheism. The Empress was much amused, but some of her councillors suggested that it might be desirable to check these expositions of doctrine. The Empress did not like to put a direct muzzle on her guest's tongue, so the following plot was contrived. Diderot was informed that a learned mathematician was in possession of an algebraical demonstration of the existence of God, and would give it him before all the Court, if he desired to hear it. Diderot gladly consented: though the name of the mathematician is not given, it was Euler. He advanced towards Diderot, and said gravely, and in a tone of perfect conviction:

$$Monsieur, \; \frac{'a + b^n}{n} = x, \; donc \; Dieu \; existe; \; répondez!$$ Diderot,

to whom algebra was Hebrew, was embarrassed and disconcerted; while peals of laughter rose on all sides. He asked permission to return to France at once, which was granted.

An examination of the Astronomical doctrine of the Moon's rotation. By J. L. Edinburgh, 1847, 8vo.

A systematic attack of the character afterwards made with less skill and more notice by Mr. Jellinger Symons.

July 1866, J. L. appears as Mr. James Laurie, with a new pamphlet 'The Astronomical doctrines of the Moon's rotation . . . .' Edinburgh. Of all the works I have seen on the question, this is the most confident, and the sorest. A writer on astronomy said of Mr. Jellinger Symons, ' Of course he convinced no one who knew anything of the subject.' This ' ungenerous slur' on the speculator's memory appears to have been keenly felt; but its truth is admitted. Those who knew anything of the subject are ' the so-called men of science,' whose three P's were assailed; prestige, pride, and prejudice: this the author tries to effect for himself with three Q's; quibble, quirk, and quiddity. He explains that the Scribes and Pharisees would not hear Jesus, and that the lordly bishop of Rome will not cast his tiara and keys at the feet of the ' humble presbyter' who now plays the part of pope in Scotland. I do not know whom he means: but perhaps the friends of the presbyter-pope may consider this an ungenerous slur. The best proof of the astronomer is just such ' as might have been expected from the merest of blockheads'; but as the giver is of course not a blockhead, this circumstance shows how deeply blinded by prejudice he must be.

Of course the paradoxers do not persuade any persons who know their subjects : and so these Scribes and Pharisees reject the Messiah. We must suppose that the makers of this comparison are Christians : for if they thought the Messiah an euthusiast or an impostor, they would be absurd in comparing those who reject what they take for truth with others who once rejected what they take for falsehood. And if Christians, they are both irreverent and blind to all analogy. The Messiah, with His Divine mission proved by miracles which all might see who chose to look, is degraded into a prototype of James Laurie, ingeniously astronomising upon ignorant geometry and false logic, and comparing to blockheads those who expose his nonsense. Their comparison is as foolish as—supposing them Christians—it is profane : but, like errors in general, its other end points to truth. There were Pseudochrists and Antichrists ; and a Concordance would find the real forerunners of all the paradoxers. But they are not so clever as the old false prophets : there are none of whom we should be inclined to say that, if it were possible, they would deceive the very educated. Not an Egyptian among them all can make uproar enough to collect four thousand men that are murderers— of common sense—to lead out into the wilderness. Nothing, says the motto of this work, is so difficult to destroy as the errors and false facts propagated by illustrious men whose words have authority. I deny it altogether. There are things much more difficult to destroy : it is much more difficult to destroy the truths and real facts supported by such men. And again, it is much more difficult to prevent men of no authority from setting up false pretensions ; and it is much more difficult to destroy assertions of fancy speculation. Many an error of thought and learning has fallen before a gradual growth of thoughtful and learned opposition. But such things as the quadrature of the circle, &c., are never put down. And why ? Because thought can influence thought, but thought cannot influence self-conceit : learning can annihilate learning : but learning cannot annihilate ignorance. A sword may cut through an iron bar ; and the severed ends will not reunite : let it go through the air, and the yielding substance is whole again in a moment.

> Miracles *versus* Nature : being an application of certain propositions in the theory of chances to the Christian miracles. By Protimalethes. Cambridge, 1847, 8vo.

The theory, as may be supposed, is carried further than most students of the subject would hold defensible.

An astronomical Lecture.   By the Rev. R. Wilson.   Greenock, 1847, 12mo.

Against the moon's rotation on her axis.

[Handed about in the streets in 1847 : I quote the whole :] Important discovery in astronomy, communicated to the Astronomer Royal, December 21st, 1846.   That the Sun revolve round the Planets in 25748⅔ years, in consequence of the combined attraction of the planets and their satellites, and that the Earth revolve round the Moon in 18 years and 228 days.   D. T. GLAZIER [altered with a pen into GLAZION.]   Price one penny.

1847.   In the *United Service Magazine* for September, 1847, Mrs. Borron, of Shrewsbury, published some remarks tending to impeach the fact that Neptune, the planet found by Galle, really was the planet which Le Verrier and Adams had a right to claim. This was followed (September 14) by two pages, separately circulated, of ' Further Observations upon the Planets Neptune and Uranus, with a Theory of Perturbations '; and (October 19, 1848) by three pages of ' A Review of M. Leverrier's Exposition.' Several persons, when the remarkable discovery was made, contended that the planet actually discovered was an intruder ; and the future histories of the discovery must contain some account of this little after-piece.   Tim Linkinwater's theory that there is no place like London for coincidences, would have been utterly overthrown in favour of what they used to call the celestial spaces, if there had been a planet which by chance was put near the place assigned to Neptune at the time when the discovery was made.

Aerial Navigation ; containing a description of a proposed flying machine, on a new principle.   By Dædalus Britannicus. London, 1847, 8vo.

In 1842–43 a Mr. Henson had proposed what he called an aeronaut steam-engine, and a Bill was brought in to incorporate an ' Aerial Transit Company.'   The present plan is altogether different, the moving power being the explosion of mixed hydrogen and air.   Nothing came of it—not even a Bill.   What the final destiny of the balloon may be no one knows: it may reasonably be suspected that difficulties will at last be overcome. Darwin, in his ' Botanic Garden ' (1781), has the following prophecy :—

Soon shall thy arm, unconquered Steam! afar
Drag the slow barge, or drive the rapid car;
Or, on wide-waving wings expanded, bear
The flying chariot through the fields of air.

Darwin's contemporaries, no doubt, smiled pity on the poor man. It is worth note that the two true prophecies have been fulfilled in a sense different from that of the predictions. Darwin was thinking of the suggestion of Jonathan Hulls, when he spoke of dragging the slow barge: it is only very recently that the steam-tug has been employed on the canals. The car was to be driven, not drawn, and on the common roads. Perhaps, the flying chariot will be something of a character which we cannot imagine, even with the two prophecies and their fulfilments to help us.

> A book for the public. New Discovery. The causes of the circulation of the blood; and the true nature of the planetary system. London, 1848, 8vo.

Light is the sustainer of motion both in the earth and in the blood. The natural standard, the pulse of a person in health, four beats to one respiration, gives the natural second, which is the measure of the earth's progress in its daily revolution. The Greek fable of the Titans is an elaborate exposition of the atomic theory: but any attempt to convince learned classics would only meet their derision; so much does long-fostered prejudice stand in the way of truth. The author complains bitterly that men of science will not attend to him and others like him: he observes, that 'in the time occupied in declining, a man of science might test the merits.' This is, alas! too true; so well do applicants of this kind know how to stick on. But every rule has its exception: I have heard of one. The late Lord Spencer—the Lord Althorp of the House of Commons—told me that a speculator once got access to him at the Home Office, and was proceeding to unfold his way of serving the public. 'I do not understand these things,' said Lord Althorp, 'but I happen to have —— (naming an eminent engineer) upstairs; suppose you talk to him on the subject.' The discoverer went up, and in half-an-hour returned, and said, 'I am very much obliged to your Lordship for introducing me to Mr. ——; he has convinced me that I am quite wrong.' I supposed, when I heard the story—but it would not have been seemly to say it—that Lord A. exhaled candour and sense, which infected those who came within reach: he would have done so, if anybody.

A method to trisect a series of angles having relation to each other ; also another to trisect any given angle. By James Sabben. 1848 (two quarto pages).

'The consequence of years of intense thought': very likely, and very sad.

1848. The following was sent to me in manuscript. I give the whole of it :—

'*Quadrature of the Circle*.—A quadrant is a curvilinear angle traversing round and at an equal distance from a given point, called a centre, no two points in the curve being at the same angle, but irreptitiously graduating from 90 to 60. It is therefore a mean angle of 90 and 60, which is 75, because it is more than 60, and less than 90, approximately from 60 to 90, and from 90 to 60, with equal generation in each irreptitious approximation, therefore meeting in 75, and which is the mean angle of the quadrant.

Or, suppose a line drawn from a given point at 90, and from the same point a line at 60. Let each of these lines revolve on this point toward each other at an equal ratio. They will become one line at 75, and bisect the curve, which is one-sixth of the entire circle. The result, taking 16 as a diameter, gives an area of 201 072400, and a circumference of 50·2681.

The original conception, its natural harmony, and the result, to my own mind is a demonstrative truth, which I presume it right to make known, though perhaps at the hazard of unpleasant if not uncourteous remarks.'

I have added punctuation : the handwriting and spelling are those of an educated person ; the word *irreptitious* is indubitable. The whole is a natural curiosity.

> The quadrature and exact area of the circle demonstrated. By Wm. Peters. 8vo. *n. d.* (circa 1848).
>
> Suggestions as to the necessity for a revolution in philosophy ; and prospectus for the establishment of a new quarterly, to be called the *Physical Philosopher and Heterodox Review*. By Q. E. D. 8vo. 1848.

These works are by one author, who also published, as appears by advertisement,

'Newton rescued from the precipitancy of his followers through a century and a half,' and ' Dangers along a coast by correcting (as it is called) a ship's reckoning by bearings of the land at night fall, or in a fog, nearly out of print. Subscriptions are requested for a new edition.'

The area of a circle is made four-fifths of the circumscribed square : proved on an assumption which it is purposed to explain in a longer essay. The author, as Q. E. D., was in controversy with the *Athenæum* journal, and criticised a correspondent, D., who wrote against a certain class of discoverers. He believed the common theories of hydrostatics to be wrong, and one of his questions was—

'Have you ever taken into account anent gravity and gravitation the fact that a five grain cube of cork will of itself half sink in the water, whilst it will take 20 grains of brass, which will sink of itself, to pull under the other half ? Fit this if you can, friend D., to your notions of gravity and specific gravity, as applied to the construction of a universal law of gravitation.'

This the *Athenæum* published—but without some Italics, for which the editor was sharply reproved, as a sufficient specimen of the *quod erat* D. *monstrandum* : on which the author remarks— ' D,—Wherefore the e caret ? is it D apostrophe ? D', D'M, D'Mo, D'Monstrandum ; we cannot find the *wit* of it.' This I conjecture to contain an illusion to the name of the supposed author ; but whether De Mocritus, De Mosthenes, or De Moivre was intended, I am not willing to decide.

> The Scriptural Calendar and Chronological Reformer, for the statute year 1849. Including a review of recent publications on the Sabbath question. London, 1849, 12mo.

This is the almanac of a sect of Christians who keep the Jewish Sabbath, having a chapel at Mill Yard, Goodman's Fields. They wrote controversial works, and perhaps do so still ; but I never chanced to see one.

> Geometry *versus* Algebra ; or the trisection of an angle geometrically solved. By W. Upton, B.A. Bath (circa 1849). 8vo.

The author published two tracts under this title, containing different alleged proofs : but neither gives any notice of the change. Both contain the same preface, complaining of the British Association for refusing to examine the production. I suppose that the author, finding his first proof wrong, invented the second, of which the Association never had the offer ; and, feeling sure that they would have equally refused to examine the second, thought it justifiable to present that second as the one which they had refused. Mr. Upton has discovered that the common way of finding the circumference is wrong, would set it right if he had leisure, and, in the mean time, has solved the problem of the duplication of the cube.

*The trisector of an angle, if he demand attention from any mathematician, is bound to produce, from his construction, an expression for the sine or cosine of the third part of any angle, in terms of the sine or cosine of the angle itself, obtained by help of no higher than the square root.* The mathematician knows that such a thing cannot be; but the trisector virtually says it can be, and is bound to produce it, to save time. This is the misfortune of most of the solvers of the celebrated problems, that they have not knowledge enough to present those consequences of their results by which they can be easily judged. Sometimes they have the knowledge, and quibble out of the use of it. In many cases a person makes an honest beginning and presents what he is sure is a solution. By conference with others he at last feels uneasy, fears the light, and puts self-love in the way of it. Dishonesty sometimes follows. The speculators are, as a class, very apt to imagine that the mathematicians are in fraudulent confederacy against them: I ought rather to say that each one of them consents to the mode in which the rest are treated, and fancies conspiracy against himself. The mania of conspiracy is a very curious subject. I do not mean these remarks to apply to the author before me.

One of Mr. Upton's trisections, if true, would prove the truth of the following equation :—

$$3 \cos \frac{\theta}{3} = 1 + \sqrt{(4 - \sin^2 \theta)}$$

which is certainly false.

In 1852 I examined a terrific construction, at the request of the late Dr. Wallich, who was anxious to persuade a poor countryman of his that trisection of the angle was waste of time. One of the principles was, that 'magnitude and direction determine each other.' The construction was equivalent to the assertion that, $\theta$ being any angle, the cosine of its third part is

$$\sin 3\theta \cdot \cos \frac{5\theta}{2} + \sin^2 \theta \sin \frac{5\theta}{2}$$

divided by the square root of

$$\sin^2 3\theta \cos \frac{^2 5\theta}{2} + \sin^4 \theta + \sin 3\theta \cdot \sin 5\theta \cdot \sin^2 \theta$$

This is from my rough notes, and I believe it is correct. It is so nearly true, unless the angle be very obtuse, that common drawing, applied to the construction, will not detect the error.

There are many formulæ of this kind : and I have several times found a speculator who has discovered the corresponding construction, has seen the approximate success of his drawing—often as great as absolute truth could give in graphical practice,—and has then set about his demonstration, in which he always succeeds to his own content.

There is a trisection of which I have lost both cutting and reference : I think it is in the *United Service Journal.* I could not detect any error in it, though certain there must be one. At least I discovered that two parts of the diagram were incompatible unless a certain point lay in line with two others, by which the angle to be trisected—and which was trisected—was bound to be either 0° or 180°.

Aug. 22, 1866. Mr. Upton sticks to his subject. He has just published ' The Uptonian Trisection. Respectfully dedicated to the schoolmasters of the United Kingdom.' It seems to be a new attempt. He takes no notice of the sentence I have put in italics : nor does he mention my notice of him, unless he mean to include me among those by whom he has been ' ridiculed and sneered at ' or ' branded as a brainless heretic.' I did neither one nor the other : I thought Mr. Upton a paradoxer to whom it was likely to be worth while to propound the definite assertion now in italics ; and Mr. Upton does not find it convenient to take issue on the point. He prefers general assertions about algebra. So long as he cannot meet algebra on the above question, he may issue as many ' respectful challenges' to the mathematicians as he can find paper to write : he will meet with no attention.

There is one trisection which is of more importance than that of the angle. It is easy to get half the paper on which you write for margin ; or a quarter ; but very troublesome to get a third. Show us how, easily and certainly, to fold the paper into three, and you will be a real benefactor to society.

Early in the century there was a Turkish trisector of the angle, Hussein Effendi, who published two methods. He was the father of Ameen Bey, who was well known in England thirty years ago as a most amiable and cultivated gentleman and an excellent mathematician. He was then a student at Cambridge ; and he died, years ago, in command of the army in Syria. Hussein Effendi was instructed in mathematics by Ingliz Selim Effendi, who translated a work of Bonnycastle into Turkish. This Englishman was Richard Baily, brother of Francis Baily the astronomer, who emigrated to Turkey in his youth, and adopted the manners of the Turks, but whether their religion also I never heard, though I should suppose he did.

I now give the letters from the agricultural labourer and his friend, described in page 9. They are curiosities; and the history of the quadrature can never be well written without some specimens of this kind:—

'Doctor Morgan, Sir. Permit me to address you

Brute Creation may perhaps enjoy the faculty of beholding visible things with a more penitrating eye than ourselves. But Spiritual objects are as far out of their reach as though they had no being

Nearest therefore to the brute Creation are those men who Suppose themselves to be so far governed by external objects as to believe nothing but what they See and feel And Can accomedate to their Shallow understanding and Imaginations

My Dear Sir Let us all Consult ourselves by the wise proverb..

I believe that evry man' merit & ability aught to be appreciated and valued In proportion to its worth & utility

In whatever State or Circumstances they may fortunately or unfortunately be placed

And happy it is for evry man to know his worth and place

When a Gentleman of your Standing in Society Clad with those honors Can not understand or Solve a problem That is explicitly explained by words and Letters and mathematacally operated by figuers He had best consult the wise proverd

Do that which thou Canst understand and Comprehend for thy good.

I would recommend that Such Gentleman Change his business

And appropriate his time and attention to a Sunday School to Learn what he Could and keep the Litle Children form durting their Close

With Sincere feelings of Gratitude for your weakness and Inability I am

Sir your Superior in Mathematics ——'
1849 June th29.


'Dor Morgin Sir

I wrote and Sent my work to Professor —— of —— State of —— United States

I am now in the possesion of the facts that he highly approves of my work. And Says he will Insure me Reward in the States

I write this that you may understand that I have knowledge of the unfair way that I am treated In my own nati County

I am told and have reasons to believe that it is the Clergy that treat me so unjust.

I am not Desirious of hcaping Disonors upon my own nation. But

if I have to Leave this kingdom without my Just dues. The world
Shall know how I am and have been treated.

I am Sir Desirous of my

Just dues ——

1849 July 3.

July 7th, 1849.

Sir, I have been given to understand that a friend of mine one whom I
shall never be ashamed to acknowledge as such tho' lowly his origine;
nay not only not ashamed but proud of doing so for I am one of those
who esteem and respect a man according to his ability and probity,
deeming with Dr. Watts ' that the mind is the standard of the man,' has
laid before you and asked your opinion of his extraordinary performance, viz. the quadrature of the circle, he did this with the firmest belief
that you would not only treat the matter in a straightforward manner
but with the conviction that from your known or supposed knowledge
of mathematicks would have given an upright and honorable decision
upon the subject; but the question is have you done so ? Could I say
yes I would with the greatest of pleasure and have congratulated you
upon your decision whatever it might have been but I am sorry to say
that I cannot your letter is a paltry evasion, you say ' that it is a great
pity that you (Mr. ——) should have attempted this (the quadrature
of the circle) for your mathematical knowledge is not sufficient to
make you know in what the problem consists,' you don't say in what
it does consist *according to your ideas*, oh ! no nothing of the sort, you
enter into no disquisition upon the subject in order to show where you
think Mr. —— is wrong and why you have not is simply—*because
you cannot*—you know that he has done it and what is if I am not
wrongly informed *you have been heard to say so.* He has done what
you nor any other mathematician as those who call themselves such
have done. And what is the reason that you will not candidly acknowledge to him as you have to others that he has squared the circle
shall I tell you ? it is because he has performed the feat to obtain the
glory of which mathematicians have battled from time immemorial
that they might encircle their brows with a wreath of laurels far more
glorious than ever conqueror won it is simply this that it is a poor
man a humble artisan who has gained that victory that you don't like
to acknowledge it you don't like to be beaten and worse to acknowledge
that you have miscalculated, you have in short too small a soul to acknowledge that he is right.

I was asked my opinion and *I* gave it unhesitatingly in the affirmative and I am backed in my opinion not only by Mr. —— a mathematician and watchmaker residing in the boro of Southwark but by
no less an authority than the Professor of mathematics of —— College
—— —— United States Mr. —— and I presume that he at least is
your equal as an authority and Mr. —— says that the government
of the U. S. will recompense M. D. for the discovery he has made if
so what a reflection upon Old england the boasted land of freedom

the nursery of the arts and sciences that her sons are obliged to go to·
a foreign country to obtain that recompense to which they are justly
entitled

In conclusion I had to contradict an assertion you made to the
effect that ' there is not nor ever was any reward offered by the
government of this country for the discovery of the quadrature of the
circle.' I beg to inform you that there *was* but that it having been
deemed an impossibility the government has withdrawn it. I do this
upon no less an authority than the Marquis of Northampton.

I am, sir, yours ——'

Dr. Morgan.

Notes on the Kinematic Effects of Revolution and Rotation, with
reference to the Motions of the Moon and of the earth. By
Henry Perigal, Jun. Esq. London, 1846–1849, 8vo.

On the misuse of technical terms. Ambiguity of the terms *Rotation*
and *Revolution*, owing to the double meaning improperly attributed to each of the words. (No date nor place, but by Mr. Perigal, I have no doubt, and containing letters of 1849 and 1850.)

The moon controversy. Facts *v.* Definitions. By H. P., Jun.
London, 1856, 8vo. (pp. 4.)

Mr. Henry Perigal helped me twenty years ago with the
diagrams, direct from the lathe to the wood, for the article
' Trochoidal Curves,' in the *Penny Cyclopædia* : these cuts add
very greatly to the value of the article, which, indeed, could not
have been made intelligible without them. He has had many
years' experience, as an amateur turner, in combination of double
and triple circular motions, and has published valuable diagrams
in profusion. A person to whom the double circular motion is
familiar in the lathe naturally looks upon one circle moving
upon another as in *simple* motion, if the second circle be
fixed to the revolving radius, so that one and the same point
of the moving circle travels upon the fixed circle. Mr. Perigal
commenced his attack upon the moon for moving about
her axis, in the first of the tracts above, ten years before
Mr. Jellinger Symons ; but he did not think it necessary to
make it a subject for the *Times* newspaper. His familiarity
with combined motions enabled him to handle his arguments
much better than Mr. J. Symons could do : in fact, he is the
clearest assailant of the lot which turned out with Mr. J. Symons.
But he is as wrong as the rest. The assault is now, I suppose,
abandoned, until it becomes epidemic again. This it will do :
it is one of those fallacies which are very tempting. There was
a dispute on the subject in 1748, between James Ferguson

and an anonymous opponent; and I think there have been others.

A poet appears in the field (July 19, 1863) who calls himself Cyclops, and writes four octavo pages. He makes a distinction between *rotation* and *revolution*; and his doctrines and phrases are so like those of Mr. Perigal that he is a follower at least. One of his arguments has so often been used that it is worth while to cite it :—

> Would Mathematicals—forsooth—
> If true, have failed to prove its truth ?
> Would not they—if they could—submit
> Some overwhelming proofs of it ?
> But still it totters *proofless* !  Hence
> There's strong presumptive evidence
> None do—or can—such proof propound
> Because *the dogma is unsound.*
> For, were there means of doing so,
> They would have proved it long ago.

This is only one of the alternatives.  Proof requires a person who can give and a person who can receive.  I feel inspired to add the following :—

> A blind man said, As to the Sun,
> I'll take my Bible oath there's none ;
> For if there had been one to show
> They would have shown it long ago.
> How came he such a goose to be ?
> Did he not know he couldn't see ?
>                         Not he !

The absurdity of the verses is in the argument.  The writer was not so ignorant or so dishonest as to affirm that nothing had been offered by the other side as proof; accordingly, his syllogism amounts to this :—If your proposition were true, you could have given proof satisfactory to *me* ; but this you have not done, therefore, your proposition is not true.

The echoes of the moon-controversy reached Benares in 1857, in which year was there published a pamphlet 'Does the Moon Rotate ? ' in Sanscrit and English.  The arguments are much the same as those of the discussion at home.

We see that there are paradoxers in argument as well as in assertion of fact : my plan does not bring me much into contact with these ; but another instance may be useful.  Sects, whether religious or political, give themselves names which, in meaning, are claimed also by their opponents ; loyal, liberal, conservative (of good), &c. have been severally appropriated by parties. *Whig* and *Tory* are unobjectionable names: the first—which occurs in English ballad as well as in Scotland—is sour milk;[1] the second is a robber. In theology, the Greek Church is *Orthodox*, the Roman is *Catholic*, the modern Puritan is *Evangelical*, &c. The word *Christian* (*ante*, p. 147) is an instance. When words begin, they carry their meanings. The Jews, who had their Messiah to come, and the followers of Jesus of Nazareth, who took *Him* for their Messiah, were both *Christians* (which means *Messianites*): the Jews would never have invented the term to signify *Jesuans*, nor would the disciples have invented such an ambiguous term for themselves; had they done so, the Jews would have disputed it, as they would have done in later times if they had had fair play. The Jews of our day, I see by their newspapers, speak of Jesus Christ as the *Rabbi Joshua*. But the Heathens, who knew little or nothing about the Jewish hope, would naturally apply the term *Christian* to the only followers of a *Messiah* of whom they had heard. For the *Jesuans* invaded them in a missionary way; while the Jews did not attempt, at least openly, to make proselytes.

All such words as Catholic, &c., are well enough as mere nomenclature; and the world falls for the most part, into any names which parties choose to give themselves. Silly people found inferences on this concession; and, as usually happens, they can cite some of their betters, St. Augustine, a freakish arguer, or, to put it in the way of an old writer, *lectorem ne multiloquii tœdio fastidiat, Punicis quibusdam argutiis recreare solet*, asks, with triumph, to what chapel a stranger would be directed, if he inquired the way to the *Catholic assembly*? But the best exhibition of this kind in our own century is that made by the excellent Dr. John Milner, in a work (first published in 1801 or 1802) which I suppose still circulates, 'The End of Religious Controversy:' a startling title which, so far as its truth is concerned, might as well have been 'The floor of the bottomless pit.' This writer, whom every one of his readers will swear to

---

[1] In the old ballad of King Alfred and the Shepherd, when the latter is tempting the disguised king into his service, he says:

> Of whig and whey we have good store,
> And keep good pease-straw fire.

*Whig* is then a preparation of milk. But another commonly cited derivation may be suspected from the word *whiggamor* being used before *whig*, as applied to the political party; *whig* may be a contraction. Perhaps both derivations conspired: the word *whiggamor*, said to be a word of command to the horses, might contract into *whig*, and the contraction might be welcomed for its own native meaning.

have been a worthy soul, though many, even of his own sect, will not admire some of his logic, speaks as follows :—

' Letter xxv. *On the true Church being Catholic.* In treating of this third mark of the true Church, as expressed in our common creed, I feel my spirits sink within me, and I am almost tempted to throw away my pen in despair. For what chance is there of opening the eyes of candid Protestants to the other marks of the Church, if they are capable of keeping them shut to this? Every time they address the God of Truth, either in solemn worship or in private devotion [stretch of rhetoric], they are forced, each of them, to repeat : *I believe in* THE CATHOLIC *Church,* and yet if I ask any of them the question : *Are you a* CATHOLIC *?* he is sure to answer me *No, I am a* PROTESTANT! Was there ever a more glaring instance of inconsistency and self-condemnation among rational beings ! '

> John Milner, honest and true,
> Did what honest people still may do,
> If they write for the many and not for the few,
> But what by and bye they must eschew.

He *shortened his clause*; and for a reason. If he had used the whole epithet which he knew so well, any one might have given his argument a half-turn. Had he written, as he ought, ' the *Holy* Catholic Church ' and then argued as above, some sly Protestant would have parodied him with ' and yet if I ask any of them the question : *Are you* HOLY *?* he is sure to answer me *No, I am a* SINNER.' To take the adjective from the Church, and apply it to the individual partisan, is recognised slipslop, but not ground of argument. If Dr. M. had asked his Protestant whether he belonged to the Catholic *Church,* the answer would have been Yes, but not to the Roman branch. When he put his question as he did, he was rightly answered and in his own division. This leaving out words is a common practice, especially when the omitter is in authority, and cannot be exposed. A year or two ago a bishop wrote a snubbing letter to a poor parson, who had complained that he was obliged, in burial, to send the worst of sinners to everlasting happiness. The bishop sternly said ' *hope* [1]

---

[1] It will be said that when the final happiness is spoken of in ' sure and certain hope,' it is *the* Resurrection, generally ; but when afterwards application is made to the individual, simple ' hope' is all that is predicated which merely means ' wish ?' I know it : but just before the general declaration, it is declared that it *has* pleased God of his great mercy to *take unto Himself*, the soul of our dear brother : and between the ' hopes' hearty thanks are given that it *has* pleased God to deliver our dear brother out of the miseries of this wicked world, with an additional prayer that the number of the elect may shortly be accomplished. All which means, that our dear

is not *assurance*.' Could the clergyman have dared to answer, he would have said, ' No, my Lord! but " *sure and certain* hope " is as like assurance as a *minikin* man is like a dwarf.' Sad to say, a theologian must be illogical : I fell sure that if you took the clearest headed writer on logic that ever lived, and made a bishop of him, he would be shamed by his own books in a twelvemonth.

Milner's sophism is glaring : but why should Dr. Milner be wiser than St. Augustine, one of his teachers ? I am tempted to let out the true derivation of the word *Catholic, as exclusively applied to the Church of Rome.* All can find it who have access to the *Rituale* of Bonaventura Piscator (lib. i. c. 12, *de nomine Sacræ Ecclesiæ*, p. 87 of the Venice folio of 1537). I am told that there is a *Rituale* in the Index Expurgatorius, but I have not thought it worth while to examine whether this be the one : I am rather inclined to think, as I have heard elsewhere, that the book was held too dangerous for the faithful to know of it, even by a prohibition : it would not surprise me at all if Roman Christians should deny its existence.[1]

It amuses me to give, at a great distance of time, a small Rowland for a small Oliver, which I received, *de par l'Église*, so far as lay in the Oliver-carrier more than twenty years ago. The following contribution of mine to *Notes and Queries* (3rd Ser. vi.

brother is declared to be taken to God, to be in a place not so miserable as this world —a description which excludes the ' wicked place '—and to be of the elect. Yes, but it will be said again! do you not know that when this Liturgy was framed, all who were not in the road to Heaven were excommunicated, and could not have the burial service read over them. Supposing the fact to have been true in old time, which is a very spicy supposition, how does that excuse the present practice ? Have you a right *always* to say what you believe *cannot always* be true, because you think it was once *always* true ? Yes, but, choose whom you please, you cannot be *certain* He is *not* gone to Heaven. True, and choose which Bishop you please, you cannot be demonstratively *certain*, he is *not* a concealed unbeliever: may I therefore say of the whole bench, *singulatim et seriatim*, that they *are* unbelievers ? No! No! The voice of common sense, of which common logic is a part, is slowly opening the eyes of the multitude to the unprincipled reasoning of theologians. Remember 1819. What chance had Parliamentary Reform when the House of Commons thanked the Manchester sabre-men ? If you do not reform your Liturgy, it will be reformed for you, and sooner than you think! The dishonest interpretations, by defence of which even the minds of children are corrupted, and which throw their shoots into literature and commerce, will be sent to the place whence they came : and over the door of the established organization for teaching religion will be posted the following notice :—

Shift and Subterfuge, Shuffle and Dodge,
No longer here allowed to lodge !

All this ought to be written by some one who belongs to the Establishment: in him, it would be quite prudent and proper ; in me, it is kind and charitable.

---

[1] This derivation has been omitted (ED.).

p. 175, Aug. 27, 1864) will explain what I say. There had been a complaint that a contributor had used the term *Papist*, which a very excellent dignitary of the Papal system pronounced an offensive term:—

### PAPIST.

The term *papist* should be stripped of all except its etymological meaning, and applied to those who give the higher and final authority to the declaration *ex cathedrâ* of the Pope. See Dr. Wiseman's article, *Catholic Church*, in the *Penny Cyclopædia*.

What is one to do about these names? First, it is clear that offence should, when possible, be avoided: secondly, no one must be required to give a name which favours *any* assumption made by those to whom it is given, and not granted by those who give it. Thus the subdivision which calls itself distinctively *Evangelical* has no right to expect others to concede the title. Now the word *Catholic*, of course, falls under this rule; and even *Roman Catholic* may be refused to those who would restrict the word *Catholic* to themselves. *Roman Christian* is unobjectionable, since the Roman Church does not deny the name of Christian to those whom she calls heretics. No one is bound in this matter by Acts of Parliament. In many cases, no doubt, names which have offensive association are used merely by habit, sometimes by hereditary transmission. Boswell records of Johnson that he always used the words 'dissenting teacher,' refusing *minister* and *clergyman* to all but the recipients of episcopal ordination.

This distinctive phrase has been widely adopted: it occurs in the Index of 3rd S. iv. [*Notes and Queries*]. Here we find 'Platts (Rev. John), Unitarian teacher, 412;' the article indexed has 'Unitarian minister.'

This, of course is habit: an intentional refusal of the word *minister* would never occur in an index. I remember that, when I first read about Sam Johnson's little bit of exclusiveness, I said to myself: 'Teacher? Teacher? surely I remember One who is often called *teacher*, but never *minister* or *clergyman*: have not the dissenters got the best of it?

When I said that the Roman Church concedes the epithet Christians to Protestants, I did not mean that all its adherents do the same. There is, or was, a Roman newspaper, the *Tablet*, which, seven or eight years ago, was one of the most virulent of the party journals. In it I read, referring to some complaint of grievance about mixed marriages, that if *Christians* would marry *Protestants* they must take the consequences. My memory notes this well; because I recollected, when I saw it, that there was in the stable a horse fit to run in the curricle with this one. About seventeen years ago an Oxford M.A., who hated mathematics like a genuine Oxonian of the last century, was writing on education, and was compelled to give some countenance to the nasty subject. He got out cleverly; for he gave as his reason for the permission, that man is an arithmetical, geometrical, and mechanical *animal*, as well as a rational *soul*.

The *Tablet* was founded by an old pupil of mine, Mr. Frederic Lucas; who availed himself of his knowledge of me to write some severe articles—even abusive, I was told, but I never saw them— against me, for contributing to the *Dublin Review*, and poking my heretic nose into orthodox places. Dr. Wiseman, the editor, came in for his share, and ought to have got all. Who ever blamed the pig for intruding himself into the cabin when the door was left open? When Mr. Lucas was my pupil, he was of the Society of Friends—in any article but this I should say *Quaker*—and was quiet and gentlemanly, as members of that Church—in any article but this I should; from mere habit, say *sect* —usually are. This is due to his memory; for, by all I heard, when he changed his religion he ceased to be Lucas couchant, and became Lucas rampant, fanged and langued gules. (I looked into Guillim to see if my terms were right: I could not find them; but to prove I have been there, I notice that he calls a violin a *violent*. How comes the word to take this form?) I met with several Roman Christians, born and bred, who were very much annoyed at Mr. Lucas and his doings; and said some severe things about new converts needing kicking-straps.

The mention of Dr. Wiseman reminds me of another word, appropriated by Christians to themselves: *fides*; the Roman faith is *fides*, and nothing else; and the adherents are *fideles*. Hereby hangs a retort. When Dr. Wiseman was first in England, he gave a course of lectures in defence of his creed, which were thought very convincing by those who were already convinced. They determined to give him a medal, and there was a very serious discussion about the legend. Dr. Wiseman told me himself that he had answered to his subscribers that he would not have the medal at all unless—(naming some Italian authority, whom I forget) approved of the legend. At last *pro fide vindicata* was chosen: this may be read either in a Popish or heretical sense. The feminine substantive *fides* means confidence, trust, (it is made to mean *belief*), but *fidis*, with the same ablative, *fide*, and also feminine, is a *fiddle-string*.[1] If a Latin writer had had to make a legend signifying 'For the defence of the fiddle-string,' he could not have done it otherwise, in the terseness of a legend, than by writing *pro fide vindicata*. Accordingly, when a Roman Christian talks to you of the *faith*, as a thing which is his and not yours, you may say *fiddle*. I have searched Bonaventura Piscator in vain for notice of this ambiguity. But the *Greeks* said fiddle; according to Suidas, σκινδαπσος—a word meaning a four stringed instrument played with a quill—was an exclamation of contemptuous dissent. How the wits of different races jump!

I am reminded of a case of *fides vindicata*, which, being in a public letter, responding to a public invitation, was not meant to be confidential. Some of the pupils of University College, in which all subdivisions of religion are (1866; *were*, 1867) on a level, have of course changed their views in after life, and become adherents of various high churches. On the occasion of a dinner of old students of the College, convened by circular, one of these students, whether then Roman or Tractarian Christian I do not remember, not content with simply giving negative answer, or none at all, concocted a jorum of theological rebuke, and sent it to the Dinner Committee. Heyday! said one of them, this man got out of bed backwards! How is that, said the rest? Why, read his name backwards, and you will see. As thus read it was— *No grub!*

To return to *Notes and Queries*. The substitution in the (editorial) index of 'Unitarian teacher,' for the contributor's 'Unitarian minister,' struck me very much. I have seldom found such things unmeaning. But as the journal had always been free from editorial sectarianisms,—and very apt to check the contributorial,—I could not be sure in this case. True it was, that the editor and publisher had been changed more than a year before; but this was not of much force. Though one swallow does not make a summer, I have generally found it show that summer is coming. However, thought I to myself, if this be Little Shibboleth, we shall have Big Shibboleth by-and-bye. At last it came. About a twelvemonth afterwards, (3rd S. vii. p. 36) the following was the *editorial* answer to the question

---

[1] The words are of the same root, and hence our word *fiddle*. Some suppose this root means a *rope*, which, as that to which you trust, becomes, in one divergence, confidence itself—just as *rock*, and other words, come to mean reliance—and in another, a little string.

when the establishment was first called the 'Church of England and Ireland:'—

'That unmeaning clause, "The United Church of England and Ireland," which occurs on the title-page of *The Book of Common Prayer*, was first used at the commencement of the present century. The authority for this phrase is the fifth article of the Union of 1800: "That the Churches of England and Ireland be united into one *Protestant* (!) episcopal Church, to be called 'The United Church of England and Ireland.' " Of course, churchmen are not responsible for the theology of Acts of Parliament, especially those passed during the dark ages of the Georgian era.'

That is to say, the journal gives its adhesion to the party which —under the assumed title of *the* Church of England—claims for the endowed corporation for the support of religion rights which Parliament cannot control, and makes it, in fact, a power above the State. The State has given an inch: it calls this corporation by the name of the 'United *Church* of England and Ireland,' as if neither England nor Ireland had any other Church. The corporation, accordingly aspires to an ell. But this the nation will only give with the aspiration prefixed. To illustrate my allusion in a delicate way to polite ears, I will relate what happened in a Johnian lecture-room at Cambridge, some fifty years ago, my informant being present. A youth of undue aspirations was giving a proposition, and at last said, 'Let E F be produced to 'L:' Not quite so far, Mr. ——, said the lecturer, quietly, to the great amusement of the class, and the utter astonishment of the aspirant, who knew no more than a Tractarian the tendency of his construction.

This word *Church* is made to have a very mystical meaning. The following dialogue between Ecclesiastes and Hæreticus, which I cannot vouch for, has often taken place in spirit, if not in letter:—E. The word *Church* (ἐκκλησια) is never used in the New Testament except generally or locally for that holy and mystical body to which the sacraments and the ordinances of Christianity are entrusted. H. Indeed! E. It is beyond a doubt (here he quoted half a dozen texts in support). H. Do you mean that any doctrine or ordinance which was solemnly practised by the ἐκκλησια is binding upon you and me? E. Certainly, unless we would be cut off from the congregation of the faithful. H. Have you a couple of hours to spare? E. What for? H. If you have, I propose we spend them in crying, Great is Diana of the Ephesians! E. What do you mean? H. You ought to know the solemn service of the ἐκκλησια

(Acts xix. 32, 41), at Ephesus; which any one might take to be true Church, by the more part not knowing wherefore they were come together, and which was dismissed, after one of the most sensible sermons ever preached, by the Recorder.  E.  I see your meaning : it is true, there is that one exception!  H.  Why, the Recorder's sermon itself contains another, the ἔννομος ἐκκλησια, legislative assembly.  E.  Ah! the New Testament can only be interpreted by the Church!  H.  I see! the Church interprets itself into existence out of the New Testament, and then interprets the New Testament out of existence into itself!

I look upon all the Churches as fair game which declare of me that *absque dubio in æternum peribo*; not for their presumption towards God, but for their personal insolence towards myself. I find that their sectaries stare when I say this.  Why! they do not speak of you *in particular*!  These poor reasoners seem to think that there could be no meaning, as against me, unless it should be propounded that 'without doubt he shall perish everlastingly, especially A. De Morgan.'  But I hold, with the schoolmen, that '*Omnis* homo est animal' in conjunction with 'Sortes est homo' amounts to '*Sortes* est animal.'  But they do not mean it *personally*!  Every universal proposition is personal to every instance of the subject.  If this be not conceded, then I retort, in their own sense and manner, 'Whosoever would serve God, before all things he must not pronounce God's decision upon his neighbour.  Which decision, except everyone leave to God himself, without doubt he is a bigoted noodle.'

The reasoning habit of the educated community, in four cases out of five, permits universal propositions to be stated at one time, and denied, *pro re nata*, at another.  'Before we proceed to consider any question involving physical principles, we should set out with *clear ideas* of the naturally possible and impossible.' The eminent man who said this, when wanting it for his views of mental education (!) never meant it for more than what was in hand, never assumed it in the researches which will give him to posterity!  I have heard half-a-dozen defences of his having said this, not one of which affirmed the truth of what was said. A worthy clergyman wrote that if A. B. had said a certain thing the point in question would have been established.  It was shown to him that A.B. *had* said it, to which the reply was a refusal to admit the point because A. B. said it in a second pamphlet and in answer to objections.  And I might give fifty such instances with very little search.  Always assume more than you want;

because you cannot tell how much you may want : put what is over into the didn't-mean-that basket, or the extreme case what-not.

Something near forty years of examination of the theologies on and off—more years very much on than quite off—have given me a good title—to myself, I ask no one else for leave— to make the following remarks : A conclusion has *premises*, facts or doctrines from proof or authority, and *mode of inference*. There may be invention or falsehood of premise, with good logic ; and there may be tenable premise, followed by bad logic ; and there may be both false premise *and* bad logic. The Roman system has such a powerful manufactory of premises, that bad logic is little wanted ; there is comparatively little of it. The doctrine-forge of the Roman Church is one glorious compound of everything that could make Heraclitus sob and Democritus snigger. But not the only one. The Protestants, in tearing away from the Church of Rome, took with them a fair quantity of the results of the Roman forge, which they could not bring themselves to give up. They had more in them of Martin than of Jack. But they would have no premises, except from the New Testament ; though some eked out with a few general Councils. The consequence is that they have been obliged to find such a logic as would bring the conclusions they require out of the canonical books. And a queer logic it is ; nothing but the Roman forge can be compared with the Protestant loom. The picking, the patching, the piecing, which goes to the Protestant *termini ad quem*, would be as remarkable to the general eye, as the Roman manufacture of *termini a quo*, if it were not that the world at large seizes the character of an asserted fact better than that of a mode of inference, A grand step towards the deifica- cation of a lady, made by alleged revelation 1800 years after her death, is of glaring evidence : two or three additional shiffle- shuffles towards defence of saying the Athanasian curse in church and unsaying it out of church, are hardly noticed. Swift has bungled his satire where he makes Peter a party to finding out what he wants, *totidem syllabis* and *totidem literis*, when he cannot find it *totidem verbis*. This is Protestant method : the Roman plan is *viam faciam*; the Protestant plan is *viam inveniam*. The public at large begins to be conversant with the ways of *wriggling out*, as shown in the interpretations of the damnatory parts of the Athanasian Creed, the phrases of the Burial Service, &c. The time will come when the same public will begin to see the ways of *wriggling in*. But one thing at a time :

neither Papal Rome nor Protestant Rome was built—nor will be pulled down—in a day.

The distinction above drawn between the two great antitheses of Christendom may be illustrated as follows. Two sets of little general dealers lived opposite to one another : all sold milk. Each vaunted its own produce : one set said that the stuff on the other side the way was only chalk and water; the other said that the opposites sold all sorts of filth, of which calves' brain was the least nasty. Now the fact was that both sets sold milk, and from the same dairy : but adulterated with different sorts of dirty water : and both honestly believed that the mixture was what they were meant to sell and ought to sell. The great difference between them, about which the apprentices fought each other like Trojans, was that the calves' brain men poured milk into the water, and the chalk men poured water into the milk. The Greek and Roman sects on one side, the Protestant sects on the other, must all have *churches* : the Greek and Roman sects pour the New Testament into their churches ; the Protestant sects pour their churches into the New Testament. The Greek and Roman insist upon the New Testament being no more than part and parcel of their churches : the Protestant insist upon their churches being as much part and parcel of the New Testament. All dwell vehemently upon the doctrine that there must be milk somewhere ; and each says—I have it. The doctrine is true : and can be verified by anyone who can and will go to the dairy for himself. Him will the several traders declare to have no milk at all. They will bring their own wares, and challenge a trial : they want nothing but to name the judges. To vary the metaphor, those who have looked at Christianity in open day, know that all who see it through painted windows shut out much of the light of heaven and colour the rest; it matters nothing that the stains are shaped into what are meant for saints and angels.

But there is another side of the question. To decompose any substance, it must be placed between the poles of the battery. Now theology is but one pole ; philosophy is the other. No one can make out the combinations of our day unless he read the writings both of the priest and the philosopher : and if any one should hold the first word offensive, I tell him that I mean *both* words to be *significant*. In reading these writings, he will need to bring both wires together to find out what it is all about. Time was when most priests were very explicit about the fate of philosophers, and most philosophers were very candid about their

opinion of priests. But though some extremes of the old sorts
still remain, there is now, in the middle, such a fusion of the two
pursuits that a plain man is wofully puzzled. The theologian
writes a philosophy which seems to tell us that the New Testament is a system of psychology; and the philosopher writes a
Christianity which is utterly unintelligible as to the question
whether the Resurrection be a fact or a transcendental allegory.
What between the theologian who assents to the Athanasian
denunciation in what seems the sense of no denunciation, and
the philosopher who parades a Christianity which looks like no
revelation, there is a maze which threatens to have the only
possible clue in the theory that everything is something else, and
nothing is anything at all. But this is a paradox far beyond
my handling : it is a Budget of itself.

Religion and Philosophy, the two best gifts of Heaven, set up
in opposition to each other at the revival of letters ; and never
did competing tradesmen more grossly misbehave. Bad wishes
and bad names flew about like swarms of wasps. The Athanasian
curses were intended against philosophers ; who, had they been
a corporation, with state powers to protect them, would have
formulized a *per contra*. But the tradesmen are beginning to
combine : they are civil to each other ; too civil by half. I speak
especially of Great Britain. Old theology has run off to ritualism,
much lamenting, with no comfort except the discovery that the
cloak Paul left at Troas was a chasuble. Philosophy, which
always had a little sense sewed up in its garments—to pay for its
funeral ?—has expended a trifle in accommodating itself to the
new system. But the two are poles of a battery ; and a question
arises.

> If Peter Piper picked a peck of pepper,
> Where is the peck of pepper Peter Piper picked ?

If Religion and Philosophy be the two poles of a battery, whose
is the battery Religion and Philosophy have been made the poles
of ? Is the change in the relation of the wires any presumption
of a removal of the managers ? We know pretty well who
handled the instrument : has he resigned or been [1] turned out ?
Has he been put under restriction ? A fool may ask more
questions than twenty sages can answer : but there is hope ; for

---

[1] The notion that the Evil Spirit is a functionary liable to be dismissed for not
attending to his duty, is, so far as my reading goes, utterly unknown in theology.
My first wrinkle on the subject was the remark of the Somersetshire farmer upon
Palmer the poisoner—'Well ! if the Devil don't take he, he didn't ought to be allowed
to be devil no longer.'

T

twenty sages cannot ask more questions than one reviewer can answer. I should like to see the opposite sides employed upon the question, What are the *commoda*, and what the *pericula*, of the current approximation of Religion and Philosophy?

All this is very profane and irreverent! It has always been so held by those whose position demands such holding. To describe the Church as it is passes for assailing the Church as it ought to be with all who cannot do without it. In Bedlam a poor creature who fancied he was St. Paul, was told by another patient that he was an impostor; the first maniac lodged a complaint against the second for calling St. Paul an impostor, which, he argued, with much appearance of sanity, ought not to be permitted in a well regulated madhouse. Nothing could persuade him that he had missed the question, which was whether *he* was St. Paul. The same thing takes place in the world *at large*. And especially must be noted the refusal to permit to the *profane* the millionth part of the licence assumed by the *sacred*. I give a sound churchman the epitaph on St. John Long; the usual pronunciation of whose name must be noted—

> Behold! ye quacks, the vengeance strong
> On deeds like yours impinging:
> For here below lies St. John Long
> Who now must be *long singeing*.

How shameful to pronounce this of the poor man! What, Mr. Orthodox! may I not do in joke to one pretender what you do in earnest—unless you quibble—to all the millions of the Greek Church, and a great many others. Enough of you and your reasoning! Go and square the circle!

The few years which end with 1867 have shown, not merely the intermediate fusion of Theology and Philosophy of which I have spoken, but much concentration of the two extremes, which looks like a gathering of forces for some very hard fought Armageddon. Extreme theology has been aiming at a high Church in England, which is to show a new front to all heresy: and extreme philosophy is contriving a physical organisation which is to *think*, and to show that mind is a consequence of matter, or thought a recreation of brain. The physical speculators begin with a possible hypothesis, in which they aim at explanation : and so the bold aspirations of the author of the 'Vestiges' find standing-ground in the variation of species by 'natural selection.' Some relics—so supposed—of extremely ancient men are brought to help the general cause. Only distant hints are given that by possibility it may end in the formation of all living organisms from a very few, if not from one. The better heads abovementioned know that their theory, if true, does not bear upon morals. The formation of solar systems from a nebular hypothesis, followed by organisations gradually emerging from some curious play of particles, nay, the very evolution of mind and thought from such an apparatus, are all as consistent with a Personal creative power to whom homage and obedience are due, and who has declared himself, as with a blind Nature of Things. A pure materialist, as to all things visible, may be even a bigotted Christian : this is not frequent, but it is possible. There is a proverb which says, A pig may fly, but it isn't a likely bird. But when the psychological speculator comes in, he often undertakes to draw inferences from the physical conclusions, by joining on his tremendous apparatus of à priori knowledge. He deduces that he can *do without* a God : he can deduce all things without any such necessity. With Occam and Newton he will have no more causes than are necessary to explain phenomena *to him* : and if by pure head-work combined with results of physical observation he can construct his universe, he must be a very *unphilosophical* man who would encumber himself with a useless Creator! There is something tangible about my method, says he ; yours is vague. He requires it to be granted that his system is *positive* and that your's is *impositive.* So reasoned the stage coachman when the railroads began to depose him—' If you're upset in a stage-coach, why, there you are! but if you're upset on the railroad, where are you ? ' The answer lies in another question, Which is most positive knowledge, God deduced from man and his history, or the postulates of the few who think they can reason à *priori* on the tacit assumption of unlimited command of data ?

We are not yet come to the existence of a school of philosophers who explicitly deny a Creator : but we are on the way, though common sense may interpose. There are always straws which show the direction of the wind. I have before me the printed letter of a medical man—to whose professional ability I have good testimony—who finds the vital principle in highly rarefied oxygen. With the usual logic of such thinkers, he dismisses the ' eternal personal identity ' because ' If soul, spirit, mind, which are merely modes of sensation, be the attribute or function of nerve-tissue, it cannot possibly have any existence apart from its material organism!' How does he know this *impossibility* ? If all the mind *we* know be from nerve-tissue, how does it appear that mind in other planets may not be another thing? Nay, when we come to *possibilities*, does not his own system give a queer one? If highly rarefied oxygen be vital power, more highly rarefied oxygen may be more vital and more powerful. Where is this to stop? Is it *impossible* that a finite quantity, rarefied *ad infinitum*, may be an Omnipotent? Perhaps the true Genesis, when written, will open with 'In the beginning was an imperial quart of oxygen at 60° of Fahrenheit, and the pressure of the atmosphere; and this oxygen was infinitely rarefied; and this oxygen became God.' For myself, my aspirations as to this system are Manichæan. The quart of oxygen is the Ormuzd, or good principle: another quart, of hydrogen, is the Ahriman, or evil principle! My author says that his system explains Freewill and Immortality so obviously that it is difficult to read previous speculations with becoming gravity. My deduction explains the conflict of good and evil with such clearness that no one can henceforward read the New Testament with becoming reverence. The surgeon whom I have described is an early bud which will probably be nipped by the frost and wither on the ground: but there is a good crop coming. Material pneuma is destined to high functions; and man is to read by gas-light.

> The solar system truly solved; demonstrating by the perfect harmony of the planets, founded on the four universal laws, the Sun to be an electrical space; and a source of every natural production displayed throughout the solar system. By James Hopkins. London, 1849, 8vo.

The author says :—

'I am satisfied that I have given the true *laws* constituting the *Sun* to be *space*; and I call upon those disposed to maintain the contrary, to give true *laws* showing him to be a body: until such can be satisfactorily established, I have an undoubted claim to the credit of my theory,—That the Sun is an *Electric Space*, fed and governed by the planets, which have the property of attracting heat from it; and the means of supplying the necessary *pabulum* by their degenerated air driven off towards the central space—the wonderful alembic in which it becomes transmuted to the revivifying necessities of continuous action; and the central space or Sun being perfectly electric, has the counter property of repulsing the bodies that attract it. How wonderful a conception! How beautiful, how magnificent an arrangement!

' O Centre! O Space! O Electric Space! '

1849. *Joseph Ady* is entitled to a place in this list of discoverers: his great fault, like that of some others, lay in pushing his method too far. He began by detecting unclaimed dividends, and disclosing them to their right owners, exacting his fee before he made his communication. He then generalized into trying to get fees from all of the *name* belonging to a dividend; and he gave mysterious hints of danger impending. For instance, he would write to a clergyman that a legal penalty was hanging over him; and when the alarmed divine forwarded the sum required for disclosure, he was favoured with an extract from some old statute or canon, never repealed, forbidding a clergyman to be a member of a corporation, and was reminded that he had insured his life in the —— Office, which had a royal charter. He was facetious, was Joseph: he described himself in his circulars as 'personally known to Sir Peter Laurie and all other aldermen'; which was nearly true, as he had been before most of them on charges of false pretence; but I believe he was nearly always within the law. Sir James Duke, when Lord Mayor, having particularly displeased him by a decision, his circulars of 1849 contain the following :—

'Should you have cause to complain of any party, Sir J. Duke has contrived a new law of evidence, viz., write to him, he will consider your letter sufficient proof, and make the parties complained of pay without judge or jury, and will frank you from every expense.'

I strongly suspect that Joseph Ady believed in himself.

He sometimes issued a second warning, of a Sibylline character :—

'Should you find cause to complain of anybody, my voluntary referee, the Rt. Hon. Sir Peter Laurie, Kt., perpetual Deputy Lord Mayor, will see justice done you without any charge whatever : he and his toady, — —— ——. The accursed of Moses can hang any man : thus, by catching him alone and swearing Naboth spake evil against God and the King. Therefore (!) I admit no strangers to a personal conference without a prepayment of 20s. each. Had you attended to my former notice you would have received twice as much : neglect this and you will lose all.'

Zadkiel's Almanac for 1849. Nineteenth number.
Raphael's Prophetic Almanac for 1849. Twenty-ninth number.
Reasons for belief in judicial astrology, and remarks on the dangerous character of popish priestcraft. London, 1849, 12mo.
Astronomy in a nutshell : or the leading problems of the solar system solved by simple proportion only, on the theory of magnetic attraction. By Lieut. Morrison, R.N. London (*s. a.*) 12mo.

Lieut. Morrison is Zadkiel Tao Sze, and declares himself in real earnest an astrologer. There are a great many books on astrology, but I have not felt interest enough to preserve many of them which have come in my way. If anything ever had a fair trial, it was astrology. The idea itself is natural enough. A human being, set down on this earth, without any tradition, would probably suspect that the heavenly bodies had something to do with the guidance of affairs. I think that any one who tries will ascertain that the planets do not prophesy : but if he should find to the contrary, he will of course go on asking. A great many persons class together belief in astrology and belief in apparitions: the two things differ in precisely the way in which a science of observation differs from a science of experiment. Many make the mistake which M. le Marquis made when he came too late, and hoped M. Cassini would do the eclipse over again for his ladies. The apparition chooses its own time, and comes as seldom or as often.as it pleases, be it departed spirit, nervous derangement, or imposition. Consequently it can only be observed, and not experimented upon. But the heavens, if astrology be true, are prophesying away day and night all the year round, and about every body. Experiments can be made, then, except only on rare phenomena, such as eclipses : anybody may choose his time and his question. This is the great difference : and experiments were made, century after century. If astrology had been true, it must have lasted in an ever-improving state. If it be true, it is a truth, and a useful truth, which had experience and prejudice both in its favour, and yet lost ground as soon as astronomy, its working tool, began to improve.

1850. A letter in the handwriting of an educated man, dated from a street in which it must be taken that educated persons live, is addressed to the Secretary of the Astronomical Society about a matter on which the writer says ' his professional pursuit will enable him to give a satisfactory reply.' In a question before a court of law it is sworn on one side that the moon was shining at a certain hour of a certain night on a certain spot in London ; on the other side it is affirmed that she was clouded. The Secretary is requested to decide. This is curious, as the question is not astrological. Persons still send to Greenwich, now and then, to have their fortunes told. In one case, not very many years ago, a young gentleman begged to know who his wife was to be, and what fee he was to remit.

Sometimes the astronomer turns conjurer for fun, and his prophecies are fulfilled. It is related of Flamsteed that an old woman came to know the whereabouts of a bundle of linen which had strayed. Flamsteed drew a circle, put a square into it, and gravely pointed out a ditch, near her cottage, in which he said it would be found. He meant to have given the woman a little good advice when she came back : but she came back in great delight, with the bundle in her hand, found in the very place. The late Baron Zach received a letter from Pons, a successful finder of comets, complaining that for a certain period he had found no comets, though he had searched diligently. Zach, a man of much sly humour, told him that no spots had been seen on the sun for about the same time—which was true,—and assured him that when the spots came back, the comets would come with them. Some time after he got a letter from Pons, who informed him with great satisfaction that he was quite right, that very large spots had appeared on the sun, and that he had found a fine comet shortly after. I do not vouch for the first story, but I have the second in Zach's handwriting. It would mend the joke exceedingly if some day a real relation should be established between comets and solar spots : of late years good reason has been shown for advancing a connexion between these spots and the earth's magnetism. If the two things had been put to Zach, he would probably have chosen the comets. Here is a hint for a paradox : the solar spots are the dead comets, which have parted with their light and heat to feed the sun, as was once suggested. I should not wonder if I were too late, and the thing had been actually maintained. My list does not contain the twentieth part of the possible whole.

The mention of coincidences suggests an everlasting source of explanations, applicable to all that is extraordinary. The great paradox of coincidence is that of Leibnitz, known as the *pre-established harmony*, or *law of coincidences*, by which, separately and independently, the body receives impressions, and the mind proceeds as if it had perceived them from without. Every sensation, and the consequent state of the soul, are independent things coincident in time by the pre-established law. The philosopher could not otherwise *account for* the connexion of mind and matter ; and he never goes by so vulgar a rule as *Whatever is, is* ; to him that which is not clear as to how, is not at all. Philosophers in general, who tolerate each other's theories much better than Christians do each other's failings, seldom revive Leibnitz's fantasy : they seem to act upon the maxim quoted by Father Eustace from the Decretals, *Facinora ostendi dum punientur, flagitia autem abscondi debent.*

The great *ghost-paradox*, and its theory of *coincidences*, will rise to the surface in the mind of everyone. But the use of the word *coincidence* is here at variance with its common meaning. When A is constantly happening, and also B, the occurrence of A and B at the same moment is the mere coincidence which may be casualty. But the case before us is that A is constantly happening, while B, when it does happen, almost always happens with A, and very rarely without it. That is to say, such is the phenomenon asserted : and all who rationally refer it to casualty, affirm that B is happening very often as well as A, but that it is not thought worthy of being recorded except when A is simultaneous. Of course A is here a death, and B the spectral appearance of the person who dies. In talking of this subject it is necessary to put out of the question all who play fast and loose with their secret convictions : these had better give us a reason, when they feel internal pressure for explanation, that there is no weathercock at Kilve ; this would do for all cases. But persons of real inquiry will see that first, experience does not bear out the asserted frequency of the spectre, without the alleged coincidence of death : and secondly, that if the crowd of purely casual spectres were so great that it is no wonder that, now and then the person should have died at or near the moment, we ought to expect a much larger proportion of cases in which the spectre should come at the moment of the death of one or another of all the cluster who are closely connected with the original of the spectre. But this, we know, is almost without example. It remains then, for all, who speculate at all, to look upon the asserted phenomenon, think what they may of it, the thing which is to be explained, as a *connexion* in time of the death, and the simultaneous appearance of the dead. Any person the least used to the theory of probabilities will see that purely casual coincidence, the *wrong spectre* being comparatively so rare that it may be said never to occur, is not within the rational field of possibility.

The purely casual coincidence, from which there is no escape except the actual doctrine of special providences, carried down to a very low point of special intention, requires a junction of the things the like of each of which is always happening. I will give three instances which have occurred to myself within the last few years : I solemnly vouch for the literal truth of every part of all three :

In August 1861, M. Senarmont, of the French Institute, wrote to me to the effect that Fresnel had sent to England, in or shortly after 1824, a paper for translation and insertion in the

*European Review*, which shortly afterwards expired. The question was what had become of that paper. I examined the Review at the Museum, found no trace of the paper, and wrote back to that effect at the Museum, adding that everything now depended on ascertaining the name of the editor, and tracing his papers : of this I thought there was no chance. I posted this letter on my way home, at a Post Office in the Hampstead Road at the junction with Edward Street, on the opposite side of which is a bookstall. Lounging for a moment over the exposed books, *sicut meus est mos*, I saw, within a few minutes of the posting of the letter, a little catch-penny book of anecdotes of Macaulay, which I bought, and ran over for a minute. My eye was soon caught by this sentence :—' One of the young fellows immediately wrote to the editor (Mr. Walker) of the *European Review.*' I thus got the clue by which I ascertained that there was no chance of recovering Fresnel's paper. Of the mention of current reviews, not one in a thousand names the editor.

In the summer of 1865 I made my first acquaintance with the tales of Nathaniel Hawthorne, and the first I read was about the siege of Boston in the War of Independence. I could not make it out : everybody seemed to have got into somebody else's place. I was beginning the second tale, when a parcel arrived : it was a lot of old pamphlets and other rubbish, as he called it, sent by a friend who had lately sold his books, had not thought it worth while to send these things for sale, but thought I might like to look at them and possibly keep some. The first thing I looked at was a sheet which, being opened, displayed ' A plan of Boston and its environs, shewing the true situation of his Majesty's army and also that of the rebels, drawn by an engineer, at Boston Oct. 1775.' Such detailed plans of current sieges being then uncommon, it is explained that ' The principal part of this plan was surveyed by Richard Williams, Lieutenant at Boston ; and sent over by the son of a nobleman to his father in town, by whose permission it was published.' 1 immediately saw that my confusion arose from my supposing that the king's troops were besieging the rebels, when it was just the other way.

April 1, 1853, while engaged in making some notes on a logical point, an idea occurred which was perfectly new to me, on the mode of conciliating the notions of *omnipresence* and *indivisibility into parts*. What it was is no matter here : suffice it that, since it was published elsewhere (in a paper on *Infinity, Camb. Phil. Trans.* vol. xi. p. 1) I have not had it produced to me. I had just finished a paragraph on the subject, when a parcel came in from a bookseller containing Heywood's 'Analysis of Kant's Critick,' 1844. On turning over the leaves I found (p. 109) the identical thought which up to this day, I only know as in my own paper, or in Kant. I feel sure I had not seen it before, for it is in Kant's first edition, which was never translated to my knowledge; and it does not appear in the later editions. Mr. Heywood gives some account of the first edition.

In the broadsheet which gave account of the dying scene of Charles II., it is said that the Roman Catholic priest was introduced by P. M. A. C. F. The chain was this : the Duchess of Portsmouth applied to the Duke of York, who may have consulted his Cordelier confessor, Mansuete, about procuring a priest, and the priest was smuggled into the king's room by the Duchess and Chiffinch. Now the letters are a verbal acrostic of *Père Mansuete a Cordelier Friar*, and a syllabic acrostic of *PortsMouth and ChifFinch*. This is a singular coincidence. Macaulay adopted the first interpretation, preferring it to the second, which I brought before him as the conjecture of a near relative of my own. But Mansuete is not mentioned in his narrative : it may well be doubted whether the writer of a broadside for English readers would use *Père* instead of *Father*. And the person who really 'reminded' the Duke of 'the duty he owed to his brother,' was the Duchess and not Mansuete. But my affair is only with the coincidence.

But there are coincidences which are really connected without the connexion being known to those who find in them matter of astonishment. Presentiments furnish marked cases : sometimes there is no mystery to those who have the clue. In the *Gentleman's Magazine* (vol. 80, part 2, p. 33) we read, the subject being presentiment of death, as follows :—' In 1778, to come nearer the recollection of survivors, at the taking of Pondicherry, Captain John Fletcher, Captain De Morgan, and Lieutenant Bosanquet, each distinctly foretold his own death on the morning of his fate.' I have no doubt of all three ; and I knew it of my grandfather long before I read the above passage. He saw that the battery he commanded was unduly exposed : I think by the sap running through the fort when produced. He represented this to the engineer officers, and to the commander-in-chief; the engineers denied the truth of the statement, the commander believed them, my grandfather quietly observed that he must make his will, and the French fulfilled his prediction. His will bore date the day of his death ; and I always thought it more remarkable than the fulfilment of the prophecy that a soldier should not consider any danger short of one like the above, sufficient reason to make his will. I suppose the other officers were similarly posted. I am told that military men very often defer making their wills until just before an action: but to face the ordinary risks intestate, and to wait until speedy death must be the all but certain consequence of a stupid mistake, is carrying the principle very far. In the matter of coincidences there are, as in other cases, two wonderful extremes with every intermediate degree. At one end we have the confident people who can attribute anything to casual coincidence; who allow Zadok Imposture and Nathan Coincidence to anoint Solomon Self-conceit king. At the other end we have those who see something *very curious* in any coincidence you please, and whose minds yearn for a deep reason. A speculator of this class happened to find that Matthew viii. 28–33 and Luke viii. 26–33 contain the same account, that of the demons entering into the swine. Very odd! chapters tallying, and verses so nearly: is the versification rightly managed? Examination is sure to show that there are monstrous inconsistencies in the mode of division, which being corrected, the verses tally as well as the chapters. And then how comes it? I cannot go on, for I have no gift at torturing a coincidence; but I would lay twopence, if I could make a bet—which I never did in all my life—that some one or more of my readers will try it. Some people say that the study of chances tends to awaken a spirit of gambling: I suspect the contrary. At any rate, I myself, the writer of a mathematical book and a comparatively popular book, have never laid a bet nor played for a stake, however small: not one single time.

It is useful to record such instances as I have given, with precision and on the solemn word of the recorder. When such a story as that of Flamsteed is told, *à priori* assures us that it could not have been: the story may have been a *ben trovato*, but not the bundle. It is also useful to establish some of the good jokes which all take for inventions. My friend Mr. J. Bellingham Inglis, before 1800, saw the tobacconist's carriage with a sample of tobacco in a shield, and the motto *Quid rides* (*N & Q.*, 3rd S. i. 245). His father was able to tell him all about it. The tobacconist was Jacob Brandon, well known to the elder Mr. Ir 'is, and the person who started the motto, the instant he was a⌄ked for such a thing, was Harry Calender of Lloyd's, a scholar and a wit. My friend Mr. H. Crabb Robinson remembers the King's Counsel (Samuel Marryat) who took the motto *Causes produce effects*, when his success enabled him to start a carriage.

The coincidences of errata are sometimes very remarkable : it may be that the misprint has a sting. The death of Sir W. Hamilton of Edinburgh was known in London on a Thursday, and the editor of the *Athenæum* wrote to me in the afternoon for a short obituary notice to appear on Saturday. I dashed off the few lines which appeared without a moment to think : and those of my readers who might perhaps think me capable of contriving errata with meaning will, I am sure, allow the hurry, the occasion, and my own peculiar relation to the departed, as sufficient reasons for believing in my entire innocence. Of course I could not see a proof : and two errata occurred. The words 'addition to Stewart' require '*for* addition to *read* edition of.' This represents what had been insisted on by the Edinburgh publisher, who, frightened by the edition of Reid, had stipulated for a simple reprint without notes. Again 'principles of logic and mathematics' required '*for* mathematics *read* metaphysics.' No four words could be put together which would have so good a title to be Hamilton's motto.

April 1850, found in the letter-box, three loose leaves, well printed and over punctuated, being

> Chapter VI. Brethren, lo I come, holding forth the word of life, for so I am commanded . . . . Chapter VII. Hear my prayer, O generations ! and walk by the way, to drink the waters of the river . . . . Chapter VIII. Hearken o earth, earth, earth, and the kings of the earth, and their armies . . . .

A very large collection might be made of such apostolic writings. They go on well enough in a misty—meant for mystical—imitation of St. Paul or the prophets, until at last some prodigious want of keeping shows the education of the writer. For example, after half a page which might pass for Irving's preaching—though a person to whom it was presented as such would say that most likely the head and tail would make something more like head and tail of it—we are astounded by a declaration from the *Holy Spirit*, speaking of himself, that he is 'not ashamed of the Gospel of Christ.' It would be long before we should find in *educated* rhapsody—of which there are specimens enough—such a thing as a person of the Trinity taking merit for moral courage enough to stand where St. Peter fell. The following declaration comes next—'I will judge between cattle and cattle, that use their tongues.'

The figure of the earth. By J. L. Murphy, of Birmingham. (London and Birmingham, 4 pages, 12mo.) (1850 ?)

Mr. Murphy invites attention and objection to some assertions, as that the earth is prolate, not oblate. 'If the philosopher's conclusion be right, then the pole is the centre of a valley (!) thirteen miles deep.' Hence it would be very warm. It is answer enough to ask—Who knows that it is not?

---

\*\*\* A paragraph in the MS. appears to have been inserted in this place by mistake. It will be found in the Appendix at the end of this volume.

---

1851. The following letter was written by one of a class of persons whom, after much experience of them, I do *not* pronounce insane. But in this case the second sentence gives a suspicion of actual delusion of the senses ; the third looks like that eye for the main chance which passes for sanity on the Stock Exchange and elsewhere :—

15th Sept. 1851.

'Gentlemen,—I pray you take steps to make known that yesterday I completed my invention which will give motion to every country on the Earth ;—to move Machinery !—the long sought in vain 'Perpetual Motion'! !—I was supported at the time by the Queen and H.R.H. Prince Albert. If, Gentlemen, you can advise me how to proceed to claim the reward, if any is offered by the Government, or how to secure the PATENT for the machine, or in any way assist me by advice in this great work, I shall most graciously acknowledge your consideration.

These are my convictions that my SEVERAL discoveries will be realised : and this great one can be at once acted upon : although at this moment it only exists in my mind, from my knowledge of certain fixed principles in nature :—the Machine I have not made, as I only completed the discovery YESTERDAY, Sunday !

I have, &c.    —— ——'

To the Directors of the
London University, Gower Street.

*The Divine Drama of History and Civilisation.* By the Rev.
James Smith, M.A. London, 1854, 8vo.

I have several books on that great paradox of our day, *Spiritualism,* but I shall exclude all but three. The bibliography of
this subject is now very large. The question is one both of
evidence and speculation ;—Are the facts true? Are they caused
by spirits? These I shall not enter upon: I shall merely recommend this work as that of a spiritualist who does not enter on
the subject, which he takes for granted, but applies his derived
views to the history of mankind with learning and thought. Mr.
Smith was a man of a very peculiar turn of thinking. He was,
when alive, the editor, or *an* editor, of the *Family Herald* : I
say when alive, to speak according to knowledge ; for, if his own
views be true, he may have a hand in it still. The answers to
correspondents, in his time, were piquant and original above any
I ever saw. I think a very readable book might be made out
of them, resembling 'Guesses at Truth :' the turn given to an
inquiry about morals, religion, or socials, is often of the highest
degree of *unexpectedness* ; the poor querist would find himself
right in a most unpalatable way.

Answers to correspondents, in newspapers, are very often the
fag ends of literature. I shall never forget the following. A
person was invited to name a rule without exception, if he could :
he answered 'A man *must* be present when he is shaved.' A
lady—what right have ladies to decide questions about shaving?
—said this was not properly a rule ; and the oracle was consulted.
The editor agreed with the lady ; he said that ' a man *must* be
present when he is shaved' is not a *rule,* but a *fact.*

[Among my anonymous communicants is one who states that
I have done injustice to the Rev. James Smith in 'referring
to him as a spiritualist,' and placing his 'Divine Drama' among
paradoxes: ' it is no paradox, nor do *spiritualistic* views mar or
weaken the execution of the design.' Quite true : for the design
is to produce and enforce 'spiritualistic views;' and leather does
not mar nor weaken a shoemaker's plan. I knew Mr. Smith
well, and have often talked to him on the subject: but more
testimony from me is unnecessary ; his book will speak for itself.
His peculiar style will justify a little more quotation than is just
necessary to prove the point. Looking at the ' battle of opinion '
now in progress, we see that Mr. Smith was a prescient :—

(P. 588.) ' From the general review of parties in England, it is evident that no country in the world is better prepared for the great Battle of Opinion. Where else can the battle be fought but where the armies are arrayed? And here they all are, Greek, Roman, Anglican, Scotch, Lutheran, Calvinist, Established and Territorial, with Baronial Bishops, and Non-established of every grade—churches with living prophets and apostles, and churches with dead prophets and apostles, and apostolical churches without apostles, and philosophies without either prophets or apostles, and only wanting one more, " the Christian Church," like Aaron's rod, to swallow up and digest them all, and then bud and flourish. As if to prepare our minds for this desirable and inevitable consummation, different parties have been favoured with a revival of that very spirit of revelation by which the Church itself was originally founded. There is a complete series of spiritual revelations in England and the United States, besides mesmeric phenomena that bear a resemblance to revelation, and thus gradually open the mind of the philosophical and infidel classes, as well as the professed believers of that old revelation which they never witnessed in living action, to a better understanding of that Law of Nature (for it is a Law of Nature) in which all revelation originates and by which its spiritual communications are regulated.'

Mr. Smith proceeds to say that there are *only* thirty-five incorporated churches in England, all formed from the New Testament except five, to each of which five he concedes a revelation of its own. The five are the Quakers, the Swedenborgians, the Southcottians, the Irvingites, and the Mormonites. Of Joanna Southcott he speaks as follows:—

(P. 592.) ' Joanna Southcott is not very gallantly treated by the gentlemen of the Press, who, we believe, without knowing anything about her, merely pick up their idea of her character from the rabble. We once entertained the same rabble idea of her; but having read her works—for we really have read them —we now regard her with great respect. However, there is a great abundance of chaff and straw to her grain; but the grain is good, and as we do not eat either the chaff or straw if we can avoid it, nor even the raw grain, but thrash it and winnow it, and grind it and bake it, we find it, after undergoing this process, not only very palatable, but a special dainty of its kind. But the husk is an insurmountable obstacle to those learned and educated gentlemen who judge of books entirely by the style and the grammar, or those who eat grain as it grows, like the cattle. Such men would reject all prological revelation; for there never was and probably never will be a revelation by voice and vision communicated in classical manner. It would be an invasion of the rights and prerogatives of Humanity, and as contrary to the Divine and the Established order of mundane government, as a field of quartern loaves or hot French rolls.'

Mr. Smith's book is spiritualism from beginning to end; and my anonymous gainsayer, honest of course, is either ignorant of the work he thinks he has read, or has a most remarkable development of the organ of imperception.]

I cut the following from a Sunday paper in 1849 :—

X. Y.—The Chaldeans began the mathematics, in which the Egyptians excelled. Then crossing the sea, by means of Thales, the Milesian, they came into Greece, where they were improved very much by Pythagoras, Anaxagoras, and Anopides of Chios. These were followed by Briso, Antipho, [two circle-squarers; where is Euclid?] and Hippocrates, but the excellence of the algebraic art was begun by Geber, an Arabian astronomer, and was carried on by Cardanus, Tartaglia, Clavius, Stevinus, Ghetaldus, Herigenius, Fran, Van Schooten [meaning Francis Van Schooten], Florida de Beaume, &c.

Bryso was a mistaken man. Antipho had the disadvantage of being in advance of his age. He had the notion of which the modern geometry has made so much, that of a circle being the polygon of an infinitely great number of sides. He could make no use of it, but the notion itself made him a sophist in the eyes of Aristotle, Eutocius, &c. Geber, an Arab astronomer, and a reputed conjurer in Europe, seems to have given his name to unintelligible language in the word *gibberish*. At one time *algebra* was traced to him; but very absurdly, though I have heard it suggested that *algebra* and *gibberish* must have had one inventor.

Any person who meddles with the circle may find himself the crane who was netted among the geese : as Antipho for one, and Olivier de Serres for another. This last gentleman ascertained, by weighing, that the area of the circle is very nearly that of the square on the side of the inscribed equilateral triangle : which it is, as near as 3·162 . . . to 3·141. . . . He did not pretend to more than approximation; but Montucla and others misunderstood him, and, still worse, misunderstood their own misunderstanding, and made him say the circle was exactly double of the equilateral triangle. He was let out of limbo by Lacroix, in a note to his edition of Montucla's History of Quadrature.

> Quadratura del cerchio, trisezione dell' angulo, et duplicazione del cubo, problemi geometricamente risolute e dimostrate dal Reverendo Arciprete di San Vito D. Domenico Angherà. Malta, 1854, 8vo.
> Equazioni geometriche, estratte dalla lettera del Rev. Arciprete . . al Professore Pullicino sulla quadratura del cerchio. Milan, 1855 or 1856, 8vo.
> Il Mediterraneo gazetta di Malta, 26 Decembre 1855, No. 909: also 911, 912, 913, 914, 936, 939.
> The Malta Times, Tuesday, 9th June 1857.
> Misura esatta del cerchio, dal Rev. D. Angherà. Malta, 1857, 12mo.
> Quadrature of the circle . . . by the Rev. D. Angherà, Archpriest of St. Vito. Malta, 1858, 12mo.

I have looked for St. Vitus in catalogues of saints, but never found his legend, though he figures as a day-mark in the oldest almanacs. He must be properly accredited, since he has an archpriest. And I pronounce and ordain, by right accruing from the trouble I have taken in this subject, that he, St. Vitus, who leads his votaries a never-ending and unmeaning dance, shall henceforth be held and taken to be the patron saint of the circle-squarer. His day is the 15th of June, which is also that of St. Modestus, with whom the said circle-squarer often has nothing to do. And he must not put himself under the first saint with a slantendicular reference to the other, as is much to be feared was done by the Cardinal who came to govern England with a title containing St. Pudentiana, who shares a day with *St. Dunstan.* The Archpriest of St. Vitus will have it that the square inscribed in a semicircle is half of the semicircle, or the circumference $3\frac{1}{5}$ diameters. He is active and able, with nothing wrong about him except his paradoxes. In the second tract named he has given the testimonials of crowned heads and ministers, &c. as follows. Louis-Napoleon gives thanks. The minister at Turin refers it to the Academy of Sciences, and hopes so much labour will be judged *degna di pregio.* The Vice-Chancellor of Oxford —a blunt Englishman—begs to say that the University has never proposed the problem, as some affirm. The Prince Regent of Baden has received the work with lively interest. The Academy of Vienna is not in a position to enter into the question. The

Academy of Turin offers the most *distinct* thanks. The Academy della Crusca attends only to literature, but gives thanks. The Queen of Spain has received the work with the highest appreciation. The University of Salamanca gives infinite thanks, and feels true satisfaction in having the book. Lord Palmerston gives thanks, by the hand of 'William San.' The Viceroy of Egypt, not being yet up in Italian, will spend his first moments of leisure in studying the book, when it shall have been translated into French : in the mean time he congratulates the author upon his victory over a problem so long held insoluble. All this is seriously published as a rate in aid of demonstration. If these royal compliments cannot make the circumference of a circle about 2 per cent. larger than geometry will have it—which is all that is wanted—no wonder that thrones are shaky.

I am informed that the legend of St. Vitus is given by Ribadeneira in his lives of the Saints, and that Baronius, in his *Martyrologium Romanum*, refers to several authors who have written concerning him. There is an account in Mrs. Jameson's 'History of Sacred and Legendary Art' (ed. of 1863, p. 544). But it seems that St. Vitus is the patron saint of *all* dances ; so that I was not so far wrong in making him the protector of the cyclometers. Why he is represented with a cock is a disputed point, which is now made clear : next after *gallus gallinaceus* himself, there is no crower like the circle-squarer.

The following is an extract from the *English Cyclopædia*, Art. TABLES :—

' 1853. William Shanks, " Contributions to Mathematics, comprising chiefly the Rectification of the Circle to 607 Places of Tables," London, 1853. (QUADRATURE OF THE CIRCLE.) Here is a *table*, because it tabulates the results of the subordinate steps of this enormous calculation as far as 527 decimals : the remainder being added as results only during the printing. For instance, one step is the calculation of the reciprocal of $601.5^{601}$ ; and the result is given. The number of pages required to describe these results is 87. Mr. Shanks has also thrown off, as chips or splinters, the values of the base of Napier's logarithms, and of its logarithms of 2, 3, 5, 10, to 137 decimals ; and the value of the modulus ·4342. . . . to 136 decimals ; with the 13th, 25th, 37th . . . up to the 721st powers of 2. These tremendous stretches of calculation—at least we so call them in our day—are useful in several respects ; they prove more than the capacity of this or that computer for labour and accuracy ; they show that there is in the community an increase of skill and courage. We say in the community : we fully believe that the unequalled turnip which every now and then appears in the newspapers is a sufficient presumption

that the average turnip is growing bigger, and the whole crop heavier.
All who know the history of the quadrature are aware that the several
increases of numbers of decimals to which $\pi$ has been carried have
been indications of a general increase in the power to calculate, and in
courage to face the labour. Here is a comparison of two different
times. In the day of Cocker, the pupil was directed to perform a
common subtraction with a voice-accompaniment of this kind: " 7 from
4 I cannot, but add 10, 7 from 14 remains 7, set down 7 and carry 1 ;
8 and 1 which I carry is 9, 9 from 2 I cannot, &c." We have before
us the announcement of the following *table*, undated, as open to
inspection at the Crystal Palace, Sydenham, in two diagrams of 7 ft.
2 in. by 6 ft. 6 in. :—" The figure 9 involved into the 912th power, and
antecedent powers or involutions, containing upwards of 73,000 figures.
Also, the proofs of the above, containing upwards of 146,000 figures.
By Samuel Fancourt, of Mincing Lane, London, and completed by him
in the year 1837, at the age of sixteen. N.B. The whole operation
performed by simple arithmetic." The young operator calculated by
successive squaring the 2nd, 4th, 8th, &c., powers up to the 512th, with
proof by division. But 511 multiplications by 9, in the short (or
10 – 1) way, would have been much easier. The 2nd, 32nd, 64th,
128th, 256th, and 512th powers are given at the back of the announcement. The powers of 2 have been calculated for many purposes. In
vol. ii. of his "Magia Universalis Naturæ et Artis," Herbipoli, 1658,
4to., the Jesuit Gaspar Schott having discovered, on some grounds of
theological magic, that the degrees of grace of the Virgin Mary were
in number the 256th power of 2, calculated that number. Whether
or no his number correctly represented the result he announced,
he certainly calculated it rightly, as we find by comparison with
Mr. Shanks.'

There is a point about Mr. Shanks' 608 figures of the value
of $\pi$ which attracts attention, perhaps without deserving it. It
might be expected that, in so many figures, the nine digits and
the cipher would occur each about the same number of times ;
that is, each about 61 times. But the fact stands thus : 3 occurs
68 times ; 9 and 2 occur 67 times each ; 4 occurs 64 times ; 1
and 6 occur 62 times each; 0 occurs 60 times; 8 occurs 58 times;
5 occurs 56 times ; and 7 occurs only 44 times. Now, if all the
digits were equally likely, and 608 drawings were made, it is 45
to 1 against the number of sevens being as distant from the
probable average (say 61) as 44 on one side or 78 on the other.
There must be some reason why the number 7 is thus deprived of
its fair share in the structure. Here is a field of speculation in
which two branches of inquirers might unite. There is but one
number which is treated with an unfairness which is incredible
as an accident : and that number is the mystic number *seven* !

If the cyclometers and the apocalyptics would lay their heads together until they come to a unanimous verdict on this phenomenon, and would publish nothing until they are of one mind, they would earn the gratitude of their race.—I was wrong : it is the Pyramid-speculator who should have been appealed to. A correspondent of my friend Prof. Piazzi Smyth notices that 3 is the number of most frequency, and that $3\frac{1}{7}$ is the nearest approximation to it in simple digits. Prof. Smyth himself, whose word on Egypt is paradox of a very high order, backed by a great quantity of useful labour, the results of which will be made available by those who do not receive the paradoxes, is inclined to see confirmation for some of his theory in these phenomena.

These paradoxes of calculation sometimes appear as illustrations of the value of a new method. In 1863, Mr. G. Suffield, M.A. and Mr. J. R. Lunn, M.A., of Clare College and of St. John's College, Cambridge, published the whole quotient of 10000 . . . divided by 7699, throughout the whole of one of the recurring periods, having 7698 digits. This was done in illustration of Mr. Suffield's method of *Synthetic division*.

Another instance of computation carried paradoxical length, in order to illustrate a method, is the solution of $x^3 - 2x = 5$, the example given of Newton's method, on which all improvements have been tested. In 1831, Fourier's posthumous work on equations showed 33 figures of solution, got with enormous labour. Thinking this a good opportunity to illustrate the superiority of the method of W. G. Horner, not then known in France, and not much known in England, I proposed to one of my classes, in 1841, to beat Fourier on this point, as a Christmas exercise. I received several answers, agreeing with each other, to 50 places of decimals. In 1848, I repeated the proposal, requesting that 50 places might be exceeded : I obtained answers of 75, 65, 63, 58, 57, and 52 places. But one answer, by Mr. W. Harris Johnston, of Dundalk, and of the Excise Office, went to 101 decimal places. To test the accuracy of this, I requested Mr. Johnston to undertake another equation, connected with the former one in a way which I did not explain. His solution verified the former one, but he was unable to see the connexion; even when his result was obtained. My reader may be as much at a loss : the two solutions are—

$$2.0945514815423265 \ldots$$
$$9.0544851845767340 \ldots$$

The results are published in the *Mathematician*, vol. iii. p. 290.

In 1851, another pupil of mine, Mr. J. Power Hicks, carried the result to 152 decimal places, without knowing what Mr. Johnston had done. The result is in the *English Cyclopædia*, article INVOLUTION AND EVOLUTION.

I remark that when I write the initial of a Christian name, the most usual name of that initial is understood. I never saw the name of W. G. Horner written at length, until I applied to a relative of his, who told me that he was, as I supposed, Wm. *George*, but that he was named after a relative of that *surname*.

The square root of 2, to 110 decimal places, was given me in 1852 by my pupil, Mr. William Henry Colvill, now (1867) Civil Surgeon at Baghdad. It was

1·41421356237309504880168872420969809807856969
71875376948073176679737990732478462107037
885038875343276415727350138462 3

Mr. James Steel of Birkenhead verified this by actual multiplication, and produced

$$2 - \frac{2580413}{10^{117}}$$

as the square.

Calcolo decidozzinale del Barone Silvio Ferrari. Turin, 1854, 4to.

This is a serious proposal to alter our numeral system and to count by twelves. Thus 10 would be twelve, 11 thirteen, &c., two new symbols being invented for ten and eleven. The names of numbers must of course be changed. There are persons who think such changes practicable. I thought this proposal absurd when I first saw it, and I think so still: but the one I shall presently describe beats it so completely in that point, that I have not a smile left for this one.

The successful and therefore probably true theory of Comets. London, 1854. (4 pp. duodecimo.)

The author is the late Mr. Peter Legh, of Norbury Booths Hall, Knutsford, who published for eight or ten years the *Ombrological Almanac*, a work of asserted discovery in meteorology. The theory of comets is that the joint attraction of the new moon and several planets in the direction of the sun, draws off the gases from the earth, and forms these cometic meteors. But how these meteors come to describe orbits round the sun, and to become capable of having their returns predicted, is not explained.

The Mormon, New York, Saturday, Oct. 27, 1855.

A newspaper headed by a grand picture of starred and striped banners, beehive, and eagle surmounting it. A scroll on each side: on the left, 'Mormon creed. Mind your own business. Brigham Young;' on the right, 'Given by inspiration of God. Joseph Smith.' A leading article on the discoveries of Prof. Orson Pratt says, 'Mormonism has long taken the lead in religion: it will soon be in the van both in science and politics.' At the beginning of the paper is Prof. Pratt's 'Law of Planetary Rotation.' The cube roots of the densities of the planets are as the square roots of their periods of rotation. The squares of the cube roots of the masses divided by the squares of the diameters are as the periods of rotation. Arithmetical verification attempted, and the whole very modestly stated and commented on. Dated G. S. L. City, Utah Ter., Aug. 1, 1855. If the creed, as above, be correctly given, no wonder the Mormonites are in such bad odour.

The two estates; or both worlds mathematically considered. London, 1855, small (pp. 16).

The author has published mathematical works with his name. The present tract is intended to illustrate mathematically a point which may be guessed from the title. But the symbols do very little in the way of illustration : thus, $x$ being the *present value* of the future estate (eternal happiness), and $a$ of all that this world can give, the author impresses it on the mathematician that, $x$ being infinitely greater than $a$, $x + a = x$, so that $a$ need not be considered. This will not act much more powerfully on a mathematician by virtue of the symbols than if those same symbols had been dispensed with : even though, as the author adds, 'It was this method of neglecting infinitely small quantities that Sir Isaac Newton was indebted to for his greatest discoveries.'

There has been a moderate quantity of well-meant attempt to enforce, sometimes motive, sometimes doctrine, by arguments drawn from mathematics, the proponents being persons unskilled in that science for the most part. The ground is very dangerous: for the illustration often turns the other way with greater power, in a manner which requires only a little more knowledge to see.

I have, in my life, heard from the pulpit or read, at least a dozen times, that all sin is infinitely great, proved as follows. The greater the being, the greater the sin of any offence against him : therefore the offence committed against an infinite being is infinitely great. Now the mathematician, of which the proposers of this argument are not aware, is perfectly familiar with quantities which increase together, and never cease increasing, but so that one of them remains finite when the other becomes infinite. In fact, the argument is a perfect *non sequitur*. Those who propose it have in their minds, though in a cloudy and indefinite form, the idea of the increase of guilt being *proportionate* to the increase of greatness in the being offended. But this it would never do to state : for by such statement not only would the argument lose all that it has of the picturesque, but the asserted premise would have no strong air of exact truth. How could any one undertake to appeal to conscience to declare that an offence against a being $4\frac{7}{10}$ times as great as another is exactly, no more and no less, $4\frac{7}{10}$ times as great an offence against the other ?

The infinite character of the offence against an infinite being is laid down in Dryden's *Religio Laici*, and is, no doubt, an old argument :—

> For, granting we have sinned, and that th' offence
> Of man is made against Omnipotence,
> Some price that bears proportion must be paid,
> And infinite with infinite be weighed.
> See then the Deist lost ; remorse for vice
> Not paid ; or, paid, inadequate in price.

Dryden, in the words ' bears proportion ' is in verse more accurate than most of the recent repeaters in prose, And this is not the only case of the kind in his argumentative poetry.

My old friend, the late Dr. Olinthus Gregory, who was a sound and learned mathematician, adopted this dangerous kind of illustration in his Letters on the Christian Religion. He argued, by parallel, from what he supposed to be the necessarily mysterious nature of the *impossible* quantity of algebra to the necessarily mysterious nature of certain doctrines of his system of Christianity. But all the difficulty and mystery of the impossible quantity is now cleared away by the advance of algebraical thought : and yet Dr. Gregory's book continues to be sold, and no doubt the illustration is still accepted as appropriate.

The mode of argument used by the author of the tract above

named has a striking defect. He talks of reducing this world and the next to 'present value,' as an actuary does with successive lives or next presentations. Does value make interest? and if not, why? And if it do, then the present value of an eternity is *not* infinitely great. Who is ignorant that a perpetual annuity at five per cent. is worth only twenty years' purchase? This point ought to be discussed by a person who treats heaven as a deferred perpetual annuity. I do not ask him to do so, and would rather he did not; but if he *will* do it, he must either deal with the question of discount, or be asked the reason why.

When a very young man, I was frequently exhorted to one or another view of religion by pastors and others who thought that a mathematical argument would be irresistible. And I heard the following more than once, and have since seen it in print, I forget where. Since eternal happiness belonged to the particular views in question, a benefit infinitely great, then, even if the probability of their arguments were small, or even infinitely small, yet the product of the chance and benefit, according to the usual rule, might give a result which no one ought in prudence to pass over. They did not see that this applied to all systems as well as their own. I take this argument to be the most perverse of all the perversions I have heard or read on the subject: there is some high authority for it, whom I forget.

The moral of all this is, that such things as the preceding should be kept out of the way of those who are not mathematicians, because they do not understand the argument; and of those who are, because they do.

[The high authority referred to above is Pascal, an early cultivator of mathematical probability, and obviously too much enamoured of his new pursuit. But he conceives himself bound to wager on one side or the other. To the argument (*Pensées*, ch. 7) that ' le juste est de ne point parier,' he answers, ' Oui : mais il faut parier: vous êtes embarqué; et ne parier point que Dieu est, c'est parier qu'il n'est pas.' Leaving Pascal's argument to make its way with a person who, *being a sceptic*, is yet positive that the issue is salvation or perdition, if a God there be,—for the case as put by Pascal requires this,—I shall merely observe that a person who elects to believe in God, as the best chance of gain, is not one who, according to Pascal's creed, or any other worth naming, will really secure that gain. I wonder whether Pascal's curious imagination ever presented to him in sleep his convert, in the future state, shaken out of a red-hot dice-box upon a red-hot hazard-table, as perhaps he might have been, if Dante had been the later of the two. The original idea is due to the elder Arnobius, who, as cited by Bayle, speaks thus:—

'Sed et ipse [Christus] quæ pollicetur, non probat. Ita est. Nulla enim, ut dixi, futurorum potest existere comprobatio. Cum ergo hæc sit conditio futurorum, ut teneri et comprehendi nullius possint anticipationis attactu ; nonne purior ratio est, ex duobus incertis, et in ambigua expectatione pendentibus, id potius credere, quod aliquas spes ferat, quam omnino quod nullas ? In illo enim periculi nihil est, si quod dicitur imminere, cassum fiat et vacuum : in hoc damnum est maximum, id est salutis amissio, si cum tempus advenerit aperiatur non fuisse mendacium.'

Really Arnobius seems to have got as much out of the notion, in the third century, as if he had been fourteen centuries later, with the arithmetic of chances to help him.]

The Sentinel, vol. ix. no. 27. London, Saturday, May 26, 1855.

This is the first London number of an Irish paper, Protestant in politics. It opens with ' Suggestions on the subject of a *Novum Organum Moralium*,' which is the application of algebra and the differential calculus to morals, socials, and politics. There is also a leading article on the subject, and some applications in notes to other articles. A separate publication was afterwards made, with the addition of a long Preface ; the author being a clergyman who I presume must have been the editor of the *Sentinel.*

> Suggestions as to the employment of·a *Novum Organum Mora-lium.* Or, thoughts on the nature of the Differential Calculus, and on the application of its principles to metaphysics, with a view to the attainment of demonstration and certainty in moral, political and ecclesiastical affairs. By Tresham Lames Gregg, Chaplain of St. Mary's, within the church of St. Nicholas intra muros, Dublin. London 1859, 8vo. (pp. xl + 32).

I have a personal interest in this system, as will appear from the following extract from the newspaper :—

'We were subsequently referred to De Morgan's " Formal Logic " and Boole's " Laws of Thought," both very elaborate works, and greatly in the direction taken by ourselves. That the writers amazingly surpass us in learning we most willingly admit, but we venture to pronounce of both their learned treatises, that they deal with the subject in a mode that is scholastic to an excess . . . That their works have been for a considerable space of time before the world and effected nothing, would argue that they have overlooked the vital nature of the theme. . . On the whole, the writings of De Morgan and Boole go to the full justification of our principle without in any wise so trenching upon our ground as to render us open to reproach in claiming our Calculus as a great discovery. . . But we renounce any paltry jealousy as to a matter so vast. If De Morgan and Boole have had a priority in the case, to them we cheerfully shall resign the glory and honour. If such be the truth, they have neither done justice to the discovery, nor to themselves [quite true]. They have, under the circumstances, acted like 'the foolish man, who roasteth not that which he taketh in hunting.' . . It will be sufficient for us, however, to be the Columbus of these great Americi, and popularise what they found, *if* they found it. We, as from the mountain top, will then become *their* trumpeters, and cry glory to De Morgan and glory to Boole, under Him who is the source of all glory, the only good and wise, to Whom be glory for ever ! *If* they be our predecessors in this matter, they have, under Him, taken moral questions out of the category of probabilities, and rendered them perfectly certain. In that case, let their books be read by those who may doubt the principles this day laid before the world as a great discovery, by our newspaper. Our cry shall be ευρηκασι ! Let us hope that they will join us, and henceforth keep their right [*sic*] from under their bushel.'

For myself, and for my old friend Mr. Boole, who I am sure would join me, I disclaim both priority, simultaneity, and posteriority, and request that nothing may be trumpeted from the mountain top except our abjuration of all community of thought or operation with this *Novum Organum*.

To such community we can make no more claim than Americus could make to being the forerunner of Columbus who popularised his discoveries. We do not wish for any ευρηκασι, and not even for ἑυρηκασι. For self and Boole, I point out what would have convinced either of us that this house is divided against itself.

A being the apostolic element, δ the doctrinal element, and X the body of the faithful, the church is A δ X, we are told. Also, that if A become negative, or the Apostolicity become Diabolicity [my words]; or if δ become negative, and doctrine become heresy; or if X become negative, that is, if the faithful become unfaithful; the church becomes negative, 'the very opposite of what it ought to be.' For self and Boole, I admit this. But—which is not noticed—if A and δ should *both* become negative, diabolical origin and heretical doctrine, then the church, A δ X, is still positive, what it ought to be, unless X be also negative, or the people unfaithful to it, in which case it is a bad church. Now, self and Boole—though I admit I have not asked my partner—are of opinion that a diabolical church with false doctrine does harm when the people are faithful, and can do good only when the people are unfaithful. We may be wrong, but this is what we *do* think. Accordingly, we have caught nothing, and can therefore roast nothing of our own: I content myself with roasting a joint of Mr. Gregg's larder.

These mathematical vagaries have uses which will justify a large amount of quotation: and in a score of years this may perhaps be the only attainable record. I therefore proceed.

After observing that by this calculus juries (heaven help them! say I) can calculate damages ' almost to a nicety,' and further that it is made abundantly evident that *c e x* is ' the general expression for an individual,' it is noted that the number of the Beast is not given in the Revelation in words at length, but as χξϛ'. On this the following remark is made :—

' Can it be possible that we have in this case a specimen given to us of the arithmetic of heaven, and an expression revealed, which indicates by its function of addibility, the name of the church in question, and of each member of it; and by its function of multiplicability the doctrine, the mission, and the members of the great Synagogue of Apostacy ? We merely propound these questions ;—we do not pretend to solve them.'

After a translation in blank verse —a very pretty one- -of the 18th Psalm, the author proceeds as follows, to render it into differential calculus :—

' And the whole tells us just this, that David did what he could. He augmented those elements of his constitution which were (*exceptis excipiendis*) subject to himself, and the Almighty then augmented his personal qualities, and his vocational *status*. Otherwise, to throw the matter into the expression of our notation, the variable *e* was augmented, and *c x* rose proportionally. The law of the variation, according to our theory, would be thus expressed. The resultant was David the king *c e x* [*c*=*r* ?] (who had been David the shepherd boy), and from the conditions of the theorem we have

$$\frac{d u}{d e} = c e \frac{d x}{d e} + e x \frac{d c}{d e} \quad x + c x$$

which, in the terms of ordinary language, just means, the increase of David's educational excellence or qualities—his piety, his prayerfulness, his humility, obedience, &c.—was so great, that when multiplied by his original talent and position, it produced a product so great as to be equal in its amount to royalty, honour, wealth, and power, &c. : in short, to all the attributes of majesty.'

The ' solution of the family problem ' is of high interest. It is to determine the effect on the family in general from a change [of conduct] in one of them. The person chosen is one of the maid-servants.

'Let $c \; e \; x$ be the father; $c_1 e_1 x_1$ the mother, &c. The family then consists of the maid's master, her mistress, her young master, her young mistress, and fellow servant. Now the master's calling (or $c$) is to exercise his share of control over this servant, and mind the rest of his business : call this remainder $a$, and let his calling generally, or all his affairs, be to his maid-servant as $m : y$, *i. e.*, $y = \dfrac{mz}{c}$; . . . . and this expression will represent his relation to the servant. Consequently,

$$c \, e \, x = \left( a + \frac{mz}{c} \right) e \, x \; ; \text{ otherwise } \left( a + \frac{mz}{c} \right) e \, x$$

is the expression for the father when viewed as the girl's master.'

I have no objection to repeat so far; but I will not give the formula for the maid's relation to her young master; for I am not quite sure that all young masters are to be trusted with it. Suffice it that the son will be affected directly as his influence over her, and inversely as his vocational power : if then he should have some influence and no vocational power, the effect on him would be infinite. This is dismal to think of. Further, the formula brings out that if one servant improve, the other must deteriorate, and *vice versâ*. This is not the experience of most families : and the author remarks as follows :—

' That is, we should venture to say, a very beautiful result, and we may say it yielded us no little astonishment. What our calculation might lead to we never dreamt of; that it should educe a conclusion so recondite that our unassisted power never could have attained to, and which, if we could have conjectured it, would have been at best the most distant probability, that conclusion being itself, as it would appear, the quintessence of truth, afforded us a measure of satisfaction that was not slight.'

That the writings of Mr. Boole and myself 'go to the full justification of' this 'principle,' is only true in the sense in which the Scotch use, or did use, the word *justification*.

[The last number of this Budget had stood in type for months, waiting until there should be a little cessation of correspondence more connected with the things of the day. I had quite forgotten what it was to contain; and little thought, when I read the proof, that my allusions to my friend Mr. Boole, then in life and health, would not be printed till many weeks after his death. Had I remembered what my last number contained, I should have added my expression of regret and admiration to the numerous obituary testimonials, which this great loss to science has called forth.

The system of logic alluded to in the last number of this series is but one of many proofs of genius and patience combined. I might legitimately have entered it among my *paradoxes*, or things counter to general opinion : but it is a paradox which, like that of Copernicus, excited admiration from its first appearance. That the symbolic processes of algebra, invented as tools of numerical calculation, should be competent to express every act of thought, and to furnish the grammar and dictionary of an all-containing system of logic, would not have been believed until it was proved. When Hobbes, in the time of the Commonwealth, published his ' Computation or Logique,' he had a remote glimpse of some of the points which are placed in the light of day by Mr. Boole. The unity of the forms of thought in all the applications of reason, however remotely separated, will one day be matter of notoriety and common wonder : and Boole's name will be remembered in connexion with one of the most important steps towards the attainment of this knowledge.]

The Decimal System as a whole. By Dover Statter. London and Liverpool, 1856, 8vo.

The proposition is to make everything decimal. The day, now 24 hours, is to be made 10 hours. The year is to have ten months, Unusber, Duober, &c. Fortunately there are ten commandments, so there will be neither addition to, nor deduction from, the moral law. But the twelve apostles! Even rejecting Judas, there is a whole apostle of difficulty. These points the author does not touch.

The first book of Phonetic Reading. London, Fred. Pitman, Phonetic Depot, 20, Paternoster Row, 1856, 12mo.

The Phonetic Journal. Devoted to the propagation of phonetic reading, phonetic longhand, phonetic shorthand, and phonetic printing. No. 46. Saturday, 15 November 1856. Vol. 15.

I write the titles of a couple out of several tracts which I have by me. But the number of publications issued by the promoters of this spirited attempt is very large indeed. The attempt itself has had no success with the mass of the public. This I do not regret. Had the world found that the change was useful, I should have gone contentedly with the stream ; but not without regretting our old language. I admit the difficulties which our unpronounceable spelling puts in the way of learning to read : and I have no doubt that, as affirmed, it is easier to teach children phonetically, and afterwards to introduce them to our common system, than to proceed in the usual way. But by the usual way I mean proceeding by letters from the very beginning. If, which I am sure is a better plan, children be taught at the commencement very much by *complete words*, as if they were learning Chinese, and be gradually accustomed to resolve the known words into letters, a fraction, perhaps a considerable one, of the advantage of the phonetic system is destroyed. It must be remembered that a phonetic system can only be an approximation. The differences of pronunciation existing among educated persons are so great, that, on the phonetic system, different persons ought to spell differently.

But the phonetic party have produced something which will immortalize their plan : I mean their *shorthand*, which has had a fraction of the success it deserves. All who know anything of shorthand must see that nothing but a phonetic system can be worthy of the name : and the system promulgated is skilfully done. Were I a young man I should apply myself to it systematically. I believe this is the only system in which books were ever published. I wish some one would contribute to a public journal a brief account of the dates and circumstances of the phonetic movement, not forgetting a list of the books published in shorthand.

A child beginning to read by himself may owe terrible dreams and waking images of horror to our spelling, as I did when six years old. In one of the common poetry-books there is an admonition against confining little birds in cages, and the child is asked what if a great giant, amazingly strong, were to take you away, shut you up,

> And feed you with vic-tu-als you ne-ver could bear.

The book was hyphened for the beginner's use ; and I had not the least idea that *vic-tu-als* were *vittles* : by the sound of the word I judged they must be of iron ; and it entered into my soul.

The worst of the phonetic shorthand books is that they nowhere, so far as I have seen, give *all* the symbols, in every stage of advancement, together, in one or following pages. It is symbols and talk, more symbols and more talk, &c. A universal view of the signs ought to begin the works.

Ombrological Almanac.  Seventeenth year.  An essay on Anemo-
logy and Ombrology.  By Peter Legh, Esq.  London, 1856,
12mo.

Mr. Legh, already mentioned, was an intelligent country
gentleman, and a legitimate speculator.  But the clue was not
reserved for him.

The proof that the three angles of a triangle are equal to two
right angles looked for in the inflation of the circle.  By Gen.
Perronet Thompson.  London, 1856, 8vo.  (pp. 4.)

Another attempt, the third, at this old difficulty, which cannot
be put into few words of explanation.

Comets considered as volcanoes, and the cause of their velocity
and other phenomena thereby explained.  London (*circa* 1856),
8vo.

The title explains the book better than the book explains the
title.

1856.  A stranger applied to me to know what the ideas
of a friend of his were worth upon the magnitude of the earth.
The matter being one involving points of antiquity, I mentioned
various persons whose speculations he seemed to have ignored;
among others, Thales.  The reply was, ' I am instructed by the
author to inform you that he is perfectly acquainted with the
works of Thales, Euclid, Archimedes, . . . ' I had some thought
of asking whether he had used the Elzevir edition of Thales,
which is known to be very incomplete, or that of Prof. Niemand
with the lections, Nirgend, 1824, 2 vols. folio ; just to see whether
the last would not have been the very edition he had read.  But
I refrained, in mercy.

The moon is the image of the Earth, and is not a solid body.  By
T^he Longitude.  (Private Circulation.)  In five parts.  London,
1856, 1857, 1857 ; Calcutta, 1858, 1858, 8vo.

The earth is ' brought to a focus '; it describes a ' looped '
orbit round the sun.  The eclipse of the sun is thus explained :
' At the time of eclipses, the image is more or less so directly
before or behind the earth that, in the case of new moon, bright
rays of the sun fall and bear upon the spot where the figure of
the earth is brought to a focus, that is, bear upon the image of

the earth, when a darkness beyond is produced reaching to the earth, and the sun becomes more or less eclipsed.' How the earth is 'brought to a focus' we do not find stated. Writers of this kind always have the argument that some things which have been ridiculed at first have been finally established. Those who put into the lottery had the same kind of argument ; but were always answered by being reminded how many blanks there were to one prize. I am loath to pronounce against anything : but it does force itself upon me that the author of these tracts has drawn a blank.

*Times*, April 6 or 7, 1856. The moon has no rotary motion.

A letter from Mr. Jellinger Symons, inspector of schools, which commenced a controversy of many letters and pamphlets. This dispute comes on at intervals, and will continue to do so. It sometimes arises from inability to understand the character of simple rotation, geometrically; sometimes from not understanding the mechanical doctrine of rotation.

> Lunar Motion. The whole argument stated, and illustrated by diagrams; with letters from the Astronomer Royal. By Jellinger C. Symons. London, 1856, 8vo.

The Astronomer Royal endeavoured to disentangle Mr. J. C. Symons, but failed. Mr. Airy can correct the error of a ship's compasses, because he can put her head which way he pleases : but this he cannot do with a speculator.

Mr. Symons, in this tract, insinuated that the rotation of the moon is one of the silver shrines of the craftsmen. To see a thing so clearly as to be satisfied that all who say they do not see it are telling wilful falsehood, is the nature of man. Many of all sects find much comfort in it, when they think of the others ; many unbelievers solace themselves with it against believers ; priests of old time founded the right of persecution upon it, and of our time, in some cases, the right of slander : many of the paradoxers make it an argument against students of science. But I must say for men of science, for the whole body, that they are fully persuaded of the honesty of the paradoxers. The simple truth is, that all those I have mentioned, believers, unbelievers, priests, paradoxers, are not so sure they are right in their points of difference that they can safely allow themselves to be persuaded of the honesty of opponents. Those who know demonstration are differently situated. I suspect a train might be laid for the formation of a better habit in this way. We know that Suvaroff taught his Russians at Ismail not to fear the Turks by accustoming them to charge bundles of faggots dressed in turbans, &c.

> At which your wise men sneered in phrases witty,
> He made no answer—but he took the city!

Would it not be a good thing to exercise boys, in pairs, in the following dialogue:—Sir, you are quite wrong!—Sir, I am sure you honestly think so! This was suggested by what used to take place at Cambridge in my day. By statute, every B.A. was obliged to perform a certain number of disputations, and the *father* of the college had to affirm that it had been done. Some were performed in earnest : the rest were huddled over as follows. Two candidates occupied the places of the respondent and the opponent : *Recte statuit Newtonus,* said the respondent: *Recte non statuit Newtonus,* said the opponent. This was repeated the requisite number of times, and counted for as many *acts* and *opponencies.* The parties then changed places, and each unsaid what he had said on the other side of the house : I remember thinking that it was capital drill for the House of Commons, if any of us should ever get there. The process was repeated with every pair of candidates.

The real disputations were very severe exercises. I was badgered for two hours with arguments given and answered in Latin,—or what we called Latin—against Newton's first section, Lagrange's derived functions, and Locke on innate principles. And though I *took off* everything, and was pronounced by the moderator to have disputed *magno honore,* I never had such a strain of thought in my life. For the inferior opponents were made as sharp as their betters by their tutors, who kept lists of queer objections, drawn from all quarters. The opponents used to meet the day before to compare their arguments, that the same might not come twice over. But, after I left Cambridge, it became the fashion to invite the respondent to be present, who therefore learnt all that was 'to be brought against him. This made the whole thing a farce : and the disputations were abolished.

The Doctrine of the Moon's Rotation, considered in a letter to the Astronomical Censor of the *Athenæum.* By Jones L. Mac-Elshender. Edinburgh, 1856, 8vo.

This is an appeal to those cultivated persons who will read it ' to overrule the *dicta* of judges who would sacrifice truth and justice to professional rule, or personal pique, pride, or prejudice'; meaning, the great mass of those who have studied the subject. But how? Suppose the 'cultivated persons' were to side with the author, would those who have conclusions to draw and applications to make consent to be wrong because the 'general body of intelligent men,' who make no special study of the subject, are against them? They would do no such thing: they would request the general body of intelligent men to find their own astronomy, and welcome. But the truth is, that this intelligent body knows better: and no persons know better that they know better than the speculators themselves.

But suppose the general body were to combine, in opposition to those who have studied. Of course all my list must be admitted to their trial; and then arises the question whether both sides are to be heard. If so, the general body of the intelligent must hear all the established side have to say: that is, they must become just as much of students as the inculpated orthodox themselves. And will they not then get into *professional rule,* pique, pride, and prejudice, as the others did? But if, which I suspect, they are intended to judge just as they are, they will be in a rare difficulty. All the paradoxers are of like pretensions: they cannot, as a class, be right, for each one contradicts a great many of the rest. There will be the puzzle which silenced the crew of the cutter in Marryat's novel of the Dog Fiend. 'A tog is a tog,' said Jansen.—'Yes,' replied another, 'we all know a dog is a dog; but the question is—Is *this* dog a dog?' And this question would arise upon every dog of them all.

Zetetic Astronomy: Earth not a globe. 1857 (Broadsheet).

Though only a travelling lecturer's advertisement, there are so many arguments and quotations that it is a little pamphlet. The lecturer gained great praise from provincial newspapers for his ingenuity in proving that the earth is a flat, surrounded by ice. Some of the journals rather incline to the view: but the *Leicester Advertiser* thinks that the statements 'would seem very seriously to invalidate some of the most important conclusions of modern astronomy,' while the *Norfolk Herald* is clear that 'there must be a great error on one side or the other.' This broadsheet is printed at Aylesbury in 1857, and the lecturer calls himself *Parallax*: but at Trowbridge, in 1849, he was S. Goulden. In this last advertisement is the following announcement:—' A paper on the above subjects was read before the Council and

Members of the Royal Astronomical Society, Somerset House, Strand, London (Sir John F. W. Herschel, President), Friday, Dec. 8, 1848.' No account of such a paper appears in the *Notice* for that month : I suspect that the above is Mr. S. Goulden's way of representing the following occurrence : —Dec. 8, 1848, the Secretary of the Astronomical Society (De Morgan by name) said, at the close of the proceedings,—'Now, gentlemen, if you will promise not to tell the Council, I will read something for your amusement ': and he then read a few of the arguments which had been transmitted by the lecturer. The fact is worth noting that from 1849 to 1857, arguments on the roundness or flatness of the earth did itinerate. I have no doubt they did much good : for very few persons have any distinct idea of the evidence for the rotundity of the earth. The *Blackburn Standard* and *Preston Guardian* (Dec. 12 and 16, 1849) unite in stating that the lecturer ran away from his second lecture at Burnley, having been rather too hard pressed at the end of his first lecture to explain why the large hull of a ship disappeared before the sails. The persons present and waiting for the second lecture assuaged their disappointment by concluding that the lecturer had slipped off the icy edge of his flat disk, and that he would not be seen again till he peeped up on the opposite side.

But, strange as it may appear, the opposer of the earth's roundness has more of a case—or less of a want of case—than the arithmetical squarer of the circle. The evidence that the earth is round is but cumulative and circumstantial : scores of phenomena ask, separately and independently, what other explanation can be imagined except the sphericity of the earth. The evidence for the earth's figure is tremendously powerful of its kind ; but the proof that the circumference is 3·14159265 . . . times the diameter is of a higher kind, being absolute mathematical demonstration.

The Zetetic system still lives in lectures and books ; as it ought to do, for there is no way of teaching a truth comparable to opposition. The last I heard of it was in lectures at Plymouth, in October, 1864. Since this time a prospectus has been issued of a work entitled ' The Earth not a Globe ;' but whether it has been published I do not know. The contents are as follows :—

' The Earth a Plane—How circumnavigated.—How time is lost or gained.—Why a ship's hull disappears (when outward bound) before the mast head.—Why the Polar Star sets when we proceed Southward, &c.—Why a pendulum vibrates with less velocity at the Equator than at the Pole.—The allowance for rotundity *supposed* to be made by surveyors, not made in practice.—Measurement of Arcs of the Meridian unsatisfactory.—Degrees of Longitude North and South of the Equator considered.—Eclipses and Earth's form considered.—The Earth no motion on axis or in orbit.—How the Sun moves above the Earth's surface concentric with the North Pole.—Cause of Day and Night, Winter and Summer; the long alternation of light and darkness at the Pole.—Cause of the Sun rising and setting.—Distance of the Sun from London, 4,028 miles—How measured.—*Challenge to Mathematicians.* —Cause of Tides.—Moon self-luminous, NOT a reflector.—Cause of Solar and Lunar eclipses.—Stars *not worlds*; their distance.—Earth, the *only material* world; its true position in the universe; its condition and ultimate destruction by fire (2 Peter iii.), &c.'

I wish there were geoplatylogical lectures in every town in England (*platylogical,* in composition, need not mean *babbling*). The late Mr. Henry Archer would, if alive, be very much obliged to me for recording his vehement denial of the roundness of the earth: he was excited if he heard any one call it a globe. I cannot produce his proof from the Pyramids, and from some caves in Arabia. He had other curious notions, of course: I should no more believe that a flat earth was a man's only paradox, than I should that Dutens, the editor of Leibnitz, was eccentric only in supplying a tooth which he had lost by one which he found in an Italian tomb, and fully believed that it had once belonged to Scipio Africanus, whose family vault was discovered, it is supposed, in 1780. Mr. Archer is of note as the suggester of the perforated border of the postage-stamps, and, I think, of the way of doing it; for this he got 4,000*l.* reward. He was a civil engineer.

(*August* 28, 1865.) The 'Zetetic Astronomy' has come into my hands. When, in 1851, I went to see the Great Exhibition, I heard an organ played by a performer who seemed very desirous to exhibit one particular stop. 'What do you think of that stop?' I was asked.—'That depends on the name of it,' said I.—'Oh! what can the name have to do with the sound? "that which we call a rose," &c.'—'The name has everything to do with it: if it be a flute-stop, I think it very harsh; but if it be a railway-whistle-stop, I think it very sweet' So as to this book: if it be childish, it is clever; if it be mannish, it is unusually foolish. The flat earth, floating tremulously on the sea; the sun moving always over the flat, giving day when near enough, and night when too far off; the self-luminous moon, with a semi-transparent invisible moon, created to give her an eclipse now and then; the new law of perspective, by which the vanishing of the hull before the masts, usually thought to prove the earth globular, really proves it flat;—all these and other things are well fitted to form exercises for a person who is learning the elements of astronomy. The manner in which the sun dips into the sea, especially in tropical climates, upsets the whole. Mungo Park, I think, gives an African hypothesis which explains phenomena better than this. The sun dips into the western ocean, and the people there cut him in pieces, fry him in a pan, and then join him together again, take him round the underway, and set him up in the east. I hope this book will be read, and that many will be puzzled by it: for there are many whose notions of astronomy deserve no better fate. There is no subject on which there is so little accurate conception as that of the motions of the heavenly bodies. The author, though confident in the extreme, neither impeaches the honesty of those whose opinions he assails, nor allots them any future inconvenience: in these points he is worthy to live on a globe, and to revolve in twenty-four hours.

(*October*, 1866.) A follower appears, in a work dedicated to the preceding author: it is 'Theoretical Astronomy examined and exposed by Common Sense.' The author has 128 well-stuffed octavo pages. I hope he will not be the last. He prints the newspaper accounts of his work: the *Church Times* says—not seeing how the satire might be retorted—'We never began to despair of Scripture until we discovered that "Common Sense" had taken up the cudgels in its defence.' This paper considers our author as the type of a *Protestant*. The author himself, who gives a summary of his arguments in verse, has one couplet which is worth quoting :—

How is't that sailors, bound to sea, with a 'globe_' would never start,
But in its place will always take *Mercator's* LEVEL *chart*!

To which I answer :—

Why, really Mr. Common Sense, you've never got so far
As to think Mercator's planisphere shows countries as they are ;
It won't do to measure distances ; it points out how to steer,
But this distortion's not for you ; another is, I fear.
The earth must be a cylinder, if seaman's charts be true,
Or else the boundaries, right and left, are one as well as two;
They contradict the notion that we dwell upon a plain,
For straight away, without a turn, will bring you home again.
There are various plane projections; and each one has its use :
I wish a milder word would rhyme—but really you're a goose !

The great wish of persons who expose themselves as above, is to be argued with, and to be treated as reputable and refutable opponents, 'Common Sense' reminds us that no amount of 'blatant ridicule' will turn right into wrong. He is perfectly correct: but then no amount of bad argument will turn wrong into right. These two things balance; and we are just where we were: but you should answer our arguments, for whom, I ask? Would reason convince this kind of reasoner? The issue is a short and a clear one. If these parties be what I contend they are, then ridicule is made for them: if not, for what or for whom? If they be right, they are only passing through the appointed trial of all good things. Appeal is made to the future: and my Budget is intended to show samples of the long line of heroes who have fallen without victory, each of whom had his day of confidence and his prophecy of success. Let the future decide: they say roundly that the earth is flat; I say flatly that it is round.

The paradoxers all want reason, and not ridicule: they are all accessible, and would yield to conviction. Well then, let them reason with one another! They divide into squads, each with a subject, and as many different opinions as persons in each squad. If they be really what they say they are, the true man of each set can put down all the rest, and can come crowned with glory and girdled with scalps, to the attack on the orthodox misbelievers. But they know, to a man, that the rest are not fit to be reasoned with: they pay the regulars the compliment of believing that the only chance lies with them. They think in their hearts, each one for himself, that ridicule is of fit appliance to the rest.

Miranda. A book divided into three parts, entitled Souls, Numbers, Stars, on the Neo-Christian Religion. . . Vol. i. London, 1858, 1859, 1860. 8vo.

The name of the author is Filopanti. He announces himself as the 49th and last Emanuel: his immediate predecessors were Emanuel Washington, Emanuel Newton, and Emanuel Galileo. He is to collect nations into one family. He knows the transmigrations of the whole human race. Thus Descartes became William III. of England: Roger Bacon became Boccaccio. But Charles IX., in retribution for the massacre of St. Bartholomew, was hanged in London under the name of Barthélemy for the murder of Collard: and many of the Protestants whom he killed as King of France were shouting at his death before the Old Bailey.

A Letter to the members of the Anglo-Biblical Institute, dated Sept. 7, 1858, and signed 'Herman Heinfetter.' (Broadsheet.)

This gentleman is well known to the readers of the *Athenæum*, in which, for nearly twenty years, he has inserted, as advertisements, long arguments in favour of Christians keeping the Jewish Sabbath, beginning on Friday Evening. The present letter maintains that, by the force of the definite article, the *days* of creation may not be consecutive, but may have any time—millions of years—between them. This ingenious way of reconciling the author of Genesis and the indications of geology is worthy to be added to the list, already pretty numerous. Mr. Heinfetter has taken such pains to make himself a public agitator, that I do not feel it to be any invasion of private life if I state that I have heard he is a large corn-dealer. No doubt he is a member of the congregation whose almanac has already been described.

The great Pyramid. Why was it built? And who built it? By John Taylor, 1859, 12mo.

This work is very learned, and may be referred to for the history of previous speculations. It professes to connect the dimensions of the Pyramid with a system of metrology which is supposed to have left strong traces in the systems of modern times ; showing the Egyptians to have had good approximate knowledge of the dimensions of the earth, and of the quadrature of the circle. These are points on which coincidence is hard to distinguish from intention. Sir John Herschel noticed this work, and gave several coincidences, in the *Athenæum*, Nos. 1696 and 1697, April 28 and May 5, 1860 : and there are some remarks by Mr. Taylor in No. 1701, June 2, 1860.

Mr. Taylor's most recent publication is—

The battle of the Standards : the ancient, of four thousand years, against the modern, of the last fifty years—the less perfect of the two. London, 1864, 12mo.

This is intended as an appendix to the work on the Pyramid. Mr. Taylor distinctly attributes the original system to revelation, of which he says the Great Pyramid is the record. We are advancing, he remarks, towards the end of the Christian Dispensation, and he adds that it is satisfactory to see that we retain the standards which were given by unwritten revelation 700 years before Moses. This is lighting the candle at both ends ; for

myself, I shall not undertake to deny or affirm either what is
said about the dark past or what is hinted about the dark future.
My old friend Mr. Taylor is well known as the author of the
argument which has convinced many, even most, that Sir Philip
Francis was Junius: pamphlet, 1813; supplement, 1817; second
edition ' The Identity of Junius with a distinguished living character established,' London, 1818, 8vo. He told me that Sir
Philip Francis, in a short conversation with him, made only this
remark, ' You may depend upon it you are quite mistaken :' the
phrase appears to me remarkable; it has an air of criticism on
the book, free from all personal denial. He also mentioned that
a hearer told him that Sir Philip said, speaking of writers on the
question,—' Those fellows, for half-a-crown, would prove that
Jesus Christ was Junius.'

Mr. Taylor implies, I think, that he is the first who started the
suggestion that Sir Philip Francis was Junius, which I have no
means either of confirming or refuting. If it be so [and I now
know that Mr. Taylor himself never heard of any predecessor],
the circumstance is very remarkable: it is seldom indeed that
the first proposer of any solution of a great and vexed question
is the person who so nearly establishes his point in general
opinion as Mr. Taylor has done.

As to the Junius question in general, there is a little bit of the
philosophy of horse-racing which may be usefully applied. A
man who is so confident of his horse that he places him far above
any other, may nevertheless, and does, refuse to give odds against
all the field: for many small adverse chances united make a big
chance for one or other of the opponents. I suspect Mr. Taylor
has made it at least 20 to 1 for Francis against any one competitor who has been named : but what the odds may be against the
whole field is more difficult to settle. What if the real Junius
should be some person not yet named ?

Mr. Jopling, *Leisure Hour*, May 23, 1863, relies on the
porphyry coffer of the Great Pyramid, in which he finds ' the most
ancient and accurate standard of measure in existence.'

I am shocked at being obliged to place a thoughtful and
learned writer, and an old friend, before such a successor as he
here meets with. But chronological arrangement defies all other
arrangement.

(I had hoped that the preceding account would have met Mr.
Taylor's eye in print: but he died during the last summer. For
a man of a very thoughtful and quiet temperament, he had a
curious turn for vexed questions. But he reflected very long and

very patiently before he published: and all his works are valuable for their accurate learning, whichever side the reader may take.)

1859. *The Cottle Church.*—For more than twenty years printed papers have been sent about in the name of Elizabeth Cottle. It is not so remarkable that such papers should be concocted as that they should circulate for such a length of time without attracting public attention. Eighty years ago Mrs. Cottle might have rivalled Lieut. Brothers or Joanna Southcott. Long hence, when the now current volumes of our journals are well-ransacked works of reference, those who look into them will be glad to see this feature of our time : I therefore make a few extracts, faithfully copied as to type. The Italic is from the New Testament ; the Roman is the requisite interpretation :—

' Robert Cottle "*was numbered* (5196) *with the transgressors* " at the back of the Church in Norwood Cemetery, May 12, 1858—Isa. liii. 12. The Rev. J. G. Collinson, Minister of St. James's Church, Clapham, the then district church, before All Saints was built, read the funeral service *over the Sepulchre wherein never before man was laid.*

' *Hewn on the stone,* " at the mouth of the Sepulchre," is his name,— Robert Cottle, born at Bristol, June 2, 1774 ; died at Kirkstall Lodge, Clapham Park, May 6, 1858. *And that day* (May 12, 1858) *was the preparation* (day and year for " the PREPARED place for you "—Cottleites —by the widowed mother of the Father's house, at Kirkstall Lodge— John xiv. 2, 3. *And the Sabbath* (Christmas Day, Dec. 25, 1859) *drew on* (for the resurrection of the Christian body on "the third [Protestant Sun]-day "—1 Cor. xv. 35). *Why seek ye the living* (God of the New Jerusalem—Heb. xii. 22 ; Rev. iii. 12) *among the dead* (men) : *he* (the God of Jesus) *is not here* (in the grave), *but is risen* (in the person of the Holy Ghost, from the supper of " the dead in the second death " of Paganism). *Remember how he spake unto you* (in the church of the Rev. George Clayton, April 14, 1839). *I will not drink henceforth* (at this last Cottle supper) *of the fruit of this* (Trinity) *vine, until that day* (Christmas Day, 1859), *when I* (Elizabeth Cottle) *drink it new with you* (Cottleites) *in my Father's Kingdom*—John xv. *If this* (Trinitarian) *cup may not pass away from me* (Elizabeth Cottle, April 14, 1839), *except I drink it* ("new with you Cottleites, in my Father's Kingdom"), *thy will be done*—Matt. xxvi. 29, 42, 64. " Our Father which art (God) in Heaven," *hallowed be thy name, thy* (Cottle) *kingdom come, thy will be done in earth, as it is* (done) *in* (the new) *Heaven* (and new earth of the new name of Cottle—Rev. xxi. 1 ; iii. 12).

' . . . Queen Elizabeth, from A.D. 1558 to 1566. *And this* WORD *yet once more* (by a second Elizabeth—the WORD of his oath) *signifieth* (at John Scott's baptism of the Holy Ghost) *the removing of those things* (those Gods and those doctrines) *that are made* (according to the Creeds and Commandments of men) *that those things* (in the moral law of God) *which cannot be shaken* (as a rule of faith and practice) *may remain, wherefore we receiving* (from Elizabeth) *a kingdom* (of God,) *which cannot be moved* (by Satan) *let us have grace* (in his Grace of Canterbury) *whereby we may serve God acceptably* (with the acceptable sacrifice of Elizabeth's body and blood of the communion of the Holy Ghost) *with reverence* (for truth) *and godly fear* (of the unpardonable sin of blasphemy against the Holy Ghost) *for our God* (the Holy Ghost) *is a consuming fire* (to the nation that will not serve him in the Cottle Church). We cannot defend ourselves against the Almighty, and if He is our defence, no nation can invade us.

'In verse 4 the Church of St. Peter is *in prison between four quaternions of soldiers*—the Holy Alliance of 1815. Rev. vii. i. Elizabeth, *the Angel of the Lord* Jesus *appears* to the Jewish and Christian body with *the vision* of prophecy to the Rev. Geo. Clayton and his clerical brethren, April 8th, 1839. *Rhoda* was the name of her maid at Putney Terrace who used *to open the door to her Peter*, the Rev. Robert Ashton, the Pastor of " the little flock " " of 120 names together, assembled in an upper (school) room " at Putney Chapel, to which little flock she gave the revelation (Acts i. 13, 15) *of Jesus the same* King of the Jews *yesterday* at the prayer meeting, Dec. 31, 1841, *and to-day*, Jan. 1, 1842, *and for ever.* See book of Life, page 24. Matt. xviii. 19, xxi. 13—16. In verse 6 the Italian body of St. Peter *is sleeping* " in the second death " *between the two* Imperial *soldiers* of France and Austria. The Emperor of France from Jan. 1, to July 11, 1859, causes the Italian *chains of St. Peter to fall off from his* Imperial *hands.*

'*I say unto thee*, Robert Ashton, *thou art Peter*, a stone, *and upon this rock*, of truth, *will I* Elizabeth, the angel of Jesus, *build my* Cottle *Church, and the gates of hell*, the doors of St. Peter, at Rome, shall not prevail against it—Matt. xvi. 18. Rev. iii. 7—12.'

This will be enough for the purpose. When any one who pleases can circulate new revelations of this kind, uninterrupted and unattended to, new revelations will cease to be a good investment of excentricity. I take it for granted that the gentlemen whose names are mentioned have nothing to do with the circulars or their doctrines. Any lady who may happen to be intrusted with a revelation may nominate her own pastor, or any other clergyman, one of her apostles ; and it is difficult to say to what court the nominees can appeal to get the commission abrogated.

*March* 16, 1865. During the last two years the circulars have continued. It is hinted that funds are low : and two gentlemen who are represented as gone ' to Bethlehem asylum in despair' say that Mrs. Cottle ' will spend all that she hath, while Her

Majesty's Ministers are flourishing on the wages of sin.' The following is perhaps one of the most remarkable passages in the whole :—

'*Extol and magnify Him* (Jehovah, the Everlasting God, see the Magnificat and Luke i. 45, 46—68—73—79), *that rideth* (by rail and steam over land and sea, from his holy habitation at Kirkstall Lodge, Psa. lxxvii. 19, 20), *upon the* (Cottle) *heavens, as it were* (Sept. 9, 1864, see pages 21, 170), *upon an* (exercising, Psa. cxxxi. 1), *horse-*(chair, bought of Mr. John Ward, Leicester-square).'

I have pretty good evidence that there is a clergyman who thinks Mrs. Cottle a very sensible woman.

[*The Cottle Church.* Had I chanced to light upon it at the time of writing, I should certainly have given the following. A printed letter to the *Western Times*, by Mr. Robert Cottle, was accompanied by a manuscript letter from Mrs. Cottle, apparently a circular. The date was Nov^r. 1853, and the subject was the procedure against Mr. Maurice at King's College for doubting that God would punish human sins by an existence of torture lasting through years numbered by millions of millions of millions of millions (repeat the word *millions* without end,) &c. The memory of Mr. Cottle has, I think, a right to the quotation : he seems to have been no participator in the notions of his wife :—

'The clergy of the Established Church, taken at the round number of 20,000, may, in their first estate, be likened to 20,000 gold blanks, destined to become sovereigns, in succession,—they are placed between the matrix of the Mint, when, by the pressure of the screw, they receive the impress that fits them to become part of the current coin of the realm. In a way somewhat analogous this great body of the clergy have each passed through the crucibles of Oxford and Cambridge,— have been assayed by the Bishop's chaplain, touching the health of their souls, and the validity of their call by the Divine Spirit, and then the gentle pressure of a prelate's hand upon their heads ; and the words—" Receive the Holy Ghost," have, in a brief space of time, wrought a change in them, much akin to the miracle of transubstantiation—the priests are completed, and they become the current ecclesiastical coin of our country. The whole body of clergy, here spoken of, have undergone the preliminary induction of baptism and confirmation ; and all have been duly ordained, *professing* to hold one faith, and to believe in the selfsame doctrines ! In short, to be as identical as the 20,000 sovereigns, if compared one with the other. But mind is not malleable and ductile, like gold ; and all the preparations of tests, creeds, and catechisms will not insure uniformity of belief. No stamp of orthodoxy will produce the same impress on the minds of different men. Variety is manifest, and patent, upon everything mental and material.  The Almighty has not created, nor man fashioned, two things alike !  How futile, then, is the attempt to shape and mould man's apprehension of divine truth by one fallible standard of man's invention !  If proof of this be required, an appeal might be made to history and the experience of eighteen hundred years.'

This is an argument of force against the reasonableness of expecting tens of thousands of educated readers of the New Testament to find the doctrine above described in it.  The lady's argument against the doctrine itself is very striking.  Speaking of an outcry on this matter among the Dissenters against one of their body, who was the son of ' the White Stone (Rev. ii. 17), or the Roman cement-maker,' she says—

' If the doctrine for which they so wickedly fight were true, what would become of the black gentlemen for whose redemption I have been sacrificed from April 8, 1839.'

There are certainly very curious points about this revelation. There have been many surmises about the final restoration of the infernal spirits, from the earliest ages of Christianity until our own day : a collection of them would be worth making.  On reading this in proof, I see a possibility that by ' black gentlemen' may be meant the clergy.  I suppose my first interpretation must have been suggested by context : I leave the point to the reader's sagacity.

> The Problem of squaring the circle solved ; or, the circumference and area of the circle discovered.  By James Smith.  London, 1859, 8vo.
> On the relations of a square inscribed in a circle.  Read at the British Association, Sept. 1859, published in the Liverpool Courier, Oct. 8, 1859, and reprinted in broadsheet.
> The question: Are there any commensurable relations between a circle and other Geometrical figures ?  Answered by a member of the British Association . . . London, 1860, 8vo.—[This has been translated into French by M. Armand Grange, Bordeaux, 1863, 8vo.]
> The Quadrature of the Circle.  Correspondence between an eminent mathematician and James Smith, Esq.  (Member of the Mersey Docks and Harbour Board),  London, 1861, 8vo. (pp. 200).
> Letter to the . . British Association . . . by James Smith, Esq. Liverpool, 1861, 8vo.
> Letter to the . . British Association . . . by James Smith, Esq. Liverpool, 1862, 8vo.—[These letters the author promised to continue.]

A Nut to crack for the readers of Professor De Morgan's 'Budget of Paradoxes.' By James Smith, Esq. Liverpool, 1863, 8vo.
Paper read at the Liverpool Literary and Philosophical Society, reported in the Liverpool Daily Courier, Jan. 26, 1864. Reprinted as a pamphlet.
The Quadrature of the circle, or the true ratio between the diameter and circumference geometrically and mathematically demonstrated. By James Smith, Esq. Liverpool, 1865, 8vo.
[On the relations between the dimensions and distances of the Sun, Moon, and Earth ; a paper read before the Literary and Philosophical Society of Liverpool, Jan. 25, 1864. By James Smith, Esq.
The British Association in Jeopardy, and Dr. Whewell, the Master of Trinity, in the stocks without hope of escape. Printed for the authors (J. S. confessed, and also hidden under *Nauticus*). (No date, 1865).
The British Association in Jeopardy, and Professor De Morgan in the Pillory without hope of escape. London, 1866, 8vo.]

When my work appeared in numbers, I had not anything like an adequate idea of Mr. James Smith's superiority to the rest of the world in the points in which he is superior. He is beyond a doubt the ablest head at unreasoning, and the greatest hand at writing it, of all who have tried in our day to attach their names to an error. Common cyclometers sink into puny orthodoxy by his side.

The behaviour of this singular character induces me to pay him the compliment which Achilles paid Hector, to drag him round the walls again and again. He was treated with unusual notice and in the most gentle manner. The unnamed mathematician, E. M. bestowed a volume of mild correspondence upon him ; Rowan Hamilton quietly proved him wrong in a way accessible to an ordinary schoolboy ; Whewell, as we shall see, gave him the means of seeing himself wrong, even more easily than by Hamilton's method. Nothing would do ; it was small kick and silly fling at all ; and he exposed his conceit by alleging that he, James Smith, had placed Whewell in the stocks. He will therefore be universally pronounced a proper object of the severest literary punishment : but the opinion of all who can put two propositions together will be that of the many strokes I have given, the hardest and most telling are my republications of his own attempts to reason.

He will come out of my hands in the position he ought to hold, the Supreme Pontiff of cyclometers, the vicegerent of St.

Vitus upon earth, the Mamamouchi of burlesque on inference. I begin with a review of him which appeared in the *Athenœum* of May 11, 1861. Mr. Smith says I wrote it: this I neither affirm nor deny; to do either would be a sin against the editorial system elsewhere described. Many persons tell me they know me by my style; let them form a guess: I can only say that many have declared as above while fastening on me something which I had never seen nor heard of.

The Quadrature of the Circle: Correspondence between an Eminent Mathematician and James Smith, Esq. (Edinburgh, Oliver & Boyd; London, Simpkin, Marshall & Co.)

'A few weeks ago we were in perpetual motion. We did not then suppose that anything would tempt us on a circle-squaring expedition: but the circumstances of the book above named have a peculiarity which induces us to give it a few words.

Mr. James Smith, a gentleman residing near Liverpool, was some years ago seized with the *morbus cyclometricus*. The symptoms soon took a defined form: his circumference shrank into exactly $3\frac{1}{8}$ times his diameter, instead of close to $3\frac{16}{113}$, which the mathematican knows to be so near to truth that the error is hardly at the rate of a foot in 2,000 miles. This shrinking of the circumference remained until it became absolutely necessary that it should be examined by the British Association. This body, which as Mr. James Smith found to his sorrow, has some interest in 'jealously guarding the mysteries of their profession,' refused at first to entertain the question. On this Mr. Smith changed his 'tactics' and the name of his paper, and smuggled in the subject under the form of 'The Relations of a Circle inscribed in a Square'! The paper was thus forced upon the Association, for Mr. Smith informs us that he 'gave the Section to understand that he was not the man that would permit even the British Association to trifle with him.' In other words, the Association bore with and were bored with the paper, as the shortest way out of the matter. Mr. Smith also circulated a pamphlet. Some kind hearted man, who did not know the disorder as well as we do, and who appears in Mr. Smith's handsome octavo as E. M. —the initials of 'eminent mathematician'—wrote to him and offered to show him in a page that he was all wrong. Mr. Smith thereupon opened a correspondence, which is the bulk of the volume. When the correspondence was far advanced, Mr. Smith announced his intention to publish. His benevolent instructor—

we mean in intention—protested against the publication, saying, 'I do not wish to be gibbeted to the world as having been foolish enough to enter upon what I feel now to have been a ridiculous enterprise.'

For this Mr. Smith cared nothing : he persisted in the publication, and the book is before us.  Mr. Smith has had so much grace as to conceal his kind adviser's name under E. M.  And that is to say, he has divided the wrong among all who may be suspected of having attempted so hopeless a task as that of putting a little sense into his head.  He has violated the decencies of private life. Against the will of the kind-hearted man who undertook his case, he has published letters which were intended for no other purpose than to clear his poor head of a hopeless delusion.  He deserves the severest castigation ; and he will get it : his abuse of confidence will stick by him all his days.  Not that he has done his benefactor—in intention, again—any harm.  The patience with which E. M. put the blunders into intelligible form, and the perseverance with which he tried to find a cranny-hole for common reasoning to get in at, are more than respectable : they are admirable.  It is, we can assure E. M., a good thing that the nature of the circle-squarer should be so completely exposed as in this volume.  The benefit which he intended Mr. James Smith may be conferred upon others.  And we should very much like to know his name, and if agreeable to him, to publish it.  As to Mr. James Smith, we can only say this : he is not mad.  Madmen reason rightly upon wrong premises : Mr. Smith reasons wrongly upon no premises at all.

E. M. very soon found out that, to all appearance, Mr. Smith got a circle of $3\frac{1}{8}$ times the diameter by making it the supposition to set out with that there was such a circle ; and then finding certain consequences which, so it happened, were not inconsistent with the supposition on which they were made.  Error is sometimes self-consistent.  However, E. M., to be quite sure of his ground, wrote a short letter, stating what he took to be Mr. Smith's hypothesis, containing the following :—'On A C as diameter, describe the circle D, which by hypothesis shall be equal to three and one-eighth times the length of A C. . . . I beg, before proceeding further, to ask whether I have rightly stated your argument.'  To which Mr. Smith replied :—'You have stated my argument with perfect accuracy.'  Still E. M. went on, and we could not help, after the above, taking these letters as the initials of Everlasting Mercy.  At last, however, when Mr. Smith flatly denied that the area of the circle lies between those of the inscribed and circumscribed polygons, E. M. was fairly beaten, and gave up the task. Mr. Smith was left to write his preface, to talk about the certain victory of truth—which, oddly enough, is the consolation of all hopelessly mistaken men ; to compare himself with Galileo ; and to expose to the world the perverse behaviour of the Astronomer Royal, on whom he wanted to fasten a conversation, and who replied, ' It would be a waste of time, Sir, to listen to anything you could have to say on such a subject.'

Having thus disposed of Mr. James Smith, we proceed to a few remarks on the subject: it is one which a journal would never originate, but which is rendered necessary from time to time by the attempts of the autopseustic to become heteropseustic. To the mathematician we have nothing to say : the question is, what kind of assurance can be given to the world at large that the wicked mathematicians are not acting in concert to keep down their superior, Mr. James Smith, the current Galileo of the quadrature of the circle.

Let us first observe that this question does not stand alone : independently of the millions of similar problems which exist in higher mathematics, the finding of the diagonal of a square has just the same difficulty, namely, the entrance of a pair of lines of which one cannot be definitely expressed by means of the other. We will show the reader who is up to the multiplication-table how he may go on, on, on, ever nearer, never there, in finding the diagonal of a square from the side.

Write down the following rows of figures, and more, if you like, in the way described :—

$$
\begin{array}{ccccccccc}
1 & 2 & 5 & 12 & 29 & 70 & 169 & 408 & 985 \\
1 & 3 & 7 & 17 & 41 & 99 & 239 & 577 & 1393
\end{array}
$$

After the second, each number is made up of double the last increased by the last but one : thus, 5 is 1 more than twice 2, 12 is 2 more than twice 5, 239 is 41 more than twice 99. Now, take out two adjacent numbers from the upper line, and the one below the first from the lower : as

$$
\begin{array}{cc}
70 & 169 \\
99. &
\end{array}
$$

Multiply together 99 and 169, giving 16,731. If, then, you will say that 70 diagonals are exactly equal to 99 sides, you are in error about the diagonal, but an error the amount of which is not so great as the 16,731st part of the diagonal. Similarly, to say that five diagonals make exactly seven sides does not involve an error of the 84th part of the diagonal.

Now, why has not the question of *crossing the square* been as celebrated as that of *squaring the circle*? Merely because Euclid demonstrated the impossibility of the first question, while that of the second was not demonstrated, completely, until the last century.

The mathematicians have many methods, totally different from each other, of arriving at one and the same result, their celebrated approximation to the circumference of the circle. An intrepid calculator has, in our own time, carried his approximation to what they call 607 decimal places: this has been done by Mr. Shanks, of Houghton-le-Spring, and Dr. Rutherford has verified 441 of these places. But though 607 looks large, the general public will form but a hazy notion of the extent of accuracy acquired. We have seen, in Charles Knight's *English Cyclopædia*, an account of the matter which may illustrate the unimaginable, though rationally conceivable, extent of accuracy obtained.

Say that the blood-globule of one of our animalcules is a millionth of an inch in diameter. Fashion in thought a globe like our own, but so much larger that our globe is but a blood-globule in one of its animalcules: never mind the microscope which shows the creature being rather a bulky instrument. Call this the first globe *above* us. Let the first globe above us be but a blood-globule, as to size, in the animalcule of a still larger globe, which call the second globe above us. Go on in this way to the twentieth globe above us. Now go down just as far on the other side. Let the blood-globule with which we started be a globe peopled with animals like ours, but rather smaller: and call this the first globe below us. Take a blood-globule out of this globe, people it, and call it the second globe below us: and so on to the twentieth globe below us. This is a fine stretch of progression both ways. Now give the giant of the twentieth globe *above* us the 607 decimal places, and, when he has measured the diameter of his globe with accuracy worthy of his size, let him calculate the circumference of his equator from the 607 places. Bring the little philosopher from the twentieth globe *below* us with his very best microscope, and set him to see the small error which the giant must make. He will not succeed, unless his microscopes be much better for his size than ours are for ours.

Now it must be remembered by any one who would laugh at

Y

the closeness of the approximation, that the mathematician generally goes *nearer*; in fact his theorems have usually no error at all. The very person who is bewildered by the preceding description may easily forget that if there were *no error at all*, the Lilliputian of the millionth globe below us could not find a flaw in the Brobdingnagian of the millionth globe above. The three angles of a triangle, of perfect accuracy of form, are *absolutely* equal to two right angles; no stretch of progression will detect *any* error.

Now think of Mr. Lacomme's mathematical adviser (*ante*, p. 32) making a difficulty of advising a stonemason about the quantity of pavement in a circular floor!

We will now, for our non-calculating reader, put the matter in another way. We see that a circle-squarer can advance, with the utmost confidence, the assertion that when the diameter is 1,000, the circumference is accurately 3,125: the mathematician declaring that it is a trifle more than $3,141\frac{1}{2}$. If the squarer be right, the mathematician has erred by about a 200th part of the whole: or has not kept his accounts right by about 10s. in every 100l. Of course, if he set out with such an error he will accumulate blunder upon blunder. Now, if there be a process in which close knowledge of the circle is requisite, it is in the prediction of the moon's place—say, as to time of passing the meridian at Greenwich—on a given day. We cannot give the least idea of the complication of details: but common sense will tell us that if a mathematician cannot find his way round the circle without a relative error four times as big as a stockbroker's commission, he must needs be dreadfully out in his attempt to predict the time of passage of the moon. Now, what is the fact? His error is less than a second of time, and the moon takes 27 days odd to revolve. That is to say, setting out with 10s. in 100l. of error in his circumference, he gets within the fifth part of a farthing in 100l. in predicting the moon's transit. Now we cannot think that the respect in which mathematical science is held is great enough—though we find it not small—to make this go down. That respect is founded upon a notion that right ends are got by right means: it will hardly be credited that the truth can be got to farthings out of data which are wrong by shillings. Even the celebrated Hamilton of Edinburgh, who held that in mathematics there was no way of going wrong, was fully impressed with the belief that this was because error was avoided from the beginning. He never went so far as to say that a mathematician who begins wrong must end right somehow.

There is always a difficulty about the mode in which the thinking man of common life is to deal with subjects he has not studied to a professional extent. He must form opinions on matters theological, political, legal, medical, and social. If he can make up his mind to choose a guide, there is, of course, no perplexity : but on all the subjects mentioned the direction-posts point different ways. Now why should he not form his opinion upon an abstract mathematical question ? Why not conclude that, as to the circle, it is possible Mr. James Smith may be the man, just as Adam Smith was the man of things then to come, or Luther, or Galileo ? It is true that there is an unanimity among mathematicians which prevails in no other class : but this makes the chance of their all being wrong only different in degree. And more than this, is it not generally thought among us that priests and physicians were never so much wrong as when there was most appearance of unanimity among them ? To the preceding questions we see no answer except this, that the individual inquirer may as rationally decide a mathematical question for himself as a theological or a medical question, so soon as he can put himself into a position in mathematics level with that in which he stands in theology or medicine. The every-day thought and reading of common life have a certain resemblance to the thought and reading demanded by the learned faculties. The research, the balance of evidence, the estimation of probabilities, which are used in a question of medicine, are closely akin in character, however different the matter of application, to those which serve a merchant to draw his conclusions about the markets. But the mathematicians have methods of their own, to which nothing in common life bears close analogy, as to the nature of the results or the character of the conclusions. The logic of mathematics is certainly that of common life : but the data are of a different species ; they do not admit of doubt. An expert arithmetician, such as is Mr. J. Smith, may fancy that calculation, merely as such, is mathematics : but the value of his book, and in this point of view it is not small, is the full manner in which it shows that a practised arithmetician, venturing into the field of mathematical demonstration, may show himself utterly destitute of all that distinguishes the reasoning geometrical investigator from the calculator.

And, further, it should be remembered that in mathematics the power of verifying results far exceeds that which is found in anything else : and also the variety of distinct methods by which they can be attained. It follows from all this that a person who desires to be as near the truth as he can will not judge the results of mathematical demonstration to be open to his criticism, in the same degree as results of other kinds. Should he feel compelled to decide, there is no harm done : his circle may be $3\frac{1}{8}$ times its diameter, if it please him. But we must warn him that, in order to get this circle, he must, as Mr. James Smith has done, *make it at home* : the laws of space and thought beg leave respectfully to decline the order."

I will insert now at length, from the *Athenæum* of June 8, 1861, the easy refutation given by my deceased friend, with the remarks which precede.

" Mr. James Smith, of whose performance in the way of squaring the circle we spoke some weeks ago in terms short of entire acquiescence, has advertised himself in our columns, as our readers will have seen. He has also forwarded his letter to the Liverpool *Albion*, with an additional statement, which he did not make in *our* journal. He denies that he has violated the decencies of private life, since his correspondent revised the proofs of his own letters, and his ' protest had respect only to making his name public.' This statement Mr. James Smith precedes by saying that we have treated as true what we well knew to be false ; and he follows by saying that we have not read his work, or we should have known the above facts to be true. Mr. Smith's pretext is as follows. His correspondent E. M. says, ' My letters were not intended for publication, and I protest against their being published,' and he subjoins 'Therefore I must desire that my name may not be used.' The obvious meaning is that E. M. protested against the publication altogether, but, judging that Mr. Smith was determined to publish, desired that his name should not be used. That he afterwards corrected the proofs merely means that he thought it wiser to let them pass under his own eyes than to leave them entirely to Mr. Smith.

We have received from Sir W. Rowan Hamilton a proof that the circumference is more than $3\frac{1}{8}$ diameters, requiring nothing but a knowledge of four books of Euclid. We give it in brief as an exercise for our juvenile readers to fill up. It reminds us of the old days when real geometers used to think it worth while seriously to demolish pretenders. Mr. Smith's fame is now assured : Sir W. R. Hamilton's brief and easy exposure will procure him notice in connexion with this celebrated problem.

It is to be shown that the perimeter of a regular polygon of 20 sides is greater than $3\frac{1}{8}$ diameters of the circle, and still more, of course, is the circumference of the circle greater than $3\frac{1}{8}$ diameters.

1. It follows from the 4th Book of Euclid, that the rectangle under the side of a regular decagon inscribed in a circle, and that side increased by the radius, is equal to the square of the radius. But the product $791(791+1280)$ is less than $1280 \times 1280$; if then the radius be 1280 the side of the decagon is greater than 791.

2. When a diameter bisects a chord, the square of the chord is equal to the rectangle under the doubles of the segments of the diameter. But the product $125 \ (4 \times 1280-125)$ is less than $791 \times 791$. If then the bisected chord be a side of the decagon, and if the radius be still 1280, the double of the lesser segment exceeds 125.

3. The rectangle under this doubled segment and the radius is equal to the square of the side of an inscribed regular polygon of 20 sides. But the product $125 \times 1280$ is equal to $400 \times 400$; therefore, the side of the last-mentioned polygon is greater than 400, if the radius be still 1280. In other words, if the radius be represented by the new member 16, and therefore the diameter by 32, this side is greater than 5, and the perimeter exceeds 100. So that, finally, if the diameter be 8, the perimeter of the inscribed regular polygon of 20 sides, and still more the circumference of the circle, is greater than 25: that is, the circumference is more than $3\frac{1}{8}$ diameters."

The last work in the list was thus noticed in the *Athenæum*, May 27, 1865.

"Mr. James Smith appears to be tired of waiting for his place in the Budget of Paradoxes, and accordingly publishes a long letter to Prof. De Morgan, with various prefaces and postscripts. The letter opens by a hint that the Budget appears at very long intervals, and 'apparently without any sufficient reason for it.' As Mr. Smith hints that he should like to see Mr. De Morgan, whom he calls an 'elephant of mathematics,' 'pumping his brains' 'behind the scenes'—an odd thing for an elephant to do, and an odd place to do it in—to get an answer, we think he may mean to hint that the Budget is delayed until the pump has worked successfully. Mr. Smith is informed that we have had the whole manuscript of the Budget, excepting only a final summing-up, in our hands since October, 1863. [This does not

refer to the Supplement.] There has been no delay : we knew
from the beginning that a series of historical articles would be
frequently interrupted by the things of the day. Mr. James
Smith lets out that he has never been able to get a private line
from Mr. De Morgan in answer to his communications : we should
have guessed it. He says, ' The Professor is an old bird and not
to be easily caught, and by no efforts of mine have I been able,
up to the present moment, either to induce or twit him into a
discussion. . . . ' Mr. Smith curtails the proverb : old birds
are not to be caught with *chaff*, nor with *twit*, which seems to be
Mr. Smith's word for his own chaff, and, so long as the first letter
is sounded, a very proper word. Why does he not try a little
grain of sense ? Mr. Smith evidently thinks that, in his character
as an elephant, the Professor has not pumped up brain enough to
furnish forth a bird. In serious earnest, Mr. Smith needs no
answer. In one thing he excites our curiosity : what is meant by
demonstrating ' geometrically *and* mathematically ? "

I now proceed to my original treatment of the case.

Mr. James Smith will, I have no doubt, be the most uneclipsed
circle-squarer of our day. He will not owe this distinction to his
being an influential and respected member of the commercial
world of Liverpool, even though the power of publishing which
his means give him should induce him to issue a whole library
upon one paradox. Neither will he owe it to the pains taken
with him by a mathematician, who corresponded with him until
the joint letters filled an octavo volume. Neither will he owe it
to the notice taken of him by Sir William Hamilton, of Dublin,
who refuted him in a manner intelligible to an ordinary student
of Euclid, which refutation he calls a remarkable paradox easily
explainable, but without explaining it. What he will owe it to I
proceed to show.

Until the publication of the ' Nut to Crack ' Mr. James Smith
stood among circle-squarers in general. I might have treated
him with ridicule, as I have done others : and he says that he
does not doubt he shall come in for his share at the tail end of
my Budget. But I can make a better job of him than so, as
Locke would have phrased it : he is such a very striking example
of something I have said on the use of logic that I prefer to make
an example of his writings. On one point indeed he well deserves
the *scutica*, if not the *horribile flagellum*. He tells me that he
will bring his solution to me in such a form as shall compel me
to admit it as *un fait accompli* [*une faute accomplie ?*] or leave

myself open to the humiliating charge of mathematical ignorance and folly.   He has also honoured me with some private letters. In the first of these he gives me a 'piece of information,' after which he cannot imagine that I, 'as an honest mathematician,' can possibly have the slightest hesitation in admitting his solution.   There is a tolerable reservoir of modest assurance in a man who writes to a perfect stranger with what he takes for an argument, and gives an oblique threat of imputation of dishonesty in case the argument be not admitted without hesitation; not to speak of the minor charges of ignorance and folly.   All this is blind self-confidence, without mixture of malicious meaning; and I rather like it : it makes me understand how Sam Johnson came to say of his old friend Mrs. Cobb,—'I love Moll Cobb for her impudence.'   I have now done with my friend's *suaviter in modo*, and proceed to his *fortiter in re* : I shall show that he *has* convicted himself of ignorance and folly, with an honesty and candour worthy of a better value of $\pi$.

Mr. Smith's method of proving that every circle is $3\frac{1}{8}$ diameters is to assume that it is so,—'if you dislike the term datum, then, by hypothesis, let 8 circumferences of a circle be exactly equal to 25 diameters,'—and then to show that every other supposition is thereby made absurd.   The right to this assumption is enforced in the 'Nut' by the following analogy :—

'I think you (!) will not dare (!) to dispute my right to this hypothesis, when I can prove by means of it that every other value of $\pi$ will lead to the grossest absurdities ; unless indeed, you are prepared to dispute the right of Euclid to adopt a false line hypothetically for the purpose of a " *reductio ad absurdum* " demonstration, in pure geometry.'

Euclid assumes what he wants to *disprove*, and shows that his *assumption* leads to absurdity, and so *upsets itself*.   Mr. Smith assumes what he wants to *prove*, and shows that. *his* assumption makes *other propositions* lead to absurdity.   This is enough for all who can reason.   Mr. James Smith cannot be argued with ; he has the whip-hand of all the thinkers in the world.   Montucla would have said of Mr. Smith what he said of the gentleman who squared his circle by giving 50 and 49 the same square root, *Il a perdu le droit d'être frappé de l'évidence.*

It is Mr. Smith's habit, when he finds a conclusion agreeing with its own assumption, to regard that agreement as proof of the assumption.   The following is the 'piece of information' which will settle me, if I be honest.   Assuming $\pi$ to be $3\frac{1}{8}$, he finds out

by working instance after instance that the mean proportional
between one-fifth of the area and one-fifth of eight is the radius.
That is,

$$\text{if } \pi = \frac{25}{8}, \ \sqrt{\left(\frac{\pi r^2}{5} \cdot \frac{8}{5}\right)} = r.$$

This 'remarkable general principle' may fail to establish Mr.
Smith's quadrature, even in an honest mind, if that mind should
happen to know that, $a$ and $b$ being any two numbers whatever,
we need only assume—

$$\pi = \frac{a^2}{b}, \text{ to get at } \sqrt{\left(\frac{\pi r^2}{a} \cdot \frac{b}{a}\right)} = r.$$

We naturally ask what sort of glimmer can Mr. Smith have of
the subject which he professes to treat? On this point he has
given satisfactory information. I had mentioned the old problem
of finding two mean proportionals, as a preliminary to the duplication of the cube. On this mention Mr. Smith writes as follows.
I put a few words in capitals; and I write $rq$ for the sign of the
square root, which embarrasses small type : —

'This establishes the following *infallible* rule, for finding two mean
proportionals OF EQUAL VALUE, and is more than a preliminary, to the
famous old problem of " Squaring the circle." Let any finite number,
say 20, and its fourth part $= \frac{1}{4}$ (20) $= 5$, be given numbers. Then
$rq$ (20 × 5) $= rq$ 100 $= 10$, is their mean proportional. Let this be
a given mean proportional TO FIND ANOTHER MEAN PROPORTIONAL OF EQUAL
VALUE. Then $20 \times \frac{\pi}{4} = 20 \times \dfrac{3 \cdot 125}{4} = 20 \times \cdot 78125 = 15 \cdot 625$ will
be the first number; as 25 : 16 : : $rq$ 20 : $rq$ 8·192: and $(rq\ 8\cdot 192)^2$
$\times \frac{\pi}{4} = 8\cdot192 \times \cdot78125 = 6\cdot4$ will be the second number; therefore
$rq$ (15·625 × 6·4) $= rq$ 100 $= 10$, is the required mean proportional
.... Now, my good Sir, however competent you may be to prove
every man a fool [not *every* man, Mr. Smith! only *some*; pray learn
logical quantification] who now thinks, or in times gone by has thought,
the " Squaring of the Circle" *a possibility*; I doubt, and, on the evidence afforded by your Budget, I cannot help doubting, whether you
were ever before competent to find two mean proportionals *by my unique
method.*'—(*Nut*, pp. 47, 48.) [That I never was, I solemnly declare !]

All readers can be made to see the following exposure. When
5 and 20 are given, $x$ is a mean proportional when in 5, $x$, 20, 5
is to $x$ as $x$ to 20. And $x$ must be 10. But $x$ and $y$ are *two*
mean proportionals when in 5, $x$, $y$, 20, $x$ is a mean proportional
between 5 and $y$, and $y$ is a mean proportional between $x$ and 20.
And these means are $x = 5 \sqrt[3]{4}, y = 5 \sqrt[3]{16}$. But Mr. Smith

finds *one* mean, finds it *again* in a roundabout way, and produces 10 and 10 as the two (equal!) means, in solution of the 'famous old problem.' This is enough: if more were wanted, there is more where this came from. Let it not be forgotten that Mr. Smith has found a translator abroad, two, perhaps three, followers at home, and—most surprising of all—a real mathematician to try to set him right. And this mathematician did not discover the character of the subsoil of the land he was trying to cultivate until a goodly octavo volume of letters had passed and repassed. I have noticed, in more quarters than one, an apparent want of perception of the *full* amount of Mr. Smith's ignorance : persons who have not been in contact with the non-geometrical circle-squarers have a kind of doubt as to whether anybody can carry things so far. But I am an 'old bird' as Mr. Smith himself calls me ; a Simorg, an 'all-knowing Bird of Ages' in matters of cyclometry.

The curious phenomena of thought here exhibited illustrate, as above said, a remark I have long ago made on the effect of proper study of logic. Most persons reason well enough on matter to which they are accustomed, and in terms with which they are familiar. But in unaccustomed matter, and with use of strange terms, few except those who are practised in the abstractions of pure logic can be tolerably sure to keep their feet. And one of the reasons is easily stated : terms which are not quite familiar partake of the vagueness of the X and Y on which the student of logic learns to see the formal force of a proposition independently of its material elements.

I make the following quotation from my fourth paper on logic in the *Cambridge Transactions* :—

'The uncultivated reason proceeds by a process almost entirely material. Though the necessary law of thought must determine the conclusion of the ploughboy as much as that of Aristotle himself, the ploughboy's conclusion will only be tolerably sure when the matter of it is such as comes within his usual cognizance. He knows that geese being all birds does not make all birds geese, but mainly because there are ducks, chickens, partridges, &c. A beginner in geometry, when asked what follows from " Every A is B," answers " Every B is A." That is, the necessary laws of thought, except in minds which have examined their tools, are not very sure to work correct conclusions except upon familiar matter . . . As the cultivation of the individual increases, the laws of thought which are of most usual application are applied to familiar matter with tolerable safety. But difficulty and risk of error make a new appearance with a new subject ; and this, in most cases, until new subjects are familiar things, unusual matter

common, untried nomenclature habitual; that is, until it is a habit to
be occupied upon a novelty.    It is observed that many persons reason
well in some things and badly in others ; and this is attributed to the
consequence of employing the mind too much upon one or another
subject.    But those who know the truth of the preceding remarks will
not have far to seek for what is often, perhaps most often, the true
reason . . . I maintain that logic tends to make the power of reason
over the unusual and unfamiliar more nearly equal to the power over
the usual and familiar than it would otherwise be.    The second is
increased ; but the first is almost created.'

Mr. James Smith, by bringing ignorance, folly, dishonesty into
contact with my name, in the way of conditional insinuation, has
done me a good turn : he has given me right to a freedom of
personal remark which I might have declined to take in the case
of a person who is useful and respected in matters which he
understands.

Tit for tat is logic all the world over.    By the way, what has
become of the rest of the maxim : we never hear it now.    When
I was a boy, in some parts of the country at least, it ran thus :—

> Tit for tat ;
> Butter for fat :
> If you kill my dog,
> I'll kill your cat.

He is a glaring instance of the truth of the observations quoted
above.    I will answer for it that, at the Mersey Dock Board, he
never dreams of proving that the balance at the banker's is larger
than that in the book by assuming that the larger sum is there,
and then proving that the other supposition—the smaller balance
—is, upon that assumption, an absurdity.    He never says to
another director, How can you dare to refuse me a right to assume
the larger balance, when you yourself, the other day, said,—
Suppose, for argument's sake, we had 80,000$l$. at the banker's,
though you knew the book only showed 30,000$l$.?    This is the
way in which he has supported his geometrical paradox by Euclid's
example : and this is not the way he reasons at the board ; I know
it by the character of him as a man of business which has reached
my ears from several quarters.    But in geometry and rational
arithmetic he is a smatterer, though expert at computation ; at
the board he is a trained man of business.    The language of
geometry is so new to him that he does not know what is meant
by ' two mean proportionals :' but all the phrases of commerce are
rooted in his mind.    He is most unerasably booked in the history
of the squaring of the circle, as the speculator who took a right

to assume a proposition for the destruction of other propositions, on the express ground that Euclid assumes a proposition to show that it destroys itself : which is as if the curate should demand permission to throttle the squire because St. Patrick drove the vermin to suicide to save themselves from slaughter. He is conspicuous as the speculator who, more visibly than almost any other known to history, reasoned in a circle by way of reasoning on a circle. But what I have chiefly to do with is the force of instance which he has lent to my assertion that men who have not had real training in pure logic are unsafe reasoners in matter which is not familiar. It is hard to get first-rate examples of this, because there are few who find the way to the printer until practice and reflection have given security against the grossest slips. I cannot but think that his case will lead many to take what I have said into consideration, among those who are competent to think of the great mental disciplines. To this end I should desire him to continue his efforts, to amplify and develope his great principle, that of proving a proposition by assuming it and taking as confirmation every consequence that does not contradict the assumption.

Since my Budget commenced, Mr. Smith has written me notes : the portion which I have preserved—I suppose several have been mislaid—makes a hundred and seven pages of note-paper, closely written. To all this I have not answered one word: but I think I cannot have read fewer than forty pages. In the last letter the writer informs me that he will not write at greater length until I have given him an answer, according to the ' rules of good society.' Did I not know that for every inch I wrote back he would return an ell? Surely in vain the net is spread in the eyes of anything that hath a wing. There were several good excuses for not writing to Mr. J. Smith: I will mention five. First, I distinctly announced at the beginning of this Budget that I would not communicate with squarers of the circle. Secondly, any answer I might choose to give might with perfect propriety be reserved for this article ; had the imputation of incivility been made after the first note, I should immediately have replied to this effect : but I presumed it was quite understood. Thirdly, Mr. Smith, by his publication of E. M.'s letters against the wish of the writer, had put himself out of the pale of correspondence. Fourthly, he had also gone beyond the rules of good society in sending letter after letter to a person who had shown by his silence an intention to avoid correspondence. Fifthly, these same rules of good society are contrived to be flexible or frangible in extreme cases: otherwise there would be no living under them; and good society would be bad. Father Aldrovand has laid down the necessary distinction—'I tell thee, thou foolish Fleming, the text speaketh but of promises made unto Christians, and there is in the rubric a special exemption of such as are made to Welchmen.' There is also a rubric to the rules of good society; and squarers of the circle are among those whom there is special permission not to answer: they are the wild Welchmen of geometry, who are always assailing, but never taking, the Garde Douloureuse of the circle. 'At this commentary,' proceeds the story, 'the Fleming grinned so broadly as to show his whole case of broad strong white teeth.' I know not whether the Welchman would have done the like, but I hope Mr. James Smith will: and I hope he has as good a case to show as Wilkin Flammock. For I wish him long life and long health, and should be very glad to see so much energy employed in a productive way. I hope he wishes me the same: if not, I will give him what all his judicious friends will think a good reason for doing so. His pamphlets and letters are all tied up together, and will form a curious lot when death or cessation of power to forage among book-shelves shall bring my little library to the hammer. And this time may not be far off: for I was X years old in A.D. $X^2$; not 4 in A.D. 16, nor 5 in A.D. 25, but still in one case under that law. And now I have made my own age a problem of quadrature, and Mr. J. Smith may solve it. But I protest against his method of assuming a result, and making itself prove itself: he might in this way, as sure as eggs is eggs (a corruption of X is X), make me 1,864 years old, which is a great deal too much.

*April* 5, 1864.—Mr. Smith continues to write me long letters, to which he hints that I am to answer. In his last, of 31 closely written sides of note-paper, he informs me, with reference to my obstinate silence, that though I think myself and am thought by others to be a mathematical Goliath, I have resolved to play the mathematical snail, and keep within my shell. A mathematical *snail*! This cannot be the thing so called which regulates the striking of a clock; for it would mean that I am to make Mr. Smith sound the true time of day, which I would by no means undertake upon a clock that gains 19 seconds odd in every hour by false quadrature. But he ventures to tell me that pebbles from the sling of simple truth and common sense will ultimately crack my shell, and put me *hors de combat*. The confusion of images is amusing: Goliath turning himself into a snail to avoid $\pi = 3\frac{1}{8}$, and James Smith, Esq., of the Mersey Dock Board: and put *hors de combat*—which should have been *caché*—by pebbles from a sling. If Goliath had crept into a snail-shell, David would have cracked the Philistine with his foot. There is something like modesty in the implication that the crack-shell pebble has not yet taken effect; it might have been thought that the slinger would by this time have been singing—

> And thrice [and one-eighth] I routed all my foes,
> And thrice [and one-eighth] I slew the slain.

But he promises to give the public his nut-cracker if I do not, before the Budget is concluded, 'unravel' the paradox, which is the mathematico-geometrical nut he has given me to crack. Mr. Smith is a crack man: he will crack his own nut; he will crack my shell; in the mean time he cracks himself up. Heaven send he do not crack himself into lateral contiguity with himself.

On June 27 I received a letter, in the handwriting of Mr. James Smith, signed Nauticus. I have ascertained that one of the letters to the *Athenæum* signed Nauticus is in the same handwriting. I make a few extracts:—

' . . . The important question at issue has been treated by a brace of mathematical birds with too much levity. It may be said, however, that sarcasm and ridicule sometimes succeed, where reason fails . . . Such a course is not well suited to a discussion . . . For this reason I shall for the future [this implies there has been a past, so that Nauticus is not before me for the first time] endeavour to confine myself to dry reasoning from incontrovertible premisses . . . It appears to me that so far as his theory is concerned he comes off unscathed. You might have found "a hole in Smith's circle" (have you seen a pamphlet bearing this title? [I never heard of it until now]), but after all it is quite possible the hole may have been left by design, for the purpose of entrapping the unwary.'

[On the publication of the above, the author of the pamphlet obligingly forwarded a copy to me of ' A Hole in Smith's Circle— by a Cantab : Longman and Co., 1859,' (pp. 15). ' It is pity to lose any fun we can get out of the affair,' says my almamaternal brother : to which I add that in such a case warning without joke is worse than none at all, as giving a false idea of the nature of the danger. The Cantab takes some absurdities on which I have not dwelt : but there are enough to afford a Cantab from every college his own separate hunting ground.]

Does this hint that his mode of proof, namely, assuming the thing to be proved, was a design to entrap the unwary? if so, it bangs Banagher. Was his confounding two mean proportionals with one mean proportional found twice over a trick of the same intent ? if so, it beats cockfighting.    That Nauticus is Mr. Smith appears from other internal evidence.    In 1819, Mr. J. C. Hobhouse was sent to Newgate for a libel on the House of Commons which was only intended for a libel on Lord Erskine.    The ex-Chancellor had taken Mr. Hobhouse to be thinking of him in a certain sentence ; this Mr. Hobhouse denied, adding, ' There is but one man in the country who is always thinking of Lord Erskine.'    I say that there is but one man of our day who would couple me and Mr. James Smith as ' brace of mathematical birds.'

Mr. Smith's 'theory' is unscathed by me.    Not a doubt about it : but how does he himself come off ?    I should never think of refuting a theory proved by assumption of itself.    I left Mr. Smith's $\pi$ untouched : or, if I put in my thumb and pulled out a plum, it was to give a notion of the cook, not of the dish.    The ' important question at issue ' was not the circle : it was, wholly and solely, whether the abbreviation of *James* might be spelt *Jimm.*[1]    This is personal to the verge of scurrility : but in literary controversy the challenger names the weapons, and Mr. Smith begins with charge of ignorance, folly, and dishonesty, by conditional implication.    So that the question is, not the personality of a word, but its applicability to the person designated : it is enough if, as the Latin grammar has it, *Verbum personale concordat cum nominativo.*

I may plead precedent for taking a liberty with the orthography of *Jem.*    An instructor of youth was scandalised at the abrupt and irregular—but very effective—opening of Wordsworth's little piece :—

> A simple child
> That lightly draws its breath,
> And feels its life in every limb,
> What should it know of death ?

So he mended the matter by instructing his pupils to read the first line thus :—

> A simple child, dear brother ——.

The brother, we infer from sound, was to be James, and the blank must therefore be filled up with *Jimb.*

I will notice one point of the letter, to make a little more

---

[1] The above is explained in the MS. by a paragraph referring to some anagrams, in one of which, by help of the orthography suggested, a designation for this cyclometer was obtained from the letters of his name. (ED )

distinction between the two birds. Nauticus lays down—quite correctly—that the sine of an angle is less than its circular measure. He then takes 3·1416 for 180°, and finds that 36' is ·010472. But this is exactly what he finds for the sine of 36' in tables: he concludes that either 3·1416 or the tables must be wrong. He does not know that sines, as well as $\pi$, are interminable decimals, of which the tables, to save printing, only take in a finite number. He is a six-figure man : let us go thrice again to make up nine, and we have as follows:—

Circular measure of 36' . . . . ·010471975 . . .
Sine of 36' . . . . . . . . . ·010471784 . . .
Excess of measure over sine . . . ·000000191 . . .

Mr. Smith invites me to say which is wrong, the quadrature, or the tables: I leave him to guess. He says his assertions ' arise naturally and necessarily out of the arguments of a circle-squarer:' he might just as well lay down that all the pigs went to market because it is recorded that ' *This* pig went to market.' I must say for circle-squarers that very few bring their pigs to so poor a market. I answer the above argument because it is, of all which Mr. James Smith has produced, the only one which rises to the level of a schoolboy: to meet him halfway I descend to that level.

Mr. Smith asks me to solve a problem in the *Athenæum* : and I will do it, because the question will illustrate what is *below* schoolboy level.

Let $x$ represent the circular measure of an angle of 15°, and $y$ half the sine of an angle of 30° = area of the square on the radius of a circle of diameter unity = ·25. If $x - y = xy$, firstly, what is the arithmetical value of $xy$ ? secondly, what is the angle of which $xy$ represents the circular measure ?

If $x$ represent 15° and $y$ be $\frac{1}{4}$, $xy$ represents 3° 45', whether $x-y$ be $xy$ or no. But, $y$ being $\frac{1}{4}$, $x-y$ is *not* $xy$ unless $x$ be $\frac{1}{3}$, that is, unless $12x$ or $\pi$ be 4, which Mr. Smith would not admit. How could a person who had just received such a lesson as I had given immediately pray for further exposure, furnishing the stuff so liberally as this ? Is it possible that Mr. Smith, because he signs himself Nauticus, means to deny his own very regular, legible, and peculiar hand ? It is enough to make the other members of the Liverpool Dock Board cry, Mersey on the man !

Mr. Smith says that for the future he will give up what he calls sarcasm, and confine himself, ' as far as possible,' to what he calls dry reasoning from incontrovertible premisses. If I have fairly taught him that *his* sarcasm will not succeed,-I hope he will find that his wit's end is his logic's beginning.

I now reply to a question I have been asked again and again since my last Budget appeared:—Why do you take so much trouble to expose such a reasoner as Mr. Smith ? I answer as a deceased friend of mine used to answer on like occasions—A man's capacity is no measure of his power to do mischief. Mr. Smith has untiring energy, which does something ; self-evident honesty of conviction, which does more ; and a long purse, which does most of all. He has made at least ten publications, full of figures which few readers can criticize. A great many people are staggered to this extent, that they imagine there must be the indefinite *something* in the mysterious *all this*. They are brought to the point of suspicion that the mathematicians ought not to treat ' all this ' with such undisguised contempt, at least. Now I have no fear for $\pi$ : but I do think it possible that general opinion might in time demand that the crowd of circle-squarers, &c. should be admitted to the honours of opposition ; and this would be a time-tax of five per cent., one man with another, upon those who are better employed. Mr. James Smith may be made useful, in hands which understand how to do it, towards preventing such opinion from growing. A speculator who expressly assumes what he wants to prove, and argues that all which contradicts it is absurd, *because* it cannot stand side by side with his assumption, is a case which can be exposed to all. And the best person to expose it is one who has lived in the past as well as the present, who takes misthinking from points of view which none but a student of history can occupy, and who has something of a turn for the business.

Whether I have any motive but public good must be referred to those who can decide whether a missionary chooses his pursuit solely to convert the heathen. I shall certainly be thought to have a little of the spirit of Col. Quagg, who delighted in strapping the Grace-walking Brethren. I must quote this myself: if I do not, some one else will, and then where am I ? The Colonel's principle is described as follows :—

'I licks ye because I kin, and because I like, and because ye'se critters that licks is good for. Skins ye have on, and skins I'll have off ; hard or soft, wet or dry, spring or fall. Walk in grace if ye like till pumpkins is peaches ; but licked ye must be till your toe-nails drop off and your noses bleed blue ink.' And—licked—they —were— accordingly.'

I am reminded of this by the excessive confidence with which Mr. James Smith predicted that he would treat me as Zephaniah Stockdolloger (Sam Slick calls it *slockdollager*) treated Goliah Quagg. He has announced his intention of bringing me, with a contrite heart, and clean shaved,—4159265. . . razored down to 25,—to a camp-meeting of circle-squarers. But there is this difference : Zephaniah only wanted to pass the Colonel's smithy in peace ; Mr. James Smith sought a fight with me. As soon as this Budget began to appear, he oiled his own strap, and attempted to treat me as the terrible Colonel would have treated the inoffensive brother.

He is at liberty to try again.

> The Moon-hoax ; or the discovery that the moon has a vast population of human beings. By Richard Adams Locke. New York, 1859, 8vo.

This is a reprint of the hoax already mentioned. I suppose R. A. Locke is the name assumed by M. Nicollet. The publisher informs us that when the hoax first appeared day by day in a morning paper, the circulation increased fivefold, and the paper obtained a permanent footing. Besides this, an edition of 60,000 was sold off in less than one month.

This discovery was also published under the name of A. R. Grant. Sohnke's ' Bibliotheca Mathematica ' confounds this Grant with Professor R. Grant of Glasgow, the author of the ' History of Physical Astronomy,' who is accordingly made to guarantee the discoveries in the moon. I hope Adams Locke will not merge in J. C. Adams, the co-discoverer of Neptune. Sohnke gives the titles of three French translations of the Moon hoax at Paris, of one at Bordeaux, and of Italian translations at Parma, Palermo, and Milan.

A Correspondent, who is evidently fully master of details, which he has given at length, informs me that the Moon hoax appeared first in the *New York Sun*, of which R. A. Locke was editor. It so much resembled a story then recently published by Edgar A. Poe, in a Southern paper, ' Adventures of Hans Pfaal,' that some New York journals published the two side by side. Mr. Locke, when he left the *New York Sun*, started another paper, and discovered the manuscript of Mungo Park ; but this did not deceive. The *Sun*, however, continued its career, and had a great success in an account of a balloon voyage from England to America, in seventy-five hours, by Mr. Monck Mason, Mr. Harrison Ainsworth, and others. I have no doubt that M. Nicollet was the author of the Moon hoax, written in a way which marks the practised Observatory astronomer beyond all doubt, and by evidence seen in the most minute details. Nicollet had an eye to Europe. I suspect that he took Poe's story, and made it a basis for his own. Mr. Locke, it would seem, when he attempted a fabrication for himself, did not succeed.

The Earth we inhabit, its past, present, and future. By Capt. Drayson. London, 1859, 8vo.

The earth is growing; absolutely growing larger: its diameter increases three-quarters of an inch per mile every year. The foundations of our buildings will give way in time: the telegraph cables break, and no cause ever assigned except ships' anchors, and such things. The book is for those whose common sense is unwarped, who can judge evidence as well as the ablest philosopher. The prospect is not a bad one, for population increases so fast that a larger earth will be wanted in time, unless emigration to the Moon can be managed, a proposal of which it much surprises me that Bishop Wilkins has a monopoly.

*Athenæum*, August 19, 1865. *Notice to Correspondents.*

' R. W.—If you will consult the opening chapter of the Budget of Paradoxes, you will see that the author presents only works in his own library at a given date; and this for a purpose explained. For ourselves we have carefully avoided allowing any writers to present themselves in our columns on the ground that the Budget has passed them over. We gather that Mr. De Morgan contemplates additions at a future time, perhaps in a separate and augmented work ; if so, those who complain that others of no greater claims than themselves have been ridiculed may find themselves where they wish to be. We have done what we can for you by forwarding your letter to Mr. De Morgan.'

The author of ' An Essay on the Constitution of the Earth,' published in 1844, demanded of the *Athenæum*, as an *act of fairness*, that a letter from him should be published, proving that he had as much right to be 'impaled' as Capt. Drayson. He holds, on speculative grounds, what the other claims to have proved by measurement, namely, that the earth is growing; and he believes that in time—a good long time, not *our* time— the earth and other planets may grow into suns, with systems of their own.

This gentleman sent me a copy of his work, after the commencement of my Budget ; but I have no recollection of having received it, and I cannot find it on the (nursery? quarantine?) shelves on which I keep my unestablished discoveries. Had I known of this work in time, (see the Introduction) I should of course, have impaled it (heraldically) with the other work; but the two are very different. Capt. Drayson professes to prove his point by results of observation; and I think he does not succeed. The author before me only speculates; and a speculator can get any conclusion into his premises, if he will only build or hire them of shape and size to suit. It reminds me of a statement I heard years ago, that a score of persons, or near it, were to dine inside the skull of one of the aboriginal animals, dear little creatures! Whereat I wondered vastly, nothing doubting; facts being stubborn and not easy drove, as Mrs. Gamp said. But I soon learned that the skull was not a real one, but artificially constructed by the methods—methods which have had striking verifications, too—which enable zoologists to go the whole hog by help of a toe or a bit of tail. This took off the edge of the wonder: a hundred people can dine inside an inference, if you draw it large enough. The method might happen to fail for once: for instance, the toe-bone might have been abnormalised by therian or saurian malady; and the possibility of such failure, even when of small probability, is of great alleviation. The author before me is, apparently, the sole fabricator of his own premises. With vital force in the earth and continual creation on the part of the original Creator, he expands our bit of a residence as desired. But, as the Newtoness of Cookery observed, First catch your hare. When this is done, when you *have* a growing earth, you shall dress it with all manner of proximate causes, and serve it up with a growing Moon for sauce, a growing Sun, if it please you, at the other end, and growing planets for side-dishes. Hoping this amount of impalement will be satisfactory, I go on to something else.

*The Hailesean System of Astronomy.* By John Davey Hailes (two pages duodecimo, 1860).

He offers to *take* 100,000*l.* to 1,000*l.* that he shows the sun to be less than seven millions of miles from the earth. The earth in the centre, revolving eastward, the sun revolving westward, so that they 'meet at half the circle distance in the 24 hours.' The diameter of a circle being 9839458303, the circumference is 30911,569,920.

The following written challenge was forwarded to the Council of the Astronomical Society: it will show the 'general reader'

—and help him towards earning his name—what sort of things come every now and then to our scientific bodies. I have added punctuation :—

### Challenge.

### 1,000 to 30,000.

Leverrier's name stand placed first. Do the worthy Frenchman justice.
By awarding him the medal in a trice.
Give Adams an extra—of which neck and neck the race.
Now I challenge to meet them and the F. R. S.'s all,
For good will and *one* thousand pounds to their *thirty* thousand withall,
That I produce a system, which shall measure the time,
When the Sun was vertical to Gibeon, afterward to Syene.
To meet any time in London—name your own period,
To be decided by a majority of twelve persons—a President, *odd*.
That mean, if the twelve equally divide, the President decide,
I should prefer the Bishop of London, over the meeting to preside.

JOHN DAVY HAILES.

Feb. 17, 1847.

Mr. Hailes still issues his flying sheets. The last I have met with (October 7, 1863) informs us that the latitude of England is slowly increasing, which is the true cause of the alteration in the variation of the magnet.

[Mr. Hailes continues his researches. Witness his new Hailesean system of Astronomy, displaying Joshua's miracle-time, origin of time from science, with Bible and Egyptian history. Rewards offered for astronomical problems. With magnetism, &c. &c. Astronomical challenge to all the world. Published at Cambridge, in 1865. The author agrees with Newton in one marked point *Errores quam minimi non sunt contemnendi*, says Isaac : meaning in figures, not in orthography. Mr. Hailes enters into the spirit, both positive and negative, of this dictum, by giving the distance of *Sidius* from the centre of the earth at 163,162,008 miles 10 feet 8 inches 17-28ths of an inch. Of course, he is aware that the centre of *figure* of the earth is 17·1998 inches from the centre of *gravity*. Which of the two is he speaking of?]

The Divine Mystery of Life. London [1861], 18mo. (pp. 32).

The author has added one class to zoology, which is printed in capitals, as derived from *zoé*, life, not from *zóon*, animal. That class is of *Incorporealia*, order I., *Infinitum*, of one genus without plurality, *Deus :* order II., *Finita*, angels good and evil.

The rest is all about a triune system, with a diagram. The author is not aware that ζωον is not *animal*, but *living being*. Aristotle has classed gods under ζωα, and has been called to account for it by moderns who have taken the word to mean *animal*.

> Explication du Zodiaque de Denderah, des Pyramides, et de Genèse. Par le Capitaine au longcours Justin Roblin. Caen, 1861. 8vo.

Capt. Roblin, having discovered the sites of gold and diamond mines by help of the zodiac of Denderah, offered half to the shareholders of a company which he proposed to form. One of our journals, by help of the zodiac of Esné, offered, at five francs a head, to tell the shareholders the exact amount of gold and diamonds which each would get, and to make up the amount predicted to those who got less. There are moods of the market in England in which this company could have been formed: so we must not laugh at our neighbours.

> A million's worth of property, and five hundred lives annually lost at sea by the Theory of Gravitation. A letter on the true figure of the earth, addressed to the Astronomer Royal, by Johannes von Gumpach. London, 1861, 8vo. (pp. 54).
>
> The true figure and dimensions of the earth, in a letter addressed to the Astronomer Royal. By Joh. von Gumpach. 2nd ed. entirely recast. London, 1862, 8vo. (pp. 266).
>
> Two issues of a letter published with two different title-pages, one addressed to the Secretary of the Royal Society, the other to the Secretary of the Royal Astronomical Society. It would seem that the same letter is also issued with two other titles, addressed to the British Association and the Royal Geographical Society. By Joh. von Gumpach. London, 1862, 8vo.
>
> Baby-Worlds. An essay on the nascent members of our solar household. By Joh. von Gumpach. London, 1863, 8vo.

The earth, it appears, instead of being flattened, is elongated at the poles : by ignorance of which the loss above mentioned occurs yearly. There is, or is to be, a substitute for attraction and an 'application hitherto neglected, of a recognised law of optics to the astronomical theory, showing the true orbits of the heavenly bodies to be perfectly circular, and their orbital motions to be perfectly uniform:' all irregularities being, I suppose, optical delusions. Mr. Von Gumpach is a learned man : what else, time must show.

*Perpetuum Mobile :* or Search for self-motive Power. By Henry
Dircks. London, 1861, 8vo.

A useful collection on the history of the attempts at perpetual
motion, that is, at obtaining the consequences of power without
any power to produce them. September 7, 1863, a correspondent
of the *Times* gave an anecdote of George Stephenson, which he
obtained from Robert Stephenson. A perpetual motionist wanted
to explain his method ; to which George replied—'Sir ! I shall
believe it when I see you take yourself up by the waistband, and
carry yourself about the room.' Never was the problem better
stated.

There is a paradox of which I ought to give a specimen, I mean
the *slander-paradox* ; the case of a person who takes it into his
head, upon evidence furnished entirely by the workings of his
own thoughts, that some other person has committed a foul act
of which the world at large would no more suppose him guilty
than they would suppose that the earth is a flat bordered by ice.
If I were to determine on giving cases in which the self-deluded
person imagines a conspiracy against *himself*, there would be no
end of choices. Many of the grosser cases are found at last to
be accompanied by mental disorder, and it is difficult to avoid
referring the whole class to something different from simple misuse
of the reasoning power. The first instance is one which puts in
a strong light the state of things in which we live, brought about
by our glorious freedom of thought, speech, and writing. The
Government treated it with neglect, the press with silent contempt, and I will answer for it many of my readers now hear of
it for the first time, when it comes to be enrolled among circle-
squarers and earth-stoppers, where, as the old philosophers said,
it will not gravitate, being *in proprio loco.*

1862. On new year's day, 1862, when the nation was in the
full tide of sympathy with the Queen, and regret for its own loss,
a paper called the *Free Press* published a number devoted to the
consideration of the causes of the death of the Prince Consort.
It is so rambling and inconsecutive that it takes more than one
reading to understand it. It is against the *Times* newspaper.
First, the following insinuation :—

' To the legal mind, the part of [the part taken by] the *Times* will
present a *primâ facie* case of the gravest nature, in the evident fore-
knowledge of the event, and the preparation to turn it to account
when it should have occurred. The article printed on Saturday must

have been written on Friday. That article could not have appeared had the Prince been intended to live.'

Next, it is affirmed that the *Times* intended to convey the idea that the Prince had been poisoned.

' Up to this point we are merely dealing with words which the *Times* publishes, and these can leave not a shadow of doubt that there is an intention to promulgate the idea that Prince Albert had been poisoned.'

The article then goes on with a strange olio of insinuations to the effect that the Prince was the obstacle to Russian intrigue, and that if he should have been poisoned,—which the writer strongly hints may have been the case,—some Minister under the influence of Russia must have done it. Enough for this record. *Un sot trouve toujours un plus sot qui l'admire*: who can he be in this case ?

1846. At the end of this year arose the celebrated controversy relative to the discovery of Neptune. Those who know it are well aware that Mr. Adams's now undoubted right to rank with Le Verrier was made sure at the very outset by the manner in which Mr. Airy, the Astronomer Royal, came forward to state what had taken place between himself and Mr. Adams. Those who know all the story about Mr. Airy being arrested in his progress by the neglect of Mr. Adams to answer a letter, with all the imputations which might have been thrown upon himself for laxity in the matter, know also that Mr. Airy's conduct exhibited moral courage, honest feeling, and willingness to sacrifice himself, if need were, to the attainment of the ends of private justice, and the establishment of a national claim. A writer in a magazine, in a long and elaborate article, argued the supposition—put in every way except downright assertion, after the fashion of such things—that Mr. Airy had communicated Mr. Adams's results to M. Le Verrier, with intention that they should be used. His presumption as to motive is that, had Mr. Adams been recognised, ' then the discovery must have been indisputably an *Englishman's*, and that Englishman not the Astronomer Royal.' Mr. Adams's conclusions were ' retouched in France, and sent over the year after.' The proof given is that it cannot be ' imagined ' otherwise.

' Can it then be imagined that the Astronomer Royal received such results from Mr. Adams, supported as they were by Professor Challis's valuable testimony as to their probable accuracy, and did not bring the French astronomer acquainted with them, especially as he was aware that his friend was engaged in matters bearing directly upon these results?'

The whole argument the author styles 'evidence which I consider it difficult to refute.' He ends by calling upon certain persons, of whom I am one, to 'see ample justice done.' This is the duty of every one, according to his opportunities. So when the reputed author—the article being anonymous—was, in 1849, proposed as a Fellow of the Astronomical Society, I joined—if I remember right, I originated—an opposition to his election, until either the authorship should be denied, or a proper retraction made. The friends of the author neither denied the first, nor produced the second: and they judged it prudent to withdraw the proposal. Had I heard of any subsequent repentance, I would have taken some other instance, instead of this : should I yet hear of such a thing, I will take care to notice it in the continuation of this list, which I confidently expect, life and health permitting, to be able to make in a few years. This much may be said, that the author, in a lecture on the subject, given in 1849, and published with his name, did *not* repeat the charge.

[The libel was published in the 'Mechanics' Magazine,' (vol. for 1846, pp. 604–615): and the editor supported it as follows, (vol. for 1847, p. 476). In answer to Mr. Sheepshanks's charitable hope that he had been hoaxed, he says 'Mr. Sheepshanks cannot certainly have read the article referred to . . . Severe and inculpatory it is—unjust some may deem it (though we ourselves are out of the number.) . . A "hoax" forsooth ! May we be often the dupes of such hoaxes !' He then goes on to describe the article as directed against the Astronomer Royal's alleged neglect to give Mr. Adams that 'encouragement and protection' which was his due, and *does not hint one word* about the article containing the charge of having secretly and fraudulently transmitted news of Mr. Adams's researches to France, that an Englishman might not have the honour of the discovery. Mr. Sheepshanks having called this a 'deliberate calumny,' without a particle of proof or probability to support it, the editor says 'what the reverend gentleman means by this, we are at a loss to understand.' He then proceeds *not* to remember. I repeat here, what I have said elsewhere, that the management of the journal has changed hands ; but from 1846 to 1856, it had the collar of S.S. (scientific slander). The prayer for more such things was answered (See pp. 349).]

I have said that those who are possessed with the idea of conspiracy against themselves are apt to imagine both conspirators and their bad motives and actions. A person who should take up the idea of combination against himself without feeling ill-will and originating accusations would be indeed a paradox. But such a paradox has existed. It is very well known, both in and beyond the scientific world, that the late James Ivory was subject to the impression of which I am speaking ; and the diaries and other sources of anecdote of our day will certainly, sooner or later, make it a part of his biography. The consequence will be that to his memory will be attached the unfavourable impression which the usual conduct of such persons creates ; unless it should happen that some one who knows the real state of the case puts the two sides of it properly together. Ivory was of that note in the scientific world which may be guessed from Laplace's description of him as the first geometer in Britain and one of the first in Europe. Being in possession of accurate knowledge of his peculiarity in more cases than one ; and in one case under his own hand : and having been able to make full inquiry about him, especially from my friend the late Thomas Galloway—who came after him at Sandhurst—one of the few persons with whom he was intimate :—I have decided, after full deliberation, to forestall the future biographies.

That Ivory was haunted by the fear of which I have spoken, to the fullest extent, came to my own public and official knowledge, as Secretary of the Astronomical Society. It was the duty of Mr. Epps, the Assistant Secretary, at the time when Francis Baily first announced his discovery of the Flamsteed Papers, to report to me that Mr. Ivory had called at the Society's apartments to inquire into the contents of those papers, and to express his hope that Mr. Baily was not attacking living persons under the names of Newton and Flamsteed. Mr. Galloway, to whom I communicated this, immediately went to Mr. Ivory, and succeeded, after some explanation, in setting him right. This is but one of many instances in which a man of thoroughly sound judgment in every other respect seemed to be under a complete chain of delusions about the conduct of others to himself. But the paradox is this :—I never could learn that Ivory, passing his life under the impression that secret and unprovoked enemies were at work upon his character, ever originated a charge, imputed a bad motive, or allowed himself an uncourteous expression. Some letters of his, now in my possession, referring to a private matter, are, except in the main impression on which they proceed, unobjectionable in every point : they might have been written by a cautious friend, whose object was, if possible, to prevent a difference from becoming a duel without compromising his principal's rights or character. Knowing that in some quarters the knowledge of Ivory's peculiarity is more or less connected with a notion that the usual consequences followed, I think the preceding statement due to his memory.

In such a record as the present, which mixes up the grossest speculative absurdities with every degree of what is better, an instance of another kind may find an appropriate place. The faults of journalism, when merely exposed by other journalism pass by and are no more regarded. A distinct account of an undeniable meanness, recorded in a work of amusement and reference both, may have its use : such a thing may act as a warning. An editor who is going to indulge his private grudge may be prevented from counting upon oblivion as a matter of certainty.

There are three kinds of journals, with reference to the mode of entrance of contributors. First, as a thing which has been, but which now hardly exists, there is the journal in which the editor receives a fixed sum to *find the matter*. In such a journal, every article which the editor can get a friend to give him is so much in his own pocket, which has a great tendency to lower the character of the articles; but I am not concerned with this point. Secondly, there is the journal which is supported by voluntary contributions of matter, the editor selecting. Thirdly, there is the journal in which the contributor is paid by the proprietors in a manner with which the literary editor has nothing to do.

The third class is the safe class, as its editors know : and, as a usual rule, they refuse unpaid contributions of the editorial cast. It is said that when Canning declined a cheque forwarded for an article in the *Quarterly,* John Murray sent it back with a blunt threat that if he did not take his money he could never be admitted again. The great publisher told him that if men like himself in position worked for nothing, all the men like himself in talent who could not afford it would not work for the *Quarterly.* If the above did not happen between Canning and Murray, it *must have happened* between some other two. Now journals of the second class—and of the first, if such there be—have a fault to which they alone are very liable, to say nothing of the editorial function (see the paper at the beginning, p. 11 et seq.), being very much cramped, a sort of gratitude towards effective contributors leads the journal to help their personal likes and dislikes, and to sympathise with them. Moreover, this sort of journal is more accessible than others to articles conveying personal imputation :

and when these provoke discussion, the journal is apt to take the part of the assailant to whom it lent itself in the first instance. Among the journals which went all lengths with contributors whom they valued, was the *Mechanics' Magazine* in the period 1846–56. I cannot say that matters have not mended in the last ten years : and I draw some presumption that they have mended from my not having heard, since 1856, of anything resembling former proceedings. And on actual inquiry, made since the last sentence was written, I find that the property has changed hands, the editor is no longer the same, and the management is of a different stamp. This journal is chiefly supported by voluntary articles : and it is the journal in which, as above noted, the ridiculous charge against the Astronomer Royal was made in 1849. The following instance of attempt at revenge is so amusing that I select it as the instance of the defect which I intend to illustrate ; for its puerility brings out in better relief the points which are not so easily seen in more adult attempts.

The *Mechanics' Magazine*, which by its connexion with engineering, &c., had always taken somewhat of a mathematical character, began, a little before 1846, to have more to do with abstract science. Observing this, I began to send short communications, which were always thankfully received, inserted, and well spoken of. Any one who looks for my name in that journal in 1846–49, will see nothing but the most respectful and even laudatory mention. In May 1849 occurred the affair at the Astronomical Society, and my share in forcing the withdrawal of the name of the alleged contributor to the journal. In February 1850 occurred the opportunity of payment. The *Companion to the Almanac* had to be noticed, in which, as then usual, was an article signed with my name. I shall give the review of this article entire, as a sample of a certain style, as well as an illustration of my point. The reader will observe that my name is not mentioned. This would not have done ; the readers of the Magazine would have stared to see a name of not infrequent occurrence in previous years all of a sudden fallen from the heaven of respect into the pit of contempt, like Lucifer, son of the morning. But before giving the review, I shall observe that Mr. Adams, in whose *favour* the attack on the Astronomer Royal was made, did not appreciate the favour ; and of course did not come forward to shield his champion. This gave deadly offence, as will appear from the following passage, (February 16, 1850) :—

" It was our intention to enter into a comparison of the contents of our Nautical Almanack with those of its rival, the *Connaissance des*

*Temps* ; but we shall defer it for the present. The Nautical Almanack for 1851 will contain Mr. Adams's paper ' On the Perturbation of Uranus ;' and when it comes, in due course, before the public, we are quite sure that that gentleman will expect that we shall again enter upon the subject with peculiar delight. Whilst we have a thorough loathing for mean, cowardly, crawlers—we have an especial pleasure in maintaining the claims of men who are truly grateful as well as highly-talented : Mr. Adams, therefore, will find that he cannot be disappointed—and the occasion will afford us an opportunity for making the comparison to which we have adverted."

This passage illustrates what I have said on the editorial function (p. 11). What precedes and follows has some criticism on the Government, the Astronomer Royal, &c., but reserved in allusion, oblique in sarcasm, and not fiercely uncourteous. The coarseness of the passage I have quoted shews editorial insertion, which is also shown by its blunder. The inserter is waiting for the Almanac of 1851 that he may review Mr. Adams's paper, which is to be contained in it. His own contributor, only two sentences before the insertion, had said, ' The Nautical Almanac, we believe, is published three or four years in advance.' In fact, the Almanac for 1851—with Mr. Adams's paper at the end—was published at the end of 1847 or very beginning of 1848; it had therefore been more than two years before the public when the passage quoted was written. And probably every person in the country who was fit to review Mr. Adams's paper—and most of those who were fit to read it—knew that it had been widely circulated, in revise, at the end of 1846 : my copy has written on it, ' 2nd revise, December 27, 1846, at noon,' in the hand-writing of the Superintendent of the Almanac ; and I know that there was an extensive issue of these revises, brought out by the Le-Verrier-and-Adams discussion. I now give the review of myself, (February 23, 1850):—

## "THE BRITISH ALMANACK AND COMPANION.

" The Companion to this Almanack, for some years after its first publication, annually contained scientific articles by Sir J. Lubbock and others of a high order and great interest ; we have now, however, closed the publication as a scientific one in remembrance of what it was, and not in consequence of what it is. Its list of contributors on science, has grown 'small by degrees and beautifully less,' until it has dwindled down to one—' a last rose of summer left withering alone.' The one contributor has contributed one paper ' On Ancient and Modern Usage in Reckoning.'

The learned critic's *chef d'œuvre*, is considered, by competent judges, to be an Essay on *Old Almanacks* printed a few years ago in this annual, and supposed to be written with the view of surpassing a profound memoir on the same subject by James O. Halliwell, Esq., F.R. and A.S.S., but the tremendous effort which the learned writer then made to excel many titled competitors for honours in the antique line appears to have had a sad effect upon his mental powers—at any rate, his efforts have since yearly become duller and duller; happily, at last, we should suppose, ' the ancient and modern usage in reckoning' indicates the lowest point to which the *vis inertia* of the learned writer's peculiar genius can force him.

We will give a few extracts from the article.

The learned author says, ' Those who are accustomed to settle the meaning of ancient phrases by self-examination will find some *strange* conclusions arrived at by us.' The writer never wrote a more correct sentence—it admits of no kind of dispute.

' Language and counting,' says the learned author, ' both came before the logical discussion of either. It is not allowable to argue that something is or was, because it ought to be or ought to have been. That two negatives make an affirmative, ought to be; if *no* man have done *nothing*, the man who has done nothing does not exist, and *every* man has done *something*. But in Greek, and in uneducated English, it is unquestionable that ' no man has done nothing ' is only an emphatic way of saying that no man has done *anything*; and it would be absurd to reason that it could not have been so, because it should not.'—p. 5.

' But there *is* another difference between old and new times, yet more remarkable, for we have *nothing* of it now : whereas in things indivisible we count with our fathers, and should say in buying an acre of land, that the result has no parts, and that the purchaser, till he owns all the ground, owns none, the change of possession being instantaneous. This second difference lies in the habit of considering nothing, nought, zero, cipher, or whatever it may be called, to be at the beginning of the scale of numbers. Count four days from Monday : we should now say Tuesday, Wednesday, Thursday, Friday ; formerly, it would have been Monday, Tuesday, Wednesday, Thursday. Had we asked, what at that rate is the first day from Monday, all would have stared at a phrase they had never heard. Those who were capable of extending language would have said, Why it must be Monday itself : the rest would have said, there can be no first day from Monday, for the day after is Tuesday, which must be the second day : Monday, one ; Tuesday, two.'—p. 10.

We assure our readers that the whole article is equally lucid, and its logic alike formal.

There are some exceedingly valuable foot-notes ; we give one of the most interesting, taken from the learned Mr. Halliwell's profound book on Nursery Rhymes—a celebrated production, for which it is supposed the author was made F.R.S.

' *One's nine,*
Two's some,
Three's a many,
Four's a penny,
Five's a little hundred.'

'The last line refers to five score, the so-called hundred being more usually six score. The first line, looked at etymologically, is *one is not one*, and the change of thought by which *nine*, the decimal of *one*, aims to be associated with the decimal of *plurality* is curious : '—Very.

This valuable and profound essay will very probably be transferred to the next edition of the learned Mr. Halliwell's rare work, of kindred worth, entitled ' RARA MATHEMATICA,' it will then be deservedly handed down to posterity as a covering for cheap trunks—a most appropriate archive for such a treasure."

In December, 1846, the *Mechanics' Magazine* published a libel on Airy in the matter of the discovery of Neptune. In May, 1849, one * * * was to have been brought forward for election at the Astronomical Society, and was opposed by me and others, on the ground that he was the probable author of this libel, and that he would not, perhaps could not, deny it. [N.B. I no more doubt that he was the author than I doubt that I am the author of this sentence.][1]

Accordingly, * * * was withdrawn, and a discussion took place, for which see the *Athenæum*, No. 1126, May 26, 1849, p. 544. The *Mechanics' Magazine* was very sore, but up to this day has never ventured beyond an attack on Airy, private whisperings against Adams—(see *ante*, p. 348),—and the above against myself. In due time, I doubt not my name will appear as one of the *âmes damnées* of the *Mechanics' Magazine*.[2]

First, as to Mr. Halliwell. The late Thomas Stephens Davies, excellent in geometry, and most learned in its history, was also a good hand at enmity, though not implacable. He and Mr. Halliwell, who had long before been very much one, were, at this date, very much two. I do not think T. S. Davies wrote this article ; and I think that by giving my reasons I shall do service to his memory. It must have been written at the beginning of February ; and within three days of that time T. S. Davies was making over to me, by his own free act, to be kept until claimed

---

[1] The subject of this criticism is of long past date, and as it has only been introduced by the author as an instance of faulty editorship, I have omitted the name of the writer of the libel, and a few lines of further detail.—ED.

[2] The editor of the *Mechanics' Magazine* died soon after the above was written.—ED.

by the relatives, what all who knew even his writings knew that he considered as the most precious deposit he had ever had in his keeping — Horner's papers. His letter announcing the transmission is dated February 2, 1850. This is a strong point; but there is another quite as strong. Euclid and his writings were matters on which T. S. Davies knew neither fear nor favour: he could not have written lightly about a man who stood high with him as a judge of Euclid. Now in this very letter of Feb. 2, there is a sentence which I highly value, because, as aforesaid, it is on a point on which he would never have yielded anything, to which he had paid life-long attention, and on which he had the bias of having long stood alone. In fact, knowing—and what I shall quote confirms me,—that in the matter of Euclid his hand was against every man, I expected, when I sent him a copy of my 22-column article, 'Eucleides' in Smith's *Dictionary*, to have received back a criticism, that would have blown me out of the water : and I thought it not unlikely that a man so well up in the subject might have made me feel demolished on some points. Instead of this, I got the following : ' Although on one or two minor points I do not quite accord with your views, yet as a whole and without regard to any minor points, I think you are the first who has succeeded in a delineation of Euclid as a geometer.' All this duly considered, it is utterly incredible that T. S. Davies should have written the review in question. And yet Mr. Halliwell is treated just as T. S. Davies would have treated him, as to tone and spirit. The inference in my mind is that we have here a marked instance of the joining of hatreds which takes place in journals supported by voluntary contributions of matter. Should anything ever have revived this article —and no one ever knows what might have been fished up from the forgotten mass of journals—the treatment of Mr. Halliwell would certainly have thrown a suspicion on T. S. Davies, a large and regular contributor to the Magazine. It is good service to his memory to point out what makes it incredible that he should have written so unworthy an article.

The fault is this. There are four extracts : the first three are perfectly well printed. The printing of the *Mechanics' Magazine* was very good. I was always exceedingly satisfied with the manner in which my articles appeared, without my seeing proof. Most likely these extracts were printed from my printed paper ; if not the extractor was a good copier. I know this by a test which has often served me. I use the subjunctive—'if no man *have* done nothing,' an ordinary transcriber, narrating a quotation almost always lets his own habit write *has*. The fourth extract has three alterations, all tending to make me ridiculous. *None* is altered, in two places, into *nine*, *denial* into *decimal*, and *comes* into *aims*; so that 'none, the denial of one, comes to be associated with the denial of plurality,' reads as ' nine, the decimal of one, aims to be associated with the decimal of plurality.' This is intentional ; had it been a compositor's reading of bad handwriting, these would not have been the only mistakes ; to say nothing of the corrector of the press. And both the compositor and reader would have guessed, from the first line being translated into ' one is not one,' that it must have been ' one's none,' not ' one's nine.' But it was not intended that the gem should be recovered from the unfathomed cave, and set in a Budget of Paradoxes.

We have had plenty of slander-paradox. I now give a halfpenny-worth of bread to all this sack, an instance of the paradox of benevolence, in which an individual runs counter to all the ideas of his time, and sees his way into the next century. At Amiens, at the end of the last century, an institution was endowed by a M. de Morgan, to whom I hope I am of kin, but I cannot trace it; the name is common at Amiens. It was the first of the kind I ever heard of. It is a Salle d'Asyle for childen, who are taught and washed and taken care of during the hours in which their parents must be at work. The founder was a large wholesale grocer and colonial importer, who was made a Baron by Napoleon I. for his commercial success and his charities.

---

1862. Mr. Smith replies to me, still signing himself Nauticus : I give an extract :—

'By hypothesis [what, again !] let 14° 24′ be the chord of an arc of 15° [but I wont, says 14° 24′], and consequently equal to a side of a regular polygon of 24 sides inscribed in the circle. Then 4 times 14° 24′ = 57° 36′ = the radius of the circle . . .'

That is, four times the chord of an arc is the chord of four times the arc : and the sum of four sides of a certain pentagon is equal to the fifth. This is the capital of the column, the crown of the arch, the apex of the pyramid, the watershed of the elevation. Oh ! J. S. ! J. S. ! groans Geometry—*Summum J. S. summa injuria*! The other J. S., Joseph Scaliger, as already mentioned, had his own way of denying that a straight line is always the shortest distance between two points. A parallel might be instituted, but not in half a column. And J. S. the *second* has been so tightly handled that he may now be dismissed, with an inscription for his circular shield, obtained by changing *Lexica contexat* into *Circus quadrandus* in an epigram of J. S. the *first* : —

Si quem dura manet sententia judicis, olim
Damnatum ærumnis suppliciisque caput,
Hunc neque fabrili lassent ergastula massa,
Nec rigidas vexent fossa metalla manus.
Circus quadrandus : nam—cætera quid moror ?—omnes
Pœnarum facies hic labor unus habet.

I had written as far as *damnatum* when in came the letter of Nauticus as a printed slip, with a request that I would consider the slip as a ' revised copy.' Not a word of alteration in the part I have quoted ! And in the evening came a letter desiring that I would alter a gross error ; but not the one above : this is revising without revision ! If there were cyclometers enough of this stamp, they would, as cultivation progresses—and really, with John Stuart Mill in for Westminster, it seems on the move, even though, as I learn while correcting the proof, Gladstone be out from Oxford ; for Oxford is no worse than in 1829, while Westminster is far above what she ever has been : election time excuses even such a parenthesis as this—be engaged to amuse those who can afford it with paralogism at their meals, after the manner of the other jokers who wore the caps and bells. The rich would then order their dinners with *panem et Circenses*,—up with the victuals and the circle-games—as the poor did in the days of old.

Mr. Smith is determined that half a column shall not do. Not a day without something from him : letter, printed proof, pamphlet. In what is the last at this moment of writing he tells me that part of the title of a work of his will be ' Professor De Morgan in the pillory without hope of escape.' And where will he be himself ? This I detected by an effort of reasoning which I never could have made except by following in his steps. In all matters connected with $\pi$ the letters $l$ and $g$ are closely related : this appears in the well-known formula for the time of oscillation, $\pi \sqrt{(l : g)}$. Hence $g$ may be written for $l$, but only once : do it twice, and you require the time to be $\pi \sqrt{(l^2 : g^2)}$. This may be reinforced by observing that if as a datum, or if you dislike that word, by hypothesis, the first $l$ be a $g$, it is absurd that it should be an $l$. Write $g$ for the first $l$, and we have *un fait accompli*. I shall be in pillory; and overhead, in a cloud, will sit Mr. James Smith on one stick laid across two others, under a nimbus of $3\frac{1}{8}$ diameters to the circumference—in $\pi$-glory. Oh for a drawing of this scene! Mr. De Morgan presents his compliments to Mr. James Smith, and requests the honour of an exchange of photographs.

*July* 26.—Another printed letter.—Mr. James Smith begs for a distinct answer to the following plain question : 'Have I not in this communication brought under your notice *truths* that were never before dreamed of in your geometrical and mathematical philosophy?' To which, he having taken the precaution to print the word *truths* in italics, I can conscientiously answer, Yes, you have. And now I shall take no more notice of these *truths*, until I receive something which surpasses all that has yet been done.

> The Circle secerned from the Square ; and its area gauged in terms of a triangle common to both. By Wm. Houlston, Esq. London and Jersey, 1862, 4to.

Mr. Houlston squares at about four poetical quotations in a page, and brings out $\pi = 3\cdot14213 \ldots$ . His forntispiece is a variegated diagram, having parts designated Inigo and Outigo. All which relieves the subject, but does not remove the error.

> Considerations respecting the figure of the Earth . . . By C. F. Bakewell. London, 1862, 8vo.

Newton and others think that in a revolving sphere the loose surface matter will tend to the equator : Mr. Bakewell thinks it will tend to the poles.

> On eccentric and centric force : a new theory of projection. By H. F. A. Pratt, M.D. London, 1862, 8vo.

Dr. Pratt not only upsets Newton, but cuts away the very ground he stands on : for he destroys the first law of motion, and will not have the natural tendency of matter in motion to be rectilinear. This, as we have seen, was John Walsh's notion. In a more recent work 'On Orbital Motion,' London, 1863, 8vo., Dr. Pratt insists on another of Walsh's notions, namely, that the precession of the equinoxes is caused by the motion of the solar system round a distant central sun. In this last work the author refers to a few notes, which completely destroy the theory of gravitation in terms 'perfectly intelligible as well to the un-learned as to the learned': to me they are quite unintelligible, which rather tends to confirm a notion I have long had, that I am neither one thing nor the other. There is an ambiguity of phrase which delights a writer on logic, always on the look out for specimens of *homonymia* or *æquivocatio*. The author, as a physician, is accustomed to 'appeal from mere formulæ': accord-ingly, he sets at nought the whole of the mathematics, which he does not understand. This equivocation between the formula of the physician and that of the mathematician is as good, though not so perceptible to the world at large, as that made by Mr. Briggs's friend in *Punch's* picture, which I cut out to paste into my Logic. Mr. Briggs wrote for a couple of *bruisers*, meaning to prepare oats for his horses: his friend sent him the Whitechapel Chicken and the Bayswater Slasher, with the gloves, all ready.

> On matter and ether, and the secret laws of physical change. By T. R. Birks, M.A. Cambridge, 1862, 8vo.

Bold efforts are made at molecular theories, and the one before me is ably aimed. When the Newton of this subject shall be seated in his place, books like the present will be sharply looked into, to see what amount of anticipation they have made.

> The history of the 'thorn tree and bush' from the earliest to the present time : in which is clearly and plainly shown the descent of her most gracious Majesty and her Anglo-Saxon people from the half tribe of Ephraim, and possibly from the half tribe of Manasseh ; and consequently her right and title to possess, at the present moment, for herself and for them, a share or shares of the desolate cities and places in the land of their forefathers ! By Theta, M.D. (Private circulation.) London, 1862, 8vo.

This is much about *Thorn*, and its connected words, Thor, Thoth, Theta, &c. It is a very mysterious vagary. The author of it is the person whom I have described elsewhere as having for his device the round man in the three-cornered hole, the writer of the little heap of satirical anonymous letters about the Beast and 666. By accident I discovered the writer: so that if there be any more thorns to crackle under the pot, they need not be anonymous.

Nor will they be anonymous. Since I wrote the above, I have received *onymous* letters, as *ominous* as the rest. The writer,

William Thorn, M.D., is obliged to reveal himself, since it is his object to prove that he himself is one 666. By using W for a double Vau (or 12) he cooks the number out of his own name. But he says it is the number not of a beast but of a man, and adds, 'Thereby hangs a tale!' which sounds like contradiction. He informs me that he will talk the matter over with me: but I shall certainly have nothing to say to a gentleman of his number; it is best to keep on the safe side.

In one letter I am informed that not a line should I have had, but for my 'sneer at 666,' which, therefore, I am well pleased to have given. I am also told that my name means the ' "garden of death," that place in which the tree of knowledge was plucked, and so you are like your name "dead" to the fact that you are an Israelite, like those in Ezekiel 37 ch.' Some hints are given that I shall not fare well in the next world, which anyone who reads the chapter in Ezekiel will see is quite against his comparison. The reader must not imagine that my prognosticator means *Morgan* to be a corruption of *Mortjardin*; he proves his point by Hebrew: but any philologist would tell him the true derivation of the name, and how *Glamorgan* came to get it. It will be of much comfort to those young men who have not got through to know that the tree of knowledge itself was once in the same case. And so good bye to 666 for the present, and the assumption that the enigma is to be solved by the united numeral forces of the letters of a word.

It is worthy of note that, as soon as my Budget commenced, two guardian spirits started up, fellow men as to the flesh, both totally unknown to me: they have stuck to me from first to last. James Smith, Esq., finally Nauticus, watches over my character in this world, and would fain preserve me from ignorance, folly, and dishonesty, by inclosing me in a magic circle of 3⅛ diameters in circumference. The round man in the three-cornered hole, finally William Thorn, M.D. takes charge of my future destiny, and tries to bring me to the truth by unfolding a score of meanings —all right—of 666. He hints that I, and my wife, are servants of Satan: at least he desires us both to remember that we cannot serve God and Satan; and he can hardly mean that we are serving the first, and that he would have us serve the second. As becomes an interpreter of the Apocalypse, he uses seven different seals; but not more than one to one letter. If his seals be all signet-rings, he must be what Aristophanes calls a sphragidonychargocometical fellow. But—and many thanks to him for the same—though an M.D., he has not sent me a single vial. And so much for my tree of secular knowledge and my tree of spiritual life : I dismiss them with thanks from myself and thanks from my reader. The dual of the Pythagorean system was Isis and Diana ; of the Jewish law, Moses and Aaron ; of the City of London, Gog and Magog ; of the Paradoxiad, James Smith, Esq., and William Thorn, M.D.

*September*, 1866. Mr. James Biden has favoured me with some of his publications. He is a rival of Dr. Thorn ; a prophet by name-right and crest-right. He is of royal descent through the De Bidun's. He is the *watchman* of Ezekiel : God has told him so. He is the author of *The True Church*, a phrase which seems to have a book-meaning and a mission-meaning. He shall speak for himself :—

' A crest of the Bidens has significance. It is a lion rampant between wing—swings in Scripture denote the flight of time. Thus the beasts or living creatures of the Revelations have each six wings, intimating a condition of mankind up to and towards the close of six thousand years of Bible teaching. The two wings of the crest would thus intimate power towards the expiration of 2000 years, as time is marked in the history of Great Britain.

. . . . .

' In a recent publication, *The Pestilence, Why Inflicted*, are given many reasons why the writer thinks himself to be the appointed watchman foretold by Ezekiel, chapters iii. and xxxiii. Among the reasons are many prophecies fulfilled in him. Of these it is now needful to note two as bearing especially on the subject of the reign of Darius.

' 1.—In Daniel it is said, ' Darius the Median took the kingdom, being about threescore and two years old."—Daniel v. 31.

' When "Belshazzar" the king of the Chaldeans is found wanting, Darius takes the kingdom. It is not given him by the popular voice ; he asserts his right, and this is not denied. He takes it when about sixty-two years of age. The language of Daniel is prophetic, and Darius has in another an antitype. The writer was born July 18th, 1803 ; and the claim was asserted at the close of 1865, when he was about sixty-two years of age.

' The claims which have been asserted demand a settled faith, and which could only be reached through a long course of divine teaching.'

When I was a little boy at school, one of my schoolfellows took it into his head to set up a lottery of marbles : the thing took, and he made a stony profit. Soon, one after another, every boy had his lottery, and it was, 'I won't put into yours unless you put into mine.' This knocked up the scheme. It will be the same with the prophets. Dr. Thorn, Mr. Biden, Mrs. Cottle, &c. will grow imitators, until we are all pointed out in the Bible : but A will not admit B's claim unless B admits his. For myself, as elsewhere shown, I am the first Beast in the Revelations.

Every contraband prophet gets a few followers : it is a great point to make these sequacious people into Buridan's asses, which they will become when prophets are so numerous that there is no choosing.

> An historical survey of the Astronomy of the Ancients. By the Rt. Hon. Sir G. C. Lewis. 8vo. 1862.

There are few men of our day whom I admire more than the late Sir G. Lewis : he was honest, earnest, sagacious, learned, and industrious. He probably sacrificed his life to his conjunction of literature and politics : and he stood high as a minister of state in addition to his character as a man of letters. The work above named is of great value, and will be read for its intrinsic merit, consulted for its crowd of valuable references, quoted for its aid to one side of many a discussion, and opposed for its force against the other. Its author was also a wit and a satirist. I know of three classical satires of our day which are inimitable imitations : Mr. Malden's *Pragmatized Legends*, Mr. Mansel's *Phrontisterion*, and Sir G. Cornewall Lewis's *Inscriptio Antiqua*. In this last, HEYDIDDLEDIDDLETHECATANDTHEFIDDLE &c. is treated as an Oscan inscription, and rendered into Latin by approved methods. As few readers have seen it, I give the result :—

'Hejus dedit libenter, dedit libenter. Deus propitius [est], deus [donatori] libenter favet. Deus in viarum juncturâ ovorum dape [colitur], deus mundi. Deus in litatione voluit, benigno animo, hædum, taurum intra fines [loci sacri] portandos. Deus, bis lustratus, beat fossam sacræ libationis.'

How then comes the history of astronomy among the paradoxes ? Simply because the author, so admirable when writing about what he knew, did not know what he did not know, and blundered like a circle-squarer. And why should the faults of so good a writer be recorded in such a list as the present ? For three reasons : First, and foremost, because if the exposure be not made by some one, the errors will gradually ooze out, and the work will get the character of inaccurate. Nothing hurts a book of which few can fathom the depths so much as a plain blunder or two on the surface. Secondly, because the reviews either passed over these errors or treated them too gently, rather implying their existence than exposing them. Thirdly, because they strongly illustrate the melancholy truth, that no one knows enough to write about what he does not know. The distinctness of the errors is a merit; it proceeds from the clear-headedness of the author. The suppression in the journals may be due partly to admiration of the talent and energy which lived two difficult lives at once, partly to respect for high position in public affairs, partly to some of the critics being themselves men of learning only, unable to detect the errors. But we know that action and reaction are equal and contrary. If our generation take no notice of defects, and allow them to go down undetected among merits, the next generation will discover them, will perhaps believe us incapable of detecting them, at least will pronounce our judgment good for nothing, and will form an opinion in which the merits will be underrated: so it has been, is, and will be. The best thing that can be done for the memory of the author is to remove the unsound part that the remainder may thrive. The errors do not affect the work; they occur in passages which might very well have been omitted: and I consider that, in making them conspicuous, I am but cutting away a deleterious fungus from a noble tree.

(P. 154). The periodic times of the five planets were stated by Eudoxus, as we learn from Simplicius; the following is his statement, to which the true times are subjoined, for the sake of comparison:—

|  | Statement of Eudoxus. | True time. |
|---|---|---|
| Mercury . . . | 1 year . . . . | − 87d. 23h. |
| Venus . . . . | 1 ,, . . . . | − 224d. 16h. |
| Mars . . . . | 2 ,, . . . . | 1y. 321d. 23h. |
| Jupiter . . . | 12 ,, . . . . | 11y. 315d. 14h. |
| Saturn . . . . | 30 ,, . . . . | 29y. 174d. 1h. |

Upon this determination two remarks may be made. First, the error with respect to Mercury and Venus is considerable; with respect to Mercury, it is, in round numbers, 365 instead of 88 days, more than four times too much. Aristotle remarks that Eudoxus distinguishes Mercury and Venus from the other three planets by giving them one sphere each, with the poles in common. The proximity of Mercury to the sun would render its course difficult to observe and to measure, but the cause of the large error with respect to Venus (130 days) is not apparent.

Sir G. Lewis takes Eudoxus as making the planets move round the sun; he has accordingly compared the *geocentric* periods of Eudoxus with our *heliocentric* periods. What greater blunder can be made by a writer on ancient astronomy than giving Eudoxus the Copernican system? If Mercury were a black spot in the middle of the sun it would of course move round the earth in a year, or appear to do so: let it swing a little on one side and the other of the sun, and the average period is still a year, with slight departures both ways. The same for Venus, with larger departures. Say that a person not much accustomed to the distinction might for once write down the mistake; how are we to explain its remaining in the mind in a permanent form, and being made a ground for such speculation as that of the difficulty of observing Mercury leading to a period four times what it ought to be, corrected in proof and published by an industrious and thoughtful person? Only in one way: the writer was quite out of his depth. This one case is conclusive; be it said with all respect for the real staple of the work and of the author. He knew well the difference of the systems, but not the effect of the difference: he is another instance of what I have had to illustrate by help of a very different person, that it is difficult to reason well upon matter which is not familiar.

(P. 254). Copernicus, in fact, supposed the axis of the earth to be always turned towards the Sun. [169.]   [(169). See Delambre, Hist. Astr. Mod. vol. i. p. 96]. It was reserved to Kepler to propound the hypothesis of the constant parallelism of the earth's axis to itself.

If there be one thing more prominent than another in the work of Copernicus himself, in the popular explanations of it, and in the page of Delambre cited, it is that the *parallelism of the earth's axis* is a glaring part of the theory of Copernicus. What Kepler did was to throw away, as unnecessary, the method by which Copernicus, *per fas et nefas*, secured it. Copernicus, thinking of the earth's orbital revolution as those would think who were accustomed to the *solid orbs*—and much as the stoppers of the moon's rotation do now: why do they not strengthen themselves with Copernicus?—thought that the earth's axis would always incline the same end towards the sun, unless measures were taken to prevent it. He *did* take measures: he invented a *compensating* conical motion of the axis to preserve the parallelism; and, which is one of the most remarkable points of his system, he obtained the precession of the equinoxes by giving the necessary trifle more than compensation. What stares us in the face at the beginning of the paragraph to which the author refers?

' C'est donc pour arriver à ce parallelisme, ou pour le conserver, que
Copernic a cru devoir recourir à ce mouvement égal et opposé qui
détruit l'effet qu'il attribue si gratuitement au premier, de déranger le
parallelisme.'

Parallelism at any price, is the motto of Copernicus: you need
not pay so dear, is the remark of Kepler.

The opinions given by Sir G. Lewis about the effects of modern
astronomy, which he does not understand and singularly under-
values, will now be seen to be of no authority. He fancies that—
to give an instance—for the determination of a ship's place, the
invention of chronometers has been far more important than any
improvement in astronomical theory (p. 254). Not to speak of
latitude,—though the omission is not without importance,—l e
ought to have known that longitude is found by the difference
between what o'clock it is at Greenwich, and at the ship's place,
at one absolute moment of time. Now if a chronometer were
quite perfect—which no chronometer is, be it said—and would
truly tell Greenwich mean time all over the world, it ought to
have been clear that just as good a watch is wanted for the time
at *the place of observation*, before the longitude of that place
with respect to Greenwich can be found. There is no such watch,
except the starry heaven itself: and that watch can only be read
by astronomical observation, aided by the best knowledge of the
heavenly motions.

I think I have done Sir G. Lewis's very excellent book more
good than all the reviewers put together.

I will give an old instance in which literature got into confusion about astronomy. Theophrastus, who is either the culprit
or his historian, attributes to Meton, the contriver of the lunar
calendar of nineteen years, which lasts to this day, that his
solstices were determined for him by a certain Phaeinus of Elis
on Mount Lycabettus. Nobody else mentions this astronomer:
though it is pretty certain that Meton himself made more than
one appointment with him for the purpose of observing solstices;
and we may be sure that if either were behind his time, it was
Meton. For *Phaeinus Helius* is the shining sun himself; and in
the astronomical poet Aratus we read about the nineteen years of
the shining sun—

$$\text{Ἐννεακαιδέκα κύκλα φαεινοῦ ἠελίοιο.}$$

Some man of letters must have turned Apollo into Phaeinus of
Elis; and there he is in the histories of astronomy to this day.

Salmasius will have Aratus to have meant· him, and proposes to read ἠλείοιο: he did not observe that *Phaeinus* is a very common adjective of Aratus, and that, if his conjecture were right, this Phaeinus would be the only non-mythical man in the poems of Aratus.

[When I read Sir George Lewis's book, the points which I have criticised struck me as not to be wondered at, but I did not remember why at the time. A Chancellor of the Exchequer and a writer on ancient astronomy are birds of such different trees that the second did not recal the first. In 1855 I was one of a deputation of about twenty persons who waited on Sir G. Lewis, as Chancellor of the Exchequer, on the subject of the decimal coinage. The deputation was one of much force: Mr. Airy, with myself and others, represented mathematics; William Brown, whose dealings with the United States were reckoned by yearly millions, counted duodecimally in England and decimally in America, was the best, but not the only, representative of commerce. There were bullionists, accountants, retailers, &c. Sir G. L. walked into the room, took his seat, and without waiting one moment, began to read the deputation a smart lecture on the evils of a decimal coinage; it would require alteration of all the tables, it would impede calculation, &c. &c. Of those arguments against it which weighed with many of better knowledge than his, he obviously knew nothing. The members of the deputation began to make their statements, and met with curious denials. He interrupted me with 'Surely there is no doubt that the calculations of our books of arithmetic are easier than those in the French books.' He was not aware that the *universally admitted* superiority of decimal *calculation* made many of those who prefer our system for the market and the counter cast a longing and lingering look towards decimals. My answer and the smiles which he saw around, made him give a queer puzzled look, which seemed to say, 'I may be out of my depth here!' His manner changed, and he listened. I saw both the slap-dash mode in which he dealt with subjects on which he had not thought, and the temperament which admitted suspicion when the means of knowledge came in his way. Having seen his two phases, I wonder neither at his more than usual exhibition of shallowness when shallow, nor at the intensity of the contrast when he had greater depth.]

Among the paradoxers are the political paradoxers who care not how far they go in debate, their only object being to carry the House with them for the current evening. What I have said

of editors I repeat of them.   The preservation of a very marked
instance, the association of political recklessness with cyclometrical and Apocalyptic absurdity, may have a tendency to warn,
not indeed any hardened publicman and sinner, but some young
minds which have yearnings towards politics, and are in formation
of habits.

In the debate on decimal coinage of July 12, 1855, Mr. Lowe,
then member for Kidderminster, an effective speaker and a smart
man, exhibited himself in a speech on which I wrote a comment
for the Decimal Association.   I have seldom seen a more wretched
attempt to distort the points of a public question than the whole
of this speech.   Looking at the intelligence shown by the speaker
on other occasions, it is clear that if charity, instead of believing
all things, believed only all things but one, he might tremble for
his political character; for the honesty of his intention on this
occasion might be the incredible exception.   I give a few paragraphs, with the comments :—

'In commenting on the humorous, but still argumentative speech
of Mr. Lowe, the member for Kidderminster, we may observe, in
general, that it consists of points which have been several times set
forth, and several times answered.   Mr. Lowe has seen these answers,
but does not allude to them, far less attempt to meet them.   There
are, no doubt, individuals, who show in their public speaking the
outward and visible signs of a greater degree of acuteness than they
can summon to guide their private thinking.   If Mr. Lowe be not one
of these, if the power of his mind in the closet be at all comparable to
the power of his tongue in the House, it may be suspected that his
reserve with respect to what has been put forward by the very parties
against whom he was contending, arises from one or both of two
things—a high opinion of the arguments which he ignored—a low
opinion of the generality of the persons whom he addressed.   [Both,
I doubt not].

"Did they calculate in florins ? "   In the name of common sense,
                                     how can it be objected to a system
that people do not use it before it is introduced ?   Let the decimal
system be completed, and calculation shall be made in florins ; that is,
florins shall take their proper place.   If florins were introduced *now*,
there must be a column for the odd shilling.

"He was glad that some hon.          If the hon. gentleman make
gentleman had derived benefit     this assertion of himself, it is not
from the issue of florins.   His    for us to gainsay it.   It only
only experience of their conve-   proves that he is one of that
nience was, that when he ought    class of men who are described
to have received half-a-crown, he   in the old song, of which one

had generally received a florin, and when he ought to have paid a florin, he had generally paid half-a-crown." (Hear, hear, and laughter.)

couplet runs thus :—
I sold my cow to buy me a calf;
I never make a bargain but I lose half,
With a &c. &c. &c.

But he cannot mean that Englishmen in general are so easily managed. And as to Jonathan, who is but John lengthened out a little, he would see creation whittled into chips before he would even split what may henceforth be called the Kidderminster difference. The House, not unmoved—for it laughed—with sly humour decided that the introduction of the florin had been "eminently successful and satisfactory."

The truth is, that Mr. Lowe here attacks nothing except the co-existence of the florin and half-crown. We are endeavouring to abolish the half-crown. Let Mr. Lowe join us ; and he will, if we succeed, be relieved from the pressure on his pocket which must arise from having the turn of the market always against him.

"From a florin they get to 2 2-5ths of a penny, but who ever bought anything, who ever reckoned or wished to reckon in such a coin as that ? " (Hear, hear.)

Note the sophism of expressing our coin in terms of the penny, which we abandon, instead of the florin, which we retain. Remember that this 2 2-5ths is the hundredth part of the pound, which is called, as yet, a *cent*. Nobody buys anything at a cent, because the cent is not yet introduced. Nobody reckons in cents for the same reason. Everybody wishes to reckon in cents, who wishes to combine the advantage of decimal reckoning with the preservation of the pound as the highest unit of account ; amongst others, a majority of the House of Commons, the Bank of England, the majority of London bankers, the Chambers of Commerce in various places, &c. &c. &c.

"Such a coin could never come into general circulation, because it represents nothing which corresponds with any of the wants of the people."

Does 2½d. never pass from hand to hand ? And is 2½d. so precisely the modulus of popular wants, that an alteration of 4 per cent. would make it useless ? Of all the values which 2½d. measures, from three pounds of potatoes down to certain arguments used in the House of Commons, there is not one for which a cent would not do just as well. Mr. Lowe has fallen into the misconception of the person who admired the dispensation of Providence by which large rivers are made to run through cities so great and towns so many. If the cent were to be introduced to-morrow, straightway the buns and cakes, the soda-water bottles, the short omnibus fares, the bunches of radishes, &c. &c. &c., would adapt themselves to the coin.

"If the proposed system were adopted, they would all be compelled to live in decimals for ever ; if a man dined at a public-house he would have to pay for his dinner in decimal fractions. (Hear, hear.) He objected to that, for he thought that a man ought to be able to pay for his dinner in integers." (Hear, hear, and a laugh.)

The confusion of ideas here exhibited is most instructive. The speaker is under the impression that *we* are introducing fractions : the truth is, that we only want to abandon the *more difficult* fractions which we *have got*, and to introduce *easier fractions*. Does he deny this ? Let us trace his denial to its legitimate consequences. A man ought to pay for his dinner in integers.

Now, if Mr. Lowe insists on it that our integer is the pound, he is bound to admit that the present integer is the pound, of which a shilling, &c., are fractions. The next time he has a chop and a pint of stout in the city, the waiter should say—" A pound, sir, to you," and should add, " Please to remember the waiter in integers." Mr. Lowe fancies that when he pays one and sixpence, he pays in integers, and so he does, if his integer be a penny or a sixpence. Let him bring his mind to contemplate a mil as the integer, the lowest integer, and the seven cents five mils which he would pay under the new system would be payment in integers also. But, as it happens with some others, he looks *up* the present system, with Cocker and Walkingame, and always looks *down* the proposed system. The word *decimal* is obstinately associated with *fractions*, for which there is no need. Hence it becomes so much of a bugbear, that, to parody the lines of Pope, which probably suggested one of Mr. Lowe's phrases—

> Dinner he finds too painful an endeavour,
> Condemned to pay in decimals for ever,

"The present system, however, had not yet been changed into decimal system. That change might appear very easy to accomplished mathematicians and men of science, but it was one which it would be very difficult to carry out. (Hear, hear). What would have to be done ? Every sum would have to be reduced into a vulgar fraction of a pound, and then divided by the decimal of a pound—a pleasant sum for an old applewoman to work out ! " (Hear, hear, and laughter.)

A pleasant sum even for an accomplished mathematician. What does divided by the decimal of a pound mean ? Perhaps it means *reduced* to the decimal of a pound ! Mr. Lowe supposes, as many others do, that, after the change, all calculations will be *proposed in old money*, and then *converted into new*. He cannot hit the idea that the new coins will take the place of the old. This lack of apprehension will presently appear further.

" It would not be an agreeable

Let the members be assured

task, even for some members of that House, to reduce 4½d., or nine half-pence, to mils." (Hear, hear.) that during the long period he sat in the House, he never knew more than three men in it, at one time, who had a tolerable notion of fractions. [I heard him give the names of three at the time when he spoke : they were Warburton, Pollock, and Hume. He himself was then out of Parliament.] Joseph Hume affirmed that he had never met with more than ten members who were arithmeticians. But both these gentlemen had a high standard. Mr. Lowe has given a much more damaging opinion. He evidently means that the general run of members could not do his question. It is done as follows : Since farthings gain on mils, at the rate of a whole mil in 24 farthings (24 farthings being 25 mils), it is clear that 18 farthings being three-quarters of 24 farthings, will gain three-quarters of a mil ; that is, 18 farthings are eighteen mils and three-quarters of a mil. Any number of farthings is as many mils and as many twenty-fourths of a mil. To a certain extent, we feel able to protest against the manner in which Kidderminster has treated the other constituencies. We do not hold it impossible to give the Members of the House in general a sufficient knowledge of the meaning and consequences of the *decimal* succession of units, tens, hundreds, thousands, &c. ; and we believe that there are in the House itself competent men, in number enough to teach all the rest. All that is wanted is the power of starting from the known to arrive at the unknown. Now there is one kind of decimals with which every member is acquainted—the *Chiltern Hundreds*. If public opinion would enable the competent minority to start from this in their teaching, not as a basis, but as an alternative, in three weeks the fundamentals would be acquired, and members in general would be as fit to turn 4½d. into mils, as any boys on the lower forms of a commercial school.

For a long period of years, allusion to the general ignorance of arithmetic, has been a standing mode of argument, and has always been well received : whenever one member describes others as *know-nothings*, those others cry *Hear* to the country in a transport of delight. In the meanwhile the country is gradually arriving at the conclusion that a true joke is no joke.

" The main objection was, if they went below 6d., that the new scale of coins would not be commensurate in any finite ratio with anything in this new currency of mils."

Fine words, wrongly used. The new coins are commensurable with, and in a finite ratio to, the old ones. The farthing is to the mil as 25 to 24. The speaker has something here in the bud, which we shall presently meet with in the flower ; and fallacies are more easily nipped in flower than in bud.

that nine half-pence will be, for every practical purpose, 18 mils. But now to the fact asserted. Davies Gilbert used to maintain

"No less than five of our present coins must be called in, or else — which would be worse— new values must be given to them."

"If a poor man put a penny in his pocket, it would come out a coin of different value, which he would not understand. Suppose he owed another man a penny, how was he to pay him? Was he to pay him in mils? Four mils would be too little, and five mils would be too much. The hon. gentlemen said there would be only a mil between them. That was exactly it. He believed there would be a 'mil' between them." (Much laughter.)

This dreadful change of value consists in sixpence farthing going to the half-shilling instead of sixpence. _ Whether the new farthings be called mils or not is of no consequence.

Mr. Lowe, who cannot pass a half-crown for more than a florin, or get in a florin at less than half-a-crown, has such a high faith in the sterner stuff of his fellow countrymen, that he believes any two of them would go to fisty cuffs for the 25th part of a farthing. He reasons thus :— He has often heard in the streets, "I'd fight you for the fiftieth part of a farden:" and having (that is, for a Member) a notion both of fractions and logic, he infers that those who would

fight for the 50th of a farthing would, à fortiori, fight for a 25th. His mistake arises from his not knowing that when a person offers to fight another for $\frac{1}{200}$d., he really means to fight for love; and that the stake is merely a matter of form, a feigned issue, a *pro formâ* report of progress. Do the Members of the House think they have all the forms to themselves?

"What would be the present expression for fourpence? Why, 0·166 (a laugh); for threepence? 0125; for a penny? ·004166, and so on *ad infinitum* (a laugh); for a half-penny? ·002083 *ad infinitum*. (A laugh). What would be the present expression for a farthing? Why, ·0010416 *ad infinitum*. (A laugh). And this was the system which was to cause such a saving in figures, and these were the quantities into which the poor would have to reduce the current coin of the realm. (Cheers). With every respect for decimal fractions, of which he boasted no profound knowledge, he doubted whether the poor were equal to mental

We should hardly believe all this to be uttered in earnest, if we had not known that several persons who have not Mr. Lowe's humour, nevertheless have his impressions on this point. It must, therefore, be answered: but how is this to be done seriously?

arithmetic of this kind, (hear, hear) and he hoped the adoption of the system would be deferred until there were some proof that they would be able to understand it; for, after all, this was the question of the poor, and the whole weight of the change would fall upon them. Let the rich by all means have permission to perplex themselves by any division of a pound they pleased; but do not let them, by any experiment like this, impose difficulties upon the poor, and compel men to carry ready-reckoners in their pocket to give them all these fractional quantities." (Hear, hear.)

*Dialogue between a member of Parliament and an orange-boy, three days after the introduction of the complete decimal system. The member, going down to the House, wants oranges to sustain his voice in a two hours' speech on moving that 100,000l. be placed at the disposal of Her Majesty, to supply the poor with ready-reckoners.*

*Boy.* Fine oranges! two a penny! two a penny!
*Member.* Here, boy, two! Now, how am I to pay you?
*Boy.* Give you change, your honour.
*Member.* Ah! but how? Where's your ready-reckoner?
*Boy.* I sells a better sort nor them. Mine's real Cheyny.
*Member.* But you see a farthing is now 001 4166666 *ad infinitum*, and if we multiply this by 4 ——
*Boy.* Hold hard, Guv'ner; I sees what you're arter. Now, what'll you stand if I puts you up to it? which Bill Smith he put me up in two minutes, cause he goes to the Ragged School.

*Member.* You don't mean that you do without a book!
*Boy.* Book be blowed. Come now, old un, here's summut for both on us. I got a florin, you gives me half-a-crown for it, and I larns you the new money, gives you your oranges, and calls you a brick into the bargain.
*Member (to himself).* Never had such a chance of getting off half-a-crown for value since that —— fellow Bowring carried his crochet. (*Aloud*). Well, boy, its a bargain. Now!
*Boy.* Why, look 'e here, my trump, its a farden more to the tizzy—that's what it is.
*Member.* What's that?
*Boy.* Why, you knows a sixpence when you sees it. (*Aside*). Blest if I think he does! Well, its six browns and a farden now. A lady buys two oranges, and forks out a sixpence; well, in coorse, I

hands over fippence farden astead of fippence. I always gives a farden more change, and takes according.

*Member (in utter surprise, lets his oranges tumble into the gutter).* Never mind! They won't be wanted now. *(Walks off one way. Boy makes a pass of naso-digital mesmerism, and walks off the other way).*

To the poor, who keep no books, the whole secret is ' Sixpence farthing to the half shilling, twelve pence half-penny to the shilling.' The *new twopence halfpenny*, or cent, will be at once five to the shilling.

In conclusion, we remark that three very common misconceptions run through the hon. Member's argument; and, combined in different proportions, give variety to his patterns.

First, he will have it that we design to bring the uneducated into contact with *decimal fractions*. If it be so, it will only be as M. Jourdain was brought into contact with prose. In fact, *Quoi ! quand je dis, Nicole, apportez-moi mes pantoufles, c'est de la prose ?* may be rendered—" What ! do you mean that *ten to the florin is a cent a piece* must be called decimal reckoning ?" If we had to comfort a poor man, horror-struck by the threat of *decimals*, we should tell him what manner of fractions had been inflicted upon him hitherto; nothing less awful than *quarto-duodecimo-vicesimals*, we should assure him.

Secondly, he assumes that the penny, such as it now is, will remain, as a coin of estimation, after it has ceased to be a coin of exchange ; and that the mass of the people will continue to think of prices in old pence, and to calculate them in new ones, or else in new mils. No answer is required to this, beyond the mere statement of the nature of the assumption and denial.

Thirdly, he attributes to the uneducated community a want of perception and of operative power which really does not belong to them. The evidence offered to the Committee of the House shows that no fear is entertained on this point by those who come most in contact with farthing purchasers. And this would seem to be a rule, —that is, fear of the intelligence of the lower orders in the minds of those who are not in daily communication with them, no fear at all in the minds of those who are.

A remarkable instance of this distinction happened five-and-twenty years ago. The Admiralty requested the Astronomical Society to report on the alterations which should be made in the *Nautical Almanac*, the seaman's guide-book over the ocean. The greatest alteration proposed was the description of celestial phenomena in *mean* (or clock time), instead of *apparent* (or sundial) time, till then always employed. This change would require that in a great many operations the seaman should let alone what he formerly altered by addition or subtraction, and alter by addition or subtraction what he formerly let alone ; provided always that what he formerly altered by addition he should, when he altered at all, alter by subtraction, and *vice versâ*. This was a tolerably difficult change for uneducated skippers, working by rules they had only learned by rote. The Astronomical Society appointed a Committee of forty, of whom nine were naval officers or merchant seamen [I was on this Committee]. Some men of science were much afraid of the change. They could not trust an ignorant skipper or mate to make those alterations in their routine, on the correctness of which the ship might depend. Had the Committee consisted of men of science only, the change might never have been ventured on. But the naval men laughed, and said there was nothing to fear; and on their authority the alteration was made. The upshot was, that, after the new almanacs appeared, not a word of complaint was ever heard on the matter. Had the House of Commons had to decide this question, with Mr. Lowe to quote the description given by Basil Hall (who, by the way, was one of the Committee) of an observation on which the safety of the ship depended, worked out by the light of a lantern in a gale of wind off a lee shore, this simple and useful change might at this moment have been in the hands of i s tenth Government Commission.'

[*Aug.* 14, 1866. The Committee was appointed in the spring of 1830 : it consisted of forty members. Death, of course, has been busy : there are now left Lord Shaftesbury, Mr. Babbage, Sir John Herschel, Sir Thomas Maclear (Astronomer Royal at the Cape of Good Hope), Dr. Robinson (of Armagh), Sir James South, Lord Wrottesley, and myself].

Project of a new system of arithmetic, weight, measure, and coins, proposed to be called the tonal system, with sixteen to the base. By J. W. Mystrom. Philadelphia, 1862, 8vo.

That is to say, sixteen is to take the place of ten, and to be written 10. The whole language is to be changed ; every man of us is to be sixteen-stringed Jack and every woman sixteen-stringed Jill. Our old *one, two, three,* up to sixteen, are to be (*Noll* going for nothing, which will please those who dislike the memory of *Old Noll*) replaced by An, De, Ti, Go, Su, By, Ra, Me, Ni, Ko, Hu, Vy, La, Po, Fy, Ton ; and then Ton-an, Ton-de, &c. for 17, 18, &c. The number which in the system has the symbol

$$28(13)5(11)7(14)0(15)$$

(using our present compounds instead of new types) is to be pronounced

Detam-memill-lasan-suton-hubong-ramill-posanfy.

The year is to have sixteen months, and here they are :—

Anuary, Debrian, Timander, Gostus,
Suvenary, Bylian, Ratamber, Mesudius,
Nictoary, Kolumbian, Husamber, Vyctorius,
Lamboary, Polian, Fylander, Tonborius.

Surely An-month, De-month, &c. would do as well. Probably
the wants of poetry were considered. But what are we to do with
our old poets? For example—

> It was a night of lovely June,
> High rose in cloudless blue the moon.

Let us translate—

> It was a night of lovely Nictoary,
> High rose in cloudless blue the (what, in the name of
> all that is absurd?).

And again, *Fylander* thrown into our December! What is to
become of those lines of Praed, which I remember coming out
when I was at Cambridge,—

Oh! now's the time of all the year for flowers and fun, the May-days;
To trim your whiskers, curl your hair, and sinivate the ladies.

If I were asked which I preferred, this system or that of Baron
Ferrari already mentioned, proceeding by *twelves*, I should reply,
with Candide, when he had the option given of running the
gauntlet or being shot: Les volontés sont libres, et je ne veux ni
l'un ni l'autre. We can imagine a speculator providing such a
system for Utopia as it would be in the mind of a Laputan: but
to explain how an engineer who has surveyed mankind from
Philadelphia to Rostof on the Don should for a moment entertain
the idea of such a system being actually adopted, would beat a
jury of solar-system-makers, though they were shut up from the
beginning of Anuary to the end of Tonborius. When I see such
a scheme as this imagined to be practicable, I admire the wisdom
of Providence in providing the quadrature of the circle, &c., to
open a harmless sphere of action to the possessors of the kind of
ingenuity which it displays. Those who cultivate mathematics
have a right to speak strongly on such efforts of arithmetic as
this: for, to my knowledge, persons who have no knowledge are
frequently disposed to imagine that their makers are true
brothers of the craft, a little more intelligible than the rest.

Vis inertiæ victa, or Fallacies affecting science. By James Reddie.
London, 1862, 8vo.

An attack on the Newtonian mechanics; revolution by gravitation demonstrably impossible; much to be said for the earth
being the immovable centre. A good analysis of contents at the
beginning, a thing seldom found. The author has followed up

his attack in a paper submitted to the British Association, but which it appears the Association declined to consider.   It is entitled—

> *Victoria Toto Cœlo;* or, Modern Astronomy recast.   London, 1863, 8vo.

At the end is a criticism of Sir G. Lewis's ' History of Ancient Astronomy.'

> On the definition and nature of the Science of Political Economy. By H. Dunning Macleod, Esq.   Cambridge, 1862, 8vo.

A paper read—but, according to the report, not understood—at the British Association.   There is a notion that political economy is entirely mathematical; and its negative quantity is strongly recommended for study : it contains ' the whole of the Funds, Credit, 32 parts out of 33 of the value of Land . . . . . .'   The mathematics are described as consisting of—first, number, or Arithmetic ;  secondly, the theory of dependent quantities, sub-divided into dependence by cause and effect, and dependence by simultaneous variations;  thirdly, ' independent quantities or unconnected events, which is the theory of probabilities.'   I am not ashamed, having the British Association as a co-non-intel-ligent, to say I do not understand this : there is a paradox in it, and the author should give further explanation, especially of his negative quantity.   Mr. Macleod has gained praise from great names for his political economy ;  but this, 1 suspect, must have been for other parts of his system.

> On the principles and practice of just Intonation, with a view to the abolition of temperament . . . By General Perronet Thomp-son.   Sixth Edition.   London, 1862, 8vo.

Here is General Thompson again, with another paradox : but always master of the subject, always well up in what his prede-cessors have done, and always aiming at a useful end.   He desires to abolish temperament by additional keys, and has constructed an enharmonic organ with forty sounds in the octave.   If this can be introduced, I, for one, shall delight to hear it : but there are very great difficulties in the way, greater than stood even in the way of the repeal of the bread-tax.

In a paper on the beats of organ-pipes and on temperament published some years ago, I said that equal temperament ap-peared to me insipid, and not so agreeable as the effect of the instrument when in progress towards being what is called out of tune, before it becomes offensively wrong. There is throughout that period unequal temperament, determined by accident. General Thompson, taking me one way, says I have launched a declaration which is likely to make an epoch in musical practice; a public musical critic, taking me another way, quizzes me for preferring music *out of tune.* I do not think I deserve either one remark or the other. My opponent critic, I suspect, takes *equally tempered* and *in tune* to be phrases of one meaning. But by equal temperament is meant equal distribution among all the keys of the error which an instrument *must* have, which, with twelve sounds only in the octave, professes to be fit for all the keys. I am reminded of the equal temperament which was once applied to the postmen's jackets. The coats were all made for the average man : the consequence was that all the tall men had their tails too short ; all the short men had them too long. Some one innocently asked why the tall men did not change coats with the short ones.

A diagram illustrating a discovery in the relation of circles to right-lined geometrical figures. London, 1863, 12mo.

The circle is divided into equal sectors, which are joined head and tail : but a property is supposed which is not true.

An attempt to assign the square roots of negative powers ; or what is $\sqrt{-1}$ ? By F. H. Laing. London, 1863, 8vo.

If I understand the author, $-a$ and $+a$ are the square roots of $-a^2$, as proved by multiplying them together. The author seems quite unaware of what has been done in the last fifty years.

Dual Arithmetic. A new art. By Oliver Byrne. London, 1863, 8vo.

The plan is to throw numbers into the form $a(1\cdot1)^b \ (1\cdot01)^c$ $(1\cdot001)^d$...... and to operate with this form. This is an ingenious and elaborate speculation ; and I have no doubt the author has practised his method until he could surprise any one else by his use of it. But I doubt if he will persuade others to use it. As asked of Wilkins's universal language, Where is the second man to come from ?

An effective predecessor in the same line of invention was the late Mr. Thomas Weddle, in his ' New, simple, and general method of solving numeric equations of all orders,' 4to, 1842. The Royal Society, to which this paper was offered, declined to print it: they ought to have printed an organised method, which, without subsidiary tables, showed them, in six quarto pages, the solution ($x = 8\cdot367975431$) of the equation

$$1379\cdot664x^{622} + 2686034 \times 10^{432}x^{153} - 17290224 \times 10^{518}x^{60} + 2524156 \times 10^{574} = 0.$$

The method proceeds by successive factors of the form, $a$ being the first approximation, $a \times 1\cdot b \times 1\cdot0c \times 1\cdot00d$ . . . . . In my copy I find a few corrections made by me at the time in Mr. Weddle's announcement. ' It was read before that learned body [the R. S.] and they were pleased [but] to transmit their thanks to the author. The en[dis]couragement which he received induces [obliges] him to lay the result of his enquiries in this important branch of mathematics before the public [, at his own expense; he being an usher in a school at Newcastle]. Which is most satirical, Mr. Weddle or myself? The Society, in the account which it gave of this paper, described it as a 'new and remarkably simple method' possessing 'several important advantages.' Mr. Rutherford's extended value of $\pi$ was read at the very next meeting, and was printed in the *Transactions*; and very properly: Mr. Weddle's paper was excluded, and very very improperly.

I think it may be admitted that the indisposition to look at and encourage improvements of calculation which once marked the Royal Society is no longer in existence. But not without severe lessons. They had the luck to accept Horner's now celebrated paper, containing the method which is far on the way to become universal: but they refused the paper in which Horner developed his views of this and other subjects: it was printed by T. S. Davies after Horner's death. I make myself responsible for the statement that the Society could not reject this paper, yet felt unwilling to print it, and suggested that it should be withdrawn; which was done.

But the severest lesson was the loss of *Barrett's Method*, now the universal instrument of the actuary in his highest calculations. It was presented to the Royal Society, and refused admission into the *Transactions*: Francis Baily printed it. The Society is now better informed: ' *live and learn*,' meaning ' *must live, so better learn*,' ought to be the especial motto of a corporation, and is generally acted on, more or less.

Horner's method begins to be introduced at Cambridge: it was published in 1820. I remember that when I first went to Cambridge (in 1823) I heard my tutor say, in conversation, there is no doubt that the true method of solving equations is the one which was published a few years ago in the *Philosophical Transactions.* I wondered it was not taught, but presumed that it belonged to the higher mathematics. This Horner himself had in his head : and in a sense it is true ; for all lower branches belong to the higher : but he would have stared to have been told that he, Horner, was without a European predecessor, and, in the distinctive part of his discovery was heir-at-law to the nameless Brahmin—Tartar— Antenoachian—what you please— who concocted the extraction of the square root.

It was somewhat more than twenty years after I had thus heard a Cambridge tutor show sense of the true place of Horner's method, that a pupil of mine who had passed on to Cambridge was desired by his college tutor to solve a certain cubic equation —one of an integer root of two figures. In a minute the work and answer were presented, by Horner's method. ' How ! ' said the tutor, ' this can't be, you know.' ' There is the answer, Sir ! ' said my pupil, greatly amused, for my pupils learnt, not only Horner's method, but the estimation it held at Cambridge. ' Yes ! ' said the tutor, ' there is the answer certainly ; but it *stands to reason* that a cubic equation cannot be solved in this space.' He then sat down, went through a process about ten times as long, and then said with triumph : ' There ! that is the way to solve a cubic equation ! '

I think the tutor in this case was never matched, except by the country organist. A master of the instrument went into the organ-loft during service, and asked the organist to let him *play the congregation out* ; consent was given. The stranger, when the time came, began a voluntary which made the people open their ears, and wonder who had got into the loft : they kept their places to enjoy the treat. When the organist saw this, he pushed the interloper off the stool, with ' You'll never play 'em out this side Christmas.' He then began his own drone, and the congregation began to move quietly away. ' There,' said he, ' that's the way to play 'em out ! '

I have not scrupled to bear hard on my own University, on the Royal Society, and on other respectable existences : being very much the friend of all. I will now clear the Royal Society from a very small and obscure slander, simply because I know how. This dissertation began with the work of Mr. Oliver Byrne, the dual arithmetician, &c. This writer published, in 1849, a method of calculating logarithms. First, a long list of instances in which, as he alleges, foreign discoverers have been pillaged by Englishmen, or turned into Englishmen: for example, O'Neill, so called by Mr. Byrne, the rectifier of the semi-cubical parabola claimed by the Saxons under the name of *Neal*: the grandfather of this mathematician was conspicuous enough as *Neal*; he was Archbishop of York. This list, says the writer, might be continued without end; but he has mercy, and finishes with his own case, as follows :—' About twenty years ago, I discovered this method of directly calculating logarithms. I could generally find the logarithm of any number in a minute or two without the use of books or tables. The importance of the discovery subjected me to all sorts of prying. Some asserted that I committed a table of logarithms to memory; others attributed it to a peculiar mental property; and when Societies and individuals failed to extract my secret, they never failed to traduce the inventor and the invention. Among the learned Societies, the Royal Society of London played a very base part. When I have more space and time at my disposal, I will revert to this subject again.'

Such a trumpery story as this remains unnoticed at the time; but when all are gone, a stray copy from a stall falls into hands which, not knowing what to make of it, make history of it. It is a very curious distortion. The reader may take it on my authority, that the Royal Society played no part, good or bad, nor had the option of playing a part. But I myself *pars magna fui*: and when the author has 'space and time' at his disposal, he must not take all of them; I shall want a little of both.

The mystery of being; or are ultimate atoms inhabited worlds ?
By Nicholas Odgers. Redruth and London, 1863, 8vo.

This book, as a paradox, beats quadrature, duplication, trisection, philosopher's stone, perpetual motion, magic, astrology, mesmerism, clairvoyance, spiritualism, homœopathy, hydropathy, kinesipathy, Essays and Reviews, and Bishop Colenso, all put together. Of all the suppositions I have given as actually argued, this is the one which is hardest to deny, and hardest to admit. Reserving the question—as beyond human discussion—whether our particles of carbon, &c. are *clusters* of worlds, the author produces his reasons for thinking that they are at least single worlds. Of course—though not mentioned—the possibility is to be added of the same thing being true of the particles which make up our particles, and so down, for ever : and, on the other hand, of our planets and stars as being particles in some larger universe, and so up, for ever.

Great fleas have little fleas upon their backs to bite 'em,
And little fleas have lesser fleas, and so *ad infinitum*.
And the great fleas themselves, in turn, have greater fleas to go on ;
While these again have greater still, and greater still, and so on.

I have often had the notion that all the nebulæ we see, including our own, which we call the Milky Way, may be particles of snuff in the box of a giant of a proportionately larger universe. Of course the minim of time—a million of years or whatever the geologists make it—which our little affair has lasted, is but a very small fraction of a second to the great creature in whose nose we shall all be in a few tens of thousands of millions of millions of millions of years.

All this is quite possible, and the probabilities for and against are quite out of our reach. Perhaps also all the worlds, both above and below us, are fac-similes of our own. If so, away goes free will for good and all ; unless, indeed, we underpin our system with the hypothesis that all the fac-simile bodies of different sizes are actuated by a common soul. These acute supplementary notions of mine go far to get rid of the difficulty which some have found in the common theory that the soul inhabits the body : it has been started that there is, somewhere or another, a world of souls which communicate with their bodies by wondrous filaments of a nature neither mental nor material, but of a *tertium quid* fit to be a go-between ; as it were a corporispiritual copper encased in a spiritucorporeal gutta-percha. My theory is that every soul is everywhere *in posse*, as the schoolmen said, but not anywhere *in actu*, except where it finds one of its bodies. These à *priori* difficulties being thus removed, the system of particle-worlds is reduced to a dry question of fact, and remitted to the decision of the microscope. And a grand field may thus be opened, as optical science progresses ! For the worlds are not fac-similes of ours in time : there is not a moment of *our* past, and not a moment of *our* future, but is the *present* of one or more of the particles. A will write the death of Cæsar, and B the building of the Pyramids, by actual observation of the processes with a power of a thousand millions ; C will discover the commencement of the Millennium, and D the termination of Ersch and Gruber's Lexicon, as mere physical phenomena. Against this glorious future there is a sad omen : the initials of the forerunner of this discovery are—NO !

> The History of the Supernatural in all ages and nations, and in all Churches, Christian and Pagan : demonstrating a universal faith. By Wm. Howitt. London, 2 vols. 8vo. 1863.

Mr. Howitt is a preacher of spiritualism. He cements an enormous collection of alleged facts with a vivid outpouring of exhortation, and an unsparing flow of sarcasm against the scorners of all classes. He and the Rev. J. Smith (*ante*, 1854) are the most thoroughgoing universalists of all the writers I know on spiritualism. If either can insert the small end of the wedge, he will not let you off one fraction of the conclusion that all countries, in all ages, have been the theatres of one vast spiritual display. And I suspect that this consequence cannot be avoided, if any part of the system be of truly spiritual origin. Mr. Howitt treats the philosophers either as ignorant babies, or as conscious spirit-fearers : and seems much inclined to accuse the world at large of dreading, lest by the actual presence of the other world their Christianity should imbibe a spiritual element which would unfit it for the purposes of their lives.

> From Matter to Spirit. By C. D. With a preface by A. B. London, 1863, 8vo.

This is a work on Spiritual Manifestations. The author upholds the facts for spiritual phenomena : the prefator suspends his opinion as to the cause, though he upholds the facts. The work begins systematically with the lower class of phenomena, proceeds to the higher class, and offers a theory, suggested by the facts, of the connexion of the present and future life. I agree in the main with A. B.; but can, of course, make none but horrescent reference to his treatment of the smaller philosopher. This is always the way with your paradoxers : they behave towards orthodoxy as the thresher fish behaves towards the whale. But if true, as is said, that the drubbing clears the great fish of parasites which he could not otherwise get rid of, he ought to bear no malice. This preface retorts a little of that contempt which the 'philosophical world' has bestowed with heaped measure upon those who have believed their senses, and have drawn natural, even if hasty, inferences. There is philosopher-craft as well as priestcraft, both from one source, both of one spirit. In English cities and towns, the minister of religion has been tamed : so many weapons are bared against him when he obtrudes his office in a dictatory manner that, as a rule, there is no more quiet and modest member of society than the urban clergyman. Domination over religious belief is reserved for the exclusive use of those who admit the right: the rare exception to this mode of behaviour is laughed at as a bigot, or shunned as a nuisance. But the overbearing minister of nature, who snaps you with *unphilosophical* as the clergyman once frightened you with *infidel*, is still a recognized member of society, wants taming, and will get it. He wears the priest's cast-off clothes, dyed to escape detection: the better sort of philosophers would gladly set him to square the circle.

The book just named appeared about the same time as this Budget began in the *Athenæum*. It was commonly attributed, the book to my wife, the preface to myself. Some time after, our names were actually announced by the publisher, who ought to know. It will be held to confirm this statement that I announce our having in our possession some twenty reviews of different lengths, and of all characters: who ever collects a number of reviews of a book, except the author?

A great many of these reviews settle the matter *à priori*. If there had been spirits in the matter, they would have done this, and they would not have done that. Jean Meslier said there could be no God over all, for, *if* there had been one, He would have established a universal religion. If J. M. *knew* that, J. M. was right: but if J. M. did not know that, then J. M. was on the 'high priori road,' and may be left to his course. The same to all who know what spirits would do and would not do.

A. B. very distinctly said that he knew some of the asserted facts, believed others on testimony, but did not pretend to know whether they were caused by spirits, or had some unknown and unimagined origin. This he said as clearly as I could have said it myself. But a great many persons cannot understand such a frame of mind: their own apparatus is a kind of spirit-level, and their conclusion on any subject is the little bubble, which is always at one end or the other. Many of the reviewers declare that A. B. is a secret believer in the spirit-hypothesis: and one of them wishes that he had 'endorsed his opinion more boldly.' According to this reviewer, anyone who writes 'I boldly say I am unable to choose,' contradicts himself. In truth, a person who does say it has a good deal of courage, for each side believes that he secretly favours the other; and both look upon him as a coward. In spite of all this, A. B. boldly repeats that he feels assured of many of the facts of *spiritualism*, and that he cannot pretend to affirm or deny anything about their cause.

The great bulk of the illogical part of the educated community —whether majority or minority I know not; perhaps six of one and half-a-dozen of the other—have not power to make a distinction, cannot be made to take a distinction, and of course, never attempt to shake a distinction. With them all such things are evasions, subterfuges, come-offs, loopholes, &c. They would hang a man for horse-stealing under a statute against sheep-stealing; and would laugh at you if you quibbled about the distinction between a horse and a sheep. I divide the illogical—I mean people who have not that amount of natural use of sound inference which is really not uncommon—into three classes :—First class, three varieties: the Niddy, the Noddy, and the Noodle. Second class, three varieties: the Niddy-Noddy, the Niddy-Noodle, and the Noddy-Noodle. Third class, undivided : the Niddy-Noddy-Noodle. No person has a right to be angry with me for more than one of these subdivisions.

The want of distinction was illustrated to me, when a boy, about 1820, by the report of a trial which I shall never forget : boys read newspapers more keenly than men. Every now and then a bench of country magistrates rather astonishes the town populations, accustomed to rub their brains [1] against one another. Such a story as the following would, in our day, bring down grave remarks from above : but I write of the olden (or Eldon) time, when nothing but conviction in a court of record would displace a magistrate. In that day the third-class amalgamator of distinct things was often on the bench of quarter-sessions.

An attorney was charged with having been out at night, poaching. A clear *alibi* was established ; and perjury had certainly been committed. The whole gave reason to suspect that some ill-willers thought the bench disliked the attorney so much that any conviction was certain on any evidence. The bench did dislike the attorney : but not to the extent of thinking he could snare any partridges in the fields while he was asleep in bed, except the dream-partridges which are not always protected by the dream-laws. So the chairman said, "Mr. ——, you are discharged ; but you should consider this one of the most fortunate days of your life." The attorney indignantly remonstrated, but the magistrate was right ; for he said, "Mr. ——, you have frequently been employed to defend poachers : have you been careful to impress upon them the

---

[1] Baron Zach relates that a friend of his. in a writing intended for publication, said *Un esprit doit se frotter contre un autre.* The censors struck it out. The Austrian police have a keen eye for consequences.

enormity of their practices ? " It appeared in a wrangling conversation that the magistrates saw little moral difference between poaching and being a poacher's professional defender without lecturing him on his wickedness : but they admitted with reluctance, that there was a legal distinction ; and the brain of $N^3$ could no further go. This is nearly fifty years ago ; and Westernism was not quite extinct. If the present lords of the hills and the valleys want to shine, let them publish a true history of their own order. I am just old enough to remember some of the last of the squires and parsons who protested against teaching the poor to read and write. They now write books for the working classes, give them lectures, and the like. There is now no class, as a class, more highly educated, broadly educated, and deeply educated, than those who were, in old times, best described as partridge-popping squireens. I have myself, when a boy, heard Old Booby speaking with pride of Young Booby as having too high a spirit to be confined to books : and I suspected that his dislike to teaching the poor arose in fact from a feeling that they would, if taught a little, pass his heir.

A. B. recommended the spirit-theory as an hypothesis on which to ground inquiry ; that is, as the means of suggestion for the direction of inquiry. Every person who knows anything of the progress of physics understands what is meant ; but not the reviewers I speak of. Many of them consider A. B as *adopting* the spirit-hypothesis. The whole book was written, as both the authors point out, to suggest inquiry to those who are curious ; C. D. firmly believing, A. B. as above. Neither C. D. nor A. B. make any other pretence. Both dwell upon the absence of authentications and the suppression of names as utterly preventive of anything like proof. And A. B. says that his reader ' will give him credit, if not himself a goose, for seeing that the tender of an anonymous cheque would be of equal effect, whether drawn on the Bank of England or on Aldgate Pump.' By this test a number of the reviewers are found to be geese : for they take the authors as offering proof, and insist, against the authors, on the very point on which the authors had themselves insisted beforehand.

Leaving aside imperceptions of this kind, I proceed to notice a clerical and medical review. I have lived much in the middle ages, especially since the invention of printing ; and from thence I have brought away a high respect for and grateful recollection of—the priest in everything but theology, and the physician in everything but medicine. The professional harness was unfavour-

able to all progress, except on a beaten road; the professional
blinkers prevented all but the beaten road from being seen : the
professional reins were pulled at the slightest attempt to quicken
pace, even on the permitted path ; and the professional whip was
heavily laid on at the slightest attempt to diverge.  But when
the intelligent man of either class turned his attention out of his
ordinary work, he had, in most cases, the freshness and vigour of
a boy at play, and like the boy, he felt his freedom all the more
from the contrast of school-restraint.

In the case of medicine, and physics generally, the learned
were, in some essential points, more rational than many of their
present impugners.  They pass for having put *à priori* obstacles
in the way of progress : they might rather be reproved for too
much belief in progress obtained by *à priori* means.  They
would have shouted with laughter at a dunce who—in a review I
read, but without making a note—declared that he would not
believe his senses except when what they showed him was capable
of explanation upon some known principle.  I have seen such
stuff as this attributed to the schoolmen ; but only by those who
knew nothing about them.  The following, which I wrote some
years ago, will give a notion of a distinction worth remembering.
It is addressed to the authorities of the College of Physicians.

" The ignominy of the word *empiric* dates from the ages in
which scholastic philosophy deduced physical consequences *à
priori* ;—the ages in which, because a lion is strong, rubbing with
lion's fat would have been held an infallible tonic.  In those
happy days, if a physician had given decoction of a certain bark,
only because in numberless instances that decoction had been
found to strengthen the patient, he would have been a miserable
empiric.  Not that the colleges would have passed over his returns because they were empirical: they knew better.  They
were as skilful in finding causes for facts, as facts for causes.
The president and the elects of that day would have walked out into
the forest with a rope, and would have pulled heartily at the tree
which yielded the bark : nor would they ever have left it until
they had pulled out a legitimate reason.  If the tree had resisted
all their efforts, they would have said ' Ah ! no wonder now ; the
bark of a strong tree makes a strong man.'  But if they had
managed to serve the tree as you would like to serve homœopathy
then it would have been ' We might have guessed it; all the
*virtus roborativa* has settled in the bark.'  They admitted, as we
know from Molière, the *virtus dormitiva* of opium, for no other
reason than that opium *facit dormire*.  Had the medicine not

been previously *known*, they would, strange as it may seem to modern pharmacopœists, have accorded a *virtus dormitiva* to the new *facit dormire*. On this point they have often been misapprehended. They were prone to infer *facit* from a *virtus* imagined *à priori*; and they were ready in supplying *facit* in favour of an orthodox *virtus*. They might have gone so far, for example, under pre-notional impressions, as the alliterative allopath, who, when maintenance of truth was busy opposing the progress of science called *vaccination*, declared that some of its patients coughed like cows, and bellowed like bulls; but they never refused to find *virtus* when *facit* came upon them, no matter whence. They would rather have accepted Tenterden steeple than have rejected the Goodwin Sands. They would have laughed their modern imitators to scorn : but as they are not here, we do it for them.

"The man of our day—the *à priori* philosopher—tries the question whether opium can cause sleep by finding out in the recesses of his own noddle whether the drug can have a dormitive power: Well! but did not the schoolman do the same? He did ; but mark the distinction. The schoolman had recourse to first principles, when there was no opium to try it by : our man settles the point in the same way *with a lump of opium before him*. The schoolman shifted his principles with his facts : the man of our drawing-rooms will fight facts with his principles, just as an old physician would have done in actual practice, with the rod of his *Church* at his back.

"The story about Galileo—which seems to have been either a joke made against him, or by him—illustrates this. *Nature abhors a vacuum* was the explanation of the water rising in a pump : but they found that the water would not rise more than 32 feet. They asked for explanation : what does the satirist make the schoolmen say? That the stoppage is *not* a fact, because nature abhors a vacuum? No! but that the principle should be that nature abhors a vacuum as far as 32 feet. And this is what would have been done.

"There are still among us both priests and physicians who would have belonged, had they lived three or four centuries ago, to the glorious band of whom I have spoken, the majority of the intelligent, working well for mankind out of the professional pursuit. But we have a great many who have helped to abase their classes. Go where we may, we find specimens of the lower orders of the ministry of religion and the ministry of health showing themselves smaller than the small of other pursuits. And how is this?

First, because each profession is entered upon a mere working smack of its knowledge, without any depth of education, general or professional. Not that this is the whole explanation, nor in itself objectionable : the great mass of the world must be tended, soul and body, by those who are neither Hookers nor Harveys : let such persons not venture *ultra crepidam*, and they are useful and respectable. But, secondly, there is a vast upheaving of thought from the depths of commonplace learning. I am a clergyman! Sir! I am a medical man! Sir! and forthwith the nature of things is picked to pieces, and there is a race, with the last the winner, between Philosophy mounted on Follys donkey, and Folly mounted on Philosophy's donkey. How fortunate it is for Law that her battles are fought by politicians in the Houses of Parliament. Not that it is better done : but then *politics* bears the blame."

I now come to the medical review. After a quantity of remark which has been already disposed of, the writer shows Greek learning, a field in which the old physician would have had a little knowledge. A. B., for the joke's sake, had left untranslated, as being too deep, a remarkably easy sentence of Aristotle, to the effect that what has happened was possible, for if impossible it would not have happened. The reviewer, in 'simple astonishment,'—it was simple—at the pretended incapacity—I was told by A. B. that the joke was intended to draw out a reviewer—translates :—He says that this sentence is A. B.'s summing up of the evidence of Spiritualism. Now, being a sort of *alter ego* of A. B., I do declare that he is not such a fool as to rest the evidence of Spiritualism—the *spirit explanation*—upon the occurrence of certain facts proving the possibility of those very facts. In truth, A. B. refuses to receive spiritualism, while he receives the facts : this is the gist of his whole preface, which simply admits spiritualism among the qualified candidates, and does not know what others there may be.

The reviewer speaks of Aristotle as 'that clear thinker and concise writer.' I strongly suspect that his knowledge of Aristotle was limited to the single sentence which he had translated or got translated. Aristotle is concise in *phrase*, not in book, and is powerful and profound in thought : but no one who knows that his writing, all we have of him, is the very opposite of clear, will pretend to decide that he thought clearly. As his writing, so probably was his thought ; and his books are, if not anything but clear, at least anything good but clear. Nobody thinks them clear except a person who always clears difficulties : which I have no doubt was the reviewer's habit ; that is, if he ever took the field at all. The gentleman who read Euclid, all except the As and Bs and the pictures of scratches and scrawls, is the type of a numerous class.

The reviewer finds that the word *amosgepotically*, used by A. B., is utterly mysterious and incomprehensible. He hopes his translation of the bit of Greek will shield him from imputation of ignorance : and thinks the word may be referred to the ' obscure dialect ' out of which sprung *aneroid*, *kalos geusis sauce*, and *Anaxyridian trousers*. To lump the first two phrases with the third smacks of ignorance in a Greek critic ; for ἀναξυριδια, *breeches*, would have turned up in the lexicon ; and *kalos geusis*, though absurd, is not obscure. And ἀμωσγεπως, *somehow or other*, is as easily found as ἀναξυριδια. The word *aneroid*, I admit, has puzzled better scholars than the critic : but never one who knows the unscholarlike way in which words ending in ειδης have been rendered. The *aneroid barometer* does *not* use a column of air in the same way as the old instrument. Now ἀεροειδης— properly *like* the atmosphere—is by scientific non-scholarship rendered having to do with the atmosphere ; and ἀναεροειδης—say *anaëroid*—denies having to do with the atmosphere ; a nice thing to say of an instrument which is to measure the weight of the atmosphere. One more absurdity, and we have *aneroid*, and there you are. The critic ends with a declaration that nothing in the book shakes his faith in a *Quarterly* reviewer who said that suspension of opinion, until further evidence arrives, is justifiable : a strange summing up for an article which insists upon utter rejection being unavoidable.[1] The expressed aim of both A. B. and C. D. was to excite inquiry, and get further evidence : until this is done, neither asks for a verdict.

Oh where ! and oh where ! is old Medicine's learning gone ! There *was* some in the days of yore, when Popery was on ! And it's oh ! for some Greek, just to find a word upon ! The reviewer who, lexicon in hand, can neither make out *anaxyridical, amosgepotical, kalos geusis*, nor distinguish them from *aneroid*, cannot be trusted when he says he has translated a sentence of Aristotle. He may have done it ; but, as he says of spiritualism, we must suspend our opinion until further evidence shall arrive.

We now come to the theological review. I have before alluded to the faults of logic which are Protestant necessities : but I never said that Protestant argument had *nothing but* paralogism. The

---

[1] This "utter rejection" has been repeated (1872) by the same writer.—ED.

writer before me attains this completeness: from beginning to
end he is of that confusion and perversion which, as applied to
interpretation of the New Testament, is so common as to pass
unnoticed by sermon-hearers; but which, when applied out of
church, is exposed with laughter in all subjects except theology.
I shall take one instance, putting some words in italics.

| A. B. | Theological Critic. |
|---|---|
| My state of mind, which refers the whole *either* to unseen intelligence, *or something which man has never had any conception of*, proves me to be out of the pale of the Royal Society. | . . . he proceeds to argue that he himself is outside its sacred pale because he refers all these strange phenomena to *unseen spiritual intelligence.* |

The possibility of a *yet unimagined* cause is insisted on in
several places. On this ground it is argued by A. B. that
spiritualists are 'incautious' for giving in at once to the spirit
doctrine. But, it is said, they may be justified by the philosophers, who make the flint *axes*, as they call them, to be the
works of men, because no one can see *what else they can be.* This
kind of adoption, *condemned* as a conclusion, is *approved* as a
provisional theory, suggestive of direction of inquiry: experience
having shown that inquiry directed by a *wrong* theory has led to
more good than inquiry without any theory at all. All this A. B.
has fully set forth, in several pages. On it the reviewer remarks
that ' with infinite satisfaction he tries to justify his view of the
case by urging that there is no other way of accounting for it;
after the fashion of the philosophers of our own day, who conclude
that certain flints found in the drift are the work of men, because
the geologist does not see what else they can be.' After this
twist of meaning, the reviewer proceeds to say, and A. B. would
certainly join him, ' There is no need to combat any such mode of
reasoning as this, because it would apply with equal force and
justice to any theory whatever, however fantastic, profane, or silly.'
And so, having shown how the reviewer has hung himself, I leave
him funipendulous.

One instance more, and I have done. A reviewer, not theological, speaking of the common argument that things which are
derided are not *therefore* to be rejected, writes as follows:—' It
might as well be said that they who laughed at Jenner and vaccination were, in a certain but very unsatisfactory way, witnesses to
the possible excellence of the system of St. John Long.' Of course
it *might*: and of course it *is* said by all people of common sense.
In introducing the word ' possible,' the reviewer has hit the point:
I suspect that this word was introduced during revision, to put

the sentence into fighting order, hurry preventing it being seen that the sentence was thus made to fight on the wrong side. Jenner, who was laughed at, was right; therefore, it is not impossible—that is, it is *possible*—that a derided system may be right. Mark the three gradations : *in medio tutissimus ibis.*

*Reviewer.*—If a system be derided, it is no ground of suspense that derided systems have turned out true : if it were, you would suspend your opinion about St. John Long on account of Jenner. —*Ans.* You ought to do so, as to *possibility*; and *before examination*; not with the notion that J. proves St. J. *probable*; only *possible.*

*Common Sense.*—The past emergence of truths out of derided systems proves that there is a practical certainty of like occurrence to come. But, inasmuch as a hundred speculative fooleries are started for one truth, the mind is permitted to approach the examination of any one given novelty with a bias against it of a hundred to one : and this permission is given because so it will be, leave or no leave. Every one has licence not to jump over the moon.

*Paradoxer.*—Great men have been derided, and I am derided : which proves that my system ought to be adopted. This is a summary of all the degrees in which paradoxers contend for the former derision of truths now established, giving their systems *probability.* I annex a paragraph which D [e &c.] inserted in the *Athenæum* of October 23, 1847.

## "DISCOVERERS AND DISCOVERIES.

" Aristotle once sent his servant to the cellar to fetch wine ;— and the fellow brought him back small beer. The Stagirite (who knew the difference) called him a blockhead. ' Sir,' said the man, ' all I can say is, that I found it in the cellar.' The philosopher muttered to himself that an affirmative conclusion could not be proved in the second figure,—and Mrs. Aristotle, who was by, was not less effective in her remark, that small beer was not wine because it was in the same cellar. Both were right enough : and our philosophers might take a lesson from either—for they insinuate an affirmative conclusion in the second figure. Great discoverers have been little valued by established schools,—and they are little valued. The results of true science are strange at first,—and so are their's. Many great men have opposed existing notions,—and so do they. All great men were obscure at first,— and they are obscure. Thinking men doubt,—and they doubt.

Their small beer, I grant, has come out of the same cellar as the wine; but this is not enough. If they had let it stand awhile in the old wine-casks, it might have imbibed a little of the flavour."

There are better reviews than I have noticed; which, though entirely dissenting, are unassailable on their own principles. What I have given represents five-sixths of the whole. But it must be confessed that the fraction of fairness and moderation and suspended opinion which the doctrine of *Spirit Manifestations* has met with—even in the lower reviews—is strikingly large compared what would have been the case fifty years ago. It is to be hoped that our popular and periodical literatures are giving us one thinker created for twenty geese double-feathered : if this hope be realised, we shall do! Seeing all that I see, I am not prepared to go the length of a friend of mine who, after reading a good specimen of the lower reviewing, exclaimed—Oh! if all the fools in the world could be rolled up into one fool, what a reviewer he would make!

> Calendrier Universel et Perpétuel; par le Commandeur P. J. Arson. Publié par ses Enfans (Œuvre posthume). Nice, 1863, 4to.

I shall not give any account of this curious calendar, with all its changes and symbols. But there is one proposal, which, could we alter the general notions of time—a thing of very dubious possibility—would be convenient. The week is made to wax and wane, culminating on the Sunday, which comes in the middle. Thursday, Friday, Saturday, are ascending or waxing days; Monday, Tuesday, Wednesday, are descending or waning days. Our six days, lumped together after the great distinguishing day, Sunday, are too many to be distinctly thought of together : a division of three preceding and three following the day of most note would be much more easily used. But all this comes too late. It may be, nevertheless, that some individuals may be able to adjust their affairs with advantage by referring Thursday, Friday, Saturday, to the following Sunday, and Monday, Tuesday, Wednesday, to the preceding Sunday. But M. Arson's proposal to alter the names of the days is no more necessary than it is practicable.

---

I am not to enter anything I do not possess. The reader therefore will not learn from me the feats of many a man-at-arms in these subjects. He must be content, unless he will bestir himself for himself, not to know how Mr. Patrick Cody trisects the angle at Mullinavat, or Professor Recalcati squares the circle at Milan. But this last is to be done by subscription, at five francs a head: a banker is named who guarantees restitution if the solution be not perfectly rigorous; the banker himself, I suppose, is the judge. I have heard of a man of business who settled the circle in this way: if it can be reduced to a debtor and creditor account, it can certainly be done ; if not, it is not worth doing. Montucla will give the accounts of the lawsuits which wagers on the problem have produced in France.

Neither will I enter at length upon the success of the new squarer who advertises (Nov. 1863) in a country paper that, having read that the circular ratio was undetermined, ' I thought it very strange that so many great scholars in all ages should have failed in finding the true ratio, and have been determined to try myself . . . I am about to secure the benefit of the discovery, so until then the public cannot know my new and true ratio.' I have been informed that this trial makes the diameter to the circumference as 64 to 201, giving $\pi = 3\cdot140625$ exactly. The result was obtained by the discoverer in three weeks after he first heard of the existence of the difficulty. This quadrator has since published a little slip, and entered it at Stationers' Hall. He says he has done it by actual measurement ; and I hear from a private source that he uses a disk of 12 inches diameter, which he rolls upon a straight rail. Mr. James Smith did the same at one time ; as did also his partisan at Bordeaux. We have, then, both $3\cdot125$ and $3\cdot140625$, by actual measurement. The second result is more than the first by about one part in 200. The second rolling is a very creditable one ; it is about as much below the mark as Archimedes was above it. Its performer is a joiner, who evidently knows well what he is about when he measures ; he is not wrong by 1 in 3,000.

The reader will smile at the quiet self-sufficiency with which ' I have been determined to try myself' follows the information that ' so many great scholars in all ages ' have failed. It is an admirable spirit, when accompanied by common sense and uncommon self-knowledge. When I was an undergraduate there was a little attendant in the library who gave me the following,— ' As to cleaning this library, Sir, if I have spoken to the Master once about it, I have spoken fifty times : but it is of no use ; he will not employ *littery* men ; and so I am obliged to look after it myself.'

I do not think I have mentioned the bright form of quadrature in which a square is made equal to a circle by making each side equal to a quarter of the circumference. The last squarer of this kind whom I have seen figures in the last number of the *Athenæum* for 1855 : he says the thing is no longer a *problem*, but an *axiom*. He does not know that the area of the circle is greater than that of any other figure of the same circuit. This any one might see without mathematics. How is it possible that the figure of greatest area should have any one length in its circuit unlike in form to any other part of the same length?

The feeling which tempts persons to this problem is that which, in romance, made it impossible for a knight to pass a castle which belonged to a giant or an enchanter. I once gave a lecture on the subject : a gentleman who was introduced to it by what I said remarked, loud enough to be heard by all around, ' Only prove to me that it is impossible, and I will set about it this very evening.'

This rinderpest of geometry cannot be cured, when once it has seated itself in the system : all that can be done is to apply what the learned call prophylactics to those who are yet sound. When once the virus gets into the brain, the victim goes round the flame, like a moth, first one way and then the other, beginning again where he ended, and ending where he begun : thus verifying the old line

In girum imus nocte, ecce ! et consumimur igni.

Every mathematician knows that scores of methods, differing altogether from each other in process, all end in this mysterious 3·14159 . . ., which insists on calling itself the circumference to a unit of diameter. A reader who is competent to follow processes of arithmetic may be easily satisfied that such methods do actually exist. I will give a sketch, carried out to a few figures, of three : the first two I never met with in my reading ; the third is the old method of Vieta. [I find that both the first and second methods are contained in a theorem of Euler.]

What Mr. James Smith says of these methods is worth noting. He says I have given three ' *fancy* proofs' of the value of $\pi$ : he evidently takes me to be offering demonstration. He proceeds thus :—

' His first proof is traceable to the diameter of a circle of radius 1. His second, to the side of any inscribed equilateral triangle to a circle of radius 1. His third, to a radius of a circle of diameter 1. Now, it may be frankly admitted that we can arrive at the same result by many other modes of arithmetical calculation, all of which may be shown to have some sort of relation to a circle; but, after all, these results are mere exhibitions of the properties of numbers, and have no more to do with the ratio of diameter to circumference in a circle than the price of sugar with the mean height of spring tides. (*Corr.* Oct. 21, 1865).'

I quote this because it is one of the few cases—other than absolute assumption of the conclusion—in which Mr. Smith's conclusions would be true if his premise were true. Had I given what follows as *proof*, it would have been properly remarked, that I had only exhibited properties of numbers. But I took care to tell my reader that I was only going to show him *methods* which end in 3·14159 . . . The proofs that these methods establish the value of $\pi$ are for those who will read and can understand.

1. Take any diameter, double it, take 1-3rd of that double, 2-5ths of the last, 3-7ths of the last, 4-9ths of the last, 5-11ths of the last, and so on. The sum of all is the circumference of that diameter. The following is the process when the diameter is a hundred millions : the errors arising from rejection of fractions being lessened by proceeding on a thousand millions, and striking off one figure.

| | | |
|---:|---:|---:|
| 200000000 | 31415 | 3799 |
| 66666667 | | 2817 |
| 26666667 | | 1363 |
| 11428571 | | 661 |
| 5079365 | | 321 |
| 2308802 | | 156 |
| 1065601 | | 76 |
| 497281 | | 37 |
| 234014 | | 18 |
| 110849 | | 9 |
| 52785 | | 5 |
| 25245 | | 2 |
| 12118 | | 1 |
| 5834 | | |
| | | |
| 314153799 | 31415 | 9265 |

Here 200 &c. is double of the diameter; 666 &c. is 1-3rd of 200 &c.; 266 &c. is 2-5ths of 666 &c.; 114 &c. is 3-7ths of 266 &c.; 507 &c. is 4-9ths of 114 &c.; and so on.

2. To the square root of 3 add its half. Take *half* the third part of this; half 2-5ths of the last; half 3-7ths of the last; and so on. The sum is the circumference to a unit of diameter.

Square root of 3  . . . .  1·73205081
                              ·86602540
                          ‾‾‾‾‾‾‾‾‾‾‾‾
                           2·59807621
                             ·43301270
                             ·08660254
                             1855768
                             412393
                             93726
                             21629
                             5047
                             1188
                             281
                             67
                             16
                             4
                             1
                          ‾‾‾‾‾‾‾‾‾‾‾‾
                           3·14159265

3. Take the square root of $\frac{1}{2}$; the square root of half of one more than this; the square root of half of one more than the last; and so on, until we come as near to unity as the number of figures chosen will permit. Multiply all the results together, and divide 2 by the product: the quotient is an approximation to the circumference when the diameter is unity. Taking aim at four figures, that is, working to five figures to secure accuracy in the fourth, we have ·70712 for the square root of $\frac{1}{2}$; ·92390 for the square root of half one more than ·70712; and so on, through ·98080, ·99520, ·99880, ·99970, ·99992, and ·99998. The product of the eight results is ·63667; divide 2 by this, and the quotient is 3·1413 . . ., of which four figures are correct. Had the product been ·636363 . . . instead of ·63667 . . ., the famous result of Archimedes, 22-7ths, would have been accurately true. It is singular that no cyclometer maintains that Archimedes hit it exactly.

A literary journal could hardly admit as much as the preceding, if it stood alone. But in my present undertaking it passes as the halfpennyworth of bread to many gallons of sack. Many more methods might be given, all ending in the same result, let that result mean what it may.

Now since dozens of methods, to which dozens more might be added at pleasure, concur in giving one and the same result; and since these methods are declared by all who have shown knowledge of mathematics to be *demonstrated*: it is not asking too much of a person who has just a little knowledge of the first elements that he should learn more, and put his hand upon the error, before he intrudes his assertion of the existence of error upon those who have given more time and attention to it than himself, and who are in possession, over and above many demonstrations, of many consequences verifying each other, of which he can know nothing. This is all that is required. Let any one square the circle, and persuade his friends, if he and they please : let him print, and let all read who choose. But let him abstain from intruding himself upon those who have been satisfied by existing demonstration, until he is prepared to lay his finger on the point in which existing demonstration is wrong. Let him also say what this mysterious 3·14159... really is, which comes in at every door and window, and down every chimney, calling itself the circumference to a unit of diameter. This most impudent and successful impostor holds false title-deeds in his hands, and invites examination : surely those who can find out the rightful owner are equally able to detect the forgery. All the quadrators are agreed that, be the right what it may, 3·14159... is wrong. It would be well if they would put their heads together, and say what this wrong result really means. The mathematicians of all ages have tried all manner of processes, with one object in view, and by methods which are admitted to yield demonstration in countless cases. They have all arrived at one result. A large number of opponents unite in declaring this result wrong, and all agree in two points : first, in differing among themselves ; secondly, in declining to point out what that curious result really is which the mathematical methods all agree in giving.

Most of the quadrators are not aware that it has been fully demonstrated that no two numbers whatsoever can represent the ratio of the diameter to the circumference with perfect accuracy. When therefore we are told that either 8 to 25 or 64 to 201 is the true ratio, we know that it is no such thing, without the necessity of examination. The point that is left open, as not fully demonstrated to be impossible, is the *geometrical* quadrature, the determination of the circumference by the straight line and circle, used as in Euclid. The general run of circle-squarers, hearing that the quadrature is not pronounced to be *demonstratively* impossible, imagine that the *arithmetical* quadrature is open to their ingenuity. Before attempting the arithmetical problem, they ought to acquire knowledge enough to read Lambert's demonstration (last given in Brewster's translation of Legendre's Geometry) and, if they can, to refute it. [It will be given in an Appendix.] Probably some have begun in this way, and have caught a Tartar who has refused to let them go : I have never heard of any one who, in producing his own demonstration, has laid his finger on the faulty part of Lambert's investigation. This is the answer to those who think that the mathematicians treat the arithmetical squarers too lightly, and that as some person may succeed at last, all attempts should be examined. Those who have so thought, not knowing that there is demonstration on the point, will probably admit that a person who contradicts a theorem of which the demonstration has been acknowledged for a century by all who have alluded to it as read by themselves, may reasonably be required to point out the error before he demands attention to his own result.

*Apopempsis of the Tutelaries.*—Again and again I am told that I spend too much time and trouble upon my two tutelaries : but when I come to my summing-up I shall make it appear that I have a purpose. Some say I am too hard upon them : but this is quite a mistake. Both of them beat little Oliver himself in the art and science of asking for more; but without Oliver's excuse, for I had given good allowance. Both began with me, not I with them : and both knew what they had to expect when they applied for a second helping.

On July 31, the Monday after the publication of my remarks on my 666 correspondent, I found *three* notes in separate envelopes, addressed to me at ' 7A, University College.' When I saw the three new digits I was taken rhythmopoetic, as follows—

> Here's the Doctor again with his figs, and by Heavens!
> He was always at sixes, and now he's at sevens.

To understand this fully the reader must know that the greater part of Apocalyptic interpretation has long been condensed, in my mind, into the Turkish street-cry—In the name of the Prophet! figs! I make a few extracts. The reader will observe that Dr. Thorn grumbles at his *private* letters being *publicly* ridiculed. A man was summoned for a glutolactic assault; he complained of the publication of his proceeding : I kicked &c. *in confidence,* he said.

"After reading your last, which tries in every way to hold me up to public ridicule for daring to write you privately ['that you would be d—d,' omitted by accident] one would say, Why have anything to do with such a testy person ? [Wrong word ; no testy person can manage cool and consecutive ridicule. Quære, what is this word ? Is it anything but a corruption of the obsolete word *tetchy* of the same meaning ?

Some think *touchy* is our modern form of *tetchy*, which I greatly doubt]. My answer is, the poor man is lamentably ignorant; he is not only so, but 'out of the way' [quite true; my readers know me by this time for an out-of-the-way person. What other could tackle my squad of paradoxers? What other would undertake the job?]. Can he be brought back and form one of those who in Ezekiel 37 ch. have the Spirit breathed into them and live . . . Have I any other feeling towards you except that of peace and goodwill? [Not to your distinct knowledge; but in all those who send people to 'the other place' for contempt of their interpretations, there is a lurking wish which is father to the thought; 'you *will* be d—d' and you *be* d—d' are Siamese twins]. Of course your sneer at 666 brought plain words; but when men meddle with what they do not understand (not having the double *Vahu*) they must be dealt with faithfully by those who do . . . [They must; which justifies the Budget of Paradoxes: but no occasion to send them anywhere; no preachee and floggee too, as the negro said]. Many will find the text Prov. i. 26 fully realized. [All this contains distinct assumption of a right 'of course' to declare accursed those who do not respect the writer's vagary] . . . If I could but get the א, the Ox-head, which in old Hebrew was just the Latin Digamma, F, out of your name, and could then Thau you with the Thau of Ezekiel ix. 4, the χ, then you would bear the number

| | | |
|---|---|---|
| M | 40 | of a man ! But this is too hard for me, although not so |
| O | 70 | for the Lord ! Jer. xxxii. 17 . . . And now a word : is |
| R | 100 | ridicule the right thing in so solemn a matter as the |
| G | 6 | discussion of Holy Writ? [Is food for ridicule the right |
| N | 50 | thing? Did I discuss Holy Writ? I did not : I con- |
| ·—— | | cussed profane scribble. Even the Doctor did not *discuss* : |
| | 266 | he only enunciated and denunciated out of the mass of |
| n = χ | 400 | inferences which a mystical head has found premises for |
| | | in the Bible]." |

[That ill opinions are near relations of ill wishes, will be detected by those who are on the look out. The following was taken down in a Scotch Church by Mr. Cobden, who handed it to a Roman friend of mine, for his delectation (in 1855): 'Lord, we thank thee that thou hast brought the Pope into trouble; and we pray that thou wouldst be mercifully pleased to increase the same.']

Here is a martyr who quarrels with his crown; a missionary who reviles his persecutor: send him to New Zealand, and he would disagree with the Maoris who ate him. Man of unilateral reciprocity! have you, who write to a stranger with hints that that stranger and his wife are children of perdition, the bad taste to complain of a facer in return? As James Smith—

the Attorney-wit, not the Dock-cyclometer — said, or nearly
said,

> " A pretty thing, forsooth !
> Is he to burn, all scalding hot,
> Me and my wife, and am I not
> To job him out a tooth ? "

Those who think parody vulgar will be pleased to substitute for
the above a quotation from Butler :—

> There's nothing so absurd or vain,
> Or barbarous or inhumane,
> But if it lay the least pretence
> To piety and godliness,
> Or tender-hearted conscience,
> And zeal for gospel truths profess,—
> Does sacred instantly commence,
> And all that dare but question it are straight
> Pronounced th' uncircumcised and reprobate.
> As malefactors that escape and fly
> Into a sanctuary for defence,
> Must not be brought to justice thence,
> Although their crimes be ne'er so great and high.
> And he that dares presume to do't
> Is sentenced and delivered up
> To Satan that engaged him to't.

Of all the drolleries of controversy none is more amusing than
the manner in which those who provoke a combat expect to lay
down the laws of retaliation. You must not strike this way! you
must not parry that way! If you don't take care, we shall never
meddle with you again! We were not *prepared* for such as this!
Why did we have anything to do with such a testy person? M.
Jourdain must needs show Nicole, his servant-maid, how good a
thing it was to be sure of fighting without being killed, by carte
and tierce : 'Et cela n'est il pas beau d'être assuré de son fait
quand on se bat contre quelqu'un? Là, pousse moi un peu, pour
voir. NICOLE. Eh bien! quoi? M. JOURDAIN. Tout beau. Hola!
Ho! doucement. Diantre soit la coquine! NICOLE. Vous me
dites de pousser. M. JOURDAIN. Oui; mais tu me pousses en
tierce, avant que de pousser en quarte, et tu n'as pas la patience
que je pare.'
His colleague, my secular tutelary, who also made an ana-
chronistic onset, with his repartees and his retorts, before there was
anything to fire at, takes what I give by way of subsequent provocation with a good humour which would make a convert of me

if he could afford 01659265 ... of a grain of logic. He instantly sent me his photograph for the asking, and another letter in proof. The Thor-hammerer does nothing but grumble, except when he tells a good story, which he says he had from Dr. Abernethy. A Mr. James Dunlop was popping at the Papists with a 666-rifled gun, when Dr. Chalmers quietly said, ' Why, Dunlop, you bear it yourself,' and handed him a paper on which the numerals in

| I | A | C | O | B | V | S | | D | V | N | L | O | P | V | S |
|---|---|---|---|---|---|---|---|---|---|---|---|---|---|---|---|
| 1 | 100 | | | | 5 | | | 500 | 5 | | 50 | | | 5 | |

were added up. This is almost as good as the *Filii Dei Vicarius*, the numeral letters of which also make 666. No more of these crazy—I first wrote *puerile*, but why should young cricketers be libelled ?—attempts to extract religious use from numerical vagaries, and to make God over all a proposer of *salvation conundrums* : and no more of the trumpery hints about future destiny which it is too great a compliment to call blasphemous. If the Doctor will cipher upon the letters in ἐν ᾦ μετρῳ μετρειτε μετρηθησεται ὑμιν, with *double Vahu* cubic measure, he will perhaps learn to leave off trying to frighten me into gathering grapes from thorns.

Mystical hermeneutics may be put to good use by out-of-the-way people. They may be made to call the attention of the many to a distinction well known among the learned. The books of the New Testament have been for 1,500 years divided into two classes : the *acknowledged* (ὁμολογουμενα), which it has always been paradox not to receive ; and the *controverted* (ἀντιλεγομενα), about which there has always been that difference of opinion which no scholar overlooks, however he may decide for himself after balance of evidence. Eusebius, who first (l. 3, c. 25) recorded the distinction—which was much insisted on by the early Protestants—states the books which are questioned as doubtful, but which yet are approved and acknowledged by *many* —or *the many*, it is not easy to say which he means—to be the Epistles of James and Jude, the second of Peter and the second and third of John. In other places he speaks doubtingly of the Epistle to the Hebrews. The Apocalypse he does not even admit into this class, for he proceeds as follows—I use the second edition of the English folio translation (1709), to avert suspicion of bias from myself :—

' Among the *spurious* [νοθοι] let there be ranked both the work entitled the *Acts of Paul*, and the book called *Pastor*, and the *Reve-*

*lation of Peter*: and moreover that which is called the *Epistle of Barnabas*, and that named the *Doctrines of the Apostles*: and moreover, as I said, the *Revelation of John* (if you think good), which some, as I have said, do reject, but others allow of, and admit among those books which are received as unquestionable and undoubted.'

Eusebius, though he will not admit the Apocalypse even into the *controverted* list, but gives permission to call it *spurious*, yet qualifies his permission in a manner which almost annihilates the distinctive force of νοθος, and gives the book a claim to rank (if you think good, again) in the controverted list. And this is the impression received by the mind of Lardner, who gives Eusebius fully and fairly, but when he sums up, considers his author as admitting the Apocalypse into the second list. A stick may easily be found to beat the father of ecclesiastical history. There are whole faggots in writers as opposite as Baronius and Gibbon, who are perhaps his two most celebrated sons. But we can hardly imagine him totally misrepresenting the state of opinion of those for whom and among whom he wrote. The usual plan, that of making an author take the views of his reader, is more easy in his case than in that of any other writer: for, as the riddle says, he is You-see-by-us ; and to this reading of his name he has often been subjected. Dr. Nathaniel Lardner, who, though heterodox in doctrine, tries hard to be orthodox as to the Canon, is 'sometimes apt to think' that the list should be collected and divided as in Eusebius. He would have no one of the controverted books to be allowed, by itself, to establish any doctrine. Even without going so far, a due use of early opinion and long continued discussion would perhaps prevent rational people from being induced by those who have the *double Vahu* to place the Apocalypse *above* the Gospels, which all the Bivahuites do in effect, and some are said to have done in express words. But my especial purpose is to point out that an easy way of getting rid of 665 out of 666 of the mystics is to require them to establish the Apocalypse before they begin. See if they even know so much as that there is a crowd of testimonies for and against, running through the first four centuries, which makes this book the most difficult of the whole Canon. Try this method, and you will escape beautiful, as the French say. Dean Alford, in vol. iv. p. 8. of his New Testament, gives an elaborate handling of this question. He concludes by saying that he cannot venture to refuse his consent to the tradition that the Apostle is the author. This modified adherence, or non-nonadherence, pretty well represents the feeling of orthodox Protestants, when learning and common sense come together.

I have often, in former days, had the attempt made to place the Apocalypse on my neck as containing prophecies yet unfulfilled. The preceding method prevents success; and so does the following. It may almost be taken for granted that theological system-fighters do not read the New Testament: they hunt it for detached texts; they listen to it in church in that state of quiescent nonentity which is called reverent attention: but they never read it. When it is brought forward, you must pretend to find it necessary to turn to the book itself: you must read 'The revelation . . . to show unto his servants *things which must shortly come to pass* . . . . Blessed is he that readeth . . . . *for the time is at hand.*' You must then ask your mystic whether things deferred for 1800 years were shortly to come to pass, &c.? You must tell him that the Greek *ἐν ταχει*, rendered ' shortly,' is as strong a phrase as the language has to signify *soon*. The interpreter will probably look as if he had never read this opening: the chances are that he takes up the book to see whether you have not been committing a fraud. He will then give you some exquisite evasion: I have heard it pleaded that the above was a *mere preamble*. This word *mere* is all-sufficient: it turns anything into nothing. Perhaps he will say that the argument is that of the Papists: if so, tell him that there is no Christian sect but bears true witness against some one or more absurdities in other sects.

An anonyme suggests that *ἐν ταχει* may not be ' soon,' it may be 'quickly, without reference to time when:' he continues thus, ' May not time be " at hand" when it is ready to come, no matter how long delayed?' I now understand what * * * and * * * meant when they borrowed my books and promised to return them quickly, it was 'without reference to time when.' As to time at *hand*—provided you make a long *arm*—I admire the quirk, but cannot receive it: the word is *ἐγγυς*, which is a word of *closeness*, in time, in place, in reckoning, in kindred, &c

Another gentleman is not surprised that Apocalyptic reading leads to a doubt of the ' canonicity' of the book: it ought not to rest on church testimony, but on visible miracle. He offers me, or any reader of the *Athenæum*, the 'sight of a miracle to that effect, and within forty-eight hours' journey (fare paid).' I seldom travel, and my first thought was whether my carpet-bag would be found without a regular hunt: but, on reading further, I found that it was only a concordance that would be wanted. Forty hours' collection and numerical calculation of Greek nouns would make it—should I happen to agree with the writer—many hundred millions to one that Revelation xiii is superhuman. There is but one verse (the fifth) which the writer does not see verified. I looked at this verse, and was much startled. The Budget began in October 1863: should it last until March 1867 —it is now August 1866—it is clear that I am the first Beast, and my paradoxers are the saints whom I persecute.

[The Budget *did* terminate in March 1867: I hope the gentleman will be satisfied with the resulting interpretation.]

The same opponent is surprised that I should suppose a thing which 'comes to pass' must be completed, and cannot contain what is to happen 1800 years after. All who have any knowledge of English idiom know that a thing *comes* to pass when it happens, and *came* to pass afterwards. But as the original is Greek, we must look at the Greek: it is δει γενεσθαι for 'must come to pass,' and we know that εγενετο is what is usually translated 'came to pass.' No word of more finished completion exists in Greek.

And now for a last round of biter-bit with the Thor-hammerer, of whom, as in the other case, I shall take no more notice until he can contrive to surpass himself, which I doubt his being able to do. He informs me that by changing A into ח in my name he can make a 666 of *me*; adding, 'This is too hard for me, although not so for the Lord!' Sheer nonsense! He could just as easily have directed to 'Prof. De Morgחn' as have assigned me apartment 7A in University College. It would have been seen for whom it was intended: and if not, it would still have reached me, for my colleagues have for many a year handed all out-of-the-way things over to me. There is no 7A: but 7 is the Museum of Materia Medica. I took the only hint which the address gave: I inquired for hellebore, but they told me it was not now recognized, that the old notion of its value was quite obsolete, and that they had nothing which was considered a specific in senary or septenary cases. The great platitude is the reference of such a difficulty as writing ח for A to the Almighty! Not childish, but fatuous: real childishness is delightful. I knew an infant to whom, before he could speak plain, his parents had attempted to give notions of the Divine attributes: a wise plan, many think. His father had dandled him up-side-down, ending with, There now! Papa could not dance on his head! The mannikin made a solemn face, and said, *But Dod tood!* I think the Doctor has rather mistaken the way of becoming as a little child, intended in Matt. xviii. 3 : let us hope the will may be taken for the deed.

Two poets have given images of transition from infancy to manhood : Dryden,—for the Hind is Dryden himself on all fours; and Wordsworth, in his own character of broad-nailed, featherless biped :—

> The priest continues what the nurse began,
> And thus the child imposes on the man.

> The child 's the father of the man,
> And I could wish my days to be
> Bound each to each by natural piety.

In Wordsworth's aspiration it is meant that sense and piety should grow together : in Dryden's description a combination of Mysticism And Bigotry (can this be the *double Vahu*?), personified as 'the priest,'—who always catches it on this score, though the same spirit is found in all associations,—succeeds the boguey-teaching of the nurse. Never was the contrast of smile and scowl, of light and darkness, better seen than in the two pictures. But an acrostic distinction may be drawn. When mysticism predominates over bigotry, we have the grotesque picturesque, and the natural order of words gives us *Mab*, an appropriate suggestion. But when bigotry has the upper hand, we see *Bam*, which is just as appropriate ; for bigotry nearly always deals with facts and logic so as to require the application of at least one of the minor words by which dishonesty is signified. I think that M is the Doctor's initial, and that Queen Mab tickles him in his sleep with the sharp end of a 6.

(*Monday, August* 21.) Three weeks having elapsed without notice from me of the Doctor, I receive a reminder of his existence, in which I find that as I am the Daniel who judges the Magi of Babylon, it is to be pointed out that Daniel ' bore a certain number, that of a man (beloved), Daniel, ch. 10. v. 11, and which you certainly do not.' Then, ' by Greek power,' Belteshazzar is made = 666. Here is another awkward imitation of the way of a baby child. When you have sported with the tiny creature until it runs away offended, by the time you have got into conversation again you will find the game is to be renewed : a little head peeps out from a hiding-place with ' I don't love you.' The proper rejoinder is, ' Very well ! then I 'll have pussy.' But in the case before me there is a rule of three sum to do ; as baby : pussy Dr. :: 666 : the answer required. I will work it out, if I can.

D D

The squaring of the circle and the discovery of the Beast are the two goals—and gaols also—of many unbalanced intellects, and of a few instances of a better kind. I might have said more of 666, but I am not deep in its bibliography. A work has come into my hands which contains a large number of noted cases : to some of my readers it will be a treat to see the collection ; and the sight will perhaps be of some use to those who have read controversy on the few celebrated cases which are of general notoriety. It is written by a learned decipherer, a man who really knew the history of his subject, the Rev. David Thom, of Bold Street Chapel, Liverpool, who died, I am told, a few years ago.

Anybody who reads his book will be inclined to parody a criticism which was once made on Paley's Evidences—' Well ! if there be anything in Christianity, this man is no fool.' And, if he should chance to remember it, he will be strongly reminded of a sentence in my opening chapter,—' The manner in which a paradoxer will show himself, as to sense or nonsense, will not depend upon what he maintains, but upon whether he has or has not made a sufficient knowledge of what has been done by others, *especially as to the mode of doing it*, a preliminary to inventing knowledge for himself.' And this is reinforced by the fact that Mr. Thom, though a scholar, was not conspicuous for learning, except in this his great pursuit. He was a paradoxer on other points. He reconciled Calvinism and eternal reprobation with Universalism and final salvation ; showing these two doctrines to be all one.

This gentleman must not be confounded with the Rev. John Hamilton Thom (no relation), at or near the same time, and until recently, of Renshaw Street Chapel, Liverpool, who was one of the minority in the Liverpool controversy when, nearly thirty years ago, *three* heretical Unitarian schooners exchanged shotted sermons with *thirteen* Orthodox ships of the line, and put up their challengers' dander—an American corruption of *d—d anger* —to such an extent, by quiet and respectful argument, that those opponents actually addressed a printed intercession to the Almighty for the Unitarian triad, as for ' Jews, Turks, Infidels, and Heretics.' So much for the distinction, which both gentlemen would thank me for making very clear : I take it quite for granted that a guesser at 666 would feel horrified at being taken for a Unitarian, and that a Unitarian would feel queerified at being taken for a guesser at 666. Mr. David Thom's book is ' The Number and Names of the Apocalyptic Beasts,' Part I. 1848, 8vo. : I think the second part was never published. I give the Greek and Latin solutions, omitting the Hebrew : as usual, all the Greek letters are numeral, but only M D C L X V I of the

Latin. I do not give either the decipherers or their reasons : I have not room for this ; nor would I, if I could, bias my reader for one rather than another.

D. F. Julianus Cæsar Atheus (or Aug.) ; Diocles Augustus ; Ludovicus ; Silvester Secundus ; Linus Secundus ; Vicarius Filii Dei ; Doctor et Rex Latinus ; Paulo V. Vice-Deo ; Vicarius Generalis Dei in Terris ; Ipse Catholicæ Ecclesiæ Visibile Caput ; Dux Cleri ; Una, Vera, Catholica, Infallibilis Ecclesia ; Auctoritas politica ecclesiasticaque Papalis (Latina will also do) ; Lutherus Ductor Gregis ; Calvinus tristis fidei interpres ; Dic Lux ; Ludvvic ; Will. Laud ; Λατεινος ; ἡ λατινη βασιλεια ; εκκλησια ιταλικα ; ευανϑας ; τειταν ; αρνουμε ; λαμπετις ; ὁ νικητης ; κακος ὁδηγος ; ἀληθης βλαβερος ; παλαι βασκανος ; ἀμνος ἀδικος ; ἀντεμος ; γενσηρικος ; ευινας ; Βενεδικτος ; Βονιβαζιος γ. παπα ξ. η. ε. ε. a., meaning Boniface III. Pope 68th, bishop of bishops the first ! οὐλπιος ; διος εἰμι ; ἡ ἡρας ; ἡ μισσα ἡ παπικη ; λουϑ=ρανα ; σοξονειος ; Βεζζα ἀντιϑεος (Beza) ; ἡ ἀλαζονεια βιου ; Μαομετις ; Μαομετης β. ; θεος εἰμι ἐπι γαιης ; ἰαπετος ; παπεισκος ; διοκλασιανος ; χεινα ; βρασκι ; Ιον Παυνε ; κουποκς (cowpox, s being the vau ; certainly the vaccinated have the mark of the Beast) ; Βοννεπαρτη ; N. Βοιηπαρτε ; εὐπορια ; παραδοσις ; το μεγαϑηριον.

All sects fasten this number on their opponents. It is found in *Martin Lauter*, affirmed to be the true way of writing the name, by carrying numbers through the Roman Alphabet. Some Jews, according to Mr. Thom, found it in ישו נצרי *Jesus of Nazareth*. I find on inquiry that this satire was actually put forth by some mediæval rabbis, but that it is not idiomatic : it represents quite fairly ' Jesus Nazarene,' but the Hebrew wants an article quite as much as the English wants ' the.'

Mr. David Thom's own solution hits hard at all sides : he finds a 666 for both beasts ; ἡ φρην (the mind) for the first, and ἐκκλησιαι σαρκικαι (fleshly churches) for the second. A solution which embodies all mental philosophy in one beast and all dogmatic theology in the other, is very tempting : for in these are the two great supports of Antichrist. It will not, however, mislead me, who have known the true explanation a long time. The three sixes indicate that any two of the three subdivisions, Roman, Greek, and Protestant, are, in corruption of Christianity, six of one and half a dozen of the other : the distinctions of units, tens, hundreds, are nothing but the old way (1 Samuel xviii. 7, and Concordance at *ten*, *hundred*, *thousand*) of symbolizing differences of number in the subdivisions.

It may be good to know that, even in speculations on 666,

there are different degrees of unreason. All the diviners, when they get a colleague or an opponent, at once proceed to reckon him up : but some do it in play and some in earnest. Mr. David Thom found a young gentleman of the name of St. Claire busy at the Beast number : he forthwith added the letters in στ κλαιρε and found 666 : this was good fun. But my spiritual tutelary, when he found that he could not make a beast of me, except by changing א into ה, solemnly referred the difficulty to the Almighty : this was poor earnest.

I am glad I did not notice, in time to insert it in the *Athenæum*, a very remarkable paradoxer brought forward by Mr. Thom, his friend Mr. Wapshare : it is a little too strong for the general public. In the *Athenæum* they would have seen and read it : but this book will be avoided by the weaker brethren. It is as follows :—

' God, the Elohim, was six days in creating all things, and having made MAN, he entered into his rest. He is no more seen as a Creator, as Elohim, but as Jehovah, the *Lord* of the Sabbath, and the Spirit of life in MAN, which Spirit worketh *sin in the flesh* ; for the Spirit of Love, in all flesh, is Lust, or the spirit of a beast, So Rom. vii. And which Spirit is *crucified* in the flesh. He then, as Jehovah—as the power of the Law, *in* and *over* all flesh, John viii. 44—increases that which he has made as the Elohim, and his power shall last for 6 days, or 6 periods of time, computed at a millennium of years ; and at the end of which six days, he who is the Spirit of all flesh shall manifest himself as the Holy Spirit of Almighty Love, and of all truth ; and so shall the Church have her Sabbath of Rest—all contention being at an end. This is, as well as I may now express it, my solution of the mystery in Hebrew, and in Greek, and also in Latin, I H S. For he that was lifted up *is* King of the Jews, and is the Lord of all Life, working in us, both to will and to do ; as is manifest in the Jews—they slaying him that his blood might be *good* for the healing of the nations, of all people and tongues. As the Father of all *natural* flesh, he is the Spirit of Lust, as in all *beasts* ; as the Father, or King of the Jews, he is the Devil, as he himself witnesseth in John viii., already referred to. As lifted up, he is transformed into the Spirit of Love, a light to the Gentiles, and the glory of his people Israel . . . For there is but ONE God, ONE Lord, ONE Spirit, ONE body, &c. and he who was Satan, the Spirit of life in that body, is, in Christ crucified, seen in the Spirit that is in all, and through all, and over all, God blessed for ever.'

All this seems well meant, and Mr. Thom prints it as convinced of its piety, and ' pronounces no opinion.' Mystics of all sorts ! see what you may come to, or what may come to you ! I have inserted the above for your good.

There is nothing in this world so steady as some of the paradoxers. They are like the spiders who go on spinning after they have web enough to catch all the flies in the neighbourhood, if the flies would but come. They are like the wild bees who go on making honey which they never can eat, proving *sic vos non vobis* to be a physical necessity of their own contriving. But nobody robs their hives : no, unlike the bees, they go about offering their ware to any who will take it at a gift. I had just written the last sentence (Oct. 30, 1866, 8·45 A.M.) when in comes the second note received this morning from Dr. Thorn : at 1·30 P.M. came in a third. These arise out of the above account of the Rev. D. Thom, published Oct. 27 : three notes had arrived before.

For curiosity I give one day's allowance, supposing these to be all : more may arrive before night.

29th Oct. 1866.

Dear Sir,—

In re ⊥.

So that 'Zaphnath Paaneah' may be after all the revealer of the Northern Tau,' Φανερоω—To make manifest, shew, or explain; and this may satisfy the House of Joseph in Amos 5ᶜ. While Belteshazzar =666 may be also satisfactory to the House of David, and so we may have Zech. 10ᶜ. 6ᵛ. in operation when Ezekiel 37ᶜ. 16ᵛ. has been realised ;—but there, what is the use of writing, it is al Coptic to a man who has not ⊥, The Thau of the North, the double Vahu וו. Look at Jeremiah 3ᶜ. 8ᵛ. and then to Psalm 83 for 'hidden ones' צְפוּנֶיךָ.יְהוָֹה—The Zephoni Jehovah, and say whether they have any connection with the Zephon *Thau.* The Hammer of Thor of Jeremiah 23ᶜ. 29ᵛ. as I gave you in No. 3 of my present edition.

Yours truly

LE CHEVALIER AU CIN.

*By Greek Power.*

|  |  |  |
|---|---|---|
| C | = | 20 |
| H | = | 8 |
| E | = | 5 |
| V | = | 6 |
| A | = | 1 |
| L | = | 30 |
| I | = | 10 |
| E | = | 5 |
| R | = | 100 |
| A | = | 1 |
| U | = | 400 |
| C | = | 20 |
| I | = | 10 |
| N | = | 50 |
|  |  | 666 |

There will be thousands of Morgans who will be among the wise and prudent of Hosea 14ᶜ. 9ᵛ. when the Seventh Angel sounds, let me number *that One* by Greek, Rev. 17ᶜ. 1ᵛ :—

| | | |
|---|---|---|
| S | = | 200 |
| E | = | 5 |
| × V | = | 6 |
| E | = | 5 |
| N | = | 50 |
| T | = | 300 |
| H | = | 8 |
| A | = | 1 |
| N | = | 50 |
| × G | = | 6 |
| E | = | 5 |
| L | = | 30 |
| | | —— |
| | | 666 |

V and G = 12 ought to be equal to one Gamma-dion or $^{3}+3 \times 4 = 12$, what say you?

London, October 29, 1866.

Dear Sir,—

In re ✝ versus ✠.

However pretentious the X or ✠ may be, and it is peculiarly so just now in this land; after all it is only made of two Roman V's—and so is only $= \bigvee_{\wedge}$ (10)—and therefore is not the perfect number 12 of Revel$^{n}$, but is the mark of the goddess *Decima* !

Yours truly

Wm. THORN.

Had the *one* who sent forth a pastoral (Romish) the other day, remained amongst the faithful expectants, see how he would have numbered, whereas he sold himself for the privilege of signing

✠ HENRY E. MANNING.

Shilling *versus* Franc.
Teutonic Long Hundred 120 *versus* 100 or the Decimal question.

*By English Key.*

| | | |
|---|---|---|
| H | = | 8 |
| E | = | 5 |
| N | = | 40 |
| R | = | 80 |
| Y | = | 140 |
| | | |
| E | = | 5 |
| D | = | 4 |
| W | = | 120 |
| A | = | 1 |
| R | = | 80 |
| D | = | 4 |
| | | |
| M | = | 30 |
| A | = | 1 |
| N | = | 40 |
| N | = | 40 |
| I | = | 9 |
| N | = | 40 |
| G | = | 7 |
| ✠ | = | 12 |
| | | 666 |

Can you now understand the difference between ✠ and ✠ or X? Look to my challenge.

Cutting from newspaper :—

## ITALY.

Rome (*viâ* Marseilles), October 24.

Mr. Gladstone has paid a visit to the Pope.

*By Greek Power.*

| | | |
|---|---|---|
| G | = | 6 |
| L | = | 30 |
| A | = | 1 |
| D | = | 4 |
| S | = | 200 |
| T | = | 300 |
| O | = | 70 |
| N | = | 50 |
| E | = | 5 |
| | | 666 |

And what then ✠ ?

In other letters *John Stuart Mill* is 666 if the *a* be left out;
*Chasuble* is perfect.   *John Brighte* is a *fait accompli* ; and I am
asked whether intellect can account for the final *e.*   Very easily :
this Beast is not the M. P., but another person who spells his
name differently.   But if John Sturt Mill and John Brighte
choose so to write themselves, they may.

A curious collection ; a mystical phantasmagoria !   There are
those who will try to find meaning : there are those who will try
to find purpose.

> And some they said—What are you at ?
> And some—What are you arter ?

My account of Mr. Thom and his 666 appeared on October 27 :
and on the 29th I received from the editor a copy of Mr. Thom's
sermons published in 1863 (he died Feb. 27, 1862) with best
wishes for my health and happiness.   The editor does not name
himself in the book ; but he signed his name in my copy : and
may my circumference never be more than $3\frac{1}{3}$ of my diameter if
the signature, name and writing both, were not that of my
$\odot$ $\square$ing friend Mr. James Smith !   And so I have come in contact
with him on 666 as well as on $\pi$ !   I should have nothing left to
live for, had I not happened to hear that he has a perpetual
motion on hand.   I returned thanks and kind regards : and
Miss Miggs's words—' Here's  forgivenesses of injuries !  here's
amicablenesses ! '—rang in my ears.   But I was made slightly
uncomfortable : how could the war go on after this armistice ?
Could I ever make it understood that the truce only extended to
the double Vahu and things thereunto relating ?   It was once
held by seafaring men that there was no peace with Spaniards
beyond the line : I was determined that there must be no concord
with J. S. inside the circle ; that this must be a special exception,
like Father Huddleston and old Grouse in the gun-room.   I was
not long in anxiety ; twenty-four hours after the book of sermons
there came a copy of the threatened exposure—' The British
Association in Jeopardy, and Professor De Morgan in the Pillory
without hope of escape.   By James Smith, Esq.'   London and
Liverpool, 8vo., 1866 (pp. 94).   This   exposure   consists   of
reprints from the *Athenæum* and *Correspondent* : of things new
there is but one.   In a short preface Mr. J. S. particularly recom-
mends to ' *read to the end.*'   At the end is an appendix of two
pages, in type as large as the work ; a very prominent peroration.
It is an article from the *Athenæum,* left out of its place.   In the
last sentence Mr. J. Smith, who had asked whether his character

as an honest Geometer and Mathematician was not at stake, is warned against the *fallacia plurium interrogationum*. He is told that there is not a more honest what's-his-name in the world : but that as to the counter which he calls his character as a mathematician, he is assured that it had been staked years ago, and lost. And thus truth has the last word. There is no occasion to say much about reprints. One of them is a letter [that given above] of August 25, 1865, written by Mr. J. S. to the *Correspondent*. It is one of his quadratures; and the joke is that I am made to be the writer : it appears as what Mr. J. S. hopes I shall have the sense to write in the *Athenæum* and forestall him. When I saw myself thus quoted—yes! quoted! double commas, first person—I felt as I suppose did Wm. Wilberforce when he set eyes on the affectionate benediction of the potato which waggish comrades had imposed on a raw Irish reporter as part of his speech    I felt as Martin of Galway—kind friend of the poor dumb creatures!—when he was told that the newspapers had put him in Italics. 'I appeal to you, Mr. Speaker! I appeal to the House! Did I speak in Italics? Do I ever speak in Italics?' I appeal to editor and readers, whether I ever squared the circle until a week or two ago, when I gave my charitable mode of reconciling the discrepant cyclometers.

The absurdity of the imitation of symbolic reasoning is so lusciously rich, that I shall insert it when I make up my final book.    Somebody mastered Spanish merely to read Don Quixote : it would be worth while to learn a little algebra merely to enjoy this $a$ $b$-istical attack on the windmills. The principle is, Prove something in as roundabout a way as possible, mention the circle once or twice irrelevantly in the course of your proof, and then make an act of Q. E. D. in words at length. The following is hardly caricature :—

To prove that 2 and 2 make 5. Let $a = 2$, $b = 5$ : let $c = 658$, the number of the House : let $d = 666$, the number of the Beast. Then of necessity $d = a + b + c + 1$; so that 1 is a harmonious and logical quantification of the number of which we are to take care. Now, $b$, the middle of our digital system, is, by mathematical and geometrical combination, a mean between $5 + 1$ and $2 + 2$. Let 1 be removed to be taken care of, a thing no real mathematician can refuse without serious injury to his mathematical and geometrical reputation. It follows of necessity that $2 + 2 = 5$, *quod erat demonstrumhorrendum*. If Simpkin & Marshall have not, after my notice, to account for a gross of copies more than would have gone off without me, the world is not worthy of its James Smith!

The only fault of the above is, that there is more connexion than in the process of Faber Cyclometricus : so much, in fact, that the blunders are visible. The utter irrelevance of premises to conclusion cannot be exhibited with the requisite obscurity by any one who is able to follow reasoning : it is high art displayed in a certain toning down of the *ægri somnia*, which brings them to a certain look of approach to reasoning which I can only burlesque. Mr. J. S. produces something which resembles argument much as a chimpanzee in dolour, because balked of his dinner, resembles a thinking man at his studies. My humble attempt at imitation of him is more like a monkey hanging by his tail from a tree and trying to crack a cocoa-nut by his chatter.

I could forgive Mr. J. S. anything, properly headed. I would allow him to prove—*for himself*—that the Quadrature of the Circle is the child of a private marriage between the Bull Unigenitus and the Pragmatic Sanction, claiming tithe of onions for repeal of the Mortmain Act, before the Bishops in Committee under the kitchen table : his mode of imitating reason would do this with ease. But when he puts his imitation into my mouth, to make me what *he* calls a ' real mathematician,' my soul rises in epigram against him. I say with the doll's dressmaker—such a job makes me feel like a puppet's tailor myself—' He ought to have a little pepper? just a few grains? I think the young man's tricks and manners make a claim upon his friends for a little pepper?' De Fauré and Joseph Scaliger come into my head : my reader may look back for them.

> Three circlesquarers to the manner born,
> Switzerland, France, and England did adorn,
> De Fauré in equations did surpass, (p. 89)
> Joseph at contradictions was an ass. (p. 67)
> Groaned Folly, I'm used up! What shall I do
> To make James Smith? Grinned Momus, *Join the two*!

As to my *locus pœnitentiæ*, the reader who is fit to enjoy the letter I have already alluded to will see that I have a soft and easy position ; that the thing is really a *pillowry*; and that I am, like Perrette's pot of milk,

> Bien posé sur un coussinet.

Joanna Southcott never had a follower who believed in her with more humble piety than Mr. James Smith believes in himself. After all that has happened to him, he asks me with high confidence to ' favour the writer with a proof' that I still continue of opinion that ' the best of the argument is in my jokes, and the best of the joke is in his arguments.' I will not so favour him. At the very outset I told him in plain English that he has the whiphand of all the reasoners in the world, and in plain French that *il a perdu le droit d'être frappé de l'évidence*; I might have said *pendu*. To which I now add, in plain Latin, *Sapienti pauca, indocto nihil.* The law of Chancery says that he who will have equity must do equity: the law of reasoning says that he who will have proof must see proof.

The introduction of things quite irrelevant, by way of reproach, is an argument in universal request : and it often happens that the argument so produced really tells against the producer. So common is it that we forget how boyish it is ; but we are strikingly reminded when it actually comes from a boy. In a certain police court, certain small boys were arraigned for conspiring to hoot an obnoxious individual on his way from one of their school exhibitions. This proceeding was necessary, because there seemed to be a permanent conspiracy to annoy the gentleman ; and the masters did not feel able to interfere in what took place outside the school. So the boys were arraigned ; and their friends, as silly in their way as themselves, allowed one of them to make the defence, instead of employing counsel ; and did not even give them any useful hints. The defence was as follows ; and any one who does not see how richly it sets off the defences of bigger boys in bigger matters has much to learn. The innocent conviction that there was answer in the latter part is delightful. Of course fine and recognizance followed.

A—— said the boys had received great provocation from B——. He was constantly threatening them with a horsewhip which he carried in his hand [the boy did not say what had passed to induce him to take such a weapon], and he had repeatedly insulted the master, which the boys could not stand. B—— had in his own drawing-room told him (A——) that he had drawn his sword against the master and thrown away the scabbard. B—— knew well that if he came to the college he would catch it, and then he went off through a side door—which was no sign of pluck ; and then he brought Mrs. B—— with him, thinking that her presence would protect him.

My readers may expect a word on Mr. Thom's sermons, after my account of his queer doings about 666. He is evidently an honest and devout man, much wanting in discrimination. He has a sermon about private *judgment,* in which he halts between the logical and legal meanings of the word. He loathes those who apply their private judgment to the word of God : here he means those who decide what it *ought to be.* He seems in other places aware that the theological phrase means taking right to determine what it *is.* He uses his own private judgment very freely, and is strong in the conclusion that others ought not to use theirs except as he tells them how ; he leaves all the rest of mankind free to think with him. In this he is not original : his fame must rest on his senary tripod.

Mr. James Smith's procedures are not caricature of reasoning ; they are caricature of blundering. The old way of proving that $2 = 1$ is solemn earnest compared with his demonstrations. As follows :

Let $x = 1$
Then $x^2 = x$
And $x^2 - 1 = x - 1$

Divide both sides by $x - 1$ ; then

$x + 1 = 1$ ; but $x = 1$, whence $2 = 1$

When a man is regularly snubbed, bullied, blown up, walked into, and put down, there is usually some reaction in his favour, a kind of deostracism, which cannot bear to hear him always called the blunderer. I hope it will be so in this case. There is nothing I more desire than to see *sects* of paradoxers. There are fully five thousand adults in England who ought to be the followers of some one false quadrature. And I have most hope of $3\frac{1}{8}$, because I think Mr. James Smith better fitted to be the leader of an organised infatuation than any one I know of. He wants no pity, and will get none. He has energy, means, good humour, strong conviction, character, and popularity in his own circle. And, most indispensable point of all, he sticks at nothing ;

In cœlum jusseris, ibit.

When my instructor found I did not print an acceptance of what I have quoted, he addressed me as follows (*Corr.*, Sept. 23):—

' In this life, however, we must do our duty, and, when necessary, use the rod, not in a spirit of revenge, but for the benefit of the culprit and the good of society. Now, Sir, the opportunity has been thrown in your way of slipping out of the pillory without risk of serious injury ; but, like an obstinate urchin, you have chosen to quarrel with your opportunity and remain there, and thus you compel me to deal with you as schoolmasters used to do with stupid boys in bygone days —that is to say, you force me to the use of the critic's rod, compel me to put you where little Jack Horner sat, and, as a warning to other naughty boys, to ornament you with a dunce's cap. The task I set you was a very simple one, as I shall make manifest at the proper time.'

In one or more other places, as well as this, Mr. Smith shows that he does not know the legend of little Jack Horner, whom he imagines to be put in the corner as a bad boy. This is curious ; for there had been many allusions to the story in the journal he was writing in, and the Christmas pie had become altered into the Seaforth $\pi$.

Mr. Smith is satisfied at last that—what between argument and punishment he has convinced me. He says (*Corr.*, Jan. 27, 1866) ' I tell him without hesitation that he knows the true ratio of diameter to circumference as well as I do, and if he be wise he will admit it.' I should hope I do, and better ; but there is no occasion to admit what everybody knows.

I have often wished that we could have a slight glimpse of the reception which was given to some of the old cyclometers: but we have nothing, except the grave disapprobation of historians. I am resolved to give the New Zealander a chance of knowing a little more than this about one of them at least ; and, by the fortunate entrance into life of the *Correspondent*, I am able to do it. I omit sober mathematical answers, of which there were several. The following letter is grave earnest :—

' Sir,—I have watched Mr. James Smith's writings on this subject from the first, and I did hope that, as the more he departs from truth the more easy it must be to refute him, [this by no means always true] some of your correspondents would by this time have done so. I own that I am unable to detect the fallacy of his argument; and I am quite certain that ' II ' is wrong, in No. 23, where he declares that Mr. Smith is ' ignorant of the very elements of mathematical truth.' I have observed an immense amount of geometrical reasoning on his part, and I cannot see that it is either fair or honest to deny this, which may be regarded as the ' elements ' of mathematical truth. Would it not be better for ' II ' to answer Mr. Smith, to refute his arguments, to point out their fallacies, and to save learners from error, than to plunge into gross insult and unmanly abuse ? Would it not be well, also, that Professor De Morgan should favour us with a little reasoning ?

I have hitherto seen no attempt to overthrow Mr. Smith's arguments ; I trust that this will not continue, since the subject is one of immense importance to science in general, especially to nautical science, and all that thereto belongs.    Yours, &c.,

A CAPTAIN, R.N.'

On looking at this homœopathic treatment of the $3\frac{1}{8}$ quadrature—remember, homœopathic, *similia similibus*, not infinitesimal—and at the imputation thrown upon it, I asked myself, what *is* vulgarity? No two agree, except in this, that every one sees vulgarity in what is directed against himself. Mark the world, and see if anything be so common as the description of the other side's remarks as 'vulgar attempt at wit.' 'I suppose you think that very witty:' the answer is 'No my friend! your remark shows that you feel it as wit, so that the purpose is answered; I keep my razor for something else than cutting blocks;' I am inclined to think that 'out of place' is a necessary attribute of true vulgarity. And further, it is to be noticed that nothing is unproducible—*salvo pudore*—which has classical authority, modern or ancient, in its favour 'He is a vulgar fellow; I asked him what he was upon, and what do you think he answered, My legs!'—'Well, and has he not justification? what do you find in Terence? *Quid agitur? Statur.*' I do not even blench from my principle where I find that it brings what is called 'taking a sight' within permissible forms of expression: Rabelais not only establishes its antiquity but makes it English. Our old translation [1] has it thus (book 2, ch. 19):—

'Then made the Englishman this sign. His left hand, all open, he lifted up into the air, then instantly shut into his fist the four fingers thereof; and his thumb extended at length he placed upon the tip of his nose. Presently after he lifted up his right hand all open and abased and bent it downwards, putting the thumb thereof in the very place where the little finger of the left hand did close in the fist, and the four right hand fingers he softly moved in the air. Then contrarily he did with the right hand what he had done with the left, and with the left what he had done with the right.'

An impressive sight! The making a fist of the left hand is a great addition of power, and should be followed in modern practice. The gentle sullation of the front fingers, with the clenched fist behind them, says as plainly as possible, Put *suaviter in modo* in the van, but don't forget to have *fortiter in re* in the rear.

My Budget was announced (March 23, 1867) for completion on the 30th. Mr. James Smith wrote five letters, one before the completion, four after it; the five contained 68 pages of quarto

---

[1] Lors feist l'Anglois tel signe. La main gausche toute ouverte il leva hault en l'aer, puis ferma au poing les quatres·doigtz d'icelle et le poulce estendu assit sus la pinne du nez. Soubdain après leva la dextre toute ouverte, et toute ouverte la baissa, joignant la poulce au lieu que fermait le petit doigt de la gausche, et les quatre doigtz d'icelle mouvoit lentement en l'aer. Puis au rebours feit de la dextre ce qu'il avoit faict de la gausche, et de la gausche ce que avoit faict de la dextre.

letter paper. Mr. J. S. had picked up a clerical correspondent, with whom he was in the heat of battle.—

*March* 27.—Dear Sir. Very truly yours. Duty; for my own sake; just time left to retrieve my errors; sends copy of letter to clergyman; new proof never before thought of; merest tyro would laugh if I were to stifle it, whether by rhodomontade or silent contempt; keep your temper. I shall be convinced; and if world be right in supposing me incapable of a foul act, I shall proclaim glorious discovery in the *Athenæum*.

*April* 15.—Sir, . . . My dear Sir, Your sincere tutelary. Copy of another letter to clergyman; discovery tested by logarithms; reasons such as none but a knave or a sinner can resist. Let me advise you to take counsel before it is too late! Keep your temper. Let not your *pride* get the better of your discretion! Screw up your courage, my good friend, and *resolve* to show the world that you are an *honest* man . . .

*April* 20.—Sir . . . Your very sincere and favourite tutelary. I have long played the *cur*, snapping and snarling . . .; suddenly lost my power, and become *half-starved* dog without *spirit* to bark; try if air cannot restore me; calls himself the *thistle* in allusion to my other tutelary, the *thorn*; Would I prefer his next work to be, ' A whip for the Mathematical Cur, Prof. De M.' In some previous letter, which I have mislaid, he told me his next would be ' a muzzle for the Mathematical Bull dog, Prof. De M.'

*April* 23.—Sir. Very sincerely yours. More letters to clergyman; you may as well knock your head against a stone wall to improve your intellect as attempt to controvert my proofs. [I thought so too; and tried neither].

*May* 6.—My dear Sir. Very sincerely yours. All to myself, and nothing to note.

*July* 2.—No more in this interval. All that precedes is a desperate attempt to induce me to continue my descriptions: notoriety at any price.

I dare say the matter is finished : the record of so marked an instance of self-delusion will be useful.

I append to the foregoing a letter from Dr. Whewell to Mr. James Smith. The Master of Trinity was conspicuous as a rough customer, an intellectual bully, an overbearing disputant : the character was as well established as that of Sam Johnson. But there was a marked difference. It was said of Johnson that if his pistol missed fire, he would knock you down with the butt end of it : but Whewell, in like case, always acknowledged the miss, and loaded again or not, as the case might be. He reminded me of Dennis Brulgruddery, who says to Dan, Pacify me with a good reason, and you'll find me a dutiful master. I knew him from the time when he was my teacher at Cambridge, more than forty years. As a teacher, he was anything but dictatorial, and he was perfectly accessible to proposal of objections. He came in contact with me in his slashing way twice in our after joint lives, and on both occasions he acknowledged himself overcome, by that change of manner, and apologetic mode of continuance, which I had seen him employ towards others under like circumstances.

I had expressed my wish to have a *thermometer of probability*, with impossibility at one end, as 2 and 2 make 5, and necessity at the other, as 2 and 2 make 4, and a graduated rise of examples between them. Down came a blow: 'What! put necessary and contingent propositions together! It's absurd!' I pointed out that the two kinds of necessity are but such extremes of probability as 0 and ∞ are of number, and illustrated by an urn with 1 white and $n$ black balls, $n$ increasing without limit. It was frankly seen, and the point yielded; a large company was present.

Again, in a large party, after dinner, and politics being the subject, I was proceeding, in discussion with Mr. Whewell, with ' I think' . . .—' Ugh! *you* think!' was the answer. I repeated my phrase, and gave as a reason the words which Lord Grey had used in the House of Lords the night before (the celebrated advice to the Bishops to set their houses in order). He had not heard of this, and his manner changed in an instant: he was the rational discutient all the rest of the evening, having previously been nothing but a disputant with all the distinctions strongly marked.

I have said that Whewell was gentle with his pupils; it was the same with all who wanted teaching: it was only on an armed enemy that he drew his weapon. The letter which he wrote to Mr. J. Smith is an instance: and as it applies with perfect fidelity to the efforts of unreasoning above described, I give it here. Mr. James Smith is skilfully exposed, and felt it; as is proved by ' putting the writer in the stocks.'—

The Lodge, Cambridge, September 14th, 1862.

Sir,—I have received your explanation of your proposition that the circumference of the circle is to its diameter as 25 to 8. I am afraid I shall disappoint you by saying that I see no force in your proof: and I should hope that you will see that there is no force in it if you consider this :—In the whole course of the proof, though the word circle occurs, there is no property of the circle employed. You may do this : you may put the word *hexagon* or *dodecagon*, or any other word describing a polygon in the place of *Circle* in your proof, and the proof would be just as good as before. Does not this satisfy you that you annot have proved a property of that special figure—a circle?

Or you may do this : calculate the side of a polygon of 24 sides inscribed in a circle. I think you are a Mathematician enough to do this. You will find that if the radius of the circle be one, the side of this polygon is ·264 &c. Now, the arc which this side subtends is according to your proposition $\frac{3\cdot125}{12} = \cdot2604$, and therefore the chord is greater than its arc, which you will allow is impossible.

I shall be glad if these arguments satisfy you, and

I am, Sir, your obedient Servant,

W. WHEWELL.

In the debate of May, 1866, on Electoral Qualifications, a question arose about arithmetical capability. Mr. Gladstone asked how many members of the House could divide 1330l. 17s. 6d. by 2l. 13s. 8d. Six hundred and fifty-eight, answered one member ; the thing cannot be done, answered another. There is an old paradox to which this relates : it arises out of the ignorance of the distinction between abstract and concrete arithmetic. *Magnitude* may be divided by *magnitude* ; and the answer is *number* : how often does 12d. contain 4d. ; answer three *times*. *Magnitude* may be divided by *number*, and the answer is *magnitude* : 12d. is divided in four equal parts, what is each part ? Answer three *pence*. The honourable objector, whose name I suppress, trusting that he has mended his ways, gave the following utterance :—

" With regard to the division sum, it was quite possible to divide by a sum, but not by money. How could any one divide money by 2l. 16s. 8d. ? (Laughter.) The question might be asked, ' How many times 2s. will go into 1l. ? ' but that was not dividing by money ; it was simply dividing 20 by 2. He might be asked, ' How many times will 6s. 8d. go into a pound ? ' but it was only required to divide 240 by 80. If the right hon. gentleman were to ask the hon. member for Brighton (Professor Fawcett), or any other authority, he would receive the same answer—viz., that it was possible to divide by a sum, but not by money. (Hear.) "

I shall leave all comment for my second edition, if I publish one. I shall be sure to have something to laugh at. Anything said from a respectable quarter, or. supposed to be said, is sure to find defenders. Sam Johnson, a sound arithmetician, comparing himself, and what he alone had done in three years, with the forty French Academicians and their forty years, said it proved that an Englishman is to a Frenchman as 40 × 40 to 3, or as 1600 to 3. Boswell, who was no great hand at arithmetic, made him say that

E E

an Englishman is to a Frenchman as 3 to 1600. When I pointed this out, the supposed Johnson was defended through thick and thin in *Notes and Queries*.

I am now curious to see whether the following will find a palliator. It is from 'Tristram Shandy,' book v. chapter 3. There are two curious idioms, 'for for' and 'half in half;' but these have nothing to do with my point :—

'A blessing which tied up my father's tongue, and a misfortune which set it loose with a good grace, were pretty equal : sometimes, indeed, the misfortune was the better of the two ; for, for instance, where the pleasure of the harangue was as *ten*, and the pain of the misfortune but as *five*, my father gained half in half; and consequently was as well again off as if it had never befallen him.'

This is a jolly confusion of ideas ; and wants nothing but a defender to make it perfect. A person who invests five with a return of ten, and one who loses five with one hand and gains ten with the other, both leave off five richer than they began, no doubt. The first gains 'half in half,' more properly 'half *on* half,' that is, of the return, 10, the second 5 is gain upon the first 5 invested. 'Half *in* half' is a queer way of saying cent. per cent. If the 5*l.* invested be all the man had in the world, he comes out, after the gain, twice as well off as he began, with reference to his whole fortune. But it is very odd to say that balance of 5*l.* gain is *twice* as good as if nothing had befallen, either loss or gain. A mathematician thinks 5 an infinite number of times as great as 0. The whole confusion is not so apparent when money is in question : for money is money whether gained or lost. But though pleasure and pain stand to one another in the same algebraical relation as money gained and lost, yet there is more than algebra can take account of in the difference.

Next, Ri. Milward (Richard, no doubt, but it cannot be proved) who published Selden's Table Talk, which he had collected while serving as amanuensis, makes Selden say, 'A subsidy was counted the fifth part of a man's estate ; and so fifty subsidies is five and and forty times more than a man is worth.' For *times* read *subsidies*, which seems part of the confusion, and there remains the making all the subsidies equal to the first, though the whole of which they are to be the fifths is perpetually diminished.

Thirdly, there is the confusion of the great misomath of our own day, who discovered two quantities which he avers to be identically the same, but the greater the one the less the other. He had a truth in his mind, which his notions of quantity were inadequate to clothe in language. This erroneous phraseology has not found a defender; and I am almost inclined to say, with Falstaff, The poor abuses of the time want countenance.

'Shallow numerists,' as Cocker is made to call them, have long been at work upon the question how to *multiply* money by money. It is, I have observed, a very common way of amusing the tædium of a sea voyage : I have had more than one bet referred to me. Because an oblong of five inches by four inches contain 5 × 4 or 20 *square* inches, people say that five inches multiplied by four inches *is* twenty *square* inches : and, thinking that they have multiplied length by length, they stare when they are told that money cannot be multiplied by money. One of my betters made it an argument for the thing being impossible, that there is no *square money* : what could I do but suggest that postage-stamps should be made legal tender. Multiplication must be *repetition* : the repeating process must be indicated by *number* of times. I once had difficulty in persuading another of my betters that if you repeat five shillings as often as there are hairs in a horse's tail, you do not *multiply five shillings by a horsetail*.

I am very sorry to say that these wrong notions have found support—I think they do so no longer—in the University of Cambridge. In 1856 or 1857, an examiner was displaced by a vote of the Senate. The pretext was that he was too severe an examiner : but it was well known that great dissatisfaction had been expressed, far and wide through the Colleges, at an absurd question which he had given. He actually proposed such a fraction as

$$\frac{6s.\ 3d.}{17s.\ 4d.}$$

As common sense gained a hearing very soon, there is no occasion to say more. In 1858, it was proposed at a college examination, to divide 22557 days, 20 hours, 20 minutes, 48 seconds, by 57 minutes, 12 seconds, and also to explain the fraction

$$\frac{32l.\ 18s.\ 8d.}{62l.\ 12s.\ 9d.}$$

All paradoxy, in matters of demonstration, arises out of muddle about first principles. Who can say how much of it is to be laid at the door of the University of Cambridge, for not taking care of the elements of arithmetical thought?

The phenomena of the two ends of society, when brought together, give interesting comparisons: I mean the early beginnings of thought and literature, and our own high and finished state, as we think it. There is one very remarkable point. In the early day, the letter was matter of the closest adherence, and implied meanings were not admitted.

The blessing of Isaac meant for Esau, went to false Jacob, in spite of the imposition ; and the writer of Genesis seems to intend to give the notion that Isaac had no power to pronounce it null and void. And ' Jacob's policy, whereby he became rich '—as the chapter-heading puts it—in speckled and spotted stock, is not considered as a violation of the agreement, which contemplated natural proportions. In the story of Lycurgus the lawgiver is held to have behaved fairly when he bound the Spartans to obey his laws until he returned—intimating a short absence—he intending never to return. And Vishnoo, when he asked the usurper for three steps of territory as a dwarf, and then enlarged himself until he could bring heaven and earth under the bargain, was thought clever, certainly, but quite fair.

There is nothing of this kind recognised in our day : so far good. But there is a bad contrary : the age is apt, in interpretation, to upset the letter in favour of the view—very often the after thought—of one side only. The case of John Palmer, the improver of the mail coach system, is smothered. He was to have an office and a salary, and 2½ per cent. for life on the increased *revenue* of the Post-Office. His rights turned out so large, that Government would not pay them. For misconduct, real or pretended, they turned him out of his *office* : but his bargain as to the percentage had nothing to do with his future conduct ; it was payment for his *plan*. I know nothing, except from the debates of 1808 in the two Houses : if any one can redeem the credit of the nation, the field is open. When I was young, the old stagers spoke of this transaction sparingly, and dismissed it speedily.

The government did not choose to remember what private persons must remember, and are made to remember, if needful. When Dr. Lardner made his bargain with the publishers for the *Cabinet Cyclopœdia* he proposed that he, as editor, should have a certain sum for every hundred sold above a certain number : the publishers, who did not think there was any chance of reaching the turning sale of this stipulation, readily consented. But it turned out that Dr. Lardner saw further than they : the returns under this stipulation gave him a very handsome addition to his other receipts. The publishers stared; but they paid. They had no idea of standing out that the amount was too much for an editor; they knew that, though the editor had a percentage, they had all the rest; and they would not have felt aggrieved if he had received ten times as much. But governments, which cannot be brought to book before a sworn jury, are ruled only by public opinion. John Palmer's day was also the day of Thomas Fyshe Palmer, and the governments, in their prosecutions for sedition, knew that these would have a reflex action upon the minds of all who wrote about public affairs.

1864–65.—It often happens that persons combine to maintain and enforce an opinion; but it is, in our state of society, a paradox to unite for the sole purpose of blaming the opposite side. To invite educated men to do this, and above all, men of learning or science, is the next paradoxical thing of all. But this was done by a small combination in 1864. They got together and drew up a *declaration*, to be signed by ' students of the natural sciences,' who were to express their ' sincere regret that researches into scientific truth are perverted by some in our own times into occasion for casting doubt upon the truth and authenticity of the Holy Scriptures.' In words of ambiguous sophistry, they proceeded to request, in effect, that people would be pleased to adopt the views of churches as to the *complete* inspiration of all the canonical books. The great question whether the Word of God is *in* the Bible, or whether the Word of God is *all* the Bible, was quietly taken for granted in favour of the second view; to the end that men of science might be induced to blame those who took the first view. The first public attention was drawn to the subject by Sir John Herschel, who in refusing to sign the writ sent to him, administered a rebuke in the *Athenæum*, which would have opened most eyes to see that the case was hopeless. The words of a man whose *suaviter in modo* makes his *fortiter in re* cut blocks with a razor are worth preserving:—

' I consider the act of calling upon me publicly to avow or disavow, to approve or disapprove, in writing, any religious doctrine or statement, however carefully or cautiously drawn up (in other words, to append my name to a religious manifesto) to be an infringement of that social forbearance which guards the freedom of religious opinion in this country with especial sanctity . . . I consider this movement simply mischievous, having a direct tendency (by putting forward a new Shibboleth, a new verbal test of religious partisanship) to add a fresh element of discord to the already too discordant relations of the Christian world . . . But no nicety of wording, no artifice of human language, will suffice to discriminate the hundredth part of the shades of meaning in which the most world-wide differences of thought on such subjects may be involved ; or prevent the most gently worded and apparently justifiable expression of regret, so embodied, from grating on the feelings of thousands of estimable and well-intentioned men with all the harshness of controversial hostility.'

Other doses were administered by Sir J. Bowring, Sir W. Rowan Hamilton, and myself. The signed declaration was promised for Christmas, 1864 : but nothing presentable was then ready ; and it was near Midsummer, 1865, before it was published. Persons often incautiously put their names without seeing the *character* of a document, because they coincide in its *opinions*. In this way, probably, fifteen respectable names were procured before printing ; and these, when committed, were hawked as part of an application to ' solicit the favour ' of other signatures. It is likely enough no one of the fifteen saw that the declaration was, not *maintenance* of their own opinion, but *regret* (a civil word for *blame*) that others should think *differently*.

When the list appeared, there were no fewer than 716 names! But analysis showed that this roll was not a specimen of the mature science of the country. The collection was very miscellaneous : 38 were designated as ' students of the College of Chemistry,' meaning young men who attended lectures in that college. But as all the Royal Society had been applied to, a test results as follows. Of Fellows of the Royal Society,· 600 in number, 62 gave their signatures ; of writers in the *Philosophical Transactions*, 166 in number, 19 gave their signatures. Roughly speaking, then, only one out of ten could be got to express disapprobation of the free comparison of the results of science with the statements of the canonical books. And I am satisfied that many of these thought they were signing only a declaration of difference of opinion, not of blame for that difference. The number of persons is not small who, when it comes to signing printed documents, would put their names to a declaration that the coffee-pot ought to be taken downstairs, meaning that the teapot ought to be brought up-stairs. And many of them would defend it. Some would say that the two things are not contradictory ; which, with a snort or two of contempt, would be very effective. Others would, in the candid and quiet tone, point out that it is all one, because coffee is usually taken before tea, and it keeps the table clear to send away the coffee-pot before the teapot is brought up.

The original signatures were decently interred in the Bodleian Library : and the advocates of scattering indefinite blame for indefinite sins of opinion among indefinite persons are, I understand, divided in opinion about the time at which the next attempt shall be made upon men of scientific studies : some are for the Greek Calends, and others for the Roman Olympiads. But, with their usual love of indefiniteness, they have determined that the choice shall be argued upon the basis that which comes first cannot be settled, and is of no consequence.

I give the declaration entire, as a curiosity : and parallel with it I give a substitute which was proposed in the *Athenæum*, as worthy to be signed both by students of theology, and by students of science, especially in past time. When a new attempt is made, it will be worth while to look at both :—

|  *Declaration.* | *Proposed Substitute.* |
|---|---|
| WE, the undersigned Students of the Natural Sciences, desire to express our sincere regret, that researches into scientific truth are perverted by some in our own times into occasion for casting doubt upon the Truth and Authenticity of the Holy Scriptures. | WE, the undersigned Students of Theology and of Nature, desire to express our sincere regret, that common notions of religious truth are perverted by some in our own times into occasion for casting reproach upon the advocates of demonstrated or highly probable scientific theories. |
| We conceive that it is impossible for the Word of God, as written in the book of nature, and God's Word written in Holy Scripture, to contradict one another, however much they may appear to differ. | We conceive that it is impossible for the Word of God, as correctly read in the Book of Nature, and the Word of God, as truly interpreted out of the Holy Scripture, to contradict one another, however much they may appear to differ. |
| We are not forgetful that Physical Science is not complete, but is only in a condition of progress, and that at present our finite reason enables us only to see as through a glass darkly, | We are not forgetful that neither theological interpretation nor physical knowledge is yet complete, but that both are in a condition of progress ; and that at present our finite reason enables us only to see both one and the other as through a glass darkly [the writers of the original de- |

and we confidently believe, that a time will come when the two records will be seen to agree in every particular.

We cannot but deplore that Natural Science should be looked upon with suspicion by many who do not make a study of it, merely on account of the unadvised manner in which some are placing it in opposition to Holy Writ.

We believe that it is the duty of every Scientific Student to investigate nature simply for the purpose of elucidating truth,

and that if he finds that some of his results appear to be in contradiction to the Written Word, or rather to his own *interpretations* of it, which may be erroneous, he should not presumptuously affirm that his own conclusions must be right, and the statements of Scripture wrong ;

rather, leave the two side by side

till it shall please God to allow us to see the manner in which they may be reconciled ;
and, instead of insisting upon

claration have distinctively applied to physical science the phrase by which St. Paul denotes the imperfections of theological vision, which they tacitly assume to be quite perfect],
and we confidently believe that a time will come when the two records will be seen to agree in every particular.

We cannot but deplore that Religion should be looked upon with suspicion by some, and Science by others, of the students of either who do not make a study of the other, merely on account of the unadvised manner in which some are placing Religion in opposition to Science, and some are placing Science in opposition to Religion.

We believe that it is the duty of every theological student to investigate the Scripture, and of every scientific student to investigate Nature, simply for the purpose of elucidating truth.

And if either should find that some of his results appear to be in contradiction, whether to Scripture or to Nature, or rather to his own *interpretation* of one or the other, which may be erroneous, he should not affirm as with certainty that his own conclusion must be right, and the other interpretation wrong :
but should leave the two side by side for further inquiry into both, until it shall please God to allow us to arrive at the manner in which they may be reconciled.

In the mean while, instead of insisting, and least of all with acrimony or injurious statements about others, upon the seeming the seeming differences between Science and the Scriptures, it would be as well to rest in faith upon the points in which they agree.

differences between Science and the Scriptures, it would be a thousand times better to rest in faith as to our future state, in hope as to our coming knowledge, and in charity as to our present differences.

The distinctness of the fallacies is creditable to the composers, and shows that scientific habits tend to clearness, even to sophistry. Nowhere does it so plainly stand out that the *Written Word* means the sense in which the accuser takes it, while the sense of the other side is *their interpretation.* The infallible church on one side, arrayed against heretical pravity on the other, is seen in all subjects in which men differ. At school there were various games in which one or another advantage was the right of those who first called for it. In adult argument the same thing is often attempted : we often hear—I cried *Church* first !

I end with the answer which I myself gave to the application : its revival may possibly save me from a repetition of the like. If there be anything I hate more than another it is the proposal to place any persons, especially those who allow freedom to me, under any abridgment of their liberty to think, to infer, and to publish. If they break the law, take the law ; but do not make the law : ἀγοραιοι ἀγονται ἐγκαλειτωσαν ἀλληλοις. I would rather be asked to take shares in an argyrosteretic company (with limited liability) for breaking into houses by night on fork and spoon errands. I should put aside this proposal with *nothing but laughter.* It was a joke against Sam Rogers that his appearance was very like that of a corpse. The *John Bull* newspaper— suppose we now say Theodore Hook—averred that when he hailed a coach one night in St. Paul's Churchyard, the jarvey said, ' Ho ! ho ! my man ; I'm not going to be taken in that way : go back to your grave ! ' This is the answer I shall make for the future to any relics of a former time who shall want to call me off the stand for their own purposes. What obligation have I to admit that they belong to our world ?

## "SCRIPTURE AND SCIENCE.

### "*The Writ De Hæretico Commiserando.*

Nov. 14, 1864.

"THIS document was sent to me four days ago.    It ' solicits the favour '—I thought at first it was a grocer's supplication for tea and sugar patronage—of my signature to expression of ' sincere regret ' that some persons unnamed—general warrants are illegal —differ from what I am supposed—by persons whom it does not concern—to hold abcut Scripture and Science in their real or alleged discrepancies.

"No such favour from me : for three reasons.    First, I agree with Sir J. Herschel that the solicitation is an intrusion to be publicly repelled.    Secondly, I do *not* regret that others should differ from me, think what I may : those others are as good as I, and as well able to think, and as much entitled to their conclusions.    Thirdly, even if I did regret, I should be ashamed to put my name to bad chemistry made to do duty for good reasoning.    The declaration is an awkward attempt to saturate sophism with truism ; but the sophism is left largely in excess.

" I owe the inquisitors a grudge for taking down my conceit of myself.    For two months I have crowed in my own mind over my friend Sir J. Herschel, fancying that the promoters instinctively knew better than to bring their fallacies before a writer on logic. Ah ! my dear Sir John ! thought I, if you had shown yourself to be well up in *Barbara Celarent*, and had ever and anon astonished the natives with the distinction between *simpliciter* and *secundum quid*, no autograph-hunters would have baited a trap with *non sequitur* to catch your signature.    What can I say now ?    I hide my diminished head, diminished by the horns which I have been compelled to draw in.

" Those who make personal solicitation for support to an opinion about religion are bound to know their men.    The king had a right to Brother Neale's money, because Brother Neale offered it. Had he put his hand into purse after purse by way of finding out all who were of Brother Neale's mind, he would have been justly met by a rap on the knuckles whenever he missed his mark.

" The kind of test before me is the utmost our time will allow of that inquisition into opinion which has been the curse of Christianity ever since the State took Providence under its protection.

The writ *de hæreticò commiserando* is little more than the smell of the empty cask : and those who issue it may represent the old woman with her

> O suavis anima, quale in te dicam bonum
> Antehac fuisse ; tales cum sint reliquiæ.

It is no excuse that the illegitimate bantling is a very little one. Its parents may think themselves hardly treated when they are called lineal successors of Tony Fire-the-faggot : but, degenerate though they be, such is their ancestry.   Let every allowance be made for them : but their unholy fire must be trodden out ; so long as a spark is left, nothing but fuel is wanted to make a blaze.   If this cannot be done, let the flame be confined to theology, though even there it burns with diminished vigour : and let charity, candour, sense, and ridicule, be ready to play upon it whenever there is any chance of its extending to literature or science.

" What would be the consequence if this test-signing absurdity were to grow ?   Deep would call unto deep ; counter-declaration would answer declaration, each stronger than the one before. The moves would go on like the dispute of two German students, of whom each is bound to a sharper retort on a graduated scale, until at last comes *dummer junge !*—and then they must fight. There is a gentleman in the upper fifteen of the signers of the writ—the hawking of whose names appears to me very bad taste —whom I met in cordial co-operation for many a year at a scientific board.   All I knew about his religion was that he, as a clergyman, must in some sense or other receive the 39 Articles : —all that he could know about mine was that I was some kind of heretic, or so reputed.   If we had come to signing opposite manifestoes, turn-about, we might have found ourselves in the lowest depths of party discussion at our very council-table.   I trust the list of subscribers to the declaration, when it comes to be published, will show that the bulk of those who have really added to our knowledge have seen the thing in its true light.

" The promoters—I say nothing about the subscribers—of the movement will, I trust, not feel aggrieved at the course I have taken or the remarks I have made.   Walter Scott says that before we judge Napoleon by the temptation to which he yielded, we ought to remember how much he may have resisted :   I invite them to apply this rule to myself ; they can have no idea of the feeling with which I contemplate all attempts to repress freedom of inquiry, nor of the loathing with which I recoil from the proposal to be art and part. They have asked me to give a public opinion upon a certain point. It is true that they have had the kindness to tender both the opinion they wish me to form, and the shape in which they would have it appear : I will let them draw me out, but I will not let them take me in. If they will put an asterisk to my name, and this letter to the asterisk, they are welcome to my signature. As I do not expect them to relish this proposal, I will not solicit the favour of its adoption. But they have given a right to think, for they have asked me to think ; to publish, for they have asked me to allow them to publish ; to blame them, for they have asked me to blame their betters. Should they venture to find fault because my direction of disapproval, publicly given, is half a revolution different from theirs, they will be known as having presented a loaded document at the head of a traveller in the highway of discussion, with—Your signature or your silence ! "

---

The paradox being the proposition of something which runs counter to what would generally be thought likely, may present itself in many ways. There is a *fly-leaf paradox*, which puzzled me for many years, until I found a probable solution. I frequently saw, in the blank leaves of old books, learned books, Bibles of a time when a Bible was very costly, &c., the name of an owner who, by the handwriting and spelling, must have been an illiterate person or a child, followed by the date of the book itself. Accordingly, this uneducated person or young child seemed to be the first owner, which in many cases was not credible. Looking one day at a Barker's Bible of 1599, I saw an inscription in a child's writing, which certainly belonged to a much later date. It was ' Martha Taylor, her book, giuen me by Granny Scott to keep for her sake.' With this the usual verses, followed by 1599, the date of the book. But it so chanced that the blank page opposite the title, on which the above was written, was a verso of the last leaf of a prayer book, which had been bound before the Bible ; and on the recto of this leaf was a colophon, with the date 1632. It struck me immediately that uneducated persons and children, having seen dates written under names, and not being quite up in chronology, did frequently finish off with the date of the book, which stared them in the face.

Always write in your books. You may be a silly person—for though your reading my book is rather a contrary presumption, yet it is not conclusive—and your observations may be silly or irrelevant, but you cannot tell what use they may be of long after you are gone where Budgeteers cease from troubling.

I picked up the following book, printed by J. Franklin at Boston, during the period in which his younger brother Benjamin was his apprentice. And as Benjamin was apprenticed very early, and is recorded as having learnt the mechanical art very rapidly, there is some presumption that part of it may be his work, though he was but thirteen at the time. As this set of editions of Hodder (by Mose) is not mentioned, to my knowledge, I give the title in full :—

Hodder's Arithmetick : or. that necessary art made most easy : Being explained in a way familiar to the capacity of any that desire to learn it in a little time. By James Hodder, Writing-master. The Five and twentieth edition, revised, augmented, and above a thousand faults amended, by Henry Mose, late servant and successor to the author. Boston : printed by J. Franklin, for S. Phillips, N. Buttolph, B. Elliot, D. Henchman, G. Phillips, J. Elliot, and E. Negus, booksellers in Boston, and sold at their shops. 1719.

The book is a very small octavo, the type and execution are creditable, the woodcut at the beginning is clumsy. It is a perfect copy, page for page, of the English editions of Mose's Hodder, of which the one called seventeenth is of London, 1690. There is not a syllable to show that the edition above described might not be of Boston in England. Presumptions, but not very strong ones, might be derived from the name of *Franklin*, and from the large number of booksellers who combined in the undertaking. It chanced, however, that a former owner had made the following note in my copy :—

Wednessday, July $y^e$ 14, 1796, att ten in $y^e$ forenoon we sail$^d$ from Boston, came too twice, once in King Rode, and once in $y^e$ Narrows. Sail$^d$ by $y^e$ lighthouse in $y^e$ even$^g$.

No ordinary map would decide these points : so I had to apply to my friend Sir Francis Beaufort, and the charts at the Admiralty decided immediately for Massachusetts.

The French are able paradoxers in their spelling of foreign names. The Abbé Sabatier de Castres, in 1772, gives an account of an imaginary dialogue between Swif, Adisson, Otwai, and Bolingbrocke. I had hoped that this was a thing of former days, like the literal roasting of heretics; but the charity which hopeth all things must hope for disappointments. Looking at a recent work on the history of the Popes, I found referred to, in the matter of Urban VIII. and Galileo, references to the works of two Englishmen, the Rev. Win Worewel and the Rev. Raden Powen. [Wm. Whewell and Baden Powell].

I must not forget the 'moderate computation' paradox. This is the way by which large figures are usually obtained. Anything surprisingly great is got by the 'lowest computation,' anything as surprisingly small by the 'utmost computation'; and these are the two great subdivisions of 'moderate computation.' In this way we learn that 70,000 persons were executed in one reign, and 150,000 persons burned for witchcraft in one century. Sometimes this computation is very close. By a card before me it appears that all the Christians, including those dispersed in heathen countries, those of Great Britain and Ireland excepted, are 198,728,000 people, and pay their clergy 8,852,000*l*. But 6,400,000 people pay the clergy of the Anglo-Irish Establishment 8,896,000*l*.; and 14,600,000 of other denominations pay 1,024,000*l*. When I read moderate computations, I always think of Voltaire and the 'mémoires du fameux évêque de Chiapa, par lesquels il paraît qu'il avait égorgé, ou brulé, ou noyé dix millions d'infidèles en Amérique pour les convertir. Je crus que cet évêque exaggérait; mais quand on réduisait ces sacrifices à cinq millions de victimes, cela serait encore admirable.'

My Budget has been arranged by authors. This is the only plan, for much of the remark is personal: the peculiarities of the paradoxer are a large part of the interest of the paradox. As to subject-matter, there are points which stand strongly out; the quadrature of the circle, for instance. But there are others which cannot be drawn out so as to be conspicuous in a review of writers: as one instance, I may take the *centrifugal force*.

When I was about nine years old I was taken to hear a course of lectures, given by an itinerant lecturer in a country town, to get as much as I could of the second half of a good, sound, philosophical omniscience. The first half (and sometimes more) comes by nature. To this end I smelt chemicals, learned that they were different kinds of *gin*, saw young wags try to kiss the girls under the excuse of what was called *laughing gas*—which I was sure was not to blame for more than five per cent. of the requisite assurance—and so forth. This was all well so far as it went; but there was also the excessive notion of creative power exhibited in the millions of miles of the solar system, of which power I wondered they did not give a still grander idea by expressing the distances in inches.     But even this was nothing to the ingenious contrivance of the centrifugal force.     'You have heard what I have said of the wonderful centripētal force, by which Divine Wisdom has retained the planets in their orbits round the Sun. But, ladies and gentlemen, it must be clear to you that if there were no other force in action, this centripētal force would draw our earth and the other planets into the Sun, and universal ruin would ensue.     To prevent such a catastrophe, the same wisdom has implanted a centrifūgal force of the same amount, and directly opposite,' &c.     I had never heard of Alfonso X. of Castile, but I ventured to think that if Divine Wisdom had just let the planets alone it would come to the same thing, with equal and opposite troubles saved.     The paradoxers deal largely in speculation conducted upon the above explanation.     They provide external agents for what they call the centrifugal force.     Some make the sun's rays keep the planets off, without a thought about what would become of our poor eyes if the *push* of the light which falls on the earth were a counterpoise to all its gravitation.     The true explanation cannot be given here, for want of room.

Sometimes a person who has a point to carry will assert a singular fact or prediction for the sake of his point; and this paradox has almost obtained the sole use of the name.     Persons who have reputation to care for should beware how they adopt this plan, which now and then eventuates a spanker, as the American editor said.     Lord Byron, in ' English Bards, &c.,' (1809) ridiculing Cambridge poetry, wrote as follows :—

> But where fair Isis rolls her purer wave,
> The partial muse delighted loves to lave ;
> On her green banks a greener wreath she wove,
> To crown the bards that haunt her classic grove ;
> Where Richards wakes a genuine poet's fires,
> And modern Britons glory in their sires.[1]

There is some account of the Rev. Geo. Richards, Fellow of Oriel and Vicar of Bampton, (M.A. in 1791) in the ' Living Authors,' by Watkins and Shoberl (1816).     In Rivers's ' Living Authors,' of 1798, which is best fitted for citation, as being published before Lord Byron wrote, he is spoken of in high terms.     The ' Aboriginal Britons ' was an Oxford (special) prize poem, of 1791. Charles Lamb mentions Richards as his school-fellow at Christ's

---

[1] The ' Aboriginal Britons,' an excellent poem, by Richards. (Note by Byron.)

Hospital, 'author of the " Aboriginal Britons," the most spirited
of the Oxford Prize Poems : a pale, studious Grecian.'

As I never heard of Richards as a poet, I conclude that his
fame is defunct, except in what may prove to be a very ambiguous
kind of immortality, conferred by Lord Byron. The awkwardness
of a case which time has broken down is increased by the eulogist
himself adding so powerful a name to the list of Cambridge poets,
that his college has placed his statue in the library, more con-
spicuously than that of Newton in the chapel ; and this although
the greatness of poetic fame had some serious drawbacks in the
moral character of some of his writings. And it will be found on
inquiry that Byron, to get his instance against Cambridge, had
to go back eighteen years, passing over seven intermediate productions, of which he had either never heard, or which he would
not cite as waking a genuine poet's fires.

The conclusion seems to be that the ' Aboriginal Britons ' is
a remarkable youthful production, not equalled by subsequent
efforts.

To enhance the position in which the satirist placed himself,
two things should be remembered. First, the glowing and
justifiable terms in which Byron had spoken,—a hundred and
odd lines before he found it convenient to say no Cambridge
poet could compare with Richards,—of a Cambridge poet who
died only three years before Byron wrote, and produced greatly
admired works while actually studying in the University. The
fame of Kirke White still lives ; and future literary critics may
perhaps compare his writings and those of Richards, simply by
reason of the curious relation in which they are here placed
alongside of each other. And it is much to Byron's credit that,
in speaking of the deceased Cambridge poet, he forgot his own
argument and its exigencies, and proved himself only a paradoxer
*pro re nata*.

Secondly, Byron was very unfortunate in another passage of the
same poem :—

> What varied wonders tempt us as they pass !
> The cow-pox, tractors, galvanism, and gas.
> In turns appear, to make the vulgar stare,
> Till the swoln bubble bursts—and all is air !

Three of the bubbles have burst to mighty ends. The metallic
tractors are disused ; but the force which, if anything, they put
in action, is at this day, under the name of mesmerism, used, pro-

hibited, respected, scorned, assailed, defended, asserted, denied, declared utterly obscure, and universally known. It was hard lines to select four candidates for oblivion not one of whom got in. I shall myself, I am assured, be some day cited for laughing at the great discovery of —— : the blank is left for my reader to fill up in his own way; but I think I shall not be so unlucky in four different ways.

The narration before the fact, as prophecy has been called, sometimes quite as true as the narration after the fact, is very ridiculous when it is wrong. Why, the pre-narrator could not know; the post-narrator might have known. A good collection of unlucky predictions might be made : I hardly know one so fit to go with Byron's as that of the Rev. Daniel Rivers, already quoted, about Johnson's biographers. Peter Pindar may be excused, as personal satire was his object, for addressing Boswell and Mrs. Piozzi as follows :—

> Instead of adding splendour to his name,
> Your books are downright gibbets to his fame ;
> You never with posterity can thrive,
> 'Tis by the Rambler's death alone you live.

But Rivers, in prose narrative, was not so excusable. He says :—
' As admirers of the learning and moral excellence of their hero, we glow at almost every page with indignation that his weaknesses and his failings should be disclosed to public view . . . Johnson, after the lustre he had reflected on the name of Thrale . . . was to have his memory tortured and abused by her detested itch for scribbling. More injury, we will venture to affirm, has been done to the fame of Johnson by this Lady and her late biographical helpmate, than his most avowed enemies have been able to effect : and if his character becomes unpopular with some of his successors, it is to those gossiping friends he is indebted for the favour.'

Poor dear old Sam ! the best known dead man alive ! clever, good-hearted, logical, ugly bear ! Where would he have been if it had not been for Boswell and Thrale, and their imitators? What would biography have been if Boswell had not shown how to write a life ?

Rivers is to be commended for not throwing a single stone at Mrs. Thrale's second marriage. This poor lady begins to receive a little justice. The literary world seems to have found out that a blue-stocking dame who keeps open house for a set among them has a right, if it so please her, to marry again without

taking measures to carry on the cake-shop. I was before my age
in this respect: as a boy-reader of Boswell, and a few other
things that fell in my way, I came to a clearness that the conduct
of society towards Mrs. Piozzi was *blackguard*. She wanted
nothing but what was in that day a woman's only efficient protection, a male relation with a brace of pistols, and a competent
notion of using them.

Byron's mistake about Hallam in the Pindar story may be
worth placing among absurdities. For elucidation, suppose that
some poet were now to speak—

> Of man's first disobedience, and the fruit
> Eve gave to Adam in his birthday suit—

and some critic were to call it nonsense, would that critic be
laughing at Milton? Payne Knight, in his *Taste*, translated part
of Gray's *Bard* into Greek. Some of his lines are

> Θερμὰ δ' ὁ τέγγων δάκρυα στοναχαῖς
> οὖλον μέλος φοβερᾷ
> ἤειδε φωνᾷ.

Literally thus:—

> Wetting warm tears with groans,
> Continuous chant with fearful
> Voice he sang.

On which Hallam remarks: 'The twelfth line [our first] is
nonsense.' And so it is, a poet can no more wet his tears with
his groans than wet his ale with his whistle. Now this first line
is from Pindar, but is only part of the sense; in full it is:—

> Θερμὰ δὲ τέγγων δάκρυα στοναχαῖς
> ὄρθιον φώνασε.

Pindar's τέγγων must be Englished by *shedding*, and he stands
alone in this use. He says, 'shedding warm tears, he cried out
loud, with groans.' Byron speaks of—

> Classic Hallam, much renowned for Greek:

and represents him as criticising *the Greek* of all Payne's lines,
and not discovering that 'the lines' were Pindar's until after
publication. Byron was too much of a scholar to make this

blunder himself: he either accepted the facts from report, or else took satirical licence. And why not? If you want to laugh at a person, and he will not give occasion, whose fault is it that you are obliged to make it? Hallam did criticise some of Payne Knight's Greek ; but with the caution of his character, he remarked that possibly some of these queer phrases might be ' critic-traps ' justified by some one use of some one author. I remember well having a Latin essay to write at Cambridge, in which I took care to insert a few monstrous and unusual idioms from Cicero : a person with a Nizolius, and without scruples may get scores of them. So when my tutor raised his voice against these oddities, I was up to him, for I came down upon him with Cicero, chapter and verse, and got round him. And so my own solecisms, many of them, passed unchallenged.

Byron had more good in his nature than he was fond of letting out : whether he was a soured misanthrope, or whether his *vein* lay that way in poetry, and he felt it necessary to fit his demeanour to it, are matters far beyond me. Mr. Crabb Robinson told me the following story more than once. He was at Charles Lamb's chambers in the Temple when Wordsworth came in, with the new *Edinburgh Review* in his hand, and fume on his countenance. ' These reviewers,' said he, ' put me out of patience ! Here is a young man—they say he is a lord—who has written a volume of poetry ; and these fellows, just because he is a lord, set upon him, laugh at him, and sneer at his writing. The young man will do something, if he goes on as he has begun. But these reviewers seem to think that nobody may write poetry, unless he lives in a garret.' Crabb Robinson told this long after to Lady Byron, who said, ' Ah ! if Byron had known that, he would never have attacked Wordsworth, He went one day to meet Wordsworth at dinner ; when he came home I said, " Well, how did the young poet get on with the old one ? " " Why, to tell you the truth," said he, " I had but one feeling from the beginning of the visit to the end, and that was—*reverense* ! "' Lady Byron told my wife that her husband had a very great respect for Wordsworth. I suppose he would have said—as the Archangel said to his Satan—' Our difference is po[li = e]tical.'

I suspect that Fielding would, if all were known, be ranked among unlucky railers at supposed paradox. In his ' Miscellanies ' (1742, 8vo.) he wrote a satire on the Chrysippus or Guinea, an animal which multiplies itself by division, like the polypus. This he supposes to have been drawn up by Petrus Gualterus, meaning the famous usurer, Peter Walter. He calls it a paper ' proper to be read before the R——l Society'; and next year 1743, a quarto reprint was made to resemble a paper in the *Philosophical Transactions.* So far as I can make out, one object is ridicule of what the zoologists said about the polypus : a reprint in the form of the *Transactions* was certainly satire on the Society, not on Peter Walter and his knack of multiplying guineas.

Old poets have recognised the quadrature of the circle as a well-known difficulty. Dante compares himself, when bewildered, to a geometer who cannot find the principle on which the circle is to be measured :—

> Quale è 'l geometra che tutto s' affige
> Per misurar lo cerchio, e non ritruôva,
> Pensando qual principio ond' egli indige.

And Quarles speaks as follows of the *summum bonum* :—

> Or is 't a tart idea, to procure
> An edge, and keep the practic soul in ure,
> Like that dear chymic dust, or puzzling quadrature ?

The poetic notion of the quadrature must not be forgotten. Aristophanes, in the *Birds,* introduces a geometer who announces his intention to *make a square circle.* Pope, in the *Dunciad,* delivers himself as follows, with a Greek pronunciation rather strange in a translator of Homer. Probably Pope recognised, as a general rule, the very common practice of throwing back the accent in defiance of quantity, seen in o'rator, au'ditor, se'nator, ca'tenary, &c.—

> Mad *Mathesis* alone was unconfined,
> Too mad for mere material chains to bind,—
> Now to pure space lifts her ecstatic stare,
> Now, running round the circle, finds it square.

The author's note explains that this ' regards the wild and fruitless attempts of squaring the circle.' The poetic idea seems to be that the geometers try to make a square circle. Disraeli quotes it as ' finds *its* square,' but the originals do not support this reading.

I have come in the way of a work, entitled 'The Grave of Human Philosophies,' (1827) translated from the French of R. de Bécourt by A. Dalmas. It supports, but I suspect not very accurately, the views of the old Hindoo books. That the sun is only 450 miles from us, and only 40 miles in diameter, may be passed over; my affair is with the state of mind into which persons of M. Bécourt's temperament are brought by a fancy. He fully grants, as certain, four millions of years as the duration of the Hindoo race, and 1956 as that of the universe. It must be admitted he is not wholly wrong in saying that our errors about the universe proceed from our ignorance of its origin, antiquity, organization, laws, and final destination. Living in an age of light, he ' avails himself of that opportunity ' to remove this veil of darkness, &c. The system of the Brahmins is the only true one : he adds that it has never before been attempted, as it could not be obtained except by him. The author requests us first, to lay aside prejudice ; next, to read all he says in the order in which he says it : we may then pronounce judgment upon a work which begins by taking the Brahmins for granted. All the paradoxers make the same requests. They do not see that compliance would bring thousands of systems before the world every year : we have scores as it is. How is a poor candid inquirer to choose Fortunately, the mind has its grand jury as well as its little one : and it will not put a book upon its trial without a *primâ facie* case in its favour. And with most of those who really search for themselves, that case is never made out without evidence of knowledge, standing out clear and strong, in the book to be examined.

There is much private history which will never come to light, *caret quia vate sacro*, because no Budgeteer comes across it. Many years ago a man of business, whose life was passed in banking, amused his leisure with quadrature, was successful of course, and bequeathed the result in a sealed book, which the legatee was enjoined not to sell under a thousand pounds. The true ratio was 3·1416 : I have the anecdote from the legatee's executor, who opened the book. That a banker should square the circle is very credible : but how could a City man come by the notion that a thousand pounds could be got for it ? A friend of mine, one of the twins of my zodiac, will spend a thousand pounds, if he have not done it already, in black and white cyclometry : but I will answer for it that he, a man of sound business notions, never entertained the idea of $\pi$ recouping him, as they now say. I speak of individual success : of course if a company were formed, especially if it were of unlimited lie-ability, the shares would be taken. No offence ; there is nothing but what a pun will either sanctify, justify, or nullify :—

It comes o'er the soul like the sweet South
That breathes upon a bank of *vile hits.*

The shares would be at a premium of $3\frac{1}{8}$ on the day after issue. If they presented me with the number of shares I deserve, for suggestion and advertisement, I should stand up for the Archpriest of St. Vitus and $3\frac{1}{5}$, with a view to a little more gold on the bridge.

I now insert a couple of reviews, one about Cyclopædias, one about epistolary collections. Should any reader wish for explanation of this insertion, I ask him to reflect a moment, and imagine me set to justify all the additions now before him! In truth these reviews are the repositories of many odds and ends : they were not made to the books; the materials were in my notes, and the books came as to a ready-made clothes shop, and found what would fit them. Many remember Curll's bequest of some very good titles which only wanted treatises written to them. Well! here were some tolerable reviews—as times go—which only wanted books fitted to them. Accordingly, some tags were made to join on the books ; and then as the reader sees.

I should find it hard to explain why the insertion is made in this place rather than another. But again, suppose I were put to make such an explanation throughout the volume. The improver who laid out grounds and always studied what he called *unexpectedness*, was asked what name he gave it for those who walked over his grounds a second time. He was silenced ; but I have an answer : It is that which is given by the very procedure of taking up my book a second time.

October 19, 1861. *The English Cyclopædia.* Conducted by Charles Knight. 22 vols. : viz., *Geography*, 4 vols. ; *Biography*, 6 vols. ; *Natural History*, 4 vols. ; *Arts and Sciences*, 8 vols. (Bradbury & Evans.)

*The Encyclopædia Britannica : a Dictionary of Arts, Sciences, and General Literature.* Eighth Edition. 21 vols. and Index. (Black.)

The two editions above described are completed at the same time : and they stand at the head of the two great branches into which pantological undertakings are divided, as at once the largest and the best of their classes.

When the works are brought together, the first thing that strikes the eye is the syllable of difference in the names. The word *Cyclopædia* is a bit of modern purism. Though ἐγκυκλο-παιδεια is not absolutely Greek of Greece, we learn from both

Pliny and Quintilian that the circle of the sciences was so called by the Greeks, and Vitruvius has thence naturalized *encyclium* in Latin. Nevertheless we admit that the initial *en* would have euphonized but badly with the word *Penny*: and the English Cyclopædia is the augmented, revised, and distributed edition of the Penny Cyclopædia. It has indeed been said that Cyclopædia should mean the education *of* a circle, just as Cyropædia is the education *of* Cyrus. But this is easily upset by Aristotle's word κυκλοφορία, motion *in* a circle, and by many other cases, for which see the lexicon.

The earliest printed Encyclopædia of this kind was perhaps the famous 'myrrour of the worlde,' which Caxton translated from the French and printed in 1480. The original Latin is of the thirteenth century, or earlier. This is a collection of very short treatises. In or shortly after 1496 appeared the 'Margarita Philosophica' of Gregory Reisch, the same we must suppose, who was confessor to the Emperor Maximilian. This is again a collection of treatises, of much more pretension : and the estimation formed of it is proved by the number of editions it went through. In 1531, appeared the little collection of *works* of Ringelberg, which is truly called an Encyclopædia by Morhof, though the thumbs and fingers of the two hands will meet over the length of its one volume. There are more small collections ; but we pass on to the first work to which the name of *Encyclopœdia* is given. This is a ponderous 'Scientiarum Omnium Encyclopædia' of Alsted, in four folio volumes, commonly bound in two ; published in 1629 and again in 1649 ; the true parent of all the Encyclopædias, or collections of treatises, or works in which that character predominates. The first great *dictionary* may perhaps be taken to be Hofman's 'Lexicon Universale' (1677); but Chambers's (so called) Dictionary (1728) has a better claim. And we support our proposed nomenclature by observing that Alsted accidentally called his work *E*ncyclopædia, and Chambers simply Cyclopædia.

We shall make one little extract from the 'myrrour,' and one from Ringelberg. Caxton's author makes a singular remark for his time ; and one well worthy of attention. The grammar rules of a language, he says, must have been invented by foreigners : 'And whan any suche tonge was perfytely had and usyd amonge any people, than other people not used to the same tonge caused rulys to be made wherby they myght lerne the same tonge . . . . and suche rulys be called the gramer of that tonge.' Ringelberg says that if the right nostril bleed, the little finger of the right

hand should be crooked, and squeezed with great force ; and the
same for the left.

We pass on to *the* Encyclopédie, commenced in 1751 ; the work
which has, in many minds, connected the word *encyclopœdist* with
that of *infidel.* Readers of our day are surprised when they look
into this work, and wonder what has become of all the irreligion.
The truth is, that the work—though denounced *ab ovo* on account
of the character of its supporters—was neither adapted, nor intended, to excite any particular remark on the subject : no work
of which D'Alembert was co-editor would have been started on
any such plan. For, first, he was a real *sceptic* : that is, doubtful,
with a mind not made up. Next, he valued his quiet more than
anything ; and would as soon have gone to sleep over an hornet's
nest as have contemplated a systematic attack upon either religion
or government. As to Diderot—of whose varied career of thought
it is difficult to fix the character of any one moment, but who is
very frequently taken among us for a pure atheist—we will quote
one sentence from the article ' *Encyclopédie*,' which he wrote
himself :—' Dans le moral, il n'y a que Dieu qui doit servir de
modèle à l'homme ; dans les arts, que la nature.'

A great many readers in our country have but a very hazy idea
of the difference between the political Encyclopædia, as we may
call it, and the Encyclopédie Méthodique, which we always take
to be meant – whether rightly or not we cannot tell—when we
hear of the ' great French Encyclopædia.' This work, which takes
much from its predecessor, professing to correct it, was begun in
1792, and finished in 1832. There are 166 volumes of text, and
6,439 plates, which are sometimes incorporated with the text,
sometimes make about 40 more volumes. This is still the monster
production of the kind ; though probably the German Cyclopædia
of Ersch and Gruber, which was begun in 1818, and is still in
progress, will beat it in size. The great French work is a collection of dictionaries ; it consists of Cyclopædias of all the separate
branches of knowledge. It is not a work, but a collection of works,
one or another department is to be bought from time to time ;
but we never heard of a complete set for sale in one lot. As ships
grow longer and longer, the question arises what limit there is to
the length. One answer is, that it will never do to try such a
length that the stern will be rotten before the prow is finished.
This wholesome rule has not been attended to in the matter
before us ; the earlier parts of the great French work were
antiquated before the whole were completed : something of the
kind will happen to that of Ersch and Gruber.

The production of a great dictionary of either of the kinds is far from an easy task. There is one way of managing the *Encyclopædia* which has been largely resorted to ; indeed, we may say that no such work has been free from it. This plan is to throw all the attention upon the great treatises, and to resort to paste and scissors, or some process of equally easy character, for the smaller articles. However it may be done, it has been the rule that the Encyclopædia of treatises should have its supplemental Dictionary of a very incomplete character. It is true that the treatises are intended to do a good deal ; and that the Index, if it be good, knits the treatises and the dictionary into one whole of reference. Still there are two stools, and between them a great deal will fall to the ground. The dictionary portion of the Britannica is not to be compared with its treatises ; the part called Miscellaneous and Lexicographical in the Metropolitana is a great failure. The defect is incompleteness. The biographical portion, for example, of the Britannica is very defective : of many names of note in literature and science, which become known to the reader from the treatises, there is no account whatever in the dictionary. So that the reader who has learnt the results of a life in astronomy, for example, must go to some other work to know when that life began and ended. This defect has run through all the editions ; it is in the casting of the work. The reader must learn to take the results at their true value, which is not small. He must accustom himself to regard the Britannica as a splendid body of treatises on all that can be called heads of knowledge, both greater and smaller ; with help from the accompanying dictionary, but not of the most complete character. Practically, we believe, this defect cannot be avoided : two plans of essentially different structure cannot be associated on the condition of each or either being allowed to abbreviate the other.

The defect of all others which it is most difficult to avoid is inequality of performance. Take any dictionary you please, of any kind which requires the association of a number of contributors, and this defect must result. We do not merely mean that some will do their work better than others; this of course : we mean that there will be structural differences of execution, affecting the relative extent of the different parts of the whole, as well as every other point by which a work can be judged. A wise editor will not attempt any strong measures of correction : he will remember that if some portions be below the rest, which is a disadvantage, it follows that some portions must be above the rest, which is an advantage. The only practical level, if level there must be, is that of mediocrity, if not of absolute worthlessness : any attempt to secure equality of strength will result in equality of weakness. Efficient development may be cut down into meagre brevity, and in this way only can apparent equality of plan be secured throughout. It is far preferable to count upon differences of execution, and to proceed upon the acknowledged expectation that the prominent merits of the work will be settled by the accidental character of the contributors ; it being held impossible that any editorial efforts can secure a uniform standard of goodness. Wherever the greatest power is found, it should be suffered to produce its natural effect. There are, indeed, critics who think that the merit of a book, like the strength of a chain, is that of its weakest part : but there are others who know that the parallel does not hold, and who will remember that the union of many writers must show exaggeration of the inequalities which almost always exist in the production of one person. The true plan is to foster all the good that can be got, and to give development in the directions in which most resources are found : a Cyclopædia, like a plant, should grow towards the light.

The Penny Cyclopædia had its share of this kind of defect or excellence, according to the way in which the measure is taken. The circumstance is not so much noticed as might be expected, and this because many a person is in the habit of using such a dictionary chiefly with relation to one subject, his own; and more still want it for the pure dictionary purpose, which does not go much beyond the meaning of the word. But the person of full and varied reference feels the differences ; and criticism makes capital of them. The Useful Knowledge Society was always odious to the organs of religious bigotry ; and one of them, adverting to the fact that geography was treated with great ability, and most unusual fullness, in the Penny Cyclopædia, announced it by making it the sole merit of the work that, with sufficient addition, it would make a tolerably good gazetteer.

Some of our readers may still have hanging about them the feelings derived from this old repugnance of a class to all that did not associate direct doctrinal teaching of religion with every attempt to communicate knowledge. I will take one more instance, by way of pointing out the extent to which stupidity can go. If there be an astronomical fact of the telescopic character which, next after Saturn's ring and Jupiter's satellites, was known to all the world, it was the existence of multitudes of double stars, treble stars, &c. A respectable quarterly of the theological cast, which in mercy we refrain from naming, was ignorant of this common knowledge,—imagined that the mention of such systems was a blunder of one of the writers in the Penny Cyclopædia, and lashed the presumed ignorance of the statement in the following words, delivered in April, 1837 :—

'We have forgotten the name of that Sidrophel who lately discovered that the fixed stars were not single stars, but appear in the heavens like soles at Billingsgate, in pairs ; while a second astronomer, under the influence of that competition in trade which the political economists tell us is so advantageous to the public, professes to show us, through his superior telescope, that the apparently single stars are really three. Before such wondrous mandarins of science, how continually must *homunculi* like ourselves keep in the background, lest we come between the wind and their nobility.'

Certainly these little men ought to have kept in the background; but they did not : and the growing reputation of the work which they assailed has chronicled them in literary history ; grubs in amber.

This important matter of inequality, which has led us so far, is one to which the Encyclopædia is as subject as the Cyclopædia ; but it is not so easily recognised as a fault. We receive the first book as mainly a collection of treatises : we know their authors, and we treat them as individuals. We see, for instance, the names of two leading writers on Optics, Brewster and Herschel. It would not at all surprise us if either of these writers should be found criticising the other by name, even though the very view opposed should be contained in the same Encyclopædia with the criticism. And in like manner, we should hold it no wonder if we found some third writer not comparable to either of those we have named. It is not so in the Cyclopædia : here we do not know the author, except by inference from a list of which we never think while consulting the work. We do not dissent from this or that author : we blame the book.

The Encyclopædia Britannica is an old friend. Though it holds a proud place in our present literature, yet the time was when it stood by itself, more complete and more clear than anything which was to be found elsewhere. There must be studious men alive in plenty who remember, when they were studious boys, what a literary luxury it was to pass a few days in the house of a friend who had a copy of this work. The present edition is a worthy successor of those which went before. The last three editions, terminating in 1824, 1842, and 1861, seem to show that a lunar cycle cannot pass without an amended and augmented edition. Detailed criticism is out of the question ; but we may notice the effective continuance of the plan of giving general historical dissertations on the progress of knowledge. Of some of these dissertations we have had to take separate notice ; and all will be referred to in our ordinary treatment of current literature.

The literary excellence of these two extensive undertakings is of the same high character. To many this will need justification : they will not easily concede to the chea pand recent work a right to stand on the same shelf with the old and tried magazine, newly replenished with the best of everything. Those who are cognizant by use of the kind of material which fills the Penny Cyclopædia will need no further evidence : to others we shall quote a very remarkable, and certainly very complete testimony. The *Cyclopædia of the Physical Sciences*, published by Dr. Nichol in 1857 (noticed by us, April 4), is one of the most original of our special dictionaries. The following is an extract from the editor's preface :—

'When I assented to Mr. Griffin's proposal that I should edit such a Cyclopædia, I had it in my mind that I might make the *scissors* eminently effective. Alas! on narrowly examining our best Cyclopædias, I found that the scissors had become blunted through too frequent and vigorous use. One great exception exists : viz. the Penny Cyclopædia of Charles Knight. The cheapest and the least pretending, it is really the most philosophical of our *scientific* dictionaries. It is not made up of a series of treatises, some good and many indifferent, but is a thorough *Dictionary*, well proportioned and generally written by the best men of the time. The more closely it is examined, the more deeply will our obligations be felt to the intelligence and conscientiousness of its projector and editor.'

After Dr. Nichol's candid and amusing announcement of his scissorial purpose, it is but fair to state that nothing of the kind was ultimately carried into effect, even upon the work in which he found so much to praise. I quote this testimony because it is of a peculiar kind.

The success of the *Penny Magazine* led Mr. Charles Knight in 1832, to propose to the Useful Knowledge Society a Cyclopædia in weekly penny numbers. These two works stamp the name of the projector on the literature of our day in very legible characters. Eight volumes of 480 pages each were contemplated ; and Mr. Long and Mr. Knight were to take the joint management. The plan embraced a popular account of Art and Science, with very brief biographical and geographical information. The early numbers of the work had some of the *Penny Magazine* character : no one can look at the pictures of the Abbot and Abbess in their robes without seeing this. By the time the second volume was completed, it was clearly seen that the plan was working out its own extension : a great development of design was submitted to, and Mr. Long became sole editor. Contributors could not be found to make articles of the requisite power in the assigned space. One of them told us that when he heard of the eight volumes, happening to want a shelf to be near at hand for containing the work as it went on, he ordered it to be made to hold twenty-five volumes easily. But the inexorable logic of facts beat him after all : for the complete work contained twenty-six volumes, and two thick volumes of Supplement.

The penny issue was brought to an end by the state of the law, which required, in 1833, that the first and last page of everything sold separately should contain the name and address of the printer. The penny numbers contained this imprint on the fold of the outer leaf : and *qui tam* informations were laid against the agents in various towns. It became necessary to call in the stock ; and the penny issue was abandoned. Monthly parts were substituted, which varied in bulk, as the demands of the plan became more urgent, and in price from one sixpence to three. The second volume of Supplement appeared in 1846, and during the fourteen years of issue no one monthly part was ever behind its time. This result is mainly due to the peculiar qualities of Mr. Long, who unites the talents of the scholar and the editor in a degree which is altogether unusual. If any one should imagine that a mixed mass of contributors is a punctual piece of machinery, let him take to editing upon that hypothesis, and he shall see what he shall see and learn what he shall learn.

The English contains about ten per cent. more matter than the Penny Cyclopædia and its Supplements ; including the third supplementary volume of 1848, which we now mention for the first time. The literary work of the two editions cost within 500*l.* of 50,000*l.* : that of the two editions of the Britannica cost 41,000*l.* But then it is to be remembered that the Britannica had matter to begin upon, which had been paid for in the former editions. Roughly speaking, it is probable that the authorship of a page of the same size would have cost nearly the same in one as in the other.

The longest articles in the Penny Cyclopædia were ' Rome ' in 98 columns and ' Yorkshire ' in 86 columns. The only article which can be called a treatise is the Astronomer Royal's ' Gravitation,' founded on the method of Newton in the eleventh section, but carried to a much greater extent. In the English Cyclopædia, the longest article of geography is 'Asia,' in 45 columns. In natural history the antelopes demand 36 columns. In biography, 'Wellington' uses up 42 columns, and his great military opponent 41 columns. In the division of Arts and Sciences, which includes much of a social and commercial character, the length of articles often depends upon the state of the times with regard to the subject. Our readers would not hit the longest article of this department in twenty guesses: it is 'Deaf and Dumb' in 60 columns. As other specimens, we may cite Astronomy, 19; Banking, 36; Blind, 24; British Museum, 35; Cotton, 27; Drama, 26; Gravitation, 50; Libraries, 50; Painting, 34; Railways, 18; Sculpture, 36; Steam, &c., 37; Table, 40; Telegraph, 30; Welsh language and literature, 39; Wool, 21; These are the long articles of special subdivisions: the words under which the Encyclopædia gives treatises are not so prominent. As in Algebra, 10; Chemistry, 12; Geometry, 8; Logic, 14; Mathematics, 5; Music, 9. But the difference between the collection of treatises and the dictionary may be illustrated thus:— though 'Mathematics' have only five columns, 'Mathematics, recent terminology of,' has eight: and this article we believe to be by Mr. Cayley, who certainly ought to know his subject, being himself a large manufacturer of the new terms which he explains. Again, though 'Music' *in genere*, as the schoolmen said, has only nine columns, 'Temperament and Tuning' has eight, and 'Chord' alone has two. And so on.

In a dictionary of this kind it is difficult to make a total clearance of *personality*: by which we mean that exhibition of peculiar opinion which is offensive to taste when it is shifted from the individual on the corporate book. The treatise of the known author may, as we have said, carry that author's controversies on its own shoulders: and even his crotchets, if we may use such a word. But the dictionary should not put itself into antagonism with general feeling, nor even with the feelings of classes. We refer particularly to the ordinary and editorial teaching of the article. If, indeed, the writer, being at issue with mankind, should confess the difference, and give abstract of his full grounds, the case is altered: the editor then, as it were, admits a correspondent to a statement of his own individual views. The dictionary portion of the Britannica is quite clear of any lapses on this point, so far as we know: the treatises and dissertations rest upon their authors. The Penny Cyclopædia was all but clear:

and great need was there that it should have been so. The Useful Knowledge Society, starting on the principle of perfect neutrality in politics and religion, was obliged to keep strict watch against the entrance of all attempt even to look over the hedge. There were two—we believe only two—instances of what we have called personality. The first was in the article 'Bunyan.' It is worth while to extract all that is said—in an article of thirty lines—about a writer who is all but universally held to be the greatest master of allegory that ever wrote:—

'His works were collected in two volumes, folio, 1736–7 : among them 'The Pilgrim's Progress' has attained the greatest notoriety. If a judgment is to be formed of the merits of a book by the number of times it has been reprinted, and the many languages into which it has been translated, no production in English literature is superior to this coarse allegory. On a composition which has been extolled by Dr. Johnson, and which in our own times has received a very high critical opinion in its favour [probably Southey], it is hazardous to venture a disapproval, and we, perhaps, speak the opinion of a small minority when we confess that to us it appears to be mean, jejune and wearisome.'

—If the unfortunate critic who thus individualized himself had been a sedulous reader of Bunyan, his power over English would not have been so *jejune* as to have needed that fearful word. This little bit of criticism excited much amusement at the time of its publication : but it was so thoroughly exceptional and individual that it was seldom or never charged on the book. The second instance occurred in the article 'Socinians.' It had been arranged that the head-words of Christian sects should be intrusted to members of the sects themselves, on the understanding that the articles should simply set forth the accounts which the sects themselves give of their own doctrines. Thus the article on the Roman Church was written by Dr. Wiseman. But the Unitarians were not allowed to come within the rule : as in other quarters, they were treated as the gypsies of Christianity. Under the head 'Socinians'—a name repudiated by themselves—an opponent was allowed not merely to state their alleged doctrines in his own way, but to apply strong terms, such as 'audacious unfairness,' to some of their doings. The protests which were made against this invasion of the understanding produced, in due time, the article 'Unitarians,' written by one of that persuasion. We need not say that these errors have been amended in the English Cyclopædia : and our chief purpose in mentioning them is to remark, that this is all we can find on the points in question against twenty-eight large volumes produced by an editor whose task was monthly, and whose issue was never delayed a single hour. How much was arrested before publication none but himself can say. We have not alluded to one or two remonstrances on questions of absolute fact, which are beside the present purpose.

Both kinds of encyclopædic works have been fashioned upon predecessors, from the very earliest which had a predecessor to be founded upon; and the undertakings before us will be themselves the ancestors of a line of successors. Those who write in such collections should be careful what they say, for no one can tell how long a misstatement may live. On this point we will give the history of a pair of epithets. When the historian De Thou died, and left the splendid library which was catalogued by Bouillaud and the brothers Dupuis (Bullialdus and Puteanus), there was a manuscript of De Thou's friend Vieta, the *Harmonicon Cœleste*, of which it is on record, under Bouillaud's hand, that he himself lent it to Cosmo de' Medici, to which must be added that M. Libri found it in the Magliabecchi Library at Florence in our own day. Bouillaud, it seems, entirely forgot what he had done. Something, probably, that Peter Dupuis said to Bouillaud, while they were at work on the catalogue, remained on his memory, and was published by him in 1645, long after; to the effect that Dupuis lent the manuscript to Mersenne, from whom it was procured by some intending plagiarist, who would not give it back. This was repeated by Sherburne, in 1675, who speaks of the work, which ' being communicated to Mersennus was, by some perfidious acquaintance of that honest-minded person, surreptitiously taken from him, and irrecoverably lost or suppressed, to the unspeakable detriment of the lettered world.' Now let the reader look through the dictionaries of the last century and the present, scientific or general, at the article ' Vieta,' and he will be amused with the constant recurrence of ' honest-minded ' Mersenne, and his 'surreptitious' acquaintance. We cannot have seen less than thirty copies of these epithets.

October 18, 1862. *Correspondence of Scientific Men of the Seventeenth Century, in the Collection of the Earl of Macclesfield.* 2 vols. (Oxford, University Press.)

Though the title-page of this collection bears date 1841, it is only just completed by the publication of its Table of Contents and Index. Without these, a work of the kind is useless for consultation, and cannot make its way.   The reason of the delay will appear: its effect is well known to us.   We have found inquirers into the history of science singularly ignorant of things which this collection might have taught them.

In the same year, 1841, the Historical Society of Science, which had but a brief existence, published a collection of letters, eighty-three in number, edited by Mr. Halliwell, of English men of science, which dovetails with the one before us, and is for the most part of a prior date.   The two should be bound up together. The smaller collection runs from 1562 to 1682 ; the larger, from 1606 to past 1700.   We shall speak of the two as the Museum collection   and   the   Macclesfield collection.   And   near   them should be placed, in every scientific library, the valuable collection published, by Mr. Edleston, for Trinity College, in 1850.

The history of these letters runs back to famous John Collins, the attorney-general of the mathematics, as he has been called, who wrote to everybody, heard from everybody, and sent copies of everybody's letter to everybody else.   He was in England what Mersenne was in France : as early as 1671, E. Bernard addresses him as ' the very Mersennus and intelligence of this age.'   John Collins was never more than accountant to the Excise Office, to which he was promoted from teaching writing and ciphering, at the Restoration : he died in 1682.   We have had a man of the same office in our own day, the late Prof. Schumacher, who made the little Danish Observatory of Altona the junction of all the lines by which astronomical information was conveyed from one country to another.   When the collision took place between Denmark and the Duchies, the English Government, moved by the Astronomical Society, instructed its diplomatic agents to represent strongly to the Danish Government, when occasion should arise, the great importance of the Observatory of Altona to the astronomical communications of the whole world.   But Schumacher had his own celebrated journal, the *Astronomische Nachrichten,* by which to work out part of his plan ; private correspondence was his supplementary assistant.   Collins had only correspondence to rely on.   Nothing is better known than that it was Collins's collection which furnished the materials put forward by the Committee of the Royal Society in 1712, as a defence of Newton against the partisans of Leibnitz.   The noted *Commercium Epistolicum* is but the abbreviation of a title which runs on with ' D. Johannis Collins et aliorum . . .'

The whole of this collection passed into the hands of William Jones, the father of the Indian Judge of the same name, who died in 1749. Jones was originally a teacher, but was presented with a valuable sinecure by the interest of George, second Earl of Macclesfield, the mover of the bill for the change of style in Britain, who died President of the Royal Society. This change of style may perhaps be traced to the union of energies which were brought into concert by the accident of a common teacher: Lord Macclesfield and Lord Chesterfield, the mover and seconder, and Daval, who drew the bill, were pupils of De Moivre. Jones, who was a respectable mathematician though not an inventor, collected the largest mathematical library of his day, and became possessor of the papers of Collins, which contained those of Oughtred and others. Some of these papers passed into the custody of the Royal Society: but the bulk was either bequeathed to, or purchased by, Lord Macclesfield; and thus they found their way to Shirburn Castle, where they still remain.

A little before 1836, this collection attracted the attention of a searching inquirer into points of mathematical history, the late Prof. Rigaud, who died in 1839. He examined the whole collection of letters, obtained Lord Macclesfield's consent to their publication, and induced the Oxford Press to bear the expense. It must be particularly remembered that there still remains at Shirburn Castle a valuable mass of non-epistolary manuscripts. So far as we can see, the best chance of a further examination and publication lies in public encouragement of the collection now before us: the Oxford Press might be induced to extend its operations if it were found that the results were really of interest to the literary and scientific world. Rigaud died before the work was completed, and the publication was actually made by one of his sons, S. Jordan Rigaud, who died Bishop of Antigua. But this publication was little noticed, for the reasons given. The completion now published consists of a sufficient table of contents, of the briefest kind, by Prof. De Morgan, and an excellent index by the Rev. John Rigaud. The work is now fairly started on its career.

If we were charged to write a volume with the title 'Small things in their connexion with great,' we could not do better than choose the small part of this collection of letters as our basis. The names, as well as the contents, are both great and small: the great names, those which are known to every mathematician who has any infusion of the history of his pursuit, are Briggs, Oughtred, Charles Cavendish, Gascoigne, Seth Ward, Wallis, Hu[y]gens, Collins, William Petty, Hooke, Boyle, Pell, Oldenburg, Brancker, Slusius, Bertit, Bernard, Borelli, Mouton,

Pardies, Fermat, Towneley, Auzout, D. Gregory, Halley, Machin, Montmort, Cotes, Jones, Saunderson, Reyneau, Brook Taylor, Maupertuis, Bouguer, La Condamine, Folkes, Macclesfield, Baker, Barrow, Flamsteed, Lord Brounker, J. Gregory, Newton and Keill. To these the Museum collection adds the names of Thomas Digges, Dee, Tycho Brahé, Harriot, Lydyat, Briggs, Warner, Tarporley, Pell, Lilly, Oldenburg, Collins, Morland.

The first who appears on the scene is the celebrated Oughtred, who is related to have died of joy at the Restoration : but it should be added, by way of excuse, that he was eighty-six years old. He is an animal of extinct race, an Eton mathematician. Few Eton men, even of the minority which knows what a sliding rule is, are aware that the inventor was of their own school and college : but they may be excused, for Dr. Hutton, so far as his Dictionary bears witness, seems not to have known it any more than they. A glance at one of his letters reminds us of a letter from the Astronomer Royal on the discovery of Neptune, which we printed March 20, 1847. Mr. Airy there contends, and proves it both by Leverrier and by Adams, that the limited publication of a private letter is more efficient than the more general publication of a printed memoir. The same may be true of a dead letter, as opposed to a dead book. Our eye was caught by a letter of Oughtred (1629), containing systematic use of contractions for the words *sine, cosine*, &c., prefixed to the symbol of the angle. This is so very important a step, simple as it is, that Euler is justly held to have greatly advanced trigonometry by its introduction. Nobody that we know of has noticed that Oughtred was master of the improvement, and willing to have taught it, if people would have learnt. After looking at his dead letter, we naturally turned to his dead book on trigonometry, and there we found the abbreviations *s, sco, t, tco, se, seco*, regularly established as part of the system of the work. But not one of those who have investigated the contending claims of Euler and Thomas Simpson has chanced to know of Oughtred's ' Trigonometrie ' : and the present revival is due to his letter, not to his book.

A casual reader, turning over the pages, would imagine that almost all the letters had been printed, either in the General Dictionary, or in Birch, &c. : so often does the supplementary remark begin with ' this letter has been printed in ——.' For ourselves we thought, until we counted, that a large majority of the letters had been given, either in whole or in part. But the positive strikes the mind more forcibly than the negative : we find that all of which any portion has been in type makes up very little more than a quarter ; the cases in which the whole letter is given being a minority of this quarter. The person who has been best ransacked is Flamsteed : of 36 letters from him, 34 had been previously given in whole or in part. Of 59 letters to and from Newton, only 17 have been culled.

The letters have been modernized in spelling, and, to some extent, in algebraical notation; it also seems that conjectural methods of introducing interpolations into the text have been necessary. For all this we are sorry : the scientific value of the collection is little altered, but its literary value is somewhat lowered. But it could not be helped: the printers could not work from the originals, and Prof. Rigaud had to copy everything himself. A fac-simile must have been the work of more time than he had to give : had he attempted it, his death would have cut short the whole undertaking, instead of allowing him to prepare everything but a preface, and to superintend the printing of one of the volumes. We may also add, that we believe we have notices of *all* the letters in the Macclesfield collection. We judge this because several which are too trivial to print are numbered and described ; and those would certainly not have been noticed if *any* omissions had been made. And we know that every letter was removed from Shirburn Castle to Oxford.

Two persons emerge from oblivion in this series of letters. The first is Michael Dary, an obscure mathematician, who was in correspondence with Newton and other stars. He was a gauger at Bristol, by the interest of Collins ; afterwards a candidate for the mathematical school at Christ's Hospital, with a certificate from Newton : he was then a gunner in the Tower, and is lastly described by Wallis as ' Mr. Dary, the tobacco-cutter, a knowing man in algebra.' In 1674, Dary writes to Newton at Cambridge, as follows :—' Although I sent you three papers yesterday, I cannot refrain from sending you this. I have had fresh thoughts this morning.' Two months afterwards poor Newton writes to Collins, ' Mr. Dary is very solicitous about mathematics': but, in spite of the persecution, he subscribes himself to Dary 'your loving friend.' Dary's *problem* is that of finding the rate of interest of an annuity of which the value and term are given. Dary's *theorem*, which he seems to have invented specially for the solution of his problem, though it is of wide range, can be exhibited to mathematical readers even in our columns. In modern language, it is that the limit of $\phi^n x$, when $n$ increases without limit, is a solution of $\phi x = x$. We have mentioned the I. Newton to whom Dary looked up ; we add a word about the one on whom he looked down. Dr. John Newton, a sedulous publisher of logarithms, tables of interest, &c., who began his career before Isaac Newton, sometimes puzzles those who do not know him, when described as I. Newton. The scientific world was of opinion that all that was valuable in one of his works was taken from Dary's private communications.

The second character above alluded to is one who carried mathematical researches a far greater length than Newton himself : the assistance which he rendered in this respect, even to Newton, has never been acknowledged in modern times : though the work before us shows that his contemporaries were fully aware of it, and never thought of concealing it. In his theory of gravitation, in which, so far as he went, we have every reason to believe he was prior to Newton, he did not extend his calculations to the distance of the moon ; his views in this matter were purely terrestrial, and led him to charge according to weight. He was John Stiles, the London and Cambridge carrier : his name is a household word in the Macclesfield Letters, and is even enshrined in the depths of Birch's quartos. Dary informs Newton—let us do his memory this justice—that he had paid John Stiles for the carriage. At the time when the railroad to Cambridge was opened, a correspondent recommended the directors, in our columns, to call an engine by the name of John Stiles, and never to let that name go off the road. We do not know whether the advice was followed : if not, we repeat it.

Little points of life and manners come out occasionally. Baker, the author of a work on algebra much esteemed at the time, wrote to Collins that their circumstances are alike, ' having a just equal number of chargeable olive-branches, and being in the same predicament and blessed condemnation with you, not more preaching than unpaid, and preaching the art of contentment to others, am forced to practise it.' But the last sentence of his letter runs as follows :—' I have sent by the bearer . . . twenty shillings, as a token to you ; desiring you to accept of it, as a small taste from Yours, Thos. Baker.' In our day, men of a station to pay parish taxes do not offer their friends hard money to buy liquor. But Flamsteed writes to Collins as follows :— ' Last week he sent us down the counterpart, which my father has sealed, and I return up to you by the carrier, with 5l. to be paid to Mr. Leneve for the writing. I have added 2s. 6d. over, which will pay the expenses and serve to drink, with him.' This would seem as odd to us as it would have seemed thirty years ago that half-a-crown should pay carriage for a deed from Derby to

London, and leave margin for a bottle of wine : in our day, the Post-office and the French treaty would just manage it between them. But Flamsteed does not limit his friend to one bottle ; he adds, 'If you expend more than the half-crown, I will make it good after Whitsuntide.' Collins does not remember exactly where he had met James Gregory, and mentions two equally likely places thus :—' Sir, it was once my good hap to meet with you in an alehouse, or in Sion College.' There is a little proof how universally the dinner-hour was twelve o'clock. Astronomers well know the method of finding time by equal altitudes of the sun before and after noon : Huyghens calls it 'le moyen de deux égales hauteurs du soleil devant et après *dîner*.'

There is one mention of 'Mr. Cocker, our famous English graver and writer, now a schoolmaster at Northampton.' This is the true Cocker : his genuine works are specimens of writing, such as engraved copy-books, including some on arithmetic, with copper-plate questions and space for the working ; also a book of forms for law-stationers, with specimens of legal handwriting. It is recorded somewhere that Cocker and another, whose name we forget, competed with the Italians in the beauty of their flourishes. This was his real fame : and in these matters he was great. The eighth edition of his book of law forms (1675), published shortly after Cocker's death, has a preface signed 'J. H.' This was John Hawkins, who became possessed of Cocker's papers—at least he said so—and subsequently forged the famous Arithmetic, a second work on Decimal Arithmetic, and an English dictionary, all attributed to Cocker. The proofs of this are set out in De Morgan's 'Arithmetical Books.' Among many other corroborative circumstances, the clumsy forger, after declaring that Cocker to his dying day resisted strong solicitation to publish his Arithmetic, makes him write in the preface an *Ille ego qui quondam* of this kind :—' I have been instrumental to the benefit of many, by virtue of those useful arts, writing and engraving ; and do *now*, with the same *wonted alacrity*, cast this my arithmetical mite into the public treasury.' The book itself is not comparable in merit to at least half-a-dozen others. How then comes Cocker to be the impersonation of Arithmetic ? Unless some one can show proof, which we have never found, that he was so before 1756, the matter is to be accounted for thus.

Arthur Murphy, the dramatist, was by taste a man of letters, and ended by being the translator of Tacitus ; though many do not know that the two are one. His friends had tried to make him a man of business ; and no doubt he had been well plied with commercial arithmetic.  His first dramatic performance, the farce of 'The Apprentice,' produced in 1756, is about an idle young man who must needs turn actor.  Two of the best known books of the day in arithmetic were those of Cocker and Wingate. Murphy chooses *Wingate* to be the name of an old merchant who delights in vulgar fractions, and *Cocker* to be his arithmetical catchword—'You read Shakspeare! get Cocker's Arithmetic! you may buy it for a shilling on any stall ; best book that ever was wrote!': and so on.  The farce became very popular, and, as we believe, was the means of elevating Cocker to his present pedestal, where Wingate would have been, if his name had had the droller sound of the two to English ears.

A notoriety of an older day turns up, Major-General Lambert. The common story is that he was banished to Guernsey, where he passed thirty years in confinement, rearing and painting flowers. But Baker, in 1678, represents him as a prisoner at Plymouth, sending equations for solution as a challenge: probably his place of confinement was varied, and his occupation also.

[General Lambert was removed to Plymouth, probably about 1668.  His daughter captured the son of the Governor of Guernsey, who therefore probably was reckoned an unsafe custodier thenceforward ; though he assured the king that he had turned the young couple out of doors, and had never given them a penny.  Great importance was attached to Lambert's safe detention : probably the remaining republicans looked upon him as to be their next Cromwell, if such a thing were to be.  There were standing orders to shoot him at once on the first appearance of any enemy before the island.  See *Notes and Queries*, 3rd S. iv. 89.]

Collins informs James Gregory that 'some of the Royal Academy wrote over to Mr. Oldenburg, who was desired to impart the same to the Council of the Royal Society, that the French King was willing to allow pensions to one or two learned Englishmen, but they never made any answer to such a proposal.' This was written in 1671, and the thing probably happened several years before.  Mr. De Morgan communicated the account of the proposal to Lord Macaulay, who replied that he did not think that any Englishman *received* a literary pension from Louis; but that there is a curious letter, about 1664, from the French Ambassador, in which he says that he has, by his master's orders, been making inquiries as to the state of learning in England, and that he is sorry to find that the best writer is *the infamous Miltonus.*  On two such independent testimonies it may be held proved that the

French King had attempted to buy a little adherence from English literature and science ; and the silent contempt of the Royal Society is an honourable fact in their history.

Another little bit of politics is as follows. Oughtred is informed that 'Mr. Foster, our Lecturer on Astronomy at Gresham College, is put out because he will not kneel down at the communion-table. A Scotsman [Mungo Murray], one that is *verbi bis minister*, is now lecturer in Mr. Foster's place.' Ward, in his work on the Gresham Professors, suppresses the reason, and the suppression lowers the character of his book. Foster was expelled in 1636, and re-elected on a vacancy in 1641, when Puritanism had gained strength.

The correspondence of Newton would require deeper sifting than could be given in such an article as the present. The first of the letters (1669) is curious, as presenting the appearance of forms belonging to the great calculus which, in this paragraph, we ought to call that of fluxions. We find, of the date February 18, 1669–70, what we believe is the earliest manifestation of that morbid part of Newton's temperament which has been so variously represented. He had solved a problem—being that which we have called Dary's—on which he writes as follows : ' The solution of the annuity problem, if it will be of any use, you have my leave to insert it into the *Philosophical Transactions,* so it be without my name to it. For I see not what there is desirable in public esteem, were I able to acquire and maintain it. It would perhaps increase my acquaintance, the thing which I chiefly study to decline.'

Three letters touch upon ' the experiment of glass rubbed to cause various motions in bits of paper underneath ' : they are supplements to the account given by Newton to the Royal Society, and printed by Birch. It was Newton, so far as appears, who added *glass* to the substances known to be electric. Soon afterwards we come to a little bit of the history of the appointment to the Mint. It has appeared from the researches of late years that Newton was long an aspirant for public employment : the only coolness which is known to have taken place between him and Charles Montague [Halifax] arose out of his imagining that his friend was not in earnest about getting him into the public service. March 14, 1696, Newton writes thus to Halley : —' And if the rumour of preferment for me in the Mint should hereafter, upon the death of Mr. Hoar [the comptroller], or any other occasion, be revived, I pray that you would endeavour to obviate it by acquainting your friends that I neither *put in* for *any* place in the Mint, nor would meddle with *Mr. Hoar's place*, were it offered to me.' This means that Mr. Hoar's place had been suggested, which Newton seems to have declined. Five days afterwards, Montague writes to Newton that he is to have the *Wardenship*. It is fair to Newton to say that in all probability this was not—or only in a smaller degree—a question of personal dignity, or of salary. It must by this time have been clear to him that the minister, though long bound to make him an object of patronage, was actually seeking him for the Mint, because he wanted both Newton's name and his talents for business— which he knew to be great—in the weighty and dangerous operation of restoring the coinage. It may have been, and probably was, the case that Newton had a tolerably accurate notion of what he would have to do, and of what degree of power would be necessary to enable him to do it in his own way.

We have said that the non-epistolary manuscripts are still unexamined. There is a chance that one of them may answer a question of two centuries' standing, which is worth answering, because it has been so often asked. About 1640, Warner, afterwards assisted by Pell, commenced a table of *antilogarithms*, of the kind which Dodson afterwards constructed anew and published. In the Museum collection there is inquiry after inquiry from Charles Cavendish, first, as to when the *Analogics*, as he called them, would be finished; next, when they would be printed. Pell answers, in 1644, that Warner left his papers to a kinsman, who had become bankrupt, and proceeds thus :—

' I am not a little afraid that all Mr. Warner's papers, and no small share of my labours therein, are seazed upon, and most unmathematically divided between the sequestrators and creditors, who (not being able to ballance the account where there appeare so many numbers, and much troubled at the sight of so many crosses and circles in the superstitious Algebra and that black art of Geometry) will, no doubt, determine once in their lives to become figure-casters, and so vote them all to be throwen into the fire, if some good body doe not reprieve them for pye-bottoms, for which purposes you know analogicall numbers are incomparably apt, if they be accurately calculated.'

Pell afterwards told Wallis that the papers had fallen into the hands of Dr. Busby, and Collins writes that they were left in the hands of Dr. Thorndike, a prebendary of Westminster; whence Rigaud seems to say that Thorndike had left them to Dr. Busby. Birch says that he procured for the Royal Society four boxes from Busby's trustees, containing papers of Warner and Pell: but there is no other tradition of such things in the Society. But in the Birch manuscripts at the British Museum, there turns up, as printed in what we call the Museum collection, a list of Warner's papers, with *Collins's* receipt to Dr. Thorndike at the bottom, and engagement to restore them on demand. The date is December 14, 1667 ; Wallis's statement being in 1693. It is possible that Busby may be a mistake altogether : he was very unlikely to have had charge of any mathematical papers : there may have been a confusion between the Prebendary of Westminster and the Head Master of Westminster School. If so, in all probability Thorndike handed the cumbrous lot over to the notorious collector of mathematical papers, blessing himself that he had got rid of them in a manner which would insure their return if he were called upon by the owners to restore them. It is much against this hypothesis that Dodson, who certainly recalculated, can say nothing more about Warner than a repetition of Wallis's story : though, had Collins kept the papers, they would probably have been in Jones's possession at the very time when Dodson, who was a friend of Jones and a user of his library, was engaged on his own computations. But even books, and still more manuscripts, are often singularly overlooked ; and it remains not very improbable that Warner's table is now at Shirburn Castle, among the unexamined manuscripts.

---

*Redit labor actus in orbem.* Among the matters which have come to me since the Budget opened, there is a pamphlet of quadrature of two pages and a half from Prof. Recalcati, already mentioned. It ends with " Quelque objection qu'on fasse touchant les raisonnements ci-dessus on tombera toujours dans l'absurde.' A civil engineer—so he says—has made the quadrature " no longer a problem, but an axiom." As follows : " Take the quadrant of a circle whose circumference is given, square the quadrant which gives the true square of the circle. Because $30 \div 4 = 7 \cdot 5 \times 7 \cdot 5 = 56 \cdot 25 =$ the positive square of a circle whose circumference is 30." Brevity, the soul of wit, is the " wings of mighty winds " to quadrature, and sends it " flying all abroad." A *surbodhicary*—something like M.A. or LL.D., I understand— at Calcutta, published in 1863 the division of an angle into any odd number of parts, demonstration and all in — when the diagram is omitted—one page, good-sized, well-leaded type, small duodecimo. But in the Preface he acknowledges " sheer inability " to execute his task. Mr. William Dean, of Todmorden, in 1863, announced $3\frac{9}{64}$ as proved both practically and geometrically : he has been already mentioned anonymously. Next I have the tract of Don Juan Larriva, published at Leiria in 1856, and dedicated to Queen Victoria. Mr. W. Peters, already mentioned, who has for some months been circulating diagrams on a card, publishes (August, 1865) 'The Circle Squared.' He agrees with the Archpriest of St. Vitus. He hints that a larger publication will depend partly on the support he receives, and partly on the castigation, for which last, of course, he looks to me. Cyclometers have their several styles of wit ; so have anti-cyclometers too, for that matter. Mr. Peters will not allow me any extra-journal being : I am essentially a quotation from the *Athenæum* ; ' A. De Morgan ' *et prœterea nihil.* If he had to pay for keeping me set up, he would find out his mistake, and would be glad to compound handsomely for a stereotype. Next comes a magnificent sheet of pasteboard, printed on both sides. Having glanced at it and detected quadrature, I began methodically at the beginning—' By Royal Command,' with the lion and unicorn, and all that comes between. Mercy on us! thought I to myself : has Her Majesty referred the question to the Judicial Committee of the Privy Council, where all the great difficulties go now-a-days, and is this proclamation the result? On reading further I was relieved by finding that the first side is entirely an advertisement of Joseph Gillott's steel pens, with engraving of his premises, and notice of novel application of his unrivalled machinery. The second side begins with 'the circle rectified ' by W. E. Walker, who finds $\pi = 3\cdot141594789624155\ldots$ This is an off-shoot from an accurate geometrical rectification, on which it is to be presumed Mr. Gillott's new machinery is founded. I have no doubt that Mr. Walker's error, which is only in the sixth place of decimals, will not hurt the pens, unless it be by the slightest possible increase of the tendency to open at the points. This arises from Mr. Walker having rectified above proof by $\cdot000002136034362\ldots$

Lastly, I, even I myself, who have long felt that I was a quadrature below par, have solved the problem by means which, in the present state of the law of libel, I dare not divulge. But the result is permitted ; and it goes far to explain all the discordances. The ratio of the circumference to the diameter is not always the same ! Not that it varies with the radius ; the geometers are right enough on that point : but it varies with the time, in a manner depending upon the difference of the true longitudes of the Sun and Moon. A friend of mine—at least until he misbehaved—insisted on the mean right ascensions : but

I served him as Abraham served his guest in Franklin's parable. The true formula is, A and $a$ being the Sun's and Moon's longitudes,

$$\pi = 3\tfrac{13}{80} + \tfrac{3}{80} \cos(A - a).$$

Mr. James Smith obtained his quadrature at full moon; the Archpriest of St. Vitus and some others at new moon. Until I can venture to publish the demonstration, I recommend the reader to do as I do, which is to adopt 3·14159 . . . . , and to think of the matter only at the two points of the lunar month at which it is correct. The *Nautical Almanac* will no doubt give these points in a short time: I am in correspondence with the Admiralty, with nothing to get over except what I must call a perverse notion on the part of the Superintendent of the *Almanac*, who suspects one correction depending on the Moon's latitude; and the Astronomer Royal leans towards another depending on the date of the Queen's accession. I have no patience with these men: what can the Moon's node or the Queen's reign possibly have to do with the ratio in question? But this is the way with all the regular men of science; Newton is to them &c. &c. &c. &c.

The following method of finding the circumference of a circle (taken from a paper by Mr. S. Drach in the *Phil. Mag.*, Jan. 1863, Suppl.) is as accurate as the use of 3·14159265. From three diameters deduct 8-thousandths and 7-millionths of a diameter; to the result add five per cent. We have then not quite enough; but the shortcoming is at the rate of about an inch and a sixtieth of an inch in 14,000 miles.

---

Though I have met with nothing but a little tract from the school of Jacob Behmen (or Böhme; I keep to the old English version of his name), yet there has been more, and of a more recent date. I am told of an 'Introduction to Theosophy [*Theo* privative, I suppose, as in theological]; or, the Science of the Mystery of Christ,' published in 1854, mostly from the writings of William Law: and also of a volume of 688 pages, of the same year, printed for private circulation, containing notes for a biography of William Law. The editor of the first work wishes to grow 'a generation of perfect Christians' by founding a Theosophic College, for which he requests the public to raise a hundred thousand pounds. There is a good account of Jacob Behmen in the *Penny Cyclopædia*. The author mentions inaccurate accounts, one of which he quotes, as follows: 'He

derived all his mystical and rapturous doctrine from Wood's
'Athenæ Oxonienses,' vol. i. p. 610, and 'Hist. et Antiq. Acad.
Oxon.,' vol. ii. p. 308.' On which the author remarks that Wood
was born after Behmen's death. There must have been a
few words which slipped out: what is meant is that Behmen
'derived his doctrine from *Robert Fludd, for whom see* Wood's
&c. &c.' Even this is absurd enough: for Behmen began to
publish in 1610, and Fludd in 1616. Fludd was a Rosicrucian,
and a mystic of a different type from Behmen. I have some of
his works, and could produce out of them paradoxes enough,
according to our ways of thinking, to fit out a host. But the
Rosicrucian system was a recognised school of its day, and Fludd,
a man of great learning, had abettors enough in all which he
advanced, and predecessors in most of it.

[A Correspondent has recently sent a short summary of the claims
of Jacob Behmen to rank higher than I have placed him. I shall
gladly insert this summary in the book I contemplate, as a statement of what is said of Behmen far less liable to suspicion of exaggeration than anything I could write. I shall add a few extracts
from Behmen himself, in support of his right to be in my list.]

'*Jacob Behmen.*—That Prof. De Morgan classes Jacob Behmen
among paradoxers can only be attributed to the fact of his being
avowedly unacquainted with the writings of that author. Perhaps you may think a few words from one who knows them well
of sufficient interest to the learned Professor, and your readers in
general, to be worthy of space in your columns. The meta-
physical system of Behmen—the most perfect and only true one
—still awaits a qualified commentator. Behmen's countryman,
Dionysius Andreas Freher, who spent the greater part of his life
in this country, and whose exposition of Behmen exists only in
MS., filling many volumes, written in English, with the exception of two, written in German, with numerous beautiful, highly
ingenious, and elaborate illustrations,—copies of some of which
are in the British Museum, but all the originals of which are in
the possession of the gentleman who is the editor of the two
works alluded to by Prof. De Morgan,—this Freher was the first
to philosophically expound Behmen's system, which was afterwards, with the help of these MSS., as it were, popularized by
William Law; but both Freher and Law confined themselves
chiefly to its theological aspect. In Behmen, however, is to be
found, not only the true ground of all theology, but also that
of all physical science. He demonstrated with a fullness, accu-

racy, completeness and certainty that leave nothing to be desired, the innermost ground of Deity and Nature ; and, confining myself to the latter, I can from my own knowledge assert, that in Behmen's writings is to be found the true and clear demonstration of every physical fact that has been discovered since his day. Thus, the science of electricity, which was not yet in existence when he wrote, is there anticipated ; and not only does Behmen describe all the now known phenomena of that force, but he even gives us the origin, generation and birth of electricity itself. Again, positive evidence can be adduced that Newton derived all his knowledge of gravitation and its laws from Behmen, with whom gravitation or attraction is, and very properly so, as he shows us, the first of the seven properties of Nature. The theory defended by Mr. Grove, at the Nottingham meeting of last year, that all the apparently distinct causes of moral and physical phenomena are but so many manifestations of one central force, and that Continuity is the law of nature, is clearly laid down, and its truth demonstrated, by Behmen, as well as the distinction between spirit and matter, and that the moral and material world is pervaded by a sublime unity. And though all this was not admitted in Behmen's days, because science was not then sufficiently advanced to understand the deep sense of our author, many of his passages, then unintelligible, or apparently absurd, read by the light of the present age, are found to contain the positive enunciation of principles at whose discovery and establishment science has only just arrived by wearisome and painful investigations. Every new scientific discovery goes to prove his profound and intuitive insight into the most secret workings of nature; and if scientific men, instead of sharing the prejudice arising from ignorance of Behmen's system, would place themselves on the vantage ground it affords, they would at once find themselves on an eminence whence they could behold all the arcana of nature. Behmen's system, in fact, shows us the *inside* of things, while modern physical science is content with looking at the *outside*. Behmen traces back every outward manifestation or development to its one central root,—to that one central energy which, as yet, is only suspected ; every link in the chain of his demonstration is perfect, and there is not one link wanting. He carries us from the outbirths of the circumference, along the radius to the centre, or point, and beyond that even to the zero, demonstrating the constitution of the zero, or nothing, with mathematical precision. C. W. H.'

And so Behmen is no subject for the Budget ! I waited until I should chance to light on one of his volumes, knowing that any volume would do, and almost any page. My first hap was on the second volume of the edition of 1664 (4to, published by M. Richardson) and opening near the beginning, a turn or two brought me to page 13, where I saw about *sulphur* and *mercurius* as follows :—

Thus SUL is the soul, in an herb it is the oil, and in man also, according to the spirit of *this* world in the third principle, which is continually generated out of the anguish of the will in the mind, and the Brimstone-worm is the Spirit, which hath the fire and *burneth* : PHUR is the sour wheel in itself which causeth that.

*Mercurius* comprehendeth all the four forms, even as the life springeth up, and yet hath not its dark beginning in the Center as the PHUR hath, but after the flash of fire, when the sour dark form is terrified, where the hardness is turned into pliant sharpness, and where the second will . (*viz.* the will of nature, which is called the Anguish) ariseth, there Mercurius hath its original. For MER is the shivering wheel, very horrible, sharp, venomous, and hostile ; which assimulateth it thus in the sourness in the flash of fire, where the sour wrathful life *ariseth*. The syllable CU is the pressing out, of the *Anxious* will of the mind, from Nature : which is climbing up, and *willeth* to be out aloft. RI is the comprehension of the flash of fire, which in MER giveth a clear sound and tune. For the flash maketh the tune, and it is the Salt-Spirit which *soundeth*, and its form (or quality) is gritty like sand, and herein arise noises, sounds, and voices, and thus CU comprehendeth the flash, and so the pressure is as a *wind* which thrusteth, and giveth a spirit to the flash, so that it liveth and burneth. Thus the syllable US is called the burning fire, which with the spirit continually driveth itself forth : and the syllable CU presseth continually upon the flash.

Shades of Tauler and Paracelsus, how strangely you do mix ! Well may Hallam call Germany the native soil of Mysticism. Had Behmen been the least of a scholar, he would not have divided *sulph-ur* and *merc-ur-i-us* as he has done :—and the inflexion *us*, that boy of all work, would have been rejected. I think it will be held that a writer from whom hundreds of pages like the above could be brought together, is fit for the Budget. If Sampson Arnold Mackay had tied his etymologies to a mystical Christology, instead of a mystical infidelity, he might have had a school of followers. The nonsense about Newton borrowing gravitation from Behmen passes only with those who know neither what Newton did, nor what was done before him.

The above reminds me of a class of paradoxers whom I wonder that I forgot ; they are without exception the greatest bores of all, because they can put the small end of their paradox into any literary conversation whatsoever. I mean the people who have heard the local pronunciation of celebrated names, and attempt not only to imitate it, but to impose on others their broken German or Arabic, or what not. They also learn the vernacular names of those who are generally spoken of in their Latin forms; at least, they learn a few cases, and hawk them as evidences of erudition. They are miserably mistaken: scholarship, as a rule, always accepts the vernacular form of a name which has vernacular celebrity. Hallam writes Behmen: his index-maker, rather superfluously, gives *Behmen* or Boehm.' And he retains Melanchthon, the name given by Reuchlin to his little kinsman Schwartzerd, because the world has adopted it: but he will none of Capnio, the name which Reuchlin fitted on to himself, because the world has not adopted it. He calls the old forms pedantry: but he sees that the rejection of well-established results of pedantry would be greater pedantry still. The paradoxers assume the question that it is more *correct* to sound a man by lame imitation of his own countrymen than as usual in the country in which the sound is to be made. Against them are, first, the world at large; next, an overpowering majority of those who know something about surnames and their history. Some thirty years ago—a fact—there appeared at the police-office a complainant who found his own law. In the course of his argument, he asked, 'What does Kitty say?'—'Who's Kitty?' said the magistrate, 'your wife, or your nurse?'—'Sir! I mean Kitty, the celebrated lawyer.'—'Oh!' said the magistrate, 'I suspect you mean Mr. Chitty, the author of the great work on pleading.' —'I do sir! but Chitty is an Italian name, and ought to be pronounced *Kitty*.' This man was a full-blown flower: but there is many a modest bud; and all ought either to blush when seen or to waste their pronunciation on the desert air.

I stand up for king Custom, or *Usus*, as Horace called him, with whom is *arbitrium* the decision, and *jus* the right, and *norma* the way of deciding, simply because he has *potestas* the power. He may admit one and another principle to advise: but Custom is not a constitutional king; he may listen to his cabinet, but he decides for himself: and if the ministry should resign, he blesses his stars and does without them. We have a glorious liberty in England of owning neither dictionary, grammar, nor spelling-book: as many as choose write by either of the three, and decide all disputed points their own way, those following them who please. Throughout this book I have called people by the names which denote them in their books, or by our vernacular names. This is the intelligible way of proceeding. I might, for instance (p. 31), have spoken of Charles de Bovelles, of Lefèvre d'Étaples, of Pèlerin, and of Etienne. But I prefer the old plan. Those who like another plan better, are welcome to substitute with a pen, when they know what to write; when they do not, it is clear that they would not have understood me if I had given modern names.

The principal advisers of King Custom are as follows. First, there is Etymology, the *chiffonnier*, or general rag-merchant, who has made such a fortune of late years in his own business that he begins to be considered highly respectable. He gives advice which is more thought of than followed, partly on account of the fearful extremes into which he runs. He lately asked some boys of sixteen, at a matriculation examination in *English*, to what branch of the Indo-Germanic family they felt inclined to refer the Pushto language, and what changes in the force of the letters took place in passing from Greek into Mœso-Gothic. Because all syllables were once words, he is a little inclined to insist that they shall be so still. He would gladly rule English with a Saxon rod, which might be permitted with a certain discretion which he has never attained : and when opposed, he defends himself with the analogies of the Aryan family until those who hear him long for the discovery of an Athanasyus. He will transport a word beyond seas—he is recorder of Rhematopolis—on circumstantial evidence which looks like mystery gone mad ; but, strange to say, something very often comes to light after sentence passed which proves the soundness of the conviction.

The next adviser is Logic, a swearing old justice of peace, quorum, and rotulorum, whose excesses brought on such a fit of the gout that for many years he was unable to move. He is now mending, and his friends say he has sown his wild oats. He has some influence with the educated subjects of Custom, and will have more, if he can learn the line at which interference ought to stop : with them he has succeeded in making an affirmative of two negatives ; but the vulgar won't never have nothing to say to him. He has always railed at Milton for writing that Eve was the fairest of her daughters ; but has never satisfactorily shown what Milton ought to have said instead.

The third adviser has more influence with the mass of the subjects of King Custom than the other two put together; his name is Fiddlefaddle, the toy-shop keeper ; and the other two put him forward to do their worst work. In return, he often uses their names without authority. He took Etymology to witness that *means* to an end must be plural : and he would have any one method to be a *mean*. But Etymology proved him wrong, King Custom referred him to his Catechism, in which is ' a means whereby we receive the same,' and Analogy—a subordinate of Etymology—asked whether he thought it a great *new* to hear that he was wrong. It was either this Fiddle-faddle, or Lindley Murray his traveller, who persuaded the Miss Slipslops, of the Ladies Seminary, to put ' The Misses Slipslop ' over the gate. Sixty years ago, this bagman called at all the girls' schools, and got many of the teachers to insist on the pupils saying ' Is it not' and ' Can I not ' for ' Isn't it ' and ' Can't I ' : of which it came that the poor girls were dreadfully laughed at by their irreverent brothers when they went home for the holidays. Had this bad adviser not been severely checked, he might by this time have proposed our saying ' The Queen's of England son,' declaring, in the name of Logic, that the prince was the Queen's son, not England's.

Lastly, there is Typography the metallurgist, an executive officer who is always at work in secret, and whose lawless mode of advising is often done by carrying his notions into effect without leave given. He it is who never ceases suggesting that the same word is not to occur in a second place within sight of the first. When the Authorized Version was first printed, he began this trick at the passage, ' Let there be light, and there was light ; ' he drew a line on the proof under the second *light*, and wrote ' *luminosity* ? ' opposite. He is strongest in the punctuations and other signs ; he has a pepper-box full of commas always by his side. He puts everything under marks of quotation which he has ever heard before. An earnest preacher, in a very moving sermon, used the phrase Alas ! and alack a day ! Typography stuck up the inverted commas because he had read the old Anglo-Indian toast, ' A lass and a lac a day ! ' If any one should have the sense to leave out of his Greek the un-meaning scratches which they call accents, he goes to a lexicon and puts them in. He is powerful in routine ; but when two routines interlace or overlap, he frequently takes the wrong one.

Subject to bad advice, and sometimes misled for a season, King Custom goes on his quiet way, and is sure to be right at last.

> Treason does never prosper : what's the reason ?
> Why, when it prospers, none dare call it treason.

Language is in constant fermentation, and all that is thrown in, so far as it is not fit to assimilate, is thrown off ; and this without any obvious struggle. In the meanwhile every one who has read good authors, from Shakspeare downward, knows what is and what is not English; and knows, also, that our language is not one and indivisible. Two very different turns of phrase may both be equally good, and as good as can be : we may be relieved of the consequences of contempt of one court by *habeas corpus* issuing out of another.

Hallam remarks that the Authorized Version of the Bible is not in the language of the time of James the First : that it is not the English of Raleigh or of Bacon. Here arises the question whether Raleigh and Bacon are the true expositors of the language of their time; and whether they were not rather the incipient promoters of a change which was successfully resisted by—among other things—the Authorized Version of the Testaments. I am not prepared to concede that I should have given to the English which would have been fashioned upon that of Bacon by imitators, such as they usually are, the admiration which is forced from me by Bacon's English from Bacon's pen. On this point we have a notable parallel. Samuel Johnson commands our admiration, at least in his matured style : but we nauseate his followers. It is an opinion of mine that the works of the leading writers of an age are seldom the proper specimens of the language of their day, when that language is in its state of progression. I judge of a language by the colloquial idiom of educated men : that is, I take this to be the best medium between the extreme cases of one who is ignorant of grammar and one who is perched upon a style. Dialogue is what I want to judge by, and plain dialogue : so I choose Robert Recorde and his pupil in the ' Castle of Knowledge,' written before 1556. When Dr. Robert gets into his altitudes of instruction, he differs from his own common phraseology as much as probably did Bacon when he wrote morals and philosophy. But every now and then I come to a little plain talk about a common thing, of which I propose to show a specimen. Anything can be made to look old by such changes as *makes* into *maketh*, with a little old spelling. I shall invert these changes, using the newer form of inflexion, and the modern spelling : with no other variation whatever.

' *Scholar.* Yet the reason of that is easy enough to be conceived, for when the day is at the longest the Sun must needs shine the more time, and so must it needs shine the less time when the day is at the shortest : this reason I have heard many men declare.

*Master.* That may be called a crabbed reason, for it goes backward like a crab. The day makes not the Sun to shine, but the Sun shining makes the day. And so the length of the day makes not the Sun to shine long, neither the shortness of the day causes not [*sic*] the Sun to shine the lesser time, but contrariwise the long shining of the Sun makes the long day, and the short shining of the Sun makes the lesser day : else answer me what makes the days long or short?

*Scholar.* I have heard wise men say that Summer makes the long days, and Winter makes the long nights.

*Master.* They might have said more wisely, that long days makes summer and short days make winter.

*Scholar.* Why, all that seems one thing to me.

*Master.* Is it all one to say, God made the earth, and the earth made God? Covetousness overcomes all men, and all men overcome covetousness?

*Scholar.* No, not so; for here the effect is turned to be the cause, and the agent is made the patient.

*Master.* So is it to say Summer makes long days, when you should say : Long days make summer.

*Scholar.* I perceive it now : but I was so blinded with the vulgar error, that if you had demanded of me further what did make the summer, I had been like to have answered that green leaves do make summer ; and the sooner by remembrance of an old saying that a year should come in which the summer should not be known but by the green leaves.

*Master.* Yet this saying does not import that green leaves do make summer, but that they betoken summer; so are they the sign and not the cause of summer.'

I have taken a whole page of our author, without omission, that the reader may see that I do not pick out sentences convenient for my purpose. I have done nothing but alter the third person of the verb and the spelling : but great is the effect thereof. We say ' the Sun shining makes the day : ' Recorde, ' the Sonne shynynge maketh the daye.' These points apart, we see a resemblance between our English and that of three hundred years ago, in the common talk of educated persons, which will allow us to affirm that the language of the authorized Bible must have been very close to that of its time. For I cannot admit that much change can have taken place in fifty years : and the language of the version represents both our common English and that of Recorde with very close approximation. Take sentences from Bacon and Raleigh, and it will be apparent that these writers will be held to differ from all three, Recorde, the version, and ourselves, by differences of the same character. But we speak of Recorde's conversation, and of our own. We conclude that it is the plain and almost colloquial character of the

Authorized Version which distinguishes it from the English of
Bacon and Raleigh, by approximating it to the common idiom of
the time.   If any one will cast an eye upon the letters of instruction written by Cecil and the Bishop of London to the translators
themselves, or to the general directions sent to them in the King's
name, he will find that these plain business compositions differ
from the English of Bacon and Raleigh by the same sort of
differences which distinguish the version itself.

The foreign word, or the word of a district, or class of people,
passes into the general vernacular; but it is long before the
specially learned will acknowledge the right of those with whom
they come in contact to follow general usage.   The rule is
simple : so long as a word is technical or local, those who know
its technical or local pronunciation may reasonably employ it.
But when the word has become general, the specialist is not very
wise if he refuse to follow the mass, and perfectly foolish if he
insist on others following him.   There have been a few who
demanded that Euler should be pronounced in the German
fashion : Euler has long been the property of the world at large ;
what does it matter how his own countrymen pronounce the
letters ?   Shall we insist on the French pronouncing *Newton*
without that final *tong* which they never fail to give him ?
They would be wise enough to laugh at us if we did.   We remember that a pedant who was insisting on all the pronuncia-
tions being retained, was met by a maxim in contradiction,
invented at the moment, and fathered upon Kaen-foo-tzee, an
authority which he was challenged to dispute.   Whom did you
speak of ? said the bewildered man of accuracy.   Learn your own
system, was the answer, before you impose it on others ; Confu-
cius says that too.

The old English has *fote, fode, loke, coke, roke,* &c., for *foot,*
&c.   And *above* rhymes in Chaucer to *remove.*   Suspecting that
the broader sounds are the older, we may surmise that *remove* and
*food* have retained their old sounds, and that *cook,* once *coke,*
would have rhymed to our *Luke,* the vowel being brought a little
nearer, perhaps, to the *o* in our present *coke,* the fuel, probably so
called as used by cooks.   If this be so, the Chief Justice *Cook* of
our lawyers, and the *Coke* (pronounced like the fuel) of the
greater part of the world, are equally wrong.   The lawyer has no
right whatever to fasten his pronunciation upon us : even leaving
aside the general custom, he cannot prove himself right, and is
probably wrong.   Those who know the village of Rokeby (pronounced Rookby) despise the world for not knowing how to name
Walter Scott's poem : that same world never asked a question

about the matter, and the reception of the parody of *Jokeby*, which soon appeared, was a sufficient indication of their notion. Those who would fasten the hodiernal sound upon us may be reminded that the question is, not what they call it now, but what it was called in Cromwell's time. Throw away general usage as a lawgiver, and this is the point which emerges. Probably *Rūke-by* would be right, with a little turning of the Italian ū towards ō of modern English.

[Some of the above is from an old review. I do not always notice such insertions : I take nothing but my own writings. A friend once said to me, 'Ah! you got that out of the *Athenæum*!' 'Excuse me,' said I, 'the *Athenæum* got that out of me!']

It is part of my function to do justice to any cyclometers whose methods have been wrongly described by any orthodox sneerers (myself included). In this character I must notice *Dethlevus Cluverius*, as the Leipzig Acts call him (probably Dethleu Cluvier), grandson of the celebrated geographer, Philip Cluvier. The grandson was a Fellow of the Royal Society, elected on the same day as Halley, November 30, 1678 : I suppose he lived in England. This man is quizzed in the Leipzig Acts for 1686; and, if Montucla insinuate rightly, by Leibnitz, who is further suspected of wanting to embroil Cluvier with his own opponent Nieuwentiit, on the matter of infinitesimals. So far good : I have nothing against Leibnitz, who though he was ironical, told us what he laughed at. But Montucla has behaved very unfairly : he represents Cluvier as placing the essence of his method in the solution of the problem *construere mundum divinæ menti analogum*, to construct a world corresponding to the divine mind. Nothing to begin with : no way of proceeding. Now, it ought to have been *ex datâ lineâ construere*, &c. : there is a given line, which is something to go on. Further, there is a way of proceeding : it is to find the product of 1, 2, 3, 4, &c. for ever. Moreover, Montucla charges Cluvier with *unsquaring* the parabola, which Archimedes had squared as tight as a glove. But he never mentions how very nearly Cluvier agrees with the Greek : they only differ by 1 divided by $3n^2$, where $n$ is the infinite number of parts of which a parabola is composed. This must have been the conceit that tickled Leibnitz, and made him wish that Cluvier and Nieuwentiit should fight it out. Cluvier, was admitted, on terms of irony, into the Leipzig Acts : he appeared on a more serious footing in London. It is very rare for one cyclometer to refute another : *les corsaires ne se battent pas* The only instance I recall is that of M. Cluvier, who (*Phil.*

*Trans.* 1686, No. 185) refuted M. Mallemont de Messange, who published at Paris in 1686. He does it in a very serious style, and shows himself a mathematician. And yet in the year in which, in the *Phil. Trans.*, he was a geometer, and one who rebukes his squarer for quoting Matthew xi. 25, in that very year he was the visionary who, in the Leipzig Acts, professed to build a world resembling the divine mind by multiplying together 1, 2, 3, 4, &c. up to infinity.

There is a very pretty opening for a paradox which has never found its paradoxer in print. The philosophers teach that the rainbow is not material : it comes from rain-drops, but those rain-drops do not *take* colour. They only *give* it, as lenses and mirrors ; and each one drop gives *all* the colours, but throws them in different directions. Accordingly, the same drop which furnishes red light to one spectator will furnish violet to another, properly placed. Enter the paradoxer whom I have to invent. The philosopher has gulled you nicely. Look into the water, and you will see the reflected rainbow : take a looking-glass held sideways, and you see another reflexion. How could this be, if there were nothing coloured to reflect ? The paradoxer's facts are true : and what are called the reflected rainbows are *other* rainbows, caused by those *other* drops which are placed so as to give the colours to the eye after reflexion, at the water or the looking-glass. A few years ago an artist exhibited a picture with a rainbow and its apparent reflexion : he simply copied what he had seen. When his picture was examined, some started the idea that there could be no reflexion of a rainbow ; they were right : they inferred that the artist had made a mistake ; they were wrong. When it was explained, some agreed and some dissented. Wanted, immediately, an able paradoxer : testimonials to be forwarded to either end of the rainbow, No. 1. No circle-squarer need apply, His Variegatedness having been pleased to adopt 3·14159 . . . from Noah downwards.

The system of Tycho Brahé, with some alteration and addition, has been revived and contended for in our own day by a Dane, W. Zytphen, who has published ' The Motion of the Sun in the Universe,' (second edition) Copenhagen, 1865, 8vo., and ' Le Mouvement Sidéral,' 1865, 8vo. I make an extract.—

' How can one explain Copernically that the velocity of the Moon must be added to the velocity of the Earth on the one place in the Earth's orbit, to learn how far the Moon has advanced from one fixed star to another ; but in another place in the orbit these velocities must be subtracted (the movements taking place in opposite directions) to attain the same result? In the Copernican and other systems, it is well known that the Moon, abstracting from the insignificant excentricity of the orbit, always in twenty-four hours performs an equally long distance. Why has Copernicus never been denominated Fundamentus or Fundator? Because he has never convinced anybody so thoroughly that this otherwise so natural epithet has occurred to the mind.'

Really the second question is more effective against Newton than against Copernicus; for it upsets gravity: the first is of great depth.

The *Correspondent* journal makes a little episode in the history of my Budget (born May, 1865, died April, 1866). It consisted entirely of letters written by correspondents. In August, a correspondent who signed ' Fair Play '—and who I was afterwards told was a lady—thought it would be a good joke to bring in the Cyclometers. Accordingly a letter was written, complaining that though Mr. Sylvester's demonstration of Newton's theorem—then attracting public attention—was duly lauded, the possibly greater discovery of the quadrature seemed to be blushing unseen, and wasting &c. It went on as follows:—

' Prof. De Morgan, who, from his position in the scientific world, might fairly afford to look favourably on less practised efforts than his own, seems to delight in ridiculing the discoverer. Science is, of course, a very respectable person when he comes out and makes himself useful in the world [it must have been a lady; each sex gives science to the other]: but when, like a monk of the Middle Ages, he shuts himself up [it must have been a lady; they always snub the bachelors] in his cloistered cell, repeating his mumpsimus from day to day, and despising the labourers on the outside, we begin to think of Galileo, Jenner, Harvey, and other glorious trios, who have been contemned . . .'

The writer then called upon Mr. James Smith to come forward. The irony was not seen; and that day fortnight appeared the first of more than thirty letters from his pen. Mr. Smith was followed by Mr. Reddie, Zadkiel, and others, on their several subjects. To some of the letters I have referred; to others I shall come. The *Correspondent* was to become a first-class scientific journal; the time had arrived at which truth had an organ: and I received formal notice that I could not stifle it by silence, nor convert it into falsehood by ridicule. When my reader sees my extracts, he will readily believe my declaration that I should have been the last to stifle a publication which was every week what James Mill would call a dose of capital for my

Budget. A few anti-paradoxers brought in common sense : but to the mass of the readers of the journal it all seemed to be the difference between Tweedledum and Tweedledee. Some said that the influx of scientific paradoxes killed the journal : but my belief is that they made it last longer than it otherwise would have done. Twenty years ago I recommended the paradoxers to combine and publish their views in a common journal : with a catholic editor, who had no pet theory, but a stern determination not to exclude anything merely for absurdity. I suspect it would answer very well. A strong title, or motto, would be wanted : not quite so coarse as was roared out in a Cambridge mob when I was an undergraduate—' No King! No Church! No House of Lords! No nothing, blast me !'—but something on that *principle*.

At the end of 1867 I addressed the following letter to the *Athenæum* :—

### PSEUDOMATH, PHILOMATH, AND GRAPHOMATH.

*December* 31, 1867.

MANY thanks for the present of Mr. James Smith's letters of Sept. 28 and of Oct. 10 and 12. He asks where you will be if you read and digest his letters : you probably will be somewhere first. He afterwards asks what the WE of the *Athenæum* will be if, finding it impossible to controvert, it should refuse to print. I answer for you, that We-We of the *Athenæum*, not being Wa-Wa the wild goose, so conspicuous in ' Hiawatha,' will leave what controverts itself to print itself, if it please.

*Philomath* is a good old word, easier to write and speak than *mathematician*. It wants the words between which I have placed it. They are not well formed; *pseudomathete* and *graphomathete* would be better : but they will do. I give an instance of each.

The *pseudomath* is a person who handles mathematics as the monkey handled the razor. The creature tried to shave himself as he had seen his master do ; but, not having any notion of the angle at which the razor was to be held, he cut his own throat. He never tried a second time, poor animal ! but the pseudomath keeps on at his work, proclaims himself clean-shaved, and all the rest of the world hairy. So great is the difference between moral and physical phenomena ! Mr. James Smith is, beyond doubt, the great pseudomath of our time. His 3⅛ is the least of a wonderful chain of discoveries. His books, like Whitbread's barrels, will one day reach from Simpkin & Marshall's to Kew, placed upright, or to Windsor laid lengthways. The Queen will run away on their near approach, as Bishop Hatto did from the rats : but Mr. James Smith will follow her were it to John o' Groats.

The *philomath*, for my present purpose, must be exhibited as giving a lesson to presumption. The following anecdote is found in Thiébault's 'Souvenirs de vingt ans de séjour à Berlin,' published in 1804. The book itself got a high character for truth. In 1807 Marshal Mollendorff answered an inquiry of the Duc de Bassano, by saying that it was the most veracious of books, written by the most honest of men. Thiébault does not claim personal knowledge of the anecdote, but he vouches for its being received as true all over the north of Europe.[1]

Diderot paid a visit to Russia at the invitation of Catherine the Second. At that time he was an atheist, or at least talked atheism : it would be easy to prove him either one thing or the other from his writings. His lively sallies on this subject much amused the Empress, and all the younger part of her Court. But some of the older courtiers suggested that it was hardly prudent to allow such unreserved exhibitions. The Empress thought so too, but did not like to muzzle her guest by an express prohibition : so a plot was contrived. The scorner was informed that an eminent mathematician had an algebraical proof of the existence of God, which he would communicate before the whole Court, if agreeable. Diderot gladly consented. The mathematician, who is not named, was Euler. He came to Diderot with the gravest air, and in a tone of perfect conviction said, " Monsieur!

$$\frac{a + b^n}{n} = x$$

donc Dieu existe ; répondez ! " Diderot, to whom algebra was Hebrew, though this is expressed in a very roundabout way by Thiébault—and whom we may suppose to have expected some verbal argument of alleged algebraical closeness, was disconcerted ; while peals of laughter sounded on all sides. Next day he asked permission to return to France, which was granted. An algebraist would have turned the tables completely, by saying, 'Monsieur! vous savez bien que votre raisonnement demande le développement de x suivant les puissances entières de n.' Goldsmith could not have seen the anecdote, or he might have been supposed to have drawn from it a hint as to the way in which the Squire demolished poor Moses.

The *graphomath* is a person who, having no mathematics, attempts to describe a mathematician. Novelists perform in this way : even Walter Scott now and then burns his fingers. His dreaming calculator, Davy Ramsay, swears 'by the bones of the immortal Napier.' Scott thought that the philomaths worshipped relics : so they do, in one sense. Look into Hutton's Dictionary for *Napier's Bones*, and you shall learn all about the little knick-knacks by which he did multiplication and division. But never a bone of his own did he contribute ; he preferred elephants' tusks. The author of ' Headlong

---

[1] This anecdote is printed at p. 251 ; but as it is used in illustration here, and is given more in detail, I have not omitted it.—ED.

Hall' makes a grand error, which is quite high science: he says that Laplace proved the precession of the equinoxes to be a periodical inequality. He should have said the variation of the obliquity. But the finest instance is the following :—Mr. Warren, in his well-wrought tale of the martyr-philosopher, was incautious enough to invent the symbols by which his *savant* satisfied himself Laplace was right on a doubtful point. And this is what he put together—

$$\sqrt{-3a^2}, \quad \square \frac{v^2}{z^2} + 9 - n = 9, n \times \log e.$$

Now, to Diderot and the mass of mankind this might be Laplace all over: and, in a forged note of Pascal, would prove him quite up to gravitation. But I know of nothing like it, except in the lately received story of the American orator, who was called on for some Latin, and perorated thus :—'Committing the destiny of the country to your hands, Gentlemen, I may without fear declare, in the language of the noble Roman poet,

> E pluribus unum,
> Multum in parvo,
> Ultima Thule,
> Sine qua non.'

But the American got nearer to Horace than the martyr-philosopher to Laplace. For all the words are in Horace, except *Thule*, which might have been there. But $\square$ is not a symbol wanted by Laplace; nor can we see how it could have been : in fact, it is not recognized in algebra. As to the junctions, &c., Laplace and Horace are about equally well imitated.

Further thanks for Mr. Smith's letters to you of Oct. 15, 18, 19, 28, and Nov. 4, 15. The last of these letters has two curious discoveries. First, Mr. Smith declares that he has *seen* the editor of the *Athenæum*: in several previous letters he mentions a name. If he knew a little of journalism he would be aware that editors are a peculiar race, obtained by natural selection. They are never seen, even by their officials; only heard down a pipe. Secondly, 'an ellipse' or oval' is composed of four arcs of circles. Mr. Smith has got hold of the construction I was taught, when a boy, for a pretty four-arc oval. But my teachers knew better than to call it an ellipse : Mr. Smith does not; but he produces from it such confirmation of $3\frac{1}{8}$ as would convince any *honest* editor.

Surely the cyclometer is a Darwinite development of a spider, who is always at circles, and always begins again when his web is brushed away. He informs you that he has been privileged to discover truths unknown to the scientific world. This we know; but he proceeds to show that he is equally fortunate in art. He goes on to say that he will make use of you to bring those truths to light, 'just as an artist makes use of a dummy for the purpose of arranging his drapery.' The painter's lay-figure is for flowing robes; the hairdresser's dummy is for curly locks. Mr. James Smith should read Sam Weller's pathetic story of the 'four wax dummies.' As to *his* use of a dummy, it is quite correct. When I was at University College, I walked one day into a room in which my Latin colleague was examining. One of the questions was, 'Give the lives and fates of Sp. Mælius and Sp. Cassius.' Umph! said I, surely all know that Spurius Mælius was whipped for adulterating flour, and that Spurius Cassius was hanged for passing bad money. Now, a robe arranged on a dummy would look just like the toga of Cassius on the gallows. Accordingly, Mr. Smith is right in the drapery-hanger which he has chosen: he has been detected in the attempt to pass bad circles. He complains bitterly that his geometry, instead of being read and understood by you, is handed over to me to be treated after my scurrilous fashion. It is clear enough that he would rather be handled in this way than not handled at all, or why does he go on writing? He must know by this time that it is a part of the institution that his 'untruthful and absurd trash' shall be distilled into mine at the rate of about $3\frac{1}{8}$ pages of the first to one column of the second. Your readers will never know how much they gain by the process, until Mr. James Smith publishes it all in a big book, or until they get hold of what he has already published. I have six pounds avoirdupois of pamphlets and letters; and there is more than half a pound of letters written to you in the last two months. Your compositor must feel aggrieved by the rejection of these clearly written documents, without erasures, and on one side only. Your correspondent has all the makings of a good contributor, except knowledge of his subject and sense to get it. He is, in fact, only a mask: of whom the fox

O quanta species, inquit, cerebrum non habet.

I do not despair of Mr. Smith on any question which does not involve that unfortunate two-stick wicket at which he persists in bowling. He has published many papers; he has forwarded them to mathematicians: and he cannot get answers; perhaps not even readers. Does he think that he would get more notice if you were to print him in your journal? Who would study his columns? Not the mathematician, we know; and he knows. Would others? His balls are aimed too wide to be blocked by any one who is near the wicket. He has long ceased to be worth the answer which a new invader may get. Rowan Hamilton, years ago, completely knocked him over; and he has never attempted to point out any error in the short and easy method by which that powerful investigator condescended to show that, be right who may, he must be wrong. There are some persons who feel inclined to think that Mr. Smith should be argued with: let those persons understand that he has been argued with, refuted, and has never attempted to stick a pen into the refutation. He stated that it was a remarkable paradox, easily explicable; and that is all. After this evasion, Mr. James Smith is below the necessity of being told that he is unworthy of answer. His friends complain that I do nothing but *chaff* him. Absurd! I winnow him; and if nothing but chaff results, whose fault is that? I am usefully employed; for he is the type of a class which ought to be known, and which I have done much to make known.

Nothing came of this until July 1869, when I received a reprint of the above letter, with a comment, described as Appendix D of a work in course of publication on the geometry of the circle. The *Athenæum* journal received the same : but the Editor, in his private capacity, received the whole work, being 'The Geometry of the Circle and Mathematics as applied to Geometry by Mathematicians, shown to be a mockery, delusion, and a snare,' Liverpool, 8vo., 1869. Mr. J. S. here appears in deep fight with Prof. Whitworth, and Mr. Wilson, the author of the alleged amendment of Euclid. How these accomplished mathematicians could be inveigled into continued discussion is inexplicable. Mr. Whitworth began by complaining of Mr. Smith's attacks upon mathematicians, continued to correspond after he was convinced that J. S. proved an arc and its chord to be equal, and only retreated when J. S. charged him with believing in $3\frac{1}{8}$, and refusing acknowledgment. Mr. Wilson was introduced to J. S. by a volunteer defence of his geometry from the assaults of the *Athenæum*. This the editor would not publish; so J. S. sent a copy to Mr. Wilson himself. Some correspondence ensued, but Mr. Wilson soon found out his man, and withdrew.

There is a little derision of the *Athenæum* and a merited punishment for 'that unscrupulous critic and contemptible mathematical twaddler, De Morgan.'

At p. 371 I mentioned Mr. Reddie, the author of *Vis Inertiæ Victa* and of *Victoria tolo cœlo*, which last is not an address to the whole heaven, either from a Roman Goddess or a British Queen, whatever a scholar may suppose. Between these Mr. Reddie has published 'The Mechanics of the Heavens,' 8vo., 1862 : this I never saw until he sent it to me, with an invitation to notice it, he very well knowing what it would catch. His speculations do battle with common notions of mathematics and of mechanics, which, to use a feminine idiom, he blasphemes so you can't think! and I suspect that if you do not blaspheme them too, *you* can't think. He appeals to the 'truly scientific,' and would be glad to have readers who have read what he controverts, *i. e.* Newton's *Principia* : I wish he may get them ; I mean I hope he may obtain them.   To none but these would an account of his speculations be intelligible : I accordingly disposed of him in a very short paragraph of description.   Now many paradoxers desire notice, even though it be disparaging.   I have letters from more than one—besides what have been sent to the Editor of the *Athenæum*—complaining that they are not laughed at ; although they deserve it, they tell me, as much as some whom I have inserted.   Mr. Reddie informs me that I have not said a single word against his books, though I have given nearly a column to sixteen-string arithmetic, and as much to animalcule universes. What need to say anything to readers of Newton against a book from which I quoted that revolution by gravitation is *demonstrably* impossible ?   It would be as useless as evidence against a man who has pleaded guilty.   Mr. Reddie derisively thanks me for ' small mercies ' ; he wrote me private letters ; he published them, and more, in the *Correspondent*.   He gave me, *pro viribus suis*, such a dressing you can't think, both for my Budget non-notice, and for reviews which he assumed me to have written. He outlawed himself by declaring (*Correspondent*, Nov. 11, 1856) that I—in a review—had made a quotation which was ' garbled, evidently on purpose to make it appear that ' he ' was advocating solely a geocentric hypothesis, which is not true.'   In fact, he did his very best to get larger 'mercy.'   And he shall have it; and at a length which shall content him, unless his mecometer be an insatiable apparatus.   But I fear that in other respects I shall no more satisfy him than the Irish drummer satisfied the poor culprit when, after several times changing the direction of the stroke at earnest entreaty, he was at last provoked to call out, ' Bad cess to ye, ye spalpeen ! strike where one will, there 's no *plasing* ye ! '

Mr. Reddie attaches much force to Berkeley's old arguments against the doctrine of fluxions, and advances objections to Newton's second section, which he takes to be new.   To me they appear ' such as have been often made,' to copy a description given in a review : though I have no doubt Mr. Reddie got them out of himself.   But the whole matter comes to this : Mr. Reddie challenged answer, especially from the British Association, and got none.   He presumes that this is because he is right, and cannot be answered : the Association is willing to risk itself upon the counter-notion that he is wrong, and need not be answered ;

because so wrong that none who could understand an answer would be likely to want one.

Mr. Reddie demands my attention to a point which had already particularly struck me, as giving the means of showing to *all* readers the kind of confusion into which paradoxers are apt to fall, in spite of the clearest instruction. It is a very honest blunder, and requires notice: it may otherwise mislead some, who may suppose that no one able to read could be mistaken about so simple a matter, let him be ever so wrong about Newton. According to his own mis-statement, in less than five months he made the Astronomer Royal abandon the theory of the solar motion in space. The announcement is made in August, 1865, as follows: the italics are not mine:—

'The third (*Victoria* . . .), although only published in September, 1863, has already had its triumph. *It is the book that forced the Astronomer Royal of England, after publicly teaching the contrary for years, to come to the conclusion, "strange as it may appear," that " the whole question of solar motion in space is at the present time in doubt and abeyance."* This admission is made in the Annual Report of the Council of the Royal Astronomical Society, published in the Society's *Monthly Notices* for February, 1864.'

It is added that solar motion is 'full of self-contradiction, which "the astronomers" simply overlooked, but which they dare not now deny after being once pointed out.'

The following is another of his accounts of the matter, given in the *Correspondent*, Nov. 18, 1865:—

'. . . You ought, when you came to put me in the "Budget," to have been aware of the Report of the Council of the Royal Astronomical Society, where it appears that Professor Airy, with a better appreciation of my demonstrations, had admitted—" strange," say the Council, " as it may appear "—that " the whole question of solar motion in space [and here Mr. Reddie omits some words] is now in *doubt and abeyance.*" You were culpable, as a public teacher of no little pretensions, if you were " unaware " of this. If aware of it, you ought not to have suppressed such an important testimony to my really having been "very successful " in drawing the teeth of the pegtops, though you thought them so firmly fixed. And if you still suppress it, in your Appendix, or when you reprint your " Budget," you will then be guilty of a *suppressio veri*, also of further injury to me, who have never injured you . . .'

Mr. Reddie must have been very well satisfied in his own mind before he ventured such a challenge, with an answer from me looming in the distance. The following is the passage of the Report of the Council, &c., from which he quotes :—

'And yet, strange to say, notwithstanding the near coincidence of all the results of the before-mentioned independent methods of investigation, the inevitable logical inference deduced by Mr. Airy is, that the whole question of solar motion in space, *so far at least as accounting for the proper motion of the stars is concerned*, [I have put in italics the words omitted by Mr. Reddie] appears to remain at this moment in doubt and abeyance.'

Mr. Reddie has forked me, as he thinks, on a dilemma : if unaware, culpable ignorance ; if aware, suppressive intention. But the thing is a *trilemma*, and the third horn, on which I elect to be placed, is surmounted by a doubly-stuffed seat. First, Mr. Airy has not changed his opinion about the *fact* of solar motion in space, but only suspends it as to the sufficiency of present means to give the amount and direction of the motion. Secondly, all that is alluded to in the Astronomical Report was said and printed before the Victoria proclamation appeared. So that the author, instead of drawing the tooth of the Astronomer Royal's pegtop, has burnt his own doll's nose.

William Herschel, and after him about six other astronomers, had aimed at determining, by the proper motions of the stars, the point of the heavens towards which the solar system is moving : their results were tolerably accordant. Mr. Airy, in 1859, proposed an improved method, and, applying it to stars of large proper motion, produced much the same result as Herschel. Mr. E. Dunkin, one of Mr. Airy's staff at Greenwich, applied Mr. Airy's method to a very large number of stars, and produced, again, nearly the same result as before. This paper was read to the Astronomical Society in *March*, 1863, was printed in abstract in the *Notice* of that month, was printed in full in the volume then current, and was referred to in the Annual Report of the Council in *February*, 1864, under the name of 'the Astronomer Royal's elaborate investigation, as exhibited by Mr. Dunkin.' Both Mr. Airy and Mr. Dunkin express grave doubts as to the sufficiency of the data : and, regarding the coincidence of all the results as highly curious, feel it necessary to wait for calculations made on better data. The report of the Council states these doubts. Mr. Reddie, who only published in *September*, 1863, happened to see the Report of February, 1864, assumes that the doubts were then first expressed, and declares that his book of September had the triumph of forcing the

Astronomer Royal to abandon the *fact* of motion of the solar system by the February following.  Had Mr. Reddie, when he saw that the Council were avowedly describing a memoir presented some time before, taken the precaution to find out *when* that memoir was presented, he would perhaps have seen that doubts of the results obtained, expressed by one astronomer in March, 1863, and by another in 1859, could not have been due to his publication of September, 1863.  And any one else would have learnt that neither astronomer doubts the *solar motion*, though both doubt the sufficiency of present means to determine its *amount* and *direction*.  This is implied in the omitted words, which Mr. Reddie—whose omission would have been dishonest if he had seen their meaning—no doubt took for pleonasm, superfluity, overmuchness.  The rashness which pushed him headlong into the quillet that *his* thunderbolt had stopped the chariot of the Sun and knocked the Greenwich Phaethon off the box, is the same which betrayed him into yet grander error— which deserves the full word, *quidlibet*—about the *Principia* of Newton.  There has been no change of opinion at all.  When a person undertakes a long investigation, his opinion is that, at a certain date, there is *primâ facie* ground for thinking a sound result may be obtained.  Should it happen that the investigation ends in doubt upon the sufficiency of the grounds, the investigator is not put in the wrong.  He knew beforehand that there was an alternative : and he takes the horn of the alternative indicated by his calculations.  The two sides of this case present an instructive contrast.  Eight astronomers produce nearly the same result, and yet the last two doubt the sufficiency of their means : compare them with the what's-his-name who rushes in where thing-em-bobs fear to tread.

I was not aware, until what I had written what precedes, that Mr. Airy had given a sufficient answer on the point.  Mr. Reddie says (*Correspondent*, Jan. 20, 1866):—

'I claim to have forced Professor Airy to give up the notion of " solar motion in space " altogether, for he admits it to be " at present in doubt and abeyance."  I first made that claim in a letter addressed to the Astronomer Royal himself in June, 1864, and in replying, very courteously, to other portions of my letter, he did not gainsay that part of it.'

Mr. Reddie is not ready at reading satire, or he never would have so missed the meaning of the courteous reply on one point, and the total silence upon another.  Mr. Airy must be one of

those peculiar persons who, when they do not think an assertion
worth notice, let it alone, without noticing it by a notification
of non-notice. He would never commit the bull of 'Sir! I will
not say a word on that subject.' He would put it thus, 'Sir! I
will only say ten words on that subject,'—and, having thus said
them, would proceed to something else. He assumed, as a matter
of form, that Mr. Reddie would draw the proper inference from
his silence : and this because he did not care whether or no the
assumption was correct.

The ' Mechanics of the Heavens,' which Mr. Reddie sends to be
noticed, shall be noticed, so far as an extract goes :—

'My connexion with this subject is, indeed, very simply explained.
In endeavouring to understand the laws of physical astronomy as
generally taught, I happened to entertain some doubt whether gravitating bodies could revolve, and having afterwards imbibed some vague
idea that the laws of the universe were chemical and physical rather
than mechanical, and somehow connected with electricity and magnetism as opposing and correlative forces—most probably suggested to
my mind, as to many others, by the transcendent discoveries made in
electro-magnetism by Professor Faraday—my former doubts about
gravitation were revived, and I was led very naturally to try and discover whether a gravitating body really could revolve ; and I became
convinced it could not, before I had ever presumed to look into the
demonstrations of the *Principia.*'

This is enough against the book, without a word from me : I
insert it only to show those who know the subject what manner of
writer Mr. Reddie is. It is clear that ' presumed ' is a slip of the
pen ; it should have been *condescended.*

Mr. Reddie represents me as dreaming over paltry paradoxes.
He is right; many of my paradoxes are paltry: he is wrong ;
I am wide awake to them. A single moth, beetle, or butterfly,
may be a paltry thing ; but when a cabinet is arranged by genus
and species, we then begin to admire the infinite variety of a
system constructed on a wonderful sameness of leading character-
istics. And why should paradoxes be denied that collective importance, paltry as many of them may individually be, which is
accorded to moths, beetles, or butterflies ? Mr. Reddie himself
sees that ' there is a method in ' my ' mode of dealing with paradoxes.' I hope I have atoned for the scantiness of my former
article, and put the demonstrated impossibility of gravitation on
that level with Hubongramillposanfy arithmetic and inhabited
atoms which the demonstrator—not quite without reason—claims
for it.

In the Introduction to a collected edition of the three works, Mr. Reddie describes his *Mechanism of the Heavens*, from which I have just quoted, as—

'a public challenge offered to the British Association and the mathematicians at Cambridge, in August, 1862, calling upon them to point to a single demonstration in the *Principia* or elsewhere, which even attempts to prove that Universal Gravitation is possible, or to show that a gravitating body could possibly revolve about a centre of attraction. The challenge was not accepted, and never will be. No such demonstration exists. And the public must judge for themselves as to the character of a so-called "certain science," which thus shrinks from rigid examination, and dares not defend itself when publicly attacked: also of the character of its teachers, who can be content to remain dumb under such circumstances.'

The above is the commonplace talk of the class, of which I proceed to speak without more application to this paradoxer than to that. It reminds one of the funny young rascals who used, in times not yet quite forgotten, to abuse the passengers, as long as they could keep up with the stage coach; dropping off at last with 'Why don't you get down and thrash us? You're afraid, you're afraid!' They will allow the public to judge for themselves, but with somewhat of the feeling of the worthy uncle in *Tom Jones*, who, though he would let young people choose for themselves, would *have them* choose wisely. They try to be so awfully moral and so ghastly satirical that they must be answered: and they are best answered in their own division. We have all heard of the way in which sailors cat's-pawed the monkeys: they taunted the dwellers in the trees with stones, and the monkeys taunted them with cocoa-nuts in return. But these were silly dendrobats: had they belonged to the British Association they would have said—No! No! dear friends; it is not in the itinerary: if you want nuts, you must climb, as we do. The public has referred the question to Time: the procedure of this great king I venture to describe, from precedents, by an adaptation of some smart anapæstic tetrameters—your anapæst is the foot for satire to halt on, both in Greek and English—which I read about twenty years ago, and with the point of which I was much tickled. Poetasters were laughed at; but Mr. Slum, whom I employed—Mr. Charles Dickens obliged me with his address—converted the idea into that of a hit at mathematicasters, as easily as he turned the Warren acrostic into Jarley. As he observed, when I settled his little account, it is cheaper than any prose, though the broom was not stolen quite ready made :—

*Forty stripes save one for the smaller Paradoxers.*

Hark to the wisdom the sages preach
Who never have learnt what they try to teach.
We are the lights of the age, they say !
We are the men, and the thinkers we !
So we build up guess-work the livelong day,
In a topsy-turvy sort of way,
Some with and some wanting *a* plus *b.*
Let the British Association fuss ;
What are theirs to the feats to be wrought by us ?
Shall the earth stand still ?   Will the round come square ?
Must Isaac's book be the nest of a mare ?
Ought the moon to be taught by the laws of space
To turn half round without right-about-face ?
Our whimsey crotchets will manage it all ;
Deep ! Deep ! posterity will them call !
Though the world, for the present, lets them fall
Down ! Down ! to the twopenny box of the stall !

Thus they—But the marplot Time stands by,
With a knowing wink in his funny old eye.
He grasps by the top an immense fool's cap,
Which he calls a philosophaster-trap :
And rightly enough, for while these little men
Croak loud as a concert of frogs in a fen,
He first singles out one, and then another,
Down goes the cap—lo ! a moment's pother,
A spirt like that which a rushlight utters
As just at the last it kicks and gutters :
When the cruel smotherer is raised again
Only snuff, and but little of that, will remain.

But though *uno avulso* thus comes every day
*Non deficit alter* is also in play :
For the vacant parts are, one and all,
Soon taken by puppets just as small ;
Who chirp, chirp, chirp, with a grasshopper glee,
We 're the lamps of the Universe, We ! We ! We !
But Time, whose speech is never long,—
He hasn't time for it—stops the song
And says—Lilliput lamps ! leave the twopenny boxes,
And shine in the Budget of Paradoxes !

When a paradoxer parades capital letters and diagrams which are
as good as Newton's to all who know nothing about it, some persons

wonder why science does not rise and triturate the whole thing. This is why : all who are fit to read the refutation are satisfied already, and can, if they please, detect the paradoxer for themselves. Those who are not fit to do this would not know the difference between the true answer and the new capitals and diagrams on which the delighted paradoxer would declare that he had crumbled the philosophers, and not they him. Trust him for having the last word : and what matters it whether he crow the unanswerable sooner or later ? There are but two courses to take. One is to wait until he has committed himself in something which all can understand, as Mr. Reddie has done in his fancy about the Astronomer Royal's change of opinion : he can then be put in his true place. The other is to construct a Budget of Paradoxes, that the world may see how the thing is always going on, and that the picture I have concocted by cribbing and spoiling a bit of poetry is drawn from life. He who wonders at there being no answer has seen one or two : he does not know that there are always fifty with equal claims, each of whom regards his being ranked with the rest as forty-nine distinct and several slanders upon himself, the great Mully Ully Gue. And the fifty would soon be five hundred if any notice were taken of them. They call mankind to witness that science *will not* defend itself, though publicly attacked in terms which might sting a pickpocket into standing up for his character : science, in return, allows mankind to witness or not, at pleasure, that it *does not* defend itself, and yet receives no injury from centuries of assault. Demonstrative reason never raises the cry of *Church in Danger*! and it cannot have any Dictionary of Heresies except a Budget of Paradoxes. Mistaken claimants are left to Time and his extinguisher, with the approbation of all thinking non-claimants : there is no need of a succession of exposures. Time gets through the job in his own workmanlike manner, as already described.

On looking back more than twenty years, I find among my cuttings the following passage, relating to a person who had signalized himself by an effort to teach comets to the conductor of the *Nautical Almanac* :—

'Our brethren of the literary class have not the least idea of the small amount of appearance of knowledge which sets up the scientific charlatan. Their world is large, and there are many who have that moderate knowledge, and perception of what is knowledge, before which extreme ignorance is detected in its first prank. There is a public of moderate cultivation, for the most part sound in its judgment, always ready in its decisions. Accordingly, all their successful pretenders have *some pretension.* It is not so in science. Those who have a right to judge are fewer and farther between. The consequence is, that many scientific pretenders have *nothing but pretension.*'

This is nearly as applicable now as then. It is impossible to make those who have not studied for themselves fully aware of the truth of what I have quoted. The best chance is collection of cases; in fact, a Budget of Paradoxes. Those who have no knowledge of the subject can thus argue from the seen to the unseen. All can feel the impracticability of the Hubongramillposanfy numeration, and the absurdity of the equality of contour of a regular pentagon and hexagon in one and the same circle. Many may accordingly be satisfied, on the assurance of those who have studied, that there is as much of impracticability, or as much of absurdity, in things which are hidden under

> Sines, tangents, secants, radius, cosines,
> Subtangents, segments, and all those signs;
> Enough to prove that he who read 'em
> Was just as mad as he who made 'em.

Not that I mean to be disrespectful to mathematical terms: they are short and easily explained, and compete favourably with those of most other subjects: for instance, with

> Horse-pleas, traverses, demurrers,
> Jeofails, imparlances, and errors,
> Averments, bars, and protestandos,
> And puis d'arreign continuandos.

From which it appears that, taking the selections made by satirists for our samples, there are, one with another, four letters more in a law term than in one of mathematics. But pleading has been simplified of late years.

All paradoxers can publish; and any one who likes may read. But this is not enough; they find that they cannot publish, or those who can find they are *not* read, and they lay their plans athwart the noses of those who, they think, ought to read. To recommend them to be content with publication, like other authors, is an affront: of this I will give the reader an amusing instance. My good nature, of which I keep a stock, though I do not use it all up in this Budget, prompts me to conceal the name.

I received the following letter, accompanied by a prospectus of a work on metaphysics, physics, astronomy, &c. The author is evidently one whom I should delight to honour :—

'Sir,—A friend of mine has mentioned your name in terms of panigeric [*sic*], as being of high standing in mathematics, and of greatly original thought. I send you the enclosed without comment; and, assuming that the bent of your mind is in free inquiry, shall feel a pleasure in showing you my portfolio, which, as a mathematician, you will acknowledge to be deeply interesting, even in an educational point of view. The work is complete, and the system so far perfected as to place it above criticism ; and, so far as regards astronomy, as will Ptolemy beyond rivalry [*sic* : no doubt some words omitted]. Believe me to be, Sir, with the profoundest respect, &c. The work is the result of thirty-five years' travel and observation, labour, expense, and self-abnegation.'

I replied to the effect that my time was fully occupied, and that I was obliged to decline discussion with many persons who have views of their own ; that the proper way is to publish, so that those who choose may read when they can find leisure. I added that I should advise a precursor in the shape of a small pamphlet, as two octavo volumes would be too much for most persons. This was sound advice ; but it is not the first, second, or third time that it has proved very unpalatable. I received the following answer, to which I take the liberty of prefixing a bit of leonine wisdom :—

Si doceas stultum, lætum non dat tibi vultum ;
Odit te multum ; vellet te scire sepultum.

'Sir,—I pray you pardon the error I unintentionally have fallen into ; deceived by the F. R. S. [I am not F. R. S.] I took you to be a man of science [omnis homo est animal, Sortes est homo, ergo Sortes est animal] instead of the mere mathematician, or human calculating-machine. Believe me, Sir, you also have mistaken your mission, as I have mine. I wrote to you as I would to any other man well up in mathematics, with the intent to call your attention to a singular fact of omission by Euclid, and other great mathematicians : and, in selecting you, I did you an honour which, from what I have just now heard, was entirely out of place. I think, considering the nature of the work set forth in the prospectus, you are guilty of both folly and presumption, in assuming the character of a patron ; for your own sense ought to have assured you that was such my object I should not have sought him in a De Morgan, who exists only by patronage of others. On the other hand, I deem it to be an unpardonable piece of presumption in offering your advice upon a subject the magnitude, importance, and real utility of which you know nothing about : by doing so you have offered me a direct insult. The system is a manual of Philosophy, a one inseparable whole of metaphysics and physio ;

embracing points the most interesting, laws the most important, doctrines the most essential to advance man in accordance with the spirit of the times. I may not live to see it in print ; for, at ——, life at best is uncertain : but, live or die, be assured, Sir, it is not my intention to debase the work by seeking patronage, or pandering to the public taste. Your advice was the less needed, seeing I am an old-established ——. I remain, &c.—P. S. You will oblige me by returning the prospectus of my work.'

My reader will, I am sure, not take this transition from the ' profoundest respect ' to the loftiest insolence for an *apocraphical* correspondence, to use a word I find in the Prospectus : on my honour it is genuine. He will be better employed in discovering whether I exist by patronizing others, or by being patronized by them. I make any one who can find it out a fair offer : I will give him my patronage if I turn out to be Bufo, on condition he gives me his, if I turn out to be Bavius. I need hardly say that I considered the last letter to be one of those to which no answer is so good as no answer.

These letters remind me in one respect of the correspondents of the newspapers. My other party wrote because a friend had pointed me out : but he would not have written if he had known what another friend told him just in time for the second letter. The man who sends his complaint to the newspaper very often says, in effect, ' Don't imagine, Sir, that I read your columns ; but a friend who sometimes does has told me . . . . ' It is worded thus : ' My attention has been directed to an article in your paper of . . . .' Many thanks to my friend's friends for not mentioning the Budget : had my friend's attention been directed to it I might have lost a striking example of the paradoxer in search of a patron. That my friend was on this scent in the first letter is revealed in the second. Language was given to man to conceal his thoughts ; but it is not every one who can do it.

Among the most valuable information which my readers will get from me is comparison of the reactions of paradoxers, when not admitted to argument, or when laughed at. Of course, they are misrepresented ; and at this they are angry, or which is the same thing, take great pains to assure the reader that they are not. So far natural, and so far good ; anything short of concession of a case which must be seriously met by counter-reasons is sure to be misrepresentation. My friend Mr. James Smith and my friend Mr. Reddie are both terribly misrepresented : they resent it by some insinuations in which it is not easy to detect whether I am a conscious smotherer of truth, or only muddle-headed and ignorant. [This was written before I received my last communication from Mr. James Smith. He tells me that I am wrong in saying that his work in which I stand in the pillory is all reprint : I have no doubt I confounded some of it with some of the manuscript or slips which I had received from my much not-agreed-with correspondent. He adds that my mistake was intentional, and that my reason is obvious to the reader. This *is* information, as the sea-serpent said when he read in the newspaper that he had a mane and tusks.]

My friend Dr. Thorn sees deeper into my mystery. By the way, he still sends an occasional touch at the old subject ; and he wants me particularly to tell my readers that the Latin numeral letters, if M be left out, give 666. And so they do : witness DCLXVI. A person who thinks of the origin of symbols will soon see that 666 is our number because we have five fingers on each hand : had we had but four, our mystic number would have been expressed by 555, and would have stood for our present 365. Had $n$ been the number on each hand, the great number would have been

$$(n+1)(4n^2 + 2n + 1)$$

With no finger on each hand, the number would have been 1 : with one finger less than none at all on each hand, it would have been 0. But what does this mean ? Here is a question for an algebraical paradoxer ! So soon as we have found out how many fingers the inhabitants of any one planet have on each hand, we have the means of knowing their number of the Beast, and thence all about them. Very much struck with this hint of discovery, I turned my attention to the means of developing it. The first point was to clear my vision of all the old cataracts. I propose the following experiment, subject of course to the consent of parties. Let Dr. Thorn Double-Vahu Mr. James Smith, and Thau Mr. Reddie : if either be deparadoxed by the treatment, I will consent to undergo it myself. Provided always that the temperature required be not so high as the Doctor hints at : if the Turkish Baths will do for this world, I am content.

The three paradoxers last named and myself have a pentasyllabic convention, under which, though we go far beyond civility, we keep within civilization. Though Mr. James Smith pronounced that I must be dishonest if I did not see his argument, which he knew I should not do [to say nothing of recent accusation] ; though Dr. Thorn declared me a competitor for fire and brimstone—and my wife, too, which doubles the joke : though

Mr. Reddie was certain I had garbled him, evidently on purpose to make falsehood appear truth ; yet all three profess respect for me as to everything but power to see truth, or candour to admit it. And on the other hand, though these were the modes of opening communication with me, and though I have no doubt that all three are proper persons of whom to inquire whether I should go up-stairs or down-stairs, &c., yet I am satisfied they are thoroughly respectable men, as to everything but reasoning. And I dare say our several professions are far more true in extent than in many which are made under more parliamentary form. We find excuses for each other : they make allowance for my being hoodwinked by Aristotle, by Newton, by the Devil ; and I permit them to feel, for I know they cannot get on without it, that their reasons are such as none but a knave or a sinner can resist. But *they* are content with cutting a slice each out of my character : neither of them is more than an uncle, a Bone-a-part ; I now come to a dreadful nephew, Bone-the-whole.

I will not give the name of the poor fellow who has fallen so far below both the *honestum* and the *utile*, to say nothing of the *decorum* or the *dulce*. He is the fourth who has taken elaborate notice of me; and my advice to him would be, *Nec quarta loqui persona laboret*. According to him, I scorn humanity, scandalize learning, and disgrace the press ; it admits of no manner of doubt that my object is to mislead the public and silence truth, at the expense of the interests of science, the wealth of the nation, and the lives of my fellow men. The only thing left to be settled is, whether this is due to ignorance, natural distaste for truth, personal malice, a wish to curry favour with the Astronomer Royal, or mere toadyism. The only accusation which has truth in it is, that I have made myself a ' public scavenger of science ': the assertion, which is the most false of all is, that the results of my broom and spade are ' shot right in between the columns of ' the *Athenæum*. I declare I never in my life inserted a word between the columns of the *Athenæum* : I feel huffed and miffed at the very supposition. I *have* made myself a public scavenger ; and why not ? Is the mud never to be collected into a heap ? I look down upon the other scavengers, of whom there have been a few—mere historical drudges ; Montucla, Hutton, &c.—as not fit to compete with me. I say of them what one crossing-sweeper said of the rest : ' They are well enough for the common thing ; but put them to a bit of fancy-work, such as sweeping round a post, and see what a mess they make of it ! ' Who can touch me at sweeping round a paradoxer ? If I complete my design of publishing a separate work, an old copy will be fished up from a stall two hundred years hence by the coming man, and will be described in an article which will end by his comparing our century with his own, and sighing out in the best New Zealand pronunciation—

Dans ces tems-là
C'était déjà comme ça!

And pray, Sir! I have been asked by more than one—do your orthodox never fall into mistake, nor rise into absurdity? They not only do both, but they admit it of each other very freely; individually, they are convinced of sin, but not of any particular sin. There is not a syndoxer among them all but draws his line in such a way as to include among paradoxers a great many whom I should exclude altogether from this work. My worst specimens are but exaggerations of what may be found, occasionally, in the thoughts of sagacious investigators. At the end of the glorious dream, we learn that there is a way to Hell from the gates of Heaven, as well as from the City of Destruction: and that this is true of other things besides Christian pilgrimage is affirmed at the end of the Budget of Paradoxes. If D'Alembert had produced *enough* of a quality to match his celebrated mistake on the chance of throwing head in two throws, he would have been in my list. If Newton had produced *enough* to match his reception of the story that Nausicaa, Homer's Phæacian princess, invented the celestial sphere, followed by his serious surmise that she got it from the Argonauts,—then Newton himself would have had an appearance entered for him, in spite of the *Principia*. In illustration, I may cite a few words from 'Tristram Shandy':—

' "A soldier," cried my uncle Toby, interrupting the Corporal, "is no more exempt from saying a foolish thing, Trim, than a man of letters."—"But not so often, an' please your honour," replied the Corporal. My uncle Toby gave a nod.'

I now proceed to die out. Some prefatory remarks will follow in time.[1] I shall have occasion to insist that all is not barren: I think I shall find, on casting up, that two out of five of my paradoxers are not to be utterly contemned. Among the better lot will be found all gradations of merit; at the same time, as was remarked on quite a different subject, there may be little to choose between the last of the saved and the first of the lost.

---

[1] These remarks were never written.—(Ed.)

The higher and better class is worthy of blame; the lower and worse class is worthy of praise. The higher men are to be reproved for not taking up things in which they could do some good: the lower men are to be commended for taking up things in which they can do no great harm. The circle problem is like Peter Peebles's lawsuit:—

' " But, Sir, I should really spoil any cause thrust on me so hastily." —" Ye cannot spoil it, Alan," said my father, " that is the very cream of the business, man,— . . . the case is come to that pass that Stair or Arniston could not mend it, and I don't think even you, Alan, can do it much harm." '

I am strongly reminded of the monks in the darker part of the Middle Ages. To a certain proportion of them, perhaps two out of five, we are indebted for the preservation of literature, and their contemporaries for good teaching and mitigation of social evils. But the remaining three were the fleas and flies and thistles and briars with whom the satirist lumps them, about a century before the Reformation :—

> Flen, flyys, and freris, populum domini male cædunt;
> Thystlis and breris crescentia gramina lædunt.
> Christe nolens guerras qui cuncta pace tueris,
> Destrue per terras breris, flen, flyys, and freris.
> Flen, flyys, and freris, foul falle hem thys fyften yeris,
> For non that her is lovit flen, flyys, ne freris.

I should not be quite so savage with my second class. Taken together, they may be made to give useful warning to those who are engaged in learning under better auspices : aye, even useful hints ; for bad things are very often only good things spoiled or misused. My plan is that of a predecessor in the time of Edward the Second :—

> Meum est propositum gentis imperitæ
> Artes frugi reddere melioris vitæ.

To this end I have spoken with freedom of books as books, of opinions as opinions, of ignorance as ignorance, of presumption as presumption ; and of writers as I judge may be fairly inferred from what they have written. Some—to whom I am therefore under great obligation—have permitted me to enlarge my plan by assaults to which I have alluded ; assaults which allow a privilege of retort, of which I have availed myself ; assaults which give my readers a right of partnership in the amusement which I myself have received.

For the present I cut and run : a Catiline, pursued by a chorus of Ciceros, with Quousque tandem ? Quamdiu nos ? Nihil ne te ? ending with, In te conferri pestem istam jam pridem oportebat, quam tu in nos omnes jamdiu machinaris ! I carry with me the reflection that I have furnished to those who need it such a magazine of warnings as they will not find elsewhere ; *a signatis cavetote* : and I throw back at my pursuers—Valete, doctores sine doctrinâ ; facite ut proximo congressu vos salvos corporibus et sanos mentibus videamus. Here ends the Budget of Paradoxes.

# APPENDIX.

I THINK it right to give the proof that the ratio of the circumference to the diameter is incommensurable. This method of proof was given by Lambert, in the *Berlin Memoirs* for 1761, and has been also given in the notes to Legendre's Geometry, and to the English translation of the same. Though not elementary algebra, it is within the reach of a student of ordinary books.

Let a continued fraction, such as

$$\cfrac{a}{b + \cfrac{c}{d + \cfrac{e}{f + \&c.,}}}$$

be abbreviated into $\dfrac{a}{b+}\ \dfrac{c}{d+}\ \dfrac{e}{f+}$ &c. : each fraction being understood as falling down to the side of the preceding sign $+$. In every such fraction we may suppose $b$, $d$, $f$, &c. positive; $a$, $c$, $e$, &c. being as required: and all are supposed integers. If this succession be continued *ad infinitum*, and if $\dfrac{a}{b}$, $\dfrac{c}{d}$, $\dfrac{e}{f}$, &c. all lie between $-1$ and $+1$, exclusive, the limit of the fraction must be incommensurable with unity; that is, cannot be $\dfrac{A}{B}$, where A and B are integers.

First, whatever this limit may be, it lies between $-1$ and $+1$. This is obviously the case with any fraction $\dfrac{p}{q + \omega}$, where $\omega$ is between $\pm 1$: for, $\dfrac{p}{q}$, being $< 1$, and $p$ and $q$ integer, cannot be brought up to $\pm$, by the value of $\omega$. Hence, if we take any of the fractions

$$\frac{a}{b}\quad \frac{a}{b+}\ \frac{c}{d},\quad \frac{a}{b+}\ \frac{c}{d+}\ \frac{e}{f},\ \&c.$$

say $\dfrac{a}{b+}\ \dfrac{c}{d+}\ \dfrac{e}{f+}\ \dfrac{g}{h}$ we have, $\dfrac{g}{h}$ being between $\pm 1$, so is $\dfrac{e}{f+}\ \dfrac{g}{h}$, so therefore is $\dfrac{c}{d+}\ \dfrac{e}{f+}\ \dfrac{g}{h}$; and so therefore is $\dfrac{a}{b+}\ \dfrac{c}{d+}\ \dfrac{e}{f+}\ \dfrac{g}{h}$.

Now, if possible, let $\dfrac{a}{b+}\ \dfrac{c}{d+}$ &c. be $\dfrac{A}{B}$ at the limit; A and B being integers. Let

$$P = A\,\dfrac{c}{d+}\ \dfrac{e}{f+}\ \&c.,\quad Q = P\,\dfrac{e}{f+}\ \dfrac{g}{h+}\ \&c.,\quad R = Q\,\dfrac{g}{h+}\ \dfrac{i}{k+}\ \&c.$$

P, Q, R, &c. being integer or fractional, as may be. It is easily shown that all must be integer: for

$$\frac{A}{B} = \frac{a}{b + \dfrac{P}{A}},\qquad \text{or, } P = a\,B - b\,A$$

$$\frac{P}{A} = \frac{c}{d + \dfrac{Q}{P}},\qquad \text{or, } Q = c\,A - d\,P$$

$$\frac{Q}{P} = \frac{e}{f + \dfrac{R}{Q}},\qquad \text{or, } R = e\,P - f\,Q$$

&c., &c. Now, since $a$, B, $b$, A, are integers, so also is P; and thence Q; and thence R, &c. But since $\dfrac{A}{B}$, $\dfrac{P}{A}$, $\dfrac{Q}{P}$, $\dfrac{R}{Q}$, &c. are all between $-1$ and $+1$, it follows that the unlimited succession of integers P, Q, R, are each less in numerical value than the preceding. Now there can be no such *unlimited* succession of *descending* integers: consequently, it is impossible that $\dfrac{a}{b+}\ \dfrac{c}{d\times}$, &c. can have a commensurable limit.

It easily follows that the continued fraction is incommensurable if $\dfrac{a}{b}$, $\dfrac{c}{d}$, &c., being at first greater than unity become and continue less than unity after some one point. Say that $\dfrac{i}{k}$, $\dfrac{l}{m}$, ... are all less than unity. Then the fraction $\dfrac{i}{k+}\ \dfrac{l}{m+}$ ... is incommensurable, as proved: let it be $\kappa$. Then $\dfrac{g}{h+\kappa}$ is incommensurable, say $\lambda$; $\dfrac{e}{f+\lambda}$ is the same, say $\mu$; also $\dfrac{c}{d+\mu}$, say $\nu$, and $\dfrac{a}{b+\nu}$, say $\rho$. But $\rho$ is the fraction $\dfrac{a}{b+}\ \dfrac{c}{d+}$ ... itself; which is therefore incommensurable.

Let $\phi z$ represent

$$1 + \frac{a}{z} + \frac{a^2}{2\,z\,(z+1)} + \frac{a^3}{2\cdot 3\cdot\,z\,(z+1)\,(z+2)} + \cdots$$

Let $z$ be positive: this series is convergent for all values of $a$, and approaches without limit to unity as $z$ increases without limit. Change $z$ into $z + 1$, and form $\phi z - \phi (z + 1)$: the following equation will result

$$\phi z - \phi (z + 1) = \frac{a}{z (z + 1)} \phi (z + 2)$$

or $a = \dfrac{a}{z} \dfrac{\phi (z + 1)}{\phi z} \cdot z + \dfrac{a}{z} \dfrac{\phi (z + 1)}{\phi z} \dfrac{a}{z + 1} \dfrac{\phi (z + 2)}{\phi (z + 1)}$

or $a = \psi z \left( z + \psi (z + 1) \right)$

$\psi z$ being $\dfrac{a}{z} \dfrac{\phi (z + 1)}{\phi. z}$; of which observe that it diminishes without limit as $z$ increases without limit. Accordingly, we have

$$\psi z = \frac{a}{z +} \quad \psi (z + 1) = \frac{a}{z +} \frac{a}{(z + 1) +} \psi (z + 2)$$

$$= \frac{a}{z +} \frac{a}{(z + 1) +} \frac{a}{(z + 2) +} \psi (z + 3), \text{ &c.}$$

And, $\psi (z + n)$ diminishing without limit, we have

$$\frac{a}{z} \cdot \frac{\phi (z + 1)}{\phi z} = \frac{a}{z +} \frac{a}{(z + 1) +} \frac{a}{(z + 2) +} \frac{a}{(z + 3) +} \ldots$$

Let $z = \frac{1}{2}$; and let $4 a = - x^2$. Then $\dfrac{a}{z} \phi (z + 1)$ is $- \dfrac{x^2}{2}$ $\left( 1 - \dfrac{x^2}{2 \cdot 3} + \dfrac{x^4}{2 \cdot 3 \cdot 4 \cdot 5} \ldots \right)$ or $- \dfrac{x}{2} \sin x$. Again $\phi z$ is $1 - \dfrac{x^2}{2} + \dfrac{x^4}{2 \cdot 3 \cdot 4}$ -or $\cos x$: and the continued fraction is

$$\frac{- \frac{1}{4} x^2}{\frac{1}{2} +} \frac{- \frac{1}{4} x^2}{\frac{3}{2} +} \frac{- \frac{1}{4} x^2}{\frac{5}{2} + \ldots} \quad \text{or} \quad - \frac{x}{2} \frac{x}{1 +} \frac{- x^2}{3 +} \frac{- x^2}{5 + \ldots}$$

whence $\tan x = \dfrac{x}{1 +} \dfrac{- x^2}{3 +} \dfrac{- x^2}{5 +} \dfrac{- x^2}{7 + \ldots}$

Or, as written in the usual way,

$$\tan x = \cfrac{x}{1 - \cfrac{x^2}{3 - \cfrac{x^2}{5 - \cfrac{x^2}{7 - \ldots}}}}$$

K K

This result may be proved in various ways : it may also be verified by calculation. To do this, remember that if

$$\frac{a_1}{b_1 +}\ \frac{a_2}{b_2 +}\ \frac{a_3}{b_3 + \ldots}\ \frac{a_n}{b_n} = \frac{\text{P}_n}{\text{Q}_n}\ ;\ \text{then}$$

$$\text{P}_1 = a_1\ \ \text{P}_2 = b_2\,\text{P}_1,\qquad \text{P}_3 = b_3\,\text{P}_2 + a_3\,\text{P}_1,\ \ \text{P}_4 = b_4\,\text{P}_3 + a_4\,\text{P}_2,\ \&\text{c}.$$
$$\text{Q}_1 = b_1\ \ \text{Q}_2 = b_2\,\text{Q}_1 + a_2,\ \ \text{Q}_3 = b_3\,\text{Q}_2 + a_3\,\text{Q}_1,\ \ \text{Q}_4 = b_4\,\text{Q}_3 + a_4\,\text{Q}_2,\ \&\text{c}.$$

in the case before us we have

$$a_1 = x,\ a_2 = -\,x^2,\ a_3 = -\,x^2,\ a_4 = -\,x^2,\ a_5 = -\,x^2,\ \&\text{c}.$$
$$b_1 = 1,\ b_2 = 3,\qquad b_3 = 5,\qquad b_4 = 7,\qquad b_5 = 9,\ \&\text{c}.$$

$$
\begin{aligned}
\text{P}_1 &= x & \text{Q}_1 &= 1\\
\text{P}_2 &= 3\,x & \text{Q}_2 &= 3 - x^2\\
\text{P}_3 &= 15\,x - x^3 & \text{Q}_3 &= 15 - 6\,x^2\\
\text{P}_4 &= 105\,x - 10\,x^3 & \text{Q}_4 &= 105 - 45\,x^2 + x^4\\
\text{P}_5 &= 945\,x - 105\,x^3 + x^5 & \text{Q}_5 &= 945 - 420\,x^2 + 15\,x^4\\
\text{P}_6 &= 10395\,x - 1260\,x^3 + 21x^5 & \text{Q}_6 &= 10395 - 4725\,x^2 + 210\,x^4 - x^6
\end{aligned}
$$

We can use this algebraically, or arithmetically. If we divide $\text{P}_n$ by $\text{Q}_n$, we shall find a series agreeing with the known series for tan $x$, as *far as* n *terms.* That series is

$$x + \frac{x^3}{3} + \frac{2\,x^5}{15} + \frac{17\,x^7}{315} + \frac{62\,x^9}{2835} + \ldots$$

Take $\text{P}_5$, and divide it by $\text{Q}_5$ in the common way, and the first five terms will be as here written. Now take $x = 1$, which means that the angle is to be one tenth of the actual unit, or, in degrees $5°\cdot729578$. We find that when $x = \cdot 1$, $\text{P}_6 = 1038\cdot24021$, $\text{Q}_6 = 10347\cdot770999$ ; whence $\text{P}_6$ divided by $\text{Q}_6$ gives $\cdot1003346711$. Now $5°\cdot729578$ is $5°\ 43'\ 46\tfrac{1}{2}''$ ; and from the old tables of Rheticus—no modern tables carry the tangents so far—the tangent of this angle is $\cdot1003347670$.

Now let $x = \tfrac{1}{4}\pi$ ; in which case tan $x = 1$. If $\tfrac{1}{4}\pi$ be commensurable with the unit, let it be $\dfrac{m}{n}$, m and n being integers : we know that $\tfrac{1}{4}\pi < 1$. We have then

$$1 = \frac{m}{n}\ \frac{\dfrac{m^2}{n^2}}{1 -}\ \frac{\dfrac{m^2}{n^2}}{3 -}\ \frac{}{5 - \ldots}\ \cdot = \frac{m}{n -}\ \frac{m^2}{3\,n -}\ \frac{m^2}{5\,n -}\ \frac{m^2}{7\,n - \ldots}$$

Now it is clear that $\dfrac{m^2}{3\,n}$, $\dfrac{m^2}{5\,n}$, $\dfrac{m^2}{7\,n}$, &c. must at last become and continue severally less than unity. The continued fraction is therefore incommensurable, and cannot be unity. Consequently $\pi^2$ cannot be commensurable : that is, $\pi$ is an incommensurable quantity, and so also is $\pi^2$.

---

I thought I should end with a grave bit of appendix, deeply mathematical : but paradox follows me wherever I go. The foregoing is—in my own language—from Dr. (now Sir David) Brewster's English edition of Legendre's Geometry, (Edinburgh, 1824, 8vo.) translated by some one who is not named. I picked up a notion, which others had at Cambridge in 1825, that the translator was the late Mr. Galbraith, then known at Edinburgh as a writer and teacher. But it turns out that it was by a very different person, and one destined to shine in quite another walk ; it was a young man named Thomas Carlyle. He prefixed, from his own pen, a thoughtful and ingenious essay on Proportion, as good a substitute for the fifth Book of Euclid as could have been given in the space ; and quite enough to show that he would have been a distinguished teacher and thinker on first principles. But he left the field immediately.

---

(The following is the passage referred to at p. 285 :—Ed.)

Michael Stifelius edited, in 1554, a second edition of the Algebra (*Die Coss.*), of Christopher Rudolf. This is one of the earliest works in which + and − are used.

Stifelius was a queer man. He has introduced into this very work of Rudolph his own interpretation of the number of the Beast. He determined to fix the character of Pope Leo : so he picked the numeral letters from LEODECIMVS, and by taking in x from LEO X. and striking out M as standing for *mysterium*, he hit the number exactly. This discovery completed his conversion to Luther, and his determination to throw off his monastic vows. Luther dealt with him as straightforwardly as with Melancthon about his astrology : he accepted the conclusions, but told him to clear his mind of all the premises about the Beast. Stifelius did not take the advice, and proceeded to settle the end of the world out of the prophet Daniel : he fixed on October, 1533. The parishioners of some cure which he held, having full faith, began to spend their savings in all kinds of good eating and drinking ; we may charitably hope this was not the way of preparing for the event which their pastor pointed out. They succeeded in making themselves as fit for Heaven as Lazarus, so far as beggary went : but when the time came, and the world lasted on, they wanted to kill their deceiver, and would have done so but for the interference of Luther.

# INDEX.

LONDON : PRINTED BY
SPOTTISWOODE AND CO., NEW-STREET SQUARE
AND PARLIAMENT STREET